Primordial Black Holes from Inflation

Primordial Black Holes from Inflation

Guest Editor

Cristiano Germani

MDPI

Basel • Beijing • Wuhan • Barcelona • Belgrade • Novi Sad • Cluj • Manchester

Guest Editor
Cristiano Germani
Institut de Ciències
del Cosmos
Universitat de Barcelona
Barcelona
Spain

Editorial Office
MDPI AG
Grosspeteranlage 5
4052 Basel, Switzerland

This is a reprint of the Special Issue, published open access by the journal *Universe* (ISSN 2218-1997), freely accessible at: https://www.mdpi.com/journal/universe/special_issues/PBHI.

For citation purposes, cite each article independently as indicated on the article page online and as indicated below:

Lastname, A.A.; Lastname, B.B. Article Title. *Journal Name* **Year**, *Volume Number*, Page Range.

ISBN 978-3-7258-7156-8 (Hbk)
ISBN 978-3-7258-7157-5 (PDF)
https://doi.org/10.3390/books978-3-7258-7157-5

Contents

About the Editor

Cristiano Germani

Cristiano Germani is an associate professor at the Universitat de Barcelona. He has worked on everything related to gravity since completing his PhD in 2003 at the Institute of Gravitation and Cosmology (University of Portsmouth). In the last few years, Cristiano has focused his research on the topic of primordial black holes (PBH), and he and his collaborators have written seminal papers on PBH statistics and their formation.

universe

MDPI

Technical Note

pbhstat: A Python Package for Calculating the Primordial Black Hole Abundance

Philippa S. Cole [1,2,*] and Jacopo Fumagalli [3,*]

1 Dipartimento di Fisica "G. Occhialini", Università degli Studi di Milano-Bicocca, Piazza della Scienza 3, 20126 Milano, Italy
2 INFN, Sezione di Milano-Bicocca, Piazza della Scienza 3, 20126 Milano, Italy
3 Departement de Física Quàntica i Astrofisica and Institut de Ciències del Cosmos (ICC), Universitat de Barcelona, Martí i Franquès 1, 08028 Barcelona, Spain
* Correspondence: philippa.cole@unimib.it (P.S.C.); jfumagalli@fqa.ub.edu (J.F.)

Abstract

We present `pbhstat`, a publicly available Python package designed to compute the mass function and total abundance of primordial black holes (PBHs) from a given primordial power spectrum. The package offers a modular framework using multiple statistical approaches, including Press–Schechter theory, peaks theory, and formalisms based on the non-linear compaction function. Currently, the implementation is limited to scenarios with nearly Gaussian initial conditions.

Keywords: primordial black holes; inflation; dark matter

1. Introduction

The possibility that dark matter consists of primordial black holes (PBHs) presents an intriguing avenue from both theoretical and observational perspectives [1–10]. We focus on the case in which PBHs are seeded by large-amplitude inflationary perturbations [11–13]. In this context, accurately determining the PBH abundance remains highly sensitive to the interplay between various contributing factors (see [14–16] for reviews on the subject). Central to this is the choice of collapse statistics, which relate the probability of black hole formation to the initial conditions and the non-linear dynamics involved. Within a given statistical framework, further ambiguities arise—such as the eventual choice of a window function, the threshold criterion that an estimator must exceed to trigger collapse, and the numerical coefficients used in the computation (inferred, for instance, from numerical simulations [17–19]). These elements, whether explicitly chosen or inherently defined by the framework, collectively have a substantial impact on the resulting PBH abundance, e.g., [20–26]. Furthermore, PBHs are challenging to search for, meaning that theoretical predictions of their mass distribution are valuable in order to better inform and refine observational approaches.

Motivated by this, we present `pbhstat` as a simple tool that allows for the consistent computation of PBH abundance across different methods for ease of comparison. We aim for this tool to be accessible to non-experts who may not work directly on PBH formation, by providing an easy way to generate PBH mass functions from pre-defined primordial power spectra using consistent statistical formalisms. At the same time, it is designed to support experts in the field, allowing them to input custom power spectra and make detailed choices within various statistical approaches. While many researchers have developed private implementations of these methods, to the best of our knowledge, they have not yet been integrated into a publicly available, unified codebase. We hope that this package will remove the need to re-implement standard techniques and serve as a platform for reproducibility and cross-checks across different approaches.

We implement three choices of statistics to describe how fluctuations collapse to form PBHs. The Press–Schechter theory [27,28] provides a relatively simple method based on the probability that the density contrast exceeds a critical threshold. It usually assumes a Gaussian distribution for the linear overdensity and integrates over all regions above this threshold to estimate the collapse fraction, offering a first-order approximation that is easy to implement. Peaks theory [29] was first applied to the PBH context in the mid-2000s in ref. [30], extending beyond Press–Schechter by modelling PBHs as forming from rare, high peaks in the curvature perturbation field. It is generally considered to be a more accurate description of PBH formation and has been explored and developed extensively in the literature (see, e.g., [20,24,31,32]). Finally, we place particular emphasis on the more involved implementation of the non-linear compaction function statistics [25,33,34]. This approach relies on using the initial conditions of the compaction function as the primary statistical measure. It focuses on identifying regions where the overdensity exhibits a pronounced peak; the compaction function attains its maximum at a characteristic smoothing scale relative to this peak, and this maximum surpasses the critical threshold necessary for the initial overdensity to collapse to form a black hole. We do not delve into the underlying physics of the statistical methodologies used and instead refer the reader to the relevant references for detailed discussions.

Under the assumption of nearly Gaussian initial conditions, all methods depend exclusively on the initial power spectrum. The latter should be motivated by a realisable inflationary scenario; see, e.g., [35] for a review on inflationary mechanisms that lead to PBH production.

Additional effects—such as quantum diffusion [36–39] and non-Gaussianities [40–50]—may play a crucial role in shaping the final abundance. Nonetheless, even when one adopts the simplifying assumption that the primordial curvature perturbations are Gaussian with a known power spectrum, the present tool provides useful first-order calculations of the mass function, illustrating the number of components that must be taken into account in different statistical approaches. We envisage implementing additional refinements in future versions of the code.

This companion to the code is organised as follows: In Section 2, we outline the main output of the code. In Section 3, we present the three statistical methods for calculating PBH abundance that we implement, along with the primordial power spectra and window functions in Section 4. We lay out the modular structure of the code in Section 5 and provide installation and quick-start usage instructions in Section 6. Finally, we comment on future additions and summarise in Section 7.

2. Output of the Code

The main output of `pbhstat` is the mass function of primordial black holes $f(M)$, which represents the fraction of energy density contained in primordial black holes per logarithmic mass interval:

$$f(M) = \frac{1}{\Omega_{\rm CDM}} \frac{d\Omega_{\rm PBH}}{d\ln M} \tag{1}$$

which satisfies

$$\int f(M) d\ln M = f_{\rm PBH} \equiv \Omega_{\rm PBH}/\Omega_{\rm CDM}, \tag{2}$$

where M is the mass of the PBH, $f_{\rm PBH}$ is the fraction of dark matter energy density contained in PBHs today, and $\Omega_{\rm CDM}$ is the energy density in cold dark matter today. Various definitions of the mass function are present in the literature, often simply related to the one above by a rescaling or numerical coefficients.

The PBH abundance today, $\Omega_{\rm PBH}$, can be obtained by integrating the contributions over all scales:

$$\Omega_{\rm PBH} = \int d\ln R \left(\frac{R_{\rm eq}}{R}\right)\beta(R), \tag{3}$$

where R is the physical scale of the horizon at the time of formation, which is related to the horizon mass, $M_{\rm H}$, as follows [51]:

$$R = \left(\frac{M_H}{M_{\rm eq}}\right)^{1/2} R_{\rm eq}, \tag{4}$$

where $\beta(R)$ is the mass fraction of PBHs at the time of formation, i.e., $\beta = \rho_{\rm PBH}/\rho_{\rm tot}|_{\rm formation}$. The factor $(M_{\rm eq}/M_H)^{1/2}$ scales β to its value at matter-radiation equality exploiting the fact that during radiation domination, $\beta \propto \rho_{PBH}/\rho_R \propto a$, and $a_{\rm eq}/a(M_H) = (M_{\rm eq}/M_H)^{1/2}$. We use $M_{eq} = 2.8 \times 10^{17} M_\odot$ and $k_{eq} = 1/R_{\rm eq} = 0.01(\Omega_m/0.31)\,{\rm Mpc}^{-1}$, neglecting for simplicity the small correction due to the evolution of the number of relativistic degrees of freedom in Equation (4). In practice, the integral over all scales in Equation (3) will be computed over horizon mass, where it is best to integrate between extrema that correspond to the scales where the primordial power spectrum has decreased by a factor of 10 from its peak amplitude. The implementation of β has a different form depending on how the perturbations are assumed to collapse. We provide three methods from the literature for calculating β that assume three different statistics, namely Press–Schechter, peaks theory, and non-linear compaction statistics. The key ingredient to $\beta(R)$ is the primordial power spectrum, which can be either analytically calculated or loaded as a bespoke spectrum in the package. A separate module smooths the power spectrum and calculates the relevant width parameters $\sigma(R)$ which are fed to $\beta(R)$ as described in Section 4. To finally arrive at the PBH mass function, one has to invert the expression for the PBH mass in terms of the horizon mass and the statistical estimator chosen (see below).

3. Choice of Statistics

In all three statistics methods, we use the compaction function (in its linear or non-linear version in ζ) as an estimator. The compaction function, denoted with C, measures the mass excess within a given areal radius [52]. In the long-wavelength approximation, and by exploiting the (approximate) spherical symmetry of high peaks in a randomly distributed Gaussian field (in this case, ζ) [29], the time-independent super-horizon compaction function can be written as [53]

$$C(R) = g(R)\left(1 - \frac{3}{8}g(R)\right), \tag{5}$$

where

$$g = -\frac{4}{3}R\partial_R\zeta(R) \tag{6}$$

which is sometimes referred to as the linear compaction function.[1] Here, $\zeta(R)$ denotes the radial profile of the comoving curvature perturbation in a coordinate system centred on one of its peaks. The compaction function C on super-horizon scales is bounded from above, with $C \leq 2/3$. This bound is reached for $g = 4/3$, which separates standard Type I fluctuations ($g \leq 4/3$) from more exotic Type II fluctuations ($g > 4/3$), with the latter being characterised by a non-monotonic areal radius. The current implementation of the code considers only Type I fluctuations. Critical collapse for Type II fluctuations is currently under investigation—see [54–57], and a dedicated statistical framework should be developed to properly explore this case. Accordingly, we bound our integral in g to $4/3$.

In practice, this consideration is relevant primarily for the non-linear statistics discussed in Section 3.3. For the statistics in Sections 3.1 and 3.2, the constant threshold is sufficiently far from 4/3 that Type II fluctuations are naturally excluded.

3.1. Press–Schechter

We implement Press–Schechter statistics in the `PressSchechterModel` class within the `collapse_stats.py` module following [58] (and references therein). Note that in the original Press–Schechter formalism for PBHs, the mass fraction is typically expressed as an integral over the density contrast δ, which is assumed to be Gaussian. When the time-independent component of the linear density contrast is averaged over a volume enclosed by a sphere of radius R, it becomes equivalent to the compaction function estimator C (see for instance [59]) on super-horizon scales. This is valid for the linear (in ζ) term of the compaction function, i.e., g, and the linear smoothed density contrast. Using g and adjusting the critical threshold C_c accordingly [58] means that we can write the subsequent expressions in terms of g in place of the original δ.

The form of β in terms of the horizon mass (which can be converted from R using Equation (4)) is

$$\beta(M_H) = 2\int_{g_c}^{4/3} dg \frac{M}{M_H} P(g) \tag{7}$$

where $P(g)$ is the probability distribution function of the initial density perturbations and the upper limit of the integral is a result of excluding Type II fluctuations. Under the Gaussian approximation, this reads as

$$P(g) = \frac{1}{\sqrt{2\pi\sigma^2(M_H)}} \exp\left(-\frac{g^2}{2\sigma^2(M_H)}\right) \tag{8}$$

where $\sigma(M_H)$ is the mass variance, which corresponds to the zero-th width parameter σ_0 in the hierarchy of moments[2]

$$\sigma_n^2 = \frac{16}{81}\int_0^\infty \frac{dk}{k}(kR)^{4+2n} W^2(k,R)\mathcal{P}_\zeta \tag{9}$$

where $(kR)^4$ accounts for the super-horizon growth of the density perturbations, described by the dimensionless primordial power spectrum of the comoving curvature perturbation $\mathcal{P}_\zeta$, defined by $\langle \zeta_{\mathbf{k}}\zeta_{\mathbf{k}'}\rangle = (2\pi)^3\delta(\mathbf{k}+\mathbf{k}')(2\pi^2/k^3)\mathcal{P}_\zeta(k)$, and $W(k,R)$ is a window function chosen to filter the perturbations over the smoothing scale R.

The window functions implemented for use with Press–Schechter are the Gaussian window function[3] and the real-space top-hat window function. A real-space top-hat window corresponds to averaging over a sharp spherical region of radius R:

$$W_{\mathrm{TH}}(kR) = 3\frac{\sin(kR) - kR\cos(kR)}{(kR)^3} \tag{10}$$

while the Gaussian window function corresponds to a weighted average over all space, where points closer to the centre are weighted more heavily, and the weight decays exponentially with distance

$$W_{\mathrm{G}}(kR) = e^{-x^2/4} \tag{11}$$

where we have defined $x \equiv kR$. From Equation (6), one can show that $\sigma_g^2 \equiv \langle gg\rangle = \sigma_0^2$ [33,34] in Equation (9), with the window function automatically given by the real-space top-hat one.

The critical scaling behaviour of the PBH mass can be written as [60]

$$M = KM_H(g - g_c)^{\gamma_{\rm cr}}, \tag{12}$$

where K, $\gamma_{\rm cr}$, and g_c are numerical coefficients that should be chosen to be consistent with the smoothing function used to calculate $\sigma(M_H)$. Standard choices based on fiducial profiles are $\{K = 4,\ g_c = 0.77\}$ for the real-space top-hat window function [59] and $\{K = 10,\ g_c = 0.28\}$ for the Gaussian window function [32]. These values of g_c are obtained by solving Equation (5) for g with $C = 0.55$ for the top-hat, and $C = 0.25$ for the Gaussian window functions, respectively, so as to account for the fact that we are using only the linear term of the compaction function [61]. $\gamma_{\rm cr}$ is implemented as an optional argument with a default value of 0.36 in the `MassVariance` class within the `mass_variance.py` module. As is customary in these statistics, we do not consider here the profile dependence of the threshold. We note that for the real-space top-hat window function, the integral over $\beta(R)$ diverges and therefore we implement a large-k cut-off as an optional argument in the `MassVariance` class, following [58]. This is placed at the point where the window function reaches its first trough, at $k = 4.49/R$. This removes the divergence from the integral, but the window function is no longer a true top-hat function.

Combining Equations (7) and (8), and trading the variable g for M by inverting Equation (12), we can write $f(M)$ as

$$f(M) = \frac{1}{\Omega_{\rm CDM}} \int {\rm d}M_H \sqrt{\frac{M_{\rm eq}}{M_H}} \frac{\left(\frac{M}{KM_H}\right)^{1/\gamma_{\rm cr}}}{\gamma_{\rm cr} M_H^2} MP(g_\bullet), \tag{13}$$

where $g_\bullet$ denotes g expressed in terms of the other variables by inverting the critical scaling relation in Equation (12):

$$g_\bullet \equiv g(M, M_H) = \left(\frac{M}{KM_H}\right)^{1/\gamma_{\rm cr}} + g_c. \tag{14}$$

3.2. Peaks Theory

We implement peaks theory statistics [29,30] in the `PeaksTheoryModel` class within the `collapse_stats.py` module in the following way. The mass fraction is given by

$$\beta(M_H) = b \int_{g_c}^{4/3} dg \frac{M}{M_H} n\left(\frac{g}{\sigma(M_H)}\right) \tag{15}$$

where b is a pre-factor that accounts for the volume over which the window function is integrated, taking $b = \{(2\pi)^{\frac{3}{2}}, 4\pi/3\}$ for Gaussian and top-hat window functions, respectively [24]. n is the number density of peaks given by

$$n(g, M_H) = \frac{1}{3^{\frac{3}{2}}(2\pi)^2} \left(\frac{\sigma_1(M_H)}{\sigma_0(M_H)}\right)^3 \left(\frac{g}{\sigma_0(M_H)}\right)^3 \exp\left(-\frac{1}{2}\left(\frac{g}{\sigma_0(M_H)}\right)^2\right), \tag{16}$$

with σ_0 and σ_1 as defined in Equation (9). Note that we have absorbed the usual R^3 factor in this equation into the dimensionless expression for σ_1; see Equation (9).

The window functions implemented for use with the `PeakTheoryModel` are the same as for the `PressSchechterModel`, namely the Gaussian window function and the real-space top-hat window function. The recommended choices for the numerical coefficients are the same as in the previous Section 3.1. The large-k cut-off flag is also compatible with this statistics choice and can be implemented in the `MassVariance` class within the `mass_variance.py` module.

Putting Equations (15) and (16) together with Equation (1), we compute the mass function with [58]

$$f(M) = \frac{b}{2}\frac{1}{\Omega_{\rm CDM}}\frac{1}{\gamma_{\rm cr}K^{\frac{1}{\gamma_{\rm cr}}}}\int dM_H\sqrt{\frac{M_{\rm eq}}{M_{\rm H}^3}}\left(\frac{M}{M_H}\right)^{1+\frac{1}{\gamma_{\rm cr}}}n(g_\bullet, M_H), \tag{17}$$

where once again $g_\bullet$ is given in Equation (14) by inverting the critical scaling relation in Equation (12).

3.3. Non-Linear Compaction Function Statistics

We implement non-linear compaction function statistics [33] in the `NonLinearModel` class within the `collapse_stats.py` module. For this method, the estimator for the PBH abundance is the full super-horizon compaction function C in Equation (5). Since PBH formation occurs when the super-horizon C at its maximum ($R = R_m$) exceeds a certain threshold C_c [21], it is useful to introduce the following variables:

$$v = Rg', \qquad w = -R^2 g'', \tag{18}$$

where primes denote derivatives with respect to the comoving radial coordinate R. In the following expressions, these quantities are understood to be evaluated at the maximum of C.

By requiring that these three conditions are satisfied—(i) the overdensity has a peak, (ii) the compaction function reaches a maximum ($v = 0$ and $w \geq 0$), and (iii) this maximum exceeds a critical threshold ($g \geq g_c$)—one can rewrite $\beta(R)$ in the following convenient form [34] (see also [25]):

$$\beta(M_H) = \frac{2\pi}{3}\int_0^\infty dw w\int_{g_c(w)}^{4/3} dg\,\frac{f\left(\frac{\chi}{\sigma_\chi}\right)}{(2\pi)^{3/2}(\sqrt{3}\sigma_1/\sigma_2)}\frac{M}{M_H}P(g, w, v = 0), \tag{19}$$

where $\chi = 2g + w$ is the trace of the Hessian at the position of the peak's centre under the condition $v = 0$. It should be noted that, here and in the following, (g, w) are evaluated at the maximum of the compaction function. The explicit form of the function f, arising from condition (i), is given in Equation (A15) of Ref. [29]. The probability distribution function $P(g, w, v = 0)$, obtained from the three conditions mentioned above, is given by

$$P(g, w, v = 0) = p(v = 0)p(g, w) = \frac{1}{\sqrt{2\pi\sigma_v^2}}p(g, w), \tag{20}$$

where $p(v = 0)$ is the (assumed) Gaussian distribution of v evaluated at $v = 0$, i.e., the probability of having a peak in the compaction function, multiplied by the conditional joint probability of the other two variables $p(g, w)$. Assuming again for simplicity that primordial fluctuations, and therefore g, are Gaussian, $p(g, w)$ becomes a bi-variate Gaussian distribution for the two correlated variables w and g with an extra correction inherited from imposing the condition $v = 0$. We then have (using $\sigma_x^2 \equiv \langle x^2\rangle$, $\sigma_{xy} \equiv \langle xy\rangle$, $\gamma_{xy} \equiv \frac{\sigma_{xy}}{\sigma_x\sigma_y}$)

$$p(w, g) = \frac{1}{2\pi\sqrt{\det\Sigma}}\exp\left(-\frac{1}{2}\vec{X}^T\Sigma^{-1}\vec{X}\right), \tag{21}$$

where

$$\vec{X}^T = (w\ g), \quad \Sigma = \begin{pmatrix}\tilde{\sigma}_w^2 & \tilde{\sigma}_{wg}^2\\ \tilde{\sigma}_{gw}^2 & \tilde{\sigma}_g^2,\end{pmatrix}, \tag{22}$$

with

$$\tilde{\sigma}_g^2 = \sigma_g^2(1-\gamma_{vg}^2), \quad \tilde{\sigma}_w^2 = \sigma_w^2(1-\gamma_{vw}^2), \quad \tilde{\sigma}_{wg}^2 = \sigma_{wg}^2 - \frac{\sigma_{wv}^2\sigma_{vg}^2}{\sigma_v^2}. \tag{23}$$

The probability distribution function can therefore be explicitly written as

$$p(g,w) = \frac{1}{\sqrt{2\pi\tilde{\sigma}_w^2(1-\tilde{\gamma}^2)}} \exp\left(-\frac{(w-\tilde{\gamma}\frac{\tilde{\sigma}_w}{\tilde{\sigma}_g}g)^2}{2\tilde{\sigma}_w^2(1-\tilde{\gamma}^2)}\right)\frac{1}{\sqrt{2\pi\tilde{\sigma}_g^2}}\exp\left(-\frac{g^2}{2\tilde{\sigma}_g^2}\right), \tag{24}$$

or analogously exchanging g with w. $\tilde{\gamma}$, the so-called Pearsol coefficient, tells us the strength of the cross-correlation between the two variables:

$$\tilde{\gamma} = \frac{\tilde{\sigma}_{wg}^2}{\tilde{\sigma}_w\tilde{\sigma}_g}. \tag{25}$$

The threshold depends on the shape of the compaction function and is characterised by the function $g_c(w)$, where the dimensionless curvature of the linear compaction function w parametrises the profile. In the current implementation, we adopt the threshold derived using the "q-approach" [62], which asymptotically approaches 4/3 for large w, corresponding to sharply peaked compaction function profiles. This method relies on fiducial C profiles expressed in terms of the normalised variable $q \equiv -R^2C''/[4C(1-\frac{3}{2}C)]$. One first inverts the universal threshold on the average compaction function [62,63] to obtain the threshold of the compaction function as a function of q:

$$C_c(q) = \frac{4}{15}e^{-1/q}\frac{q^{1-\frac{5}{2q}}}{\Gamma\left(\frac{5}{2q}\right)-\Gamma\left(\frac{5}{2q},\frac{1}{q}\right)}, \tag{26}$$

where $\Gamma(x)$ and $\Gamma(x,y)$ denote the Euler gamma function and the upper incomplete gamma function, respectively. Then, to express the threshold as a function of w, one uses the relation[4]

$$w = -R^2C'' = 4q\,C\sqrt{1-\frac{3}{2}C}. \tag{27}$$

Evaluating the right-hand side of the previous expression at the threshold C_c and substituting (26), one finds $q(w)$. Finally, this expression is substituted into the inverted relation for g as a function of C to obtain the threshold for g as a function of w:

$$g_c(w) = \frac{4}{3}\left(1-\sqrt{1-\frac{3}{2}C_c(q(w))}\right). \tag{28}$$

The steps outlined in Equations (26)–(28) are implemented in the module `stats_utils.py` to compute $g_c(w)$.

Within this method, it is customary to express the mass of the PBHs using the critical scaling given in terms of the compaction function as

$$M = \tilde{K}M_H[C(g)-C(g_c(w))]^{\gamma_{cr}}. \tag{29}$$

The term in parentheses, when expanded around g_c, reduces to Equation (12) with a different re-scaled coefficient, which we denote by $\tilde{K}$. The two numerical coefficients are optional arguments in the code, but we choose $K = 6$ and critical exponent $\gamma_{cr} = 0.36$ as default values [62].

The final abundance depends crucially on the choice of the profile-dependent threshold $g_c(w)$ and, to some extent, the critical scaling. Both elements are subject to ongoing scrutiny in the literature (e.g., [54–57]) and can be readily modified within the `stats_utils.py` module of the code in the future.

By inverting (29) and trading the variable g for M, one can derive the mass function [25]:

$$f(M) = \frac{2\pi\tilde{K}}{3}\int d\ln M_H \int_0^\infty dw \frac{w\left(\frac{M}{KM_H}\right)^{\frac{1}{\gamma_{cr}}+1}}{\gamma_{cr}(1-\frac{3}{4}g_\bullet)}\left(\frac{M_{eq}}{M_H}\right)^{1/2} P(g_\bullet, w, v=0)\frac{f\left(\frac{2g_\bullet+w}{\sigma_\chi}\right)}{(2\pi)^{3/2}(\sqrt{3}\sigma_1/\sigma_2)}. \tag{30}$$

Here, $g_\bullet$ denotes g evaluated by inverting Equation (29).

For completeness, we report the expressions for the various correlators used in this choice of statistics. All of them can be derived from the definitions of the relevant variables in Equations (6) and (18), the σ defined in Equation (9), and the standard expression for the dimensionless primordial power spectrum given just below that equation:

$$\begin{aligned}
\sigma_g^2 &= \sigma_0^2, \qquad \sigma_w^2 = \sigma_2^2 - 4\sigma_1^2 + 4\sigma_0^2, \qquad \sigma_\chi^2 = \sigma_2^2 \\
\sigma_v^2 &= \frac{16}{81}\int \frac{dk}{k}\left(3x\sin x - x^2 W(x)\right)^2 \mathcal{P}_\zeta \\
\sigma_{vg} \equiv \langle vg\rangle &= \frac{16}{81}\int_0^\infty \frac{dk}{k}\left(3\,x\sin x - x^2\,W(x)\right) W(x)\,x^2\,\mathcal{P}_\zeta \\
\sigma_{gw} &= \frac{16}{81}\int_0^\infty \frac{dk}{k}\left(x^6 - 2\,x^4\right) W^2(x)\,\mathcal{P}_\zeta \\
\sigma_{vw} &= \frac{16}{81}\int_0^\infty \frac{dk}{k}\left(3\,x\sin x - x^2\,W(x)\right)\left(x^4 - 2\,x^2\right) W(x)\mathcal{P}_\zeta,
\end{aligned} \tag{31}$$

where we have defined $x \equiv kR$. Note that, if one strictly takes as an estimator the orthodox definition of the compaction function C (as in Equation (5)), then W, playing the role of the window function in the formula above, is automatically the real-space top-hat function.

In practice, we choose the range of k-values so that only those corresponding to amplitudes of the primordial power spectrum within a factor of 10 of the peak amplitude are integrated over. In this way, the code determines an effective width of the power spectrum and sets the limits of both w (see [25]) and $M_H \sim R^2 \sim k^{-2}$ accordingly. This `threshold_factor` can be modified from its default value of 0.1 in the `collapse_stats.py` module. We emphasise that it may be particularly important to test different values if the input power spectrum is especially broad, to ensure convergence of the σ's in Equation (31) and to properly account for high values of w.

As already mentioned, if initial non-Gaussianity is assumed to be present, the expressions in this Section 3 have to be corrected with the specific form of non-Gaussianity included (see, e.g., [47–50]).

4. Power Spectra and Smoothing

We implement both pre-set options for the primordial power spectrum, as well as the possibility for a user-defined primordial power spectrum that can be defined analytically or by uploading an array of $\mathcal{P}_\mathcal{R}$ along with their corresponding wavenumbers k.

The key pre-set options are `piecewise`, defined as

$$\mathcal{P}_\mathcal{R}(k) = \begin{cases} A\left(\frac{k}{k_\star}\right)^{n_g}, & \text{if } k \le k_\star \\ A\left(\frac{k}{k_\star}\right)^{-n_d}, & \text{if } k > k_\star \end{cases} \tag{32}$$

where A is the amplitude of the peak of the power spectrum, $k_\star$ is the position of the peak in the power spectrum, n_g is the index of the growing slope and $n_d \geq 0$ is the index of the decaying slope. We also set the minimum amplitude of the power spectrum at any k to be $A = 2 \times 10^{-9}$ in order to match CMB observations at large scales. While such small amplitude perturbations will not affect PBH abundance, this setting can be overridden by adjusting the value of `min_amplitude` in the `power_spectrum.py` module.

The `lognormal` power spectrum class is defined by

$$\mathcal{P}_\mathcal{R}(k) = A \exp\left(-\frac{1}{2}\left(\frac{\ln(k/k_\star)}{\sigma_{\ln}}\right)^2\right) \tag{33}$$

where A is the amplitude of the peak, $k_\star$ the position of the peak, and $\sigma_{\ln}$ the width of the log-normal distribution. We also implement a `delta` class by setting $\sigma_{\ln}$ to be very small (e.g., 0.001). This produces a very narrow log-normal distribution which can be used to approximate a delta-function and avoids numerical issues with using a true delta-function in the same framework as the rest of the code. We note that this type of power spectrum is unphysical but it is often used for simplicity.

The `flat` power spectrum class is scale-invariant across the range of scales provided by the user, simply defined with an amplitude A.

These pre-set power spectra aim to capture the key classes of inflationary models in the literature that lead to PBH production. We emphasise that a numerical power spectrum calculated by evolving the equations of motion for the inflaton with a given inflationary potential should be used in order to calculate the precise PBH abundance from specific initial conditions.

Finally, the `custom` class allows the user to load an array of $\mathcal{P}_\mathcal{R}$ values that correspond to an array of k-values.

The values of $\mathcal{P}_\mathcal{R}$ are fed to the `MassVariance` class. The calculation of the relevant $\sigma(R)$, depending on the choice of statistics, involves convolving $\mathcal{P}_\mathcal{R}$ with the chosen window function to find the smoothed power spectrum for Press–Schechter and peaks theory methods, as well as other combinations for the non-linear statistics. These expressions are then integrated over the user-defined k-range such that the output of the `MassVariance` class is an interpolating function for the relevant $\sigma(R)$, where the R-range of validity for the interpolators are given by $R = 1/k$.

5. Structure of the Code and Installation

The `pbhstat` package is organised into modular components, each handling a distinct stage of the calculation pipeline outlined above, from reading the initial primordial power spectrum to computing the PBH mass function. The core codebase resides in the `src/` directory, structured as follows:

- `power_spectrum.py`
  Contains the `PowerSpectrum` class, which defines and evaluates the primordial curvature power spectrum. There are some pre-defined forms for the power spectrum which can be loaded, such as `flat` and `piecewise`. It also supports both analytical and tabulated user-defined spectra, which can be implemented directly into the class or loaded as an array. The class then provides interpolation over a user-specified range of wavenumbers k.
- `mass_variance.py`
  Implements the `MassVariance` class, which computes the smoothed mass variance $\sigma^2(R)$ over a range of scales, using the power spectrum and a choice of window

function. Currently, the Gaussian window function and the real-space top-hat window functions are implemented, with a large-k cut-off optional for the latter.

- `collapse_stats.py`
  Provides statistical models for PBH formation. It includes classes for the Press–Schechter formalism (`PressSchechterModel`), peaks theory (`PeaksTheoryModel`), and for non-linear compaction statistics (`NonLinearModel`). These classes set up the computation of the mass function of primordial black holes as a function of mass.
- `mass_function.py`
  Contains the `MassFunction` class, which wraps around the collapse statistics to compute the differential PBH mass function $f(M)$ and total abundance f_{PBH}.
- `stats_utils.py`
  Hosts auxiliary functions used predominantly for the `NonLinearkModel` class, such as critical threshold calculations, filtering functions, and intermediate formulae that can be pre-computed.
- `plot_utils.py`
  Provides helper routines for plotting key quantities such as the power spectrum, mass variance, and PBH mass function.
- `constants.py`
  Stores fiducial values of cosmological quantities that can be edited and will be imported for abundance calculations.

This modular architecture is designed to be user-friendly and easily adaptable with extensions and further refinements as the user sees fit.

6. Usage Instructions

6.1. Installation

To install, it is recommended to create either a Python virtual environment or a conda environment, and then run the following:

```
pip install pbhstat
```

We provide two example `jupyter` Python notebooks for quick-start examples of how to use the code when installed using `pip`. Alternatively, download the code from `github` with

```
git clone https://github.com/pipcole/pbhstat.git
```

and run

```
cd pbhstat
pip install -e .
```

{for an editable install. The required dependencies are `numpy`, `matplotlib`, `scipy`, and `tqdm`.

The code has been tested with Python version 3.9.15 on an Apple M3 Pro running macOS Sonoma 14.1 and with Python 3.12.3 on a Lenovo Yoga 7 2-in-1 14IML9 running Ubuntu 24.04 LTS as well as with Python version 3.11 in a Google Colab notebook.

6.2. Example Usage

We provide two example Python notebooks with the code release for calculating the mass function and PBH abundance from a custom primordial power spectrum and from a pre-defined piecewise power spectrum. We annotate below the step-by-step code snippets to perform this.

First, import `numpy`, `pickle`, or similar if loading own arrays from file and `pbhstat`, assuming it has been successfully installed with `pip`. Then, import the relevant classes and functions from the package. Note that the imports will need to be edited if the package is not installed with pip.

```
import numpy as np
import pickle

import pbhstat

from pbhstat.power_spectrum import PowerSpectrum
from pbhstat.mass_variance import MassVariance
from pbhstat.collapse_stats import PressSchechterModel, PeaksTheoryModel, BroadPeakModel
from pbhstat.mass_function import MassFunction
from pbhstat.plot_utils import plot_power_spectrum, plot_mass_variance, plot_mass_function
```

Next, define a power spectrum and corresponding k-values. All inverse length units should be in Mpc^{-1}. This could be a custom power spectrum either defined analytically, for example,

```
# Set up k values in [1/Mpc]
k_values = np.logspace(3, 7, 3000)

# Custom power spectrum
Ak = 0.008
Deltak = 1
kpeak = 1e5 # [1/Mpc]
P_k_custom = Ak * np.exp(-0.5 * (1/(np.sqrt(2 * np.pi) * Deltak))
            * (np.log(k_values / kpeak) / Deltak)**2)
```

or loaded from file:

```
with open("path_to_k_array.pickle", "rb") as f:
    k_values = pickle.load(f)
with open("path_to_P_k_array.pickle", "rb") as f:
    P_k_custom = pickle.load(f)
```

which should then be instantiated with

```
ps_custom = PowerSpectrum(
    shape='custom',
    k_values=k_values,
    P_k_values=P_k_custom
)
```

Alternatively, the power spectrum can be generated from one of the pre-set options, for example, a piecewise power spectrum which can be instantiated directly:

```
# Set up k values in [1/Mpc]
k_values = np.logspace(3, 7, 3000)

# Piecewise power spectrum
ps_piecewise = PowerSpectrum(
    shape='piecewise',
    amplitude=0.008,
    k_star=1e5,
    ng=4,
    nd=2,
    k_values=k_values
)
```

We will now assume that we are using ps_piecewise for the remainder of these instructions. The next step is to instantiate the mass variance (and higher moments) with a choice of window function (realtophat or gaussian) and statistics (press, peaks, or nonlinear). When choosing the real top-hat window function, there is also an option to use the large-k cut-off as described in Section 3.1—the default is False. There are some incompatible choices which will throw an exception, for example, that the non-linear statistics must be chosen together with the real top-hat window function.

```
# Instantiate mass variance with choice of window function and statistics
mv_piecewise = MassVariance(window='realtophat', power_spectrum=ps_piecewise,
                statistics='nonlinear', cutoff=False)
```

Finally, all that is left is to instantiate the mass function class with the same statistics model as the mass variance, and evaluate on a grid of M values, calculating $f_{\rm PBH}$ as well if desired. Numerical values of K, g_c, and γ can be defined by the user here as optional arguments when instantiating the mass function class, but recommended values are defaulted to.

Furthermore, correlations with v can be enabled or disabled in the non-linear compaction function statistic using the flag vcorr. By toggling vcorr, the user can assess the impact of consistently accounting for the conditional probability of $v = 0$, i.e., the probability of being at an extremum of the compaction function. In particular, setting vcorr=False implies that all $\tilde{\sigma}$ in Equation (23) reduce to their untilde versions. It has been shown analytically in [25] that for a broad spectrum, these correlations are subleading for $R \sim 1/k_{\rm IR}$, where $k_{\rm IR}$ is the infrared scale associated with the broad power spectrum.

The number of M values that $f(M)$ is calculated for can be adjusted from the default 50 with the optional argument mpbh_vals in the mass function evaluate call.

```
# Instantiate mass function
mass_function = MassFunction(mass_variance=mv_custom, statistics='nonlinear', K=6,
                vcorr=True, gamma=0.36)

# Evaluate f(M_PBH) and f_PBH
mpbh, f_mpbh = mass_function.evaluate(k_values, mpbh_vals=50)

fpbh = mass_function.fpbh(f_mpbh, mpbh)
```

Key quantities can be plotted using the routines in plot_utils.py.

```
# Plot power spectrum
plot_power_spectrum(k_values, ps_piecewise(k_values))

# Plot mass variance
# Define R values
R_values = 1/k_values

# Evaluate mass variance at k_values
sigma0sq_piecewise = mv_piecewise.evaluate(k_values)

plot_mass_variance(R_values, sigma0sq_piecewise)

# Plot mass function
plot_mass_function(mpbh, f_mpbh, fpbh_val=fpbh)
```

We plot a comparison of mass functions for different statistics choices in Figure 1 for a broad log-normal distribution and in Figure 2 for a narrow step function.

In Figure 1, the non-linear compaction function statistics yields a narrower and higher peak toward the infrared part of the spectrum compared to the other two statistical approaches considered. This behaviour has been extensively discussed in Ref. [25], and we adopt the same underlying assumptions here, particularly regarding the threshold criterion.

It was shown that for broad power spectra, the mass function is dominated by infrared modes associated with curvature profiles in which the compaction function is sharply peaked (i.e., with large values of w). In the mildly broad log-normal example considered here, these infrared modes produce PBHs that are not-so-heavy due to the suppression implied by the critical scaling relation in Equation (29), namely $M \simeq M_H w^{-2\gamma_{cr}}$ for large w, resulting in PBHs with masses significantly below the horizon mass. It is worth noting that while Press–Schechter and peaks theory lead to mass functions with the same shape, their overall amplitudes agree only because we choose the power spectrum amplitudes such that the total PBH abundance for the two cases are comparable. If instead the same value for the power spectrum amplitude is used for both methods, there is a difference of 2–3 orders of magnitude in the resulting PBH abundance as demonstrated in, e.g., ref. [64]. In Figure 2, as a benchmark for a narrow spectrum, we use a top-hat profile with width $k_{UV}/k_{IR} = 1.3$. For such narrow spectra, all statistics prefer the same comoving scale, namely $r \sim 2.74/k_p$, which corresponds to the first maximum of σ_g^2. The difference in the position of the main peak in the resulting mass function arises from the different choices of threshold and overall normalisation in the critical scaling relation. Specifically, we adopt the same parameters for Press–Schechter and peaks theory (see Section 3), namely $(K, g_c) = (4, 0.77)$. For the non-linear compaction function statistics, we instead choose $\tilde{K} = 6$, and for a narrow spectrum, $g_c \simeq 0.91$ [25].

In the case of a monochromatic power spectrum, all methods produce mass functions that are nearly identical in shape because the statistical profile of the various estimators is essentially fixed. In contrast, for broader spectra, the situation becomes significantly more complex due to the influence of different statistical realisations. This influence is more relevant for the non-linear compaction function statistics, where one scans over different realisations of the curvature of the compaction function. Furthermore, the dependence on the threshold on w provides a way to account for the influence of UV modes on IR modes. Whether this approach is sufficient is still under investigation. In contrast, for the other two statistics, this effect is incorporated by imposing an artificial cut-off at $k \sim 4.49/R$ (see discussion in Section 3.1). These are the sources of the discrepancy behind the different curves in Figure 1. As stated in the introduction, the present tool introduced in this work is also intended to explore differences between distinct statistical approaches that remain under active investigation in the literature.

Finally, we provide a simple interface so that mass functions can be overlaid with the observational constraints available in the `PBHbounds` repository [65], available here: https://github.com/bradkav/PBHbounds/ (accessed on 3 July 2025). Arrays of $f(M)$ and corresponding M values can be saved to the directory in which `PBHbounds` is located with

```
bounds_utility(mpbh, f_mpbh, 'path_to_PBHbounds_directory')
```

Then, a modified version of the default script `PlotPBHbounds.py` is available at https://github.com/pipcole/pbhstat (accessed on 3 July 2025) to be run in the directory of `PBHbounds`. This will overplot the mass function produced with `pbhstat` with the bounds (that can be specified via the command line, if not all bounds are desired) from `PBHbounds`. We emphasise that the constraints themselves depend on the form of the mass function, and therefore overlaying broad mass functions with monochromatic PBH constraints is inconsistent. We recommend that this utility should only be used for narrow mass functions, and in any case as an approximate guide. See [66] for a method for converting constraints from monochromatic to extended mass functions.

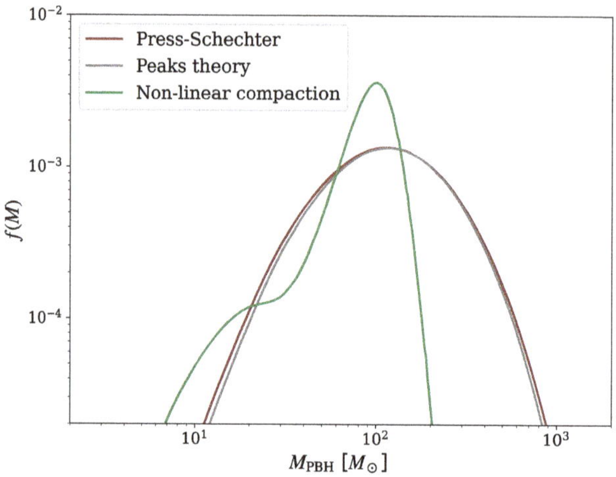

Figure 1. Comparison of mass functions calculated with the three different statistical methods. The power spectrum is chosen to be a log-normal distribution with width $\Delta = 1$ and the window function is a real-space top-hat. The amplitudes of the peak of the power spectrum are $A = \{0.00865, 0.0077, 0.009\}$ for Press–Schechter, peaks theory, and the non-linear compaction statistics, respectively. All mass functions result in $f_{\rm PBH} = 2.5 \times 10^{-3}$.

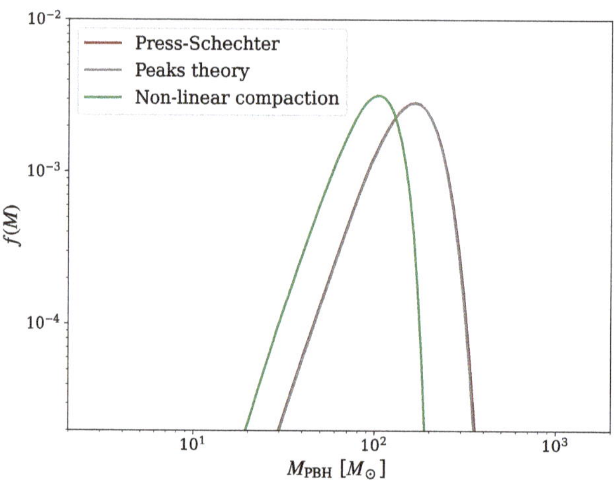

Figure 2. Comparison of mass functions calculated with the three different statistical methods. The power spectrum is chosen to be a narrow top-hat function with width in k-space equal to 1.3 and the window function is a real-space top-hat. The amplitudes of the peak of the power spectrum are $A = \{0.02795, 0.02455, 0.03118\}$ for Press–Schechter, peaks theory, and the non-linear compaction statistics, respectively. All mass functions result in $f_{\rm PBH} = 2.5 \times 10^{-3}$.

7. Summary and Outlook

`pbhstat` is a simple, user-friendly tool for computing the mass function of primordial black holes from an initial primordial power spectrum. Given the many sensitivities of this calculation to the choice of power spectrum, window function, numerical coefficients, and statistics that describe the collapse of the perturbations, we aim to provide a community tool that enables easy comparisons. Furthermore, we hope it will be useful for researchers who are non-experts on primordial black holes, who need a reasonable PBH mass function to input into their pipelines for calculating, e.g., observational constraints from pulsar timing arrays [67,68] or microlensing experiments [69,70]. The code could also provide a supporting tool for future and ongoing efforts toward a complete framework for computing gravitational wave signatures from PBHs, as is in progress within the LISA Cosmology Working Group.

The current version of the code includes the three statistical methods briefly outlined in Section 3 and assumes that the comoving curvature perturbation follows an approximately Gaussian distribution. We would like, in the future, to add the ability to deal with additional complications such as the softening of the equation of state during the QCD phase transition [71–73] (relevant for $\sim M_{\odot}$ black holes) and non-Gaussianity [40–50,74] as well as keeping the package up to date with theoretical advances as they are established. We welcome feedback and suggestions at https://github.com/pipcole/pbhstat (accessed on 3 July 2025).

Author Contributions: Conceptualization, P.S.C. and J.F.; Methodology, P.S.C. and J.F.; Software, P.S.C. and J.F.; Validation, P.S.C. and Jacopo Fumagalli; Writing—original draft, P.S.C. and J.F.; Writing—review and editing, P.S.C. and J.F. All authors have read and agreed to the published version of the manuscript.

Funding: P.C. is supported by ERC Starting Grant No. 945155–GWmining, Cariplo Foundation Grant No. 2021-0555, MUR PRIN Grant No. 2022-Z9X4XS, MUR Grant "Progetto Dipartimenti di Eccellenza 2023-2027" (BiCoQ), and the ICSC National Research Centre funded by NextGenerationEU. The research of J.F. is supported by the Grant PID2022-136224NB-C22, funded by MCIN/AEI/10.13039/501100011033/FEDER, UE, and by the Grant/2021-SGR00872.

Acknowledgments: The authors thank Andrew Gow for helpful discussions.

Conflicts of Interest: The authors declare no conflicts of interest.

Notes

1 In some works, this variable is also denoted by C_G, C_{ζ}, or δ_l.

2 Here, we defined the width parameters to be dimensionless.

3 This choice, strictly speaking, refers to a non-canonical definition of the compaction function; see [32].

4 We remind the reader that all expressions are considered at the maximum of the compaction function where $C' = g' = 0$.

References

1. Zel'dovich, Y.B.; Novikov, I.D. The Hypothesis of Cores Retarded during Expansion and the Hot Cosmological Model. *Soviet Astron. AJ (Engl. Transl.)* **1967**, *10*, 602.
2. Hawking, S. Gravitationally collapsed objects of very low mass. *Mon. Not. Roy. Astron. Soc.* **1971**, *152*, 75. [CrossRef]
3. Carr, B.J.; Hawking, S.W. Black holes in the early Universe. *Mon. Not. Roy. Astron. Soc.* **1974**, *168*, 399–415. [CrossRef]
4. Page, D.N.; Hawking, S.W. Gamma rays from primordial black holes. *Astrophys. J.* **1976**, *206*, 1–7. [CrossRef]
5. MacGibbon, J.H.; Carr, B.J. Cosmic rays from primordial black holes. *Astrophys. J.* **1991**, *371*, 447–469. [CrossRef]
6. Barrau, A.; Boudoul, G.; Donnard, J.; Grain, J.; Servant, G. Gamma-ray constraints on the density of primordial black holes. *Astron. Astrophys.* **2003**, *398*, 403–410. [CrossRef]
7. Sasaki, M.; Suyama, T.; Tanaka, T.; Yokoyama, S. Primordial black hole scenario for the gravitational wave event GW150914. *Phys. Rev. Lett.* **2016**, *117*, 061101. [CrossRef]
8. Carr, B.; Kohri, K.; Sendouda, Y.; Yokoyama, J. Constraints on primordial black holes. *Rept. Prog. Phys.* **2021**, *84*, 116902. [CrossRef] [PubMed]
9. Green, A.M.; Kavanagh, B.J. Primordial black holes as a dark matter candidate. *J. Phys. G* **2021**, *48*, 043001. [CrossRef]
10. Khlopov, M.Y. Primordial black holes. *Res. Astron. Astrophys.* **2010**, *10*, 495–528. [CrossRef]
11. Ivanov, P.; Naselsky, P.; Novikov, I. Inflation and primordial black holes as dark matter. *Phys. Rev. D* **1994**, *50*, 7173–7178. [CrossRef]
12. Carr, B.J.; Gilbert, J.H.; Lidsey, J.E. Black hole relics and inflation: Limits on blue perturbation spectra. *Phys. Rev. D* **1994**, *50*, 4853–4867. [CrossRef]

13. Garcia-Bellido, J.; Linde, A.D.; Wands, D. Density perturbations and black hole formation in hybrid inflation. *Phys. Rev. D* **1996**, *54*, 6040–6058. [CrossRef]
14. Yoo, C.M. The basics of primordial black hole formation and abundance estimation. *arXiv* **2024**, arXiv:2211.13512. [CrossRef]
15. Carr, B.; Kuhnel, F. Primordial black holes as dark matter candidates. *SciPost Phys. Lect. Notes* **2022**, *48*, 1. [CrossRef]
16. Escrivà, A.; Kuhnel, F.; Tada, Y. Primordial Black Holes. In *Black Holes in the Era of Gravitational-Wave Astronomy*; Elsevier Amsterdam, The Netherlands, 2022. [CrossRef]
17. Musco, I.; Miller, J.C.; Rezzolla, L. Computations of primordial black hole formation. *Class. Quant. Grav.* **2005**, *22*, 1405–1424. [CrossRef]
18. Escrivà, A. Simulation of primordial black hole formation using pseudo-spectral methods. *Phys. Dark Univ.* **2020**, *27*, 100466. [CrossRef]
19. Musco, I. Numerical Simulations of Primordial Black Holes. In *Primordial Black Holes*; Springer: Singapore, 2025. [CrossRef]
20. Germani, C.; Musco, I. Abundance of Primordial Black Holes Depends on the Shape of the Inflationary Power Spectrum. *Phys. Rev. Lett.* **2019**, *122*, 141302. [CrossRef] [PubMed]
21. Musco, I. Threshold for primordial black holes: Dependence on the shape of the cosmological perturbations. *Phys. Rev. D* **2019**, *100*, 123524. [CrossRef]
22. Ando, K.; Inomata, K.; Kawasaki, M. Primordial black holes and uncertainties in the choice of the window function. *Physical Review D* **2018**, *97*, 103528. [CrossRef]
23. Yoo, C.M.; Harada, T.; Garriga, J.; Kohri, K. Primordial black hole abundance from random Gaussian curvature perturbations and a local density threshold. *Prog. Theor. Exp. Phys.* **2018**, *2018*, 123E01. [CrossRef]
24. Young, S. Computation of the Abundance of Primordial Black Holes. In *Primordial Black Holes*; Springer: Singapore, 2025. [CrossRef]
25. Fumagalli, J.; Garriga, J.; Germani, C.; Sheth, R.K. Unexpected shape of the primordial black hole mass function. *Phys. Rev. D* **2025**, *111*, 123518. [CrossRef]
26. Pi, S.; Sasaki, M.; Takhistov, V.; Wang, J. Primordial Black Hole Formation from Power Spectrum with Finite-width. *arXiv* **2024**, arXiv:2501.00295. [CrossRef]
27. Press, W.H.; Schechter, P. Formation of Galaxies and Clusters of Galaxies by Self-Similar Gravitational Condensation. *Astrophys. J.* **1974**, *187*, 425–438. [CrossRef]
28. Carr, B.J. The Primordial black hole mass spectrum. *Astrophys. J.* **1975**, *201*, 1–19. [CrossRef]
29. Bardeen, J.M.; Bond, J.R.; Kaiser, N.; Szalay, A.S. The Statistics of Peaks of Gaussian Random Fields. *Astrophys. J.* **1986**, *304*, 15. [CrossRef]
30. Green, A.M.; Liddle, A.R.; Malik, K.A.; Sasaki, M. A New calculation of the mass fraction of primordial black holes. *Phys. Rev. D* **2004**, *70*, 041502. [CrossRef]
31. Young, S.; Byrnes, C.T.; Sasaki, M. Calculating the mass fraction of primordial black holes. *J. Cosmol. Astropart. Phys.* **2014**, *2014*, 45. [CrossRef]
32. Young, S.; Musso, M. Application of peaks theory to the abundance of primordial black holes. *J. Cosmol. Astropart. Phys.* **2020**, *11*, 22. [CrossRef]
33. Germani, C.; Sheth, R.K. Nonlinear statistics of primordial black holes from Gaussian curvature perturbations. *Phys. Rev. D* **2020**, *101*, 063520. [CrossRef]
34. Germani, C.; Sheth, R.K. The Statistics of Primordial Black Holes in a Radiation-Dominated Universe: Recent and New Results. *Universe* **2023**, *9*, 421. [CrossRef]
35. Özsoy, O.; Tasinato, G. Inflation and Primordial Black Holes. *Universe* **2023**, *9*, 203. [CrossRef]
36. Pattison, C.; Vennin, V.; Assadullahi, H.; Wands, D. Quantum diffusion during inflation and primordial black holes. *J. Cosmol. Astropart. Phys.* **2017**, *2017*, 46. [CrossRef]
37. Ezquiaga, J.M.; García-Bellido, J. Quantum diffusion beyond slow-roll: Implications for primordial black-hole production. *J. Cosmol. Astropart. Phys.* **2018**, *2018*, 18. [CrossRef]
38. Biagetti, M.; De Luca, V.; Franciolini, G.; Kehagias, A.; Riotto, A. The formation probability of primordial black holes. *Phys. Lett. B* **2021**, *820*, 136602. [CrossRef]
39. Vennin, V.; Wands, D. Quantum Diffusion and Large Primordial Perturbations from Inflation. In *Primordial Black Holes*; Springer Nature: Singapore, 2025; pp. 201–227. [CrossRef]
40. Bullock, J.S.; Primack, J.R. NonGaussian fluctuations and primordial black holes from inflation. *Phys. Rev. D* **1997**, *55*, 7423–7439. [CrossRef]
41. Ivanov, P. Nonlinear metric perturbations and production of primordial black holes. *Phys. Rev. D* **1998**, *57*, 7145–7154. [CrossRef]
42. Young, S.; Byrnes, C.T. Primordial black holes in non-Gaussian regimes. *J. Cosmol. Astropart. Phys.* **2013**, *08*, 52. [CrossRef]
43. Byrnes, C.T.; Copeland, E.J.; Green, A.M. Primordial black holes as a tool for constraining non-Gaussianity. *Phys. Rev. D* **2012**, *86*, 043512. [CrossRef]

44. Tada, Y.; Yokoyama, S. Primordial black holes as biased tracers. *Phys. Rev. D* **2015**, *91*, 123534. [CrossRef]
45. Atal, V.; Germani, C. The role of non-Gaussianities in Primordial Black Hole formation. *Phys. Dark Univ.* **2019**, *24*, 100275. [CrossRef]
46. Franciolini, G.; Kehagias, A.; Matarrese, S.; Riotto, A. Primordial Black Holes from Inflation and non-Gaussianity. *J. Cosmol. Astropart. Phys.* **2018**, *3*, 16. [CrossRef]
47. Kehagias, A.; Musco, I.; Riotto, A. Non-Gaussian Formation of Primordial Black Holes: Effects on the Threshold. *J. Cosmol. Astropart. Phys.* **2019**, *12*, 29. [CrossRef]
48. Riccardi, F.; Taoso, M.; Urbano, A. Solving peak theory in the presence of local non-gaussianities. *J. Cosmol. Astropart. Phys.* **2021**, *8*, 60. [CrossRef]
49. Gow, A.D.; Assadullahi, H.; Jackson, J.H.P.; Koyama, K.; Vennin, V.; Wands, D. Non-perturbative non-Gaussianity and primordial black holes. *EPL* **2023**, *142*, 49001. [CrossRef]
50. Matsubara, T.; Sasaki, M. Non-Gaussianity effects on the primordial black hole abundance for sharply-peaked primordial spectrum. *J. Cosmol. Astropart. Phys.* **2022**, *2022*, 94. [CrossRef]
51. Nakama, T.; Silk, J.; Kamionkowski, M. Stochastic gravitational waves associated with the formation of primordial black holes. *Phys. Rev. D* **2017**, *95*, 043511. [CrossRef]
52. Shibata, M.; Sasaki, M. Black hole formation in the Friedmann universe: Formulation and computation in numerical relativity. *Phys. Rev. D* **1999**, *60*, 084002. [CrossRef]
53. Harada, T.; Yoo, C.M.; Nakama, T.; Koga, Y. Cosmological long-wavelength solutions and primordial black hole formation. *Phys. Rev. D* **2015**, *91*, 084057. [CrossRef]
54. Uehara, K.; Escrivà, A.; Harada, T.; Saito, D.; Yoo, C.M. Numerical simulation of type II primordial black hole formation. *J. Cosmol. Astropart. Phys.* **2025**, *1*, 3. [CrossRef]
55. Shimada, M.; Escrivá, A.; Saito, D.; Uehara, K.; Yoo, C.M. Primordial black hole formation from type II fluctuations with primordial non-Gaussianity. *J. Cosmol. Astropart. Phys.* **2025**, *2*, 18. [CrossRef]
56. Inui, R.; Joana, C.; Motohashi, H.; Pi, S.; Tada, Y.; Yokoyama, S. Primordial black holes and induced gravitational waves from logarithmic non-Gaussianity. *J. Cosmol. Astropart. Phys.* **2025**, *3*, 21. [CrossRef]
57. Escrivà, A. The threshold for PBH formation in the type-II region and its analytical estimation. *arXiv* **2025**, arXiv:2504.05814. [CrossRef]
58. Gow, A.D.; Byrnes, C.T.; Cole, P.S.; Young, S. The power spectrum on small scales: Robust constraints and comparing PBH methodologies. *J. Cosmol. Astropart. Phys.* **2021**, *2*, 2. [CrossRef]
59. Young, S. The primordial black hole formation criterion re-examined: Parametrisation, timing and the choice of window function. *Int. J. Mod. Phys. D* **2019**, *29*, 2030002. [CrossRef]
60. Niemeyer, J.C.; Jedamzik, K. Near-Critical Gravitational Collapse and the Initial Mass Function of Primordial Black Holes. *Phys. Rev. Lett.* **1998**, *80*, 5481–5484. [CrossRef]
61. Young, S.; Musco, I.; Byrnes, C.T. Primordial black hole formation and abundance: Contribution from the non-linear relation between the density and curvature perturbation. *J. Cosmol. Astropart. Phys.* **2019**, *11*, 12. [CrossRef]
62. Escrivà, A.; Germani, C.; Sheth, R.K. Universal threshold for primordial black hole formation. *Phys. Rev. D* **2020**, *101*, 044022. [CrossRef]
63. Kehagias, A.; Perrone, D.; Riotto, A. Why the universal threshold for primordial black hole formation is universal. *Class. Quant. Grav.* **2025**, *42*, 055010. [CrossRef]
64. Wang, Q.; Liu, Y.C.; Su, B.Y.; Li, N. Primordial black holes from the perturbations in the inflaton potential in peak theory. *Phys. Rev. D* **2021**, *104*, 083546. [CrossRef]
65. Kavanagh, B.J. *Bradkav/PBHbounds: Release Version*, version = 1.0; Zenodo: Geneve, Switzerland, 2019. [CrossRef]
66. Bellomo, N.; Bernal, J.L.; Raccanelli, A.; Verde, L. Primordial Black Holes as Dark Matter: Converting Constraints from Monochromatic to Extended Mass Distributions. *J. Cosmol. Astropart. Phys.* **2018**, *1*, 4. [CrossRef]
67. Afzal, A.; Agazie, G.; Anumarlapudi, A.; Archibald, A.M.; Arzoumanian, Z.; Baker, P.T.; Bécsy, B.; Blanco-Pillado, J.J.; Blecha, L.; Boddy, K.K. The NANOGrav 15 yr Data Set: Search for Signals from New Physics. *Astrophys. J. Lett.* **2023**, *951*, L11. [CrossRef]
68. Antoniadis, J.; Arumugam, P.; Arumugam, S.; Babak, S.; Bagchi, M.; Nielsen, A.S.B.; Bassa, C.G.; Bathula, A.; Berthereau, A.; Bonetti, M.; et al. The second data release from the European Pulsar Timing Array-IV. Implications for massive black holes, dark matter, and the early Universe. *A&A* **2024**, *685*, A94. [CrossRef]
69. Niikura, H.; Takada, M.; Yasuda, N.; Lupton, R.H.; Sumi, T.; More, S.; Kurita, T.; Sugiyama, S.; More, A.; Oguri, M.; et al. Microlensing constraints on primordial black holes with Subaru/HSC Andromeda observations. *Nat. Astron.* **2019**, *3*, 524–534. [CrossRef]
70. Mróz, P.; Udalski, A.; Szymański, M.K.; Soszyński, I.; Pietrukowicz, P.; Kozłowski, S.; Poleski, R.; Skowron, J.; Ulaczyk, K.; Gromadzki, M.; et al. Limits on Planetary-mass Primordial Black Holes from the OGLE High-cadence Survey of the Magellanic Clouds. *Astrophys. J. Lett.* **2024**, *976*, L19. [CrossRef]

71. Byrnes, C.T.; Hindmarsh, M.; Young, S.; Hawkins, M.R. Primordial black holes with an accurate QCD equation of state. *J. Cosmol. Astropart. Phys.* **2018**, *2018*, 41. [CrossRef]
72. Escrivà, A.; Bagui, E.; Clesse, S. Simulations of PBH formation at the QCD epoch and comparison with the GWTC-3 catalog. *J. Cosmol. Astropart. Phys.* **2023**, *2023*, 4. [CrossRef]
73. Musco, I.; Jedamzik, K.; Young, S. Primordial black hole formation during the QCD phase transition: Threshold, mass distribution, and abundance. *Phys. Rev. D* **2024**, *109*, 083506. [CrossRef]
74. Young, S.; Musco, I.; Byrnes, C.T. Influence of large local and nonlocal bispectra on primordial black hole abundance. *J. Cosmol. Astropart. Phys.* **2022**, *2*, 32. [CrossRef]

universe

MDPI

Article

Two-Loop Corrections in Power Spectrum in Models of Inflation with Primordial Black Hole Formation

Hassan Firouzjahi

School of Astronomy, Institute for Research in Fundamental Sciences (IPM), Tehran P.O. Box 19395-5746, Iran; firouz@ipm.ir

Abstract: We calculated the two-loop corrections in the primordial power spectrum in models of single-field inflation incorporating an intermediate USR phase employed for PBH formation. Among the overall eleven one-particle irreducible Feynman diagrams, we calculated the corrections from the "double scoop" two-loop diagram involving two vertices of quartic Hamiltonians. We demonstrate herein the fractional two-loop correction in power spectrum scales, like the square of the fractional one-loop correction. We confirm our previous findings that the loop corrections become arbitrarily large in the setup where the transition from the intermediate USR to the final slow-roll phase is very sharp. This suggests that in order for the analysis to be under perturbative control against loop corrections, one requires a mild transition with a long enough relaxation period towards the final attractor phase.

Keywords: inflation; primordial black holes; loop corrections

1. Introduction

There have been intense debates in the recent literature on the nature of loop corrections in single-field models of inflation involving an intermediate ultra slow-roll (USR) phase [1–37]; for earlier works concerning the quantum loop effects in models of inflation, see [38–42]. These models have been employed to generate primordial black holes (PBHs) as candidates for the observed dark matter [43–46]; for a review on the mechanism of generating PBHs from a USR setup, see [47–51]. In the USR setup involving a flat potential, the curvature perturbation grows on superhorizon scales. The enhancement in the power spectrum allows one to use this setup to generate PBHs on desired scales. However, the rapid growth of the power spectrum during the intermediate USR stage can be problematic. More specifically, Kristiano and Yokoyama argued in [1] that the one-loop corrections originating from the small-scale USR modes can affect the CMB perturbations. Correspondingly, it was originally concluded in [1] that the analysis is not under perturbative control and the setup is not trusted for PBH formation. Following [1], the one-loop corrections in the curvature perturbation power spectrum were studied with different (conflicting) conclusions. For example, the results of [1] was criticized by Riotto [3,4], arguing that the one-loop corrections can be small if the transition to the final attractor phase is smooth enough. Similarly, in [13], employing δN formalism, it was argued that the large loop effects in the models with mild transitions are suppressed by the slow-roll parameters and the model is under perturbative control for generating PBHs. In addition, the loop corrections were studied numerically in [27] and also by employing the formalism of separate universe in [28].

In order to estimate the total one-loop corrections in the power spectrum, we need to have both the cubic and quartic Hamiltonians. The cubic action was calculated originally by Maldacena [52], but there is still no concrete result for the full quartic interaction including the USR phase; for earlier studies on quartic action, however, see [53,54]. In [11], we employed the formalism of the effective field theory (EFT) of inflation, which enabled us to calculate the cubic and the quartic Hamiltonians in reasonable ease in the decoupling limit.

Furthermore, we were able to incorporate the effects of the sharpness of the transition from the USR stage to the final slow-roll phase as well. We showed in [11] that the one-loop corrections can be dangerous in models with a sharp transition, supporting the conclusion of [1].

Based on the physical intuitions and the expectation of the decoupling of scales, it looks counterintuitive that the small USR modes can affect the long CMB scales in the first place [4,25]. Motivated by this question, the conclusion of [1] was critically revisited in [23,24], where it was claimed that the one-loop corrections are canceled. Specifically, in [23], the contributions of the boundary terms, which were not taken into account in previous works, were highlighted. On the other hand, it was argued in [24] that the loop contributions vanish after the UV limit of the momentum is considered via some $i\varepsilon$ prescription. However, in both [23,24], like many other previous works, only the cubic interactions were considered. The conclusions of [23,24] were reviewed critically in [25], highlighting the flaws in their arguments. More recently, there were new claims of loop cancellation in [55–57]; we would like to return to the results of these claims elsewhere.

As for the physical origins of the loop corrections, the non-linear coupling between the short and long modes induces a source term in the equation governing the evolution of the long mode. At the same time, the spectrum of the short modes are modulated by the long modes. The modulation effects becomes large if the power spectrum of the short modes is highly scale-dependent which is the case in the USR model. The combined effects of the non-linear coupling between the short and long modes and the modulation of the short modes by the long mode backreacts on the long mode itself, causing loop corrections, as highlighted in [3,15,25].

In light of the above discussions, in this work, we aim to calculate the corrections in power spectrum at the two-loop order. As one may expect, the analysis at two-loop order are significantly more complicated than in the one-loop case. First, we have more Feynman diagrams involving not only the cubic and quartic interactions, but also the quintic and sextic Hamiltonians. Secondly, most of these Feynman diagrams involve double- or higher-order nested in-in integrals, which make the time integral very complicated. As we shall see, there are in total eleven one-particle irreducible Feynman diagrams at the two-loop level. Out of these eleven diagrams, we consider the case of the "double scoop" diagram, which involves a double nested time integral containing two quartic Hamiltonians. We believe that the results from this case are illustrative enough, which can shed light on the structure of the two-loop corrections.

2. The Setup

In this section, we briefly review our setup. This is the same setup as employed in [1] for PBH formation. It is a single-field inflation with three distinct phases, $SR \to USR \to SR$, where the first and and the third stages are assumed to be in SR phases while the USR phase is sandwiched in between. The observed CMB perturbations leave the horizon during the first SR phase with the amplitude of power spectrum set by the COBE normalization. However, the intermediate USR phase, engineered to produce the PBHs at the desired mass scales, may start at about 30 e-folds after the CMB scales have left the Hubble horizon, and it typically lasts for 2-3 e-folds. Finally, the intermediate USR phase is followed by the second SR phase, where it is assumed that the system reaches its attractor stage.

As usual, during the SR phases, the curvature perturbation $\mathcal{R}$ is frozen after the mode leaves the horizon. However, during the USR phase, it experiences an exponential growth [58–64]. The rapid growth of the modes which become superhorizon during the intermediate USR phase is the key idea behind the enhancement in the curvature perturbation power spectrum to generate PBHs on the corresponding scales. In addition, the superhorizon growth of curvature perturbation plays important roles in violating the Maldacena non-Gaussianity consistency condition in the USR model [59–61,65–77].

Starting with the FLRW metric,

$$ds^2 = -dt^2 + a(t)^2 d\mathbf{x}^2 , \tag{1}$$

the equations governing the dynamics of the background inflaton field ϕ and the scale factor $a(t)$ during the USR stage are

$$\ddot{\phi}(t) + 3H\dot{\phi}(t) = 0 , \qquad 3M_P^2 H^2 \simeq V_0, \tag{2}$$

in which M_P is the reduced Planck mass, H is the Hubble expansion rate, and V_0 is the constant potential during the USR stage.

Since H is very nearly constant during the USR phase, then $\dot{\phi} \propto a^{-3}$. Correspondingly, the first slow-roll parameter $\epsilon \equiv -\frac{\dot{H}}{H^2}$ falls off like a^{-6}, while the second slow-roll parameter $\eta \equiv \frac{\dot{\epsilon}}{H\epsilon}$ is nearly constant, $\eta \simeq -6$. It is assumed that the USR phase is extended in the interval $\tau_s < \tau < \tau_e$, so ϵ, at the time of the end of the USR, is given by $\epsilon_e = \epsilon_i \left(\frac{\tau_e}{\tau_s}\right)^6$, where ϵ_i is the value of ϵ at the start of the USR phase. Here, τ is the conformal time, which is related to cosmic time as usual via $d\tau = dt/a(t)$, with the understanding that towards the end of inflation, $\tau \to 0$. Alternatively, working with the number of e-fold $dN = Hdt$ as the clock, the duration of the USR is determined by $\Delta N \equiv N(\tau_e) - N(\tau_s)$, yielding $\epsilon_e = \epsilon_i e^{-6\Delta N}$. For PBH formation, we typically require ΔN to be around 2 to 3 e-folds.

To simplify the analysis, we assume that the transitions $SR \to USR$ and $USR \to SR$ happen instantaneously, at $\tau = \tau_s$ and $\tau = \tau_e$, respectively. However, it may take time to end up in its attractor phase during the final SR phase. This is determined by the sharpness (which is actually the relaxation) parameter h, initially defined in [77] via

$$h = -6\sqrt{\frac{\epsilon_V}{\epsilon_e}} , \tag{3}$$

in which ϵ_V is the value of ϵ during the final slow-roll phase, which is determined as usual by the first derivative of the potential. Note that by construction, $h < 0$. In our analysis, we assume a sharp enough transition, so $|h| > 1$. For a mild transition, e.g., h, at the order of slow-roll parameters, the mode function keeps evolving during the final phase and the analysis become complicated. However, for a sharp transition, the system reaches the attractor phase quickly and the errors in our analytical results are expected to be negligible.

For a very sharp transition with $h \to -\infty$, ϵ approaches rapidly to a larger value such that towards the end of inflation, $\epsilon(\tau_0) \simeq \epsilon_V = \epsilon_e(\frac{h}{6})^2$. On the other hand, for an "instant" sharp transition, which was assumed in [1,2], one has $h = -6$. In this situation, ϵ in the third SR phase is equal to its value at the time of the end of the USR, i.e., $\epsilon_V = \epsilon_e$.

The evolution of the slow-roll parameters after the USR phase is studied in [77]. In particular, ϵ is smooth across the transition point but η experiences a jump at $\tau = \tau_e$. Prior to the transition and close to the end of the USR, $\eta = -6$, while right after the transition, we have $\eta = -6 - h$. Following [77], we can approximate η near the transition point as follows:

$$\eta = -6 - h\theta(\tau - \tau_e) \qquad \tau_e^- < \tau < \tau_e^+ , \tag{4}$$

yielding to

$$\frac{d\eta}{d\tau} = -h\delta(\tau - \tau_e) , \qquad \tau_e^- < \tau < \tau_e^+ . \tag{5}$$

As we shall see, the jump in η, highlighted by the Dirac delta function above, plays a crucial role in the loop corrections.

After presenting our background, we briefly review the perturbations in this setup. To perform the in-in analysis, we need the mode function associated to the comoving curvature perturbation $\mathcal{R}$ during the USR and afterwards. In the Fourier space, the mode function is

$$\mathcal{R}(\mathbf{x},t) = \int \frac{d^3k}{(2\pi)^3} e^{i\mathbf{k}\cdot\mathbf{x}} \hat{\mathcal{R}}_{\mathbf{k}}(t)\,, \tag{6}$$

where, as usual, the operator $\hat{\mathcal{R}}_{\mathbf{k}}(t)$ is expressed in terms of the creation and annihilation operators $a_{\mathbf{k}}$ and $a^{\dagger}_{\mathbf{k}}$ via $\hat{\mathcal{R}}_{\mathbf{k}}(t) = \mathcal{R}_k(t) a_{\mathbf{k}} + \mathcal{R}^*_k(t) a^{\dagger}_{-\mathbf{k}}$. In this notation, $\hat{\mathcal{R}}_{\mathbf{k}}$ is a quantum operator and $\mathcal{R}_k$ is the mode function. As usual, the creation and annihilation operators satisfy the standard commutation relations $[a_{\mathbf{k}}, a^{\dagger}_{\mathbf{k}'}] = (2\pi)^3\delta(\mathbf{k}-\mathbf{k}')$.

The quantum initial condition is fixed by the Bunch–Davies vacuum with the mode function

$$\mathcal{R}^{(1)}_k = \frac{H}{M_P\sqrt{4\epsilon_i k^3}}(1+ik\tau)e^{-ik\tau}\,, \qquad (\tau < \tau_s)\,. \tag{7}$$

During the intermediate USR phase, the mode function is parameterized via

$$\mathcal{R}^{(2)}_k = \frac{H}{M_P\sqrt{4\epsilon_i k^3}}\left(\frac{\tau_s}{\tau}\right)^3 \Big[\alpha^{(2)}_k (1+ik\tau)e^{-ik\tau} + \beta^{(2)}_k (1-ik\tau)e^{ik\tau}\Big]\,, \tag{8}$$

in which the coefficients $\alpha^{(2)}_k$ and $\beta^{(2)}_k$ are determined by imposing the continuity of the mode function and its time derivative at $\tau=\tau_s$, yielding

$$\alpha^{(2)}_k = 1 + \frac{3i}{2k^3\tau_s^3}(1+k^2\tau_s^2)\,, \qquad \beta^{(2)}_k = -\frac{3i}{2k^3\tau_s^3}(1+ik\tau_s)^2 e^{-2ik\tau_s}\,. \tag{9}$$

On the other hand, imposing the matching conditions at τ_e, the outgoing mode function during the final SR phase is given by [11]

$$\mathcal{R}^{(3)}_k = \frac{H}{M_P\sqrt{4\epsilon(\tau)k^3}}\Big[\alpha^{(3)}_k (1+ik\tau)e^{-ik\tau} + \beta^{(3)}_k (1-ik\tau)e^{ik\tau}\Big]\,, \tag{10}$$

in which $\alpha^{(3)}_k$ and $\beta^{(3)}_k$ are determined to be

$$\alpha^{(3)}_k = \frac{1}{8k^6\tau_s^3\tau_e^3}\Big[3h(1-ik\tau_e)^2(1+ik\tau_s)^2 e^{2ik(\tau_e-\tau_s)} - i(2k^3\tau_s^3 + 3ik^2\tau_s^2 + 3i)(4ik^3\tau_e^3 - hk^2\tau_e^2 - h)\Big]\,,$$

and

$$\beta^{(3)}_k = \frac{-1}{8k^6\tau_s^3\tau_e^3}\Big[3(1+ik\tau_s)^2(h + hk^2\tau_e^2 + 4ik^3\tau_e^3)e^{-2ik\tau_s} + ih(1+ik\tau_e)^2(3i + 3ik^2\tau_s^2 + 2k^3\tau_s^3)e^{-2ik\tau_e}\Big]\,.$$

With the mode functions given above, we can calculate the two-loop corrections in the curvature perturbations' power spectrum. We choose the convention that the momentum associated to the long CMB modes are denoted by $\mathbf{p}_1$ and $\mathbf{p}_2$, while the momentum corresponding to the short modes that run inside the loops are denoted by $\mathbf{q}$ and $\mathbf{k}$. There is the vast hierarchy $p_i \ll q, k$. In our analysis, we are interested in the loop corrections induced from the short modes, which become superhorizon during the USR phase. Therefore, we cut the momentum loop integrals in the range $q_s \le q < q_e$, where $q_s = -\frac{1}{\tau_s}$ and $q_e = -\frac{1}{\tau_e}$ are the modes which leave the horizon at $\tau=\tau_s$ and $\tau=\tau_e$, respectively. Furthermore, the duration of the USR period $\Delta N \equiv N(\tau_e) - N(\tau_s)$ is given in terms of q_s and q_e by

$$e^{-\Delta N} = \frac{\tau_e}{\tau_s} = \frac{q_s}{q_e}\,. \tag{11}$$

As mentioned previously, to generate PBHs with the desired mass scales, we require $\Delta N \sim 2-3$.

In order to simplify the analysis, we have assumed an instant transition at $\tau=\tau_e$ to the final SR phase. However, the mode functions keep evolving for $\tau > \tau_e$ before it assumes its final attractor value. This process is governed by the relaxation parameter h. For example, for an instant and sharp transition, which was employed in [1,2] with $h = -6$, at the end of inflation, $\mathcal{R}$ is smaller by a factor of $1/4$ compared to its value at $\tau = \tau_e$. The reason is that the mode function is not frozen right after the transition and it keeps evolving until it reaches its attractor value. However, for a very sharp transition corresponding to $h \to -\infty$, the mode function is assumed to freeze immediately after τ_e. This is the situation considered in [59,61], which yields $f_{NL} = \frac{5}{2}$. But, as demonstrated in [77], for a mild transition with $|h| \ll 1$, non-Gaussianity is mostly erased during the second SR phase. Motivated by these discussions, we distinguish between an instant transition and a sharp transition. For example, the assumption of an instant transition can be relaxed, and one may consider the situation where the transition takes place within some time interval [15,27]. However, this will complicate our theoretical analysis and is beyond the scope of this work.

3. Two-Loop Feynman Diagrams and Interaction Hamiltonians

In order to obtain the loop corrections in the curvature perturbation power spectrum $\mathcal{P}_\mathcal{R}$, we need the interaction Hamiltonians. Here we present the structure of two-loops Feynman diagrams and the subset of interaction Hamiltonians necessary for our two-loop calculations.

To understand the structure of Feynman diagrams associated at the two-loop level, consider a general L-loop one-particle irreducible Feynman diagram associated to the following scalar-type potential:

$$V = \sum_n g_n \phi^n \qquad (n > 2)\,, \tag{12}$$

in which g_n is the coupling (vertex) and n is the order of interaction. For example, for the cubic and quartic interactions, we have $n = 3$ and $n = 4$, respectively. Suppose that we have a Feynman diagram with L loops, P internal propagators, V_n vertices associated to each power of interaction n, and N external lines. For example, in our case of interest, $L = 2$ (two-loops) and $N = 2$ (two external lines for power spectrum). Then, using the following topological conditions [78]:

$$L = P - \sum_n V_n + 1 \tag{13}$$

and

$$N + 2P = \sum_n n V_n\,, \tag{14}$$

we obtain the following relation between L, N and V_n:

$$2L = (2 - N) + \sum_n (n-2) V_n\,. \tag{15}$$

For the loop corrections in the power spectrum with $N = 2$, this further simplifies to

$$2L = \sum_n (n-2) V_n \qquad (N = 2)\,. \tag{16}$$

In particular, for one-loop corrections, the above condition allows for only two Feynman diagrams, a single quartic vertex, and a diagram with two cubic vertices, as studied in detail in [11].

Now, in our current case of interest with two-loop corrections ($L = 2$), Equation (16) yields the following constraint:

$$4 = V_3 + 2V_4 + 3V_5 + 4V_6\,. \tag{17}$$

Based on the allowed integer solutions of the above equation, one obtains the allowed Feynman diagrams. Below are all possible allowed solutions:

$$\begin{aligned} (1) \quad &: V_6 = V_5 = V_3 = 0,\ V_4 = 2\,, \\ (2) \quad &: V_6 = V_5 = V_4 = 0,\ V_3 = 4, \\ (3) \quad &: V_6 = V_5 = 0,\ V_4 = 1, V_3 = 2, \\ (4) \quad &: V_6 = V_4 = 0,\ V_3 = V_5 = 1\,, \\ (5) \quad &: V_3 = V_4 = V_5 = 0,\ V_6 = 1\,. \end{aligned} \tag{18}$$

The above solutions yield 11 distinct two-loop diagrams as plotted in Figure 1. The Feynman diagrams in the first category involves two vertices of quartic Hamiltonian $\mathbf{H}_4$ (diagrams (**a**,**b**) in Figure 1), while the diagrams in the second category involve four vertices of the cubic Hamiltonian $\mathbf{H}_3$ (diagrams (**c**,**d**)). The diagrams in the third category contain three vertices, one from $\mathbf{H}_4$ and two from $\mathbf{H}_3$ (diagrams (**e**–**h**)). On the other hand, the diagram in the fourth category involves two vertices, one from $\mathbf{H}_3$ and one from the quintic interaction $\mathbf{H}_5$ (diagrams (**k**,**l**)). Finally, the diagram in the fifth category involves a single vertex from the sextic Hamiltonian $\mathbf{H}_6$ (diagram (**m**)). From the above discussions, we see that we need $\mathbf{H}_3, \mathbf{H}_4, \mathbf{H}_5$ and $\mathbf{H}_6$ to calculate the full two-loop corrections in $\mathcal{P}_{\mathcal{R}}$.

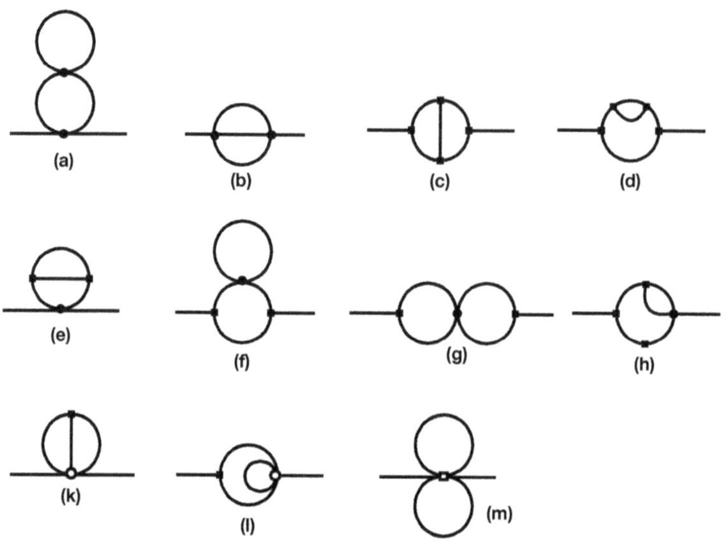

Figure 1. The one-particle irreducible Feynman diagrams for the two-loop corrections constructed from the solutions of Equation (18). The diagrams (**a**,**b**) belong to category (1) in Equation (18), diagrams (**c**,**d**) to category (2), diagrams (**e**–**h**) to category (3), diagrams (**k**,**l**) belong to categories (4), and diagram (**m**) belongs to category (5).

The cubic action for the curvature perturbations $\mathcal{R}$ and the corresponding cubic Hamiltonian were calculated in detail by Maldacena [52]. However, calculating the quartic action and the corresponding quartic Hamiltonian in this method is a very difficult task. Fortunately, the formalism of the EFT of inflation [79,80] provides a very useful alternative, in which the interaction Hamiltonians can be calculated with reasonable ease. In particular, in the decoupling limit where the gravitational backreactions can be neglected, the EFT formalism was employed to calculate the cubic and quartic Hamiltonians in [11] (see also [66] for the first work in this direction, calculating the cubic Hamiltonian). While the cubic and the quartic Hamiltonians were constructed in [11], one still needs to calculate $\mathbf{H}_5$ and $\mathbf{H}_6$ to perform the full two-loop corrections. In principle, it is possible to calculate $\mathbf{H}_5$ and $\mathbf{H}_6$ using the EFT approach, but it turns out that there are new technical complications which require careful considerations [81].

The analysis of full two-loop corrections associated to the above 11 diagrams is a demanding task. As a first step forward, we calculated the two-loop corrections from the "double scoop" diagram (**a**), which are relatively easier to technically handle. This is because this diagram involves two vertices, so one deals with double nested in-in integrals (this is also true for diagram (**b**)). However, diagrams (**c–h**) contain nested integrals with three-fold or four-fold time integrals involving $\mathbf{H}_3$ or $\mathbf{H}_4$, which are far more complicated than diagram (**a**). As we shall see, the analysis even for the simple-looking diagram (**a**) is non-trivial. Having said this, physically, one expects that the result obtained from this single diagram will shed light onto the structure of two-loop corrections, which should not be vary different than the remaining diagrams.

Here, we briefly review the results of [11], which are required to calculate the quartic Hamiltonian to calculate the loop corrections from diagram (**a**). We refer the reader to [11] for further details.

The second-order action employed to quantize the free theory is

$$S_2 = M_P^2 \int d\tau d^3x\, a^2 \epsilon H^2 \left(\pi'^2 - (\partial_i \pi)^2\right), \tag{19}$$

in which the prime represents the derivative with respect to τ. Here, $\pi(x^\mu)$ is the Goldstone boson associated to time diffeomorphism breaking, which is related to curvature perturbations $\mathcal{R}$ [79,80].

The cubic action is given by

$$S_{\pi^3} = M_P^2 H^3 \int d\tau d^3x\, \eta \epsilon a^2 \left[\pi \pi'^2 - \pi (\partial \pi)^2\right], \tag{20}$$

and the corresponding cubic interaction Hamiltonian is given by [66]

$$\mathbf{H}_3 = -M_P^2 H^3 \eta \epsilon a^2 \int d^3x \left[\pi \pi'^2 + \frac{1}{2}\pi^2 \partial^2 \pi\right]. \tag{21}$$

On the other hand, the quartic action is obtained to be

$$S_{\pi^4} = \frac{M_P^2}{2} \int d\tau d^3x\, \epsilon a H^3 (\eta^2 a H + \eta') \left[\pi^2 \pi'^2 - \pi^2 (\partial \pi)^2\right]. \tag{22}$$

In particular, note that there is the term η', which induces the delta contribution $\delta(\tau - \tau_e)$ in the interaction Hamiltonian when η undergoes a jump at $\tau = \tau_e$, as given by Equation (5).

As discussed in [11], in calculating the quartic Hamiltonian, care must be taken as the time derivative interaction $\pi' \pi^2$ in $\mathbf{H}_3$ induces an additional contribution in the quartic Hamiltonian [82,83]. As a result, one can not simply conclude that $\mathbf{H}_4 = -\mathbf{L}_4$. More specifically, the quartic Hamiltonian receives additional contribution $+M_P^2 H^4 \eta^2 \epsilon a^2\, \pi^2 \pi'^2$ from the cubic action. Combining all contributions, the total quartic Hamiltonian is given by [11]

$$\mathbf{H}_4 = \frac{M_P^2}{2} \epsilon a H^3 \int d^3x \left[(\eta^2 a H - \eta') \pi^2 \pi'^2 + (\eta^2 a H + \eta') \pi^2 (\partial \pi)^2\right]. \tag{23}$$

The interaction Hamiltonians (21) and (23) have been used in [11] to calculate the one-loop corrections to the power spectrum. We note that in obtaining the above Hamiltonian, we have ignored total time derivatives in the form $\frac{d}{dt}(f(t)\pi^4)$, where $f(t)$ is a function of the background quantity. However, as shown in [33], these boundary terms are harmless as they do not involve $\dot{\pi}$ and their contributions can be absorbed via a canonical transformation in phase space.

The quantity of interest is the curvature perturbation $\mathcal{R}$, while the above interaction Hamiltonians are given in terms of the Goldstone field π. The relation between $\mathcal{R}$ and π is non-linear. For example, to cubic order in π, they are related to each other via [53,54]

$$\begin{aligned}\mathcal{R} &= -H\pi + \big(H\pi\dot{\pi} + \frac{\dot{H}}{2}\pi^2\big) + \big(-H\pi\dot{\pi}^2 - \frac{H}{2}\ddot{\pi}\pi^2 - \dot{H}\dot{\pi}\pi^2 - \frac{\ddot{H}}{6}\pi^3\big), \\ &= -H\pi + \frac{1}{2}\frac{d}{dt}\big(H\pi^2\big) - \frac{1}{6}\frac{d^2}{dt^2}\big(H\pi^3\big). \end{aligned} \tag{24}$$

However, we calculate the loop corrections in power spectrum at the time of end of inflation $\tau = \tau_0 \to 0$ when it is assumed that the system is in the slow-roll phase and the long mode perturbations are frozen with $\dot{\pi} = \ddot{\pi} = 0$. Fortunately, one can neglect the non-linear corrections in $\mathcal{R}$ in Equation (24) in this limit and simply consider the linear relation between them,

$$\mathcal{R} = -H\pi\,, \qquad (\tau \to \tau_0). \tag{25}$$

Since the relation between $\mathcal{R}$ and π is linear at τ_0, we can simply write $\langle \mathcal{R}(\tau_0)\mathcal{R}(\tau_0)\rangle = H^2\langle \pi(\tau_0)\pi(\tau_0)\rangle$. Consequently, one can use π and $\mathcal{R}$ interchangeably in the following in-in analysis. More specifically, we will use the free mode function of $\mathcal{R}$ in the interaction picture in place of the π perturbations in the following in-in integrals.

4. Loop Corrections in Power Spectrum

Employing the perturbative in-in formalism [84], the expectation value of the quantum operator $\hat{O}[\tau_0]$ measured at the time of end of inflation τ_0 is,

$$\langle \hat{O}(\tau_0)\rangle = \Big\langle \Big[\bar{\mathrm{T}}\exp\Big(i\int_{-\infty}^{\tau_0} d\tau' H_{\rm in}(\tau')\Big)\Big]\,\hat{O}(\tau_0)\,\Big[\mathrm{T}\exp\Big(-i\int_{-\infty}^{\tau_0} d\tau' H_{\rm in}(\tau')\Big)\Big]\Big\rangle. \tag{26}$$

Here, as usual, T and $\bar{\mathrm{T}}$ represent the time ordering and anti-time ordering respectively and $H_{\rm in}(t)$ is the interaction Hamiltonian. For our case of interest here $\hat{O}(\tau_0) = \mathcal{R}_{\mathbf{p}_1}(\tau_0)\mathcal{R}_{\mathbf{p}_2}(\tau_0)$, while for the Feynman diagram (**a**) which we consider, we only need the quartic interactions so $H_{\rm in} = \mathbf{H}_4$.

In order to calculate the two-loop corrections, one requires to expand the in-in formula Equation (26) to second orders $H_{\rm in}$. For this purpose, it is more convenient to use the Weinberg commutator method associated to Equation (26). To the second order in $H_{\rm in} = \mathbf{H}_4$, we obtain [84],

$$\begin{aligned}\langle \hat{O}(\tau_0)\rangle &= i^2\int_{-\infty}^{\tau_0} d\tau_2\int_{-\infty}^{\tau_2} d\tau_1\Big\langle\Big[H_{\rm in}(\tau_1),\big[H_{\rm in}(\tau_2),\hat{O}(\tau_0)\big]\Big]\Big\rangle \qquad (27)\\ &= 2\int_{-\infty}^{\tau_0} d\tau_2\int_{-\infty}^{\tau_2} d\tau_1 \mathrm{Re}\Big[\big\langle \mathbf{H_4}(\tau_1)\hat{O}(\tau_0)\mathbf{H_4}(\tau_2)\big\rangle - \big\langle \mathbf{H_4}(\tau_1)\mathbf{H_4}(\tau_2)\hat{O}(\tau_0)\big\rangle\Big], \qquad (28)\end{aligned}$$

with $\hat{O}(\tau_0) \equiv \mathcal{R}_{\mathbf{p}_1}(\tau_0)\mathcal{R}_{\mathbf{p}_2}(\tau_0)$.

Depending on the contractions of external leg operators $\hat{O}(\tau_0)$, there are two distinct Feynman diagrams as shown by diagrams (**a**) and (**b**) in Figure 1. The diagram (**a**) corresponds to the situation in which $\hat{O}(\tau_0)$ contracts only with $\mathbf{H_4}(\tau_2)$, with no contractions to $\mathbf{H_4}(\tau_1)$. On the other hand, the diagram (**b**) corresponds to the case where $\hat{O}(\tau_0)$ contracts jointly with both $\mathbf{H_4}(\tau_2)$ and $\mathbf{H_4}(\tau_1)$. AS mentioned before, as a first try for the two-loop corrections, in this work, we only consider the "double scoop" diagram (**a**).

It turns out that it is very convenient to decompose the expectation values in terms of sub-component Wick contractions. As an example, consider $\langle \mathbf{H_4}(\tau_1)\hat{O}(\tau_0)\mathbf{H_4}(\tau_2)\rangle$ in the second line in Equation (28). It involves three forms of contractions: $\overbrace{\mathbf{H_4}(\tau_1)\mathbf{H_4}(\tau_2)}$, $\overbrace{\hat{O}(\tau_0)\mathbf{H_4}(\tau_2)}$, and $\overbrace{\mathbf{H_4}(\tau_1)\mathbf{H_4}(\tau_1)}$. Let us define

$$\overbrace{\mathbf{H_4}(\tau_1)\mathbf{H_4}(\tau_2)} \equiv h(\tau_1,\tau_2)\,, \qquad \overbrace{\hat{O}(\tau_0)\mathbf{H_4}(\tau_2)} \equiv g(\tau_2)\,, \qquad \overbrace{\mathbf{H_4}(\tau_1)\mathbf{H_4}(\tau_1)} \equiv c(\tau_1)\,. \tag{29}$$

With these definitions, considering both terms in the second line of Equation (28), we obtain

$$\langle \hat{O}(\tau_0)\rangle = -4\int_{-\infty}^{\tau_0} d\tau_2 \int_{-\infty}^{\tau_2} d\tau_1 \mathrm{Im}[g(\tau_2)]\,\mathrm{Im}\big[c(\tau_1)h(\tau_1,\tau_2)\big]\,. \tag{30}$$

Our job is now to calculate the functions $g(\tau_2), c(\tau_1)$ and $h(\tau_1,\tau_2)$ for various possible contractions.

To perform the in-in integrals, we write $\mathbf{H_4}$ in Equation (23) as follows (neglecting the non-linear relations between π and $\mathcal{R}$ as discussed before):

$$\mathbf{H_4} = A_4(\tau)\int d^3\mathbf{x}\,\mathcal{R}^2\mathcal{R}'^2 + B_4(\tau)\int d^3\mathbf{x}\,\mathcal{R}^2(\partial\mathcal{R})^2\,, \tag{31}$$

with

$$A_4(\tau) \equiv \frac{1}{2}M_P^2\eta^2\epsilon a^2\Big(1-\frac{h}{\eta^2}\delta(\tau-\tau_e)\tau_e\Big), \qquad B_4(\tau) \equiv \frac{1}{2}M_P^2\eta^2\epsilon a^2\Big(1+\frac{h}{\eta^2}\delta(\tau-\tau_e)\tau_e\Big). \tag{32}$$

In particular, note the term $\delta(\tau-\tau_e)$ appearing above, which is originated from the term η' in $\mathbf{H_4}$; see Equation (5).

Depending on which term that $\mathbf{H_4}$ is contracted with each other and with $\hat{O}(\tau_0)$, we will have four different contributions, as follows:

$$\langle\hat{O}(\tau_0)\rangle = \langle\hat{O}\rangle_{A_4A_4} + \langle\hat{O}\rangle_{A_4B_4} + \langle\hat{O}\rangle_{B_4A_4} + \langle\hat{O}\rangle_{B_4B_4}\,, \tag{33}$$

For example, for $\langle\hat{O}\rangle_{A_4A_4}$, we have

$$\langle\hat{O}\rangle_{A_4A_4} = \int_{-\infty}^{\tau_0} d\tau_2\int_{-\infty}^{\tau_2} d\tau_1 A_4(\tau_1)A_4(\tau_2)\int d^3\mathbf{x}\int d^3\mathbf{y}\Big\langle\Big[\mathcal{R}^2\mathcal{R}'^2(\mathbf{x},\tau_1),\big[\hat{O}(\tau_0),\mathcal{R}^2\mathcal{R}'^2(\mathbf{y},\tau_2)\big]\Big]\Big\rangle. \tag{34}$$

Proceeding to the Fourier space, this is cast into

$$\langle\hat{O}\rangle_{A_4A_4} = \int_{-\infty}^{\tau_0} d\tau_2\int_{-\infty}^{\tau_2} d\tau_1 A_4(\tau_1)A_4(\tau_2)\Big[\prod_i^4\int\frac{d^3\mathbf{q}_i}{(2\pi)^3}(2\pi)^3\delta^3(\sum_i\mathbf{q}_i)\Big]\Big[\prod_j^4\int\frac{d^3\mathbf{k}_i}{(2\pi)^3}(2\pi)^3\delta^3(\sum_i\mathbf{k}_i)\Big]$$
$$\times\Big\langle\Big[(\hat{\mathcal{R}}_{\mathbf{q}_1}\hat{\mathcal{R}}_{\mathbf{q}_2}\hat{\mathcal{R}}'_{\mathbf{q}_3}\hat{\mathcal{R}}'_{\mathbf{q}_4})(\tau_1),\Big[(\hat{\mathcal{R}}_{\mathbf{p}_1}\hat{\mathcal{R}}_{\mathbf{p}_2})(\tau_0),(\hat{\mathcal{R}}_{\mathbf{k}_1}\hat{\mathcal{R}}_{\mathbf{k}_2}\hat{\mathcal{R}}'_{\mathbf{k}_3}\hat{\mathcal{R}}'_{\mathbf{k}_4})(\tau_2)\Big]\Big]\Big\rangle. \tag{35}$$

After performing the contractions and imposing the constraints from the delta functions, in the soft limit where $p \ll q, k$, we end up with the following form of two-loop integrals

$$\langle\mathcal{R}_{\mathbf{p}_1}\mathcal{R}_{\mathbf{p}_2}(\tau_0)\rangle_{A_4A_4} = (2\pi)^3\delta^3(\mathbf{p}_1+\mathbf{p}_2)\int_{-\infty}^{\tau_0} d\tau_2\int_{-\infty}^{\tau_2} d\tau_1 A_4(\tau_1)A_4(\tau_2)\int\frac{d^3\mathbf{q}}{(2\pi)^3}\int\frac{d^3\mathbf{k}}{(2\pi)^3}F(\tau_1,\tau_2;k,q),$$

in which the function $F(\tau_1,\tau_2;k,q)$ is determined by different values of $c(\tau_1), g(\tau_2)$ and $h(\tau_1,\tau_2)$.

There are nine different terms in $F(\tau_1,\tau_2;k,q)$ from different contractions. We list them as follows:

$$(a_0): (-4)(2)^2\mathrm{Im}\big[\mathcal{R}_p(\tau_0)^2\mathcal{R}_p^*(\tau_2)^2\big]\mathrm{Im}\big[\mathcal{R}_k'(\tau_1)^2\mathcal{R}_k'^*(\tau_2)^2\big]|\mathcal{R}_q(\tau_1)|^2$$
$$(b_0): (-4)(2)^2\mathrm{Im}\big[\mathcal{R}_p(\tau_0)^2\mathcal{R}_p^*(\tau_2)^2\big]\mathrm{Im}\big[\mathcal{R}_k(\tau_1)^2\mathcal{R}_k'^*(\tau_2)^2\big]|\mathcal{R}'_q(\tau_1)|^2$$
$$(c_0): (-4)(2)^3\mathrm{Im}\big[\mathcal{R}_p(\tau_0)^2\mathcal{R}_p^*(\tau_2)^2\big]\mathrm{Im}\big[\mathcal{R}_k(\tau_1)\mathcal{R}_k'(\tau_1)\mathcal{R}_k'^*(\tau_2)^2\big]\mathrm{Re}\big[\mathcal{R}_q(\tau_1)\mathcal{R}'_q(\tau_1)^*\big], \tag{36}$$

and

$$
\begin{aligned}
&(d_0):(-4)(2)^4\mathrm{Im}\big[\mathcal{R}_p(\tau_0)^2\mathcal{R}_p^*(\tau_2)\mathcal{R}_p^{'*}(\tau_2)\big]\mathrm{Im}\big[\mathcal{R}_k{}'(\tau_1)^2\mathcal{R}_k^{'*}(\tau_2)\mathcal{R}_k^*(\tau_2)\big]|\mathcal{R}_q(\tau_1)|^2\\
&(e_0):(-4)(2)^4\mathrm{Im}\big[\mathcal{R}_p(\tau_0)^2\mathcal{R}_p^*(\tau_2)\mathcal{R}_p^{'*}(\tau_2)\big]\mathrm{Im}\big[\mathcal{R}_k(\tau_1)^2\mathcal{R}_k^{'*}(\tau_2)\mathcal{R}_k^*(\tau_2)\big]|\mathcal{R}'_q(\tau_1)|^2\\
&(f_0):(-4)(2)^5\mathrm{Im}\big[\mathcal{R}_p(\tau_0)^2\mathcal{R}_p^*(\tau_2)\mathcal{R}_p^{'*}(\tau_2)\big]\mathrm{Im}\big[\mathcal{R}_k(\tau_1)\mathcal{R}_k{}'(\tau_1)\mathcal{R}_k^{'*}(\tau_2)\mathcal{R}_k^*(\tau_2)\big]\mathrm{Re}\big[\mathcal{R}_q(\tau_1)\mathcal{R}'_q(\tau_1)^*\big],
\end{aligned}
\tag{37}
$$

and

$$
\begin{aligned}
&(m):(-4)(2)^2\mathrm{Im}\big[\mathcal{R}_p(\tau_0)^2\mathcal{R}_p^{'*}(\tau_2)^2\big]\mathrm{Im}\big[\mathcal{R}_k{}'(\tau_1)^2\mathcal{R}_k^*(\tau_2)^2\big]|\mathcal{R}_q(\tau_1)|^2,\\
&(n):(-4)(2)^2\mathrm{Im}\big[\mathcal{R}_p(\tau_0)^2\mathcal{R}_p^{'*}(\tau_2)^2\big]\mathrm{Im}\big[\mathcal{R}_k(\tau_1)^2\mathcal{R}_k^*(\tau_2)^2\big]|\mathcal{R}'_q(\tau_1)|^2,\\
&(r):(-4)(2)^3\mathrm{Im}\big[\mathcal{R}_p(\tau_0)^2\mathcal{R}_p^{'*}(\tau_2)^2\big]\mathrm{Im}\big[\mathcal{R}_k(\tau_1)\mathcal{R}_k{}'(\tau_1)\mathcal{R}_k^*(\tau_2)^2\big]\mathrm{Re}\big[\mathcal{R}_q(\tau_1)\mathcal{R}'_q(\tau_1)^*\big].
\end{aligned}
\tag{38}
$$

One subtle issue in the above expressions is the appearance of the term $\mathrm{Re}\big[\mathcal{R}_q(\tau_1)\mathcal{R}'_q(\tau_1)^*\big]$ in terms (c_0), (f_0), and (r_0). This term originates from the contraction of $\hat{\mathcal{R}}$ and $\hat{\mathcal{R}}'$ in $\mathcal{R}^2\mathcal{R}'^2$ in $A_4(\tau_1)$. As $\hat{\mathcal{R}}$ and $\hat{\mathcal{R}}'$ do not commute, we have symmetrized the ordering of $\hat{\mathcal{R}}$ and $\hat{\mathcal{R}}'$ in $A_4(\tau_1)$, which yields $\mathrm{Re}\big[\mathcal{R}_q(\tau_1)\mathcal{R}'_q(\tau_1)^*\big]$ in terms (c_0), (f_0), and (r_0).

Looking at the above expressions, we notice that the terms containing momentum q are separated from the terms containing the momentum k. This property is depicted in Figure 2, where the momentum q runs in the top loop attached to time τ_1, while the momentum k runs in the lower loop attached to both τ_1 and τ_2. This is a simplifying feature of the diagram (**a**) in our Feynman diagrams. Because of this separation of the two momenta, it is technically easier to calculate the corrections from diagram (**a**) compared to diagram (**b**) in Figure 1. While this is a simplification, calculating the time integrals is still a non-trivial task. This is because we have two nested integrals over τ_1 and τ_2 involving 10 factors of $\mathcal{R}(\tau)$ and its derivatives.

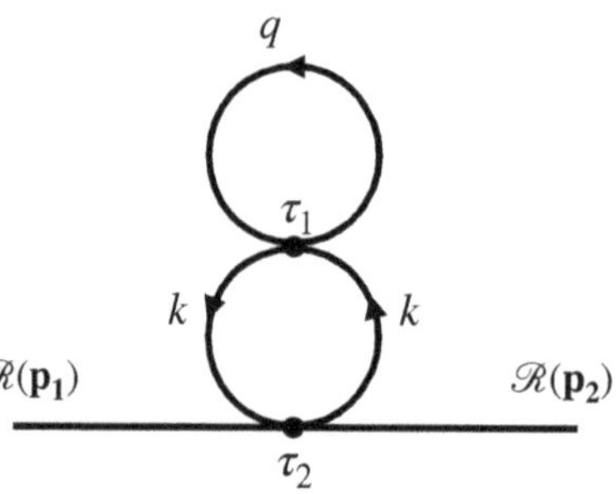

Figure 2. The arrangement of momenta inside each loop and the relative positions of τ_1 and τ_2.

The above three classes of contributions in Equations (36)–(38) are grouped by their form of the function $c(\tau_1)$, which depends only on the soft momentum p but not on the loop momenta q and k. Now, calculating these common factors in each class, we obtain

$$
\mathrm{Im}\big[\mathcal{R}_p(\tau_0)^2\mathcal{R}_p^*(\tau_2)^2\big]=\frac{-H^4\tau_s^6}{24M_P^4\epsilon_i^2h\tau_e^3\tau^3p^3}\left(h\tau_e^3+(6-h)\tau^3\right),
\tag{39}
$$

$$
\mathrm{Im}\big[\mathcal{R}_p(\tau_0)^2\mathcal{R}_p^*(\tau_2)\mathcal{R}_p^{'*}(\tau_2)\big]=\frac{H^4\tau_s^6}{16M_P^4\epsilon_i^2\tau^4p^3},
\tag{40}
$$

and

$$\mathrm{Im}\big[\mathcal{R}_p(\tau_0)^2\mathcal{R}_p'^*(\tau_2)^2\big] = \frac{H^4\tau_s^6(6\tau_s^5-\tau^5)}{40M_P^4\epsilon_i^2\tau^8 p}. \tag{41}$$

Comparing the above three expressions, we conclude that the term in Equation (41) is much more suppressed compared to the terms in Equations (39) and (40) in the soft limit, where $p \to 0$. Correspondingly, we can neglect the contributions of the terms (m), (n) and (r) in Equation (38).

We present the details of the analysis concerning the remaining three contributions $\langle\hat{O}\rangle_{A_4B_4}$, $\langle\hat{O}\rangle_{B_4A_4}$ and $\langle\hat{O}\rangle_{B_4B_4}$ in Appendix A.

Before presenting the final result, it is useful to have an estimate of the leading contributions. Let us look at the contributions of terms (a_0) and (b_0). The difference is that in (a_0), $\mathcal{R}'$ comes with k inside the expression $\mathrm{Im}\big[\mathcal{R}_k'(\tau_1)^2\mathcal{R}_k'^*(\tau_2)^2\big]$, while in (b_0), $\mathcal{R}'$ involves the momentum q simply as $|\mathcal{R}'_q(\tau_1)|^2$. On the other hand, for the superhorizon perturbations during the USR phase, we have that

$$\mathcal{R}_q'(\tau) \simeq -\frac{3}{\tau}\mathcal{R}_q(\tau) \propto \tau^{-4}. \tag{42}$$

Therefore, the leading effects in the in-in integrals on the superhorizon limit where $\tau_i \to 0$ are controlled by the contributions of $\mathcal{R}'$. Since the term (b_0) involves $|\mathcal{R}'_q(\tau_1)|^2$, it scales like τ_1^{-8}, while in $\mathrm{Im}\big[\mathcal{R}_k'(\tau_1)^2\mathcal{R}_k'^*(\tau_2)^2\big]$, the dependence cannot be steeper than this. Indeed, calculating the leading terms on superhorizon limits of $\tau_1, \tau_2 \to 0$, one can show that the ratio $(a_0)/(b_0)$ scales like $(k\tau_1)^2$, which becomes smaller than unity on superhorizon scales. Therefore, it is expected that the contribution of the term (b_0) is more dominant than the term (a_0). Indeed, calculating the in-in integrals for both terms (a_0) and (b_0), and neglecting the numerical prefactors and the common $(2\pi)^3\delta^3(\mathbf{p}_1+\mathbf{p}_2)$, we obtain that they scale as follows:

$$(a_0) : (\mathcal{P}_{\rm CMB})^3\, e^{12\Delta N}\Delta N, \qquad (b_0) : (\mathcal{P}_{\rm CMB})^3\, e^{12\Delta N}\Delta N^2 \tag{43}$$

in which $\mathcal{P}_{\rm CMB}$ is the tree-level CMB scale power spectrum:

$$\mathcal{P}_{\rm CMB} = \frac{H^2}{8\pi^2\epsilon_i M_P^2}. \tag{44}$$

For large enough ΔN, the contributions from (b_0) is typically larger than that of (a_0), as expected.

As we demonstrated in Appendix A, in total, there are 15 leading contributions (i.e., containing p^{-3}) given by $(a_0), (b_0), (c_0), (d_0), (e_0), (f_0), (a_1), (b_1), (c_1), (a_2), (b_2), (c_2), (d_2), (e_2), (f_2), (a_3)$, and (b_3) presented in Appendix A. The contributions from these 15 terms are either like the contribution from the (a_0) term scaling like $e^{12\Delta N}\Delta N$ or like that of the (b_0) term, scaling like $e^{12\Delta N}\Delta N^2$. It turns out that only the contributions from terms $(b_0), (c_0), (d_0), (e_0)$ and (f_0) have the latter form.

In the limit of large enough ΔN, and adding the contributions of terms $(b_0), (c_0), (d_0), (e_0)$, and (f_0) as the leading terms, we obtain the following fractional two-loop correction:

$$\frac{\Delta\mathcal{P}^{(2-\rm loop)}}{\mathcal{P}_{\rm CMB}} \simeq -\frac{27(23h^2+132h+1152)}{8h}\, e^{12\Delta N}\Delta N^2\mathcal{P}_{\rm CMB}^2, \tag{45}$$

where we have neglected the subleading terms involving $\Delta N e^{12\Delta N}$. The fact that the leading two-loop corrections scale as $e^{12\Delta N}\Delta N^2$ is both interesting and reassuring. In addition, the two-loop corrections scale linearly with h for $h \to -\infty$. Note that our result above is obtained assuming a sharp enough transition with $|h| > 1$, where the effects of the relaxation during the final SR phase are neglected.

It is instructive to compare the above two-loop corrections associated to diagram (**a**) with the full one-loop correction obtained in [11]:

$$\frac{\Delta\mathcal{P}^{(1-\text{loop})}}{\mathcal{P}_{\text{CMB}}} \simeq \frac{6(h^2+24h+180)}{h} e^{6\Delta N} \Delta N \mathcal{P}_{\text{CMB}}. \tag{46}$$

Comparing Equations (45) and (46), we obtain

$$\frac{\Delta\mathcal{P}^{(2-\text{loop})}}{\mathcal{P}_{\text{CMB}}} \sim \left(\frac{\Delta\mathcal{P}^{(1-\text{loop})}}{\mathcal{P}_{\text{CMB}}}\right)^2. \tag{47}$$

This is an interesting result, indicating that the fractional two-loop corrections is typically the square of the fractional one-loop correction.

For the loop effects to be under perturbative control, one requires that the successive loop corrections to be hierarchical, i.e., $\mathcal{P}_{\text{CMB}} > \Delta\mathcal{P}^{(1-\text{loop})} > \Delta\mathcal{P}^{(2-\text{loop})}$. Neglecting the numerical prefactors, from Equations (45) and (46), we obtain

$$\frac{\Delta\mathcal{P}^{(2-\text{loop})}}{\Delta\mathcal{P}^{(1-\text{loop})}} \sim \frac{\Delta\mathcal{P}^{(1-\text{loop})}}{\mathcal{P}_{\text{CMB}}} \sim e^{6\Delta N} \Delta N \mathcal{P}_{\text{CMB}}. \tag{48}$$

This shows that if the fractional one-loop correction is not small, then the two-loop corrections become significant as well, so the perturbative treatment quickly loses control. As elaborated in [11], the one-loop correction becomes significant for a sharp transition when $|h| \gg 1$, as can be seen in Equation (46). Therefore, we conclude that the two-loop corrections become significant in a model of sharp transition, which can be seen in Equation (45) as well. In order for the loop corrections to be under perturbative control, one has to either have a mild transition with h at the order of slow-roll parameters, or to take the duration of the USR phase to be reasonably short, e.g., $\Delta N \lesssim 1$. However, the latter arrangement may not be suitable for the PBH formation as we need a long enough period of the USR phase to enhance the power spectrum in the first place. For example, with $h = -6$, we need $\Delta N \lesssim 2.3$ to enhance $\mathcal{P}_{\mathcal{R}}$ by 7 orders of magnitudes compared to the CMB scale for the purpose of PBH formation, while still satisfying the perturbative bound from one-loop correction in Equation (46). Therefore, the only safe strategy for PBH formation in this setup is to employ a mild transition with $|h| \ll 1$.

There are two important comments in order. The first comment is that in this work, we studied the corrections from the diagram (**a**) in Figure 1. Naturally, one may ask how the two-loop corrections from the remaining ten diagram can be compared to the current result given in Equation (45). In a work in progress [81], we are studying the correction from diagram (**m**) in Figure 1 involving a single vertex of sextic Hamiltonian. We have confirmed that it scales like Equation (45). On the physical ground, we expect the full two-loop corrections to have the same general form as in Equation (45), i.e., scaling like $\Delta N^2 e^{12\Delta N}$. The second comment is the the issue of regularization and renormalization. In the current analysis, we have restricted the momentum in the range $q_s \leq q \leq k_e$. In principle, one should integrate over the entire range $q_{IR} < q < q_f$, in which q_{IR} is the lower IR limit, while q_f represents the mode which leaves the horizon at the time of the end of inflation. Alternatively, one may simply set $q_f \to \infty$. While the IR contributions are under control, the UV contribution will diverge, which has to be regularized and renormalized. Having said this, the peak of the power spectrum at the end of the USR, the dimensionless factor $e^{6\Delta N}\mathcal{P}_{\text{CMB}}$ fixes the overall scale of the finite term after regularization. The fact that the two-loop correction in Equation (45) is the square of the one-loop correction in Equation (46) qualitatively supports this expectation. However, it is an important open question to study the renormalization of the loop corrections in more detail.

5. Summary and Discussion

In this work, we have looked at the quantum corrections in the primordial power spectrum at two-loop orders in models of single-field inflation involving an intermediate USR phase engineered for PBH formation. This is the natural continuation of the previous works concerning the one-loop corrections. As the one-loop correction in models with sharp transition to the final SR phase can be large [1,11], it is necessary to examine the two-loop corrections for the severity of the loop contributions. As we have shown, there are eleven distinct one-particle irreducible Feynman diagrams at the two-loop orders. They require the cubic, quartic, quintic, and sextic interaction Hamiltonians.

As a first step forward, we have studied the corrections from the diagram (**a**) in Figure 1. This is because this diagram involves two vertices of quartic interaction, while their momentum integrals over q and k are separable. This brings computational simplicities. Other diagrams in Figure 1 are more complicated, either having higher-order nested integrals or the momentum integrals are not separable. On the other hand, the in-in analysis corresponding to diagram (**m**) would be easier to handle as it involves a single sextic Hamiltonian vertex. However, one has to calculate the action to a sixth order to construct $\mathbf{H_6}$, which involves additional technical complexities [81]. Our result shows that the two-loop corrections scale as the square of the one-loop corrections, i.e., like $\left(\Delta N e^{6\Delta N}\mathcal{P}_{\rm CMB}\right)^2$. This is interesting and physically expected. This result confirms the previous results [1,11] that the loop corrections can quickly lose control if the transition to the final attractor phase is very sharp and the duration of USR phase is long enough. In order for the loop corrections to be under perturbative control, and at the same time, to generate PBHs with the desired mass scales for dark matter purposes, it is necessary that the transition to the final attractor regime is mild so the loop corrections are rendered harmless.

The current work can be extended in various directions. One natural direction to proceed is to study the loop corrections from the remaining Feynman diagrams in Figure 1. While this is an interesting and yet cumbersome task, we believe that the total two-loop corrections would be similar to our current result, i.e., scaling like $\left(\Delta N e^{6\Delta N}\mathcal{P}_{\rm CMB}\right)^2$. The other direction to investigate is to study the general case of non-attractor inflation and the case of constant-roll inflation and see if the two-loop correction has a similar relation compared to the one-loop correction. Finally, an interesting direction to investigate is the question of regularization and renormalization at both one-loop and two-loop orders. We would like to come back to this important question in the future.

Funding: This work is supported partially by INSF of Iran under the grant number 4038049.

Data Availability Statement: No data are associated with this work.

Acknowledgments: We thank Jacopo Fumagalli, Haidar Sheikhahmadi, Amin Nassiri-Rad and Bahar Nikbakht for useful discussions. We thank ICCUB, University of Barcelona, for the kind hospitality during the workshop "BBH initiative: I. Primordial Black Holes" where this work was in progress.

Conflicts of Interest: The author declares no conflict of interest.

Appendix A. In-In Integrals

In this Appendix, we present the details of in-in analysis.

As explained in the main text, the two-loop corrections have the following contributions:

$$\langle \hat{O}(\tau_0)\rangle = \langle \hat{O}\rangle_{A_4A_4} + \langle \hat{O}\rangle_{A_4B_4} + \langle \hat{O}\rangle_{B_4A_4} + \langle \hat{O}\rangle_{B_4B_4}\,. \tag{A1}$$

In the main text, we have presented the results for $\langle \hat{O}\rangle_{A_4A_4}$, which we report here as well for concreteness:

$$\langle \hat{O}\rangle_{A_4A_4} = \int_{-\infty}^{\tau_0} d\tau_2 \int_{-\infty}^{\tau_2} d\tau_1 A_4(\tau_1)A_4(\tau_2) \int d^3\mathbf{x} \int d^3\mathbf{y} \Big\langle \Big[\mathcal{R}^2\mathcal{R}'^2(\mathbf{x},\tau_1), \big[\hat{O}(\tau_0), \mathcal{R}^2\mathcal{R}'^2(\mathbf{y},\tau_2)\big]\Big]\Big\rangle\,. \tag{A2}$$

Proceeding to the Fourier space, and after performing the contractions and imposing the constraints from the delta functions, in the soft limit where $p \ll q, k$, we obtain

$$\langle \mathcal{R}_{\mathbf{p}_1}\mathcal{R}_{\mathbf{p}_2}(\tau_0)\rangle_{A_4A_4} = (2\pi)^3\delta^3(\mathbf{p}_1+\mathbf{p}_2)\int_{-\infty}^{\tau_0} d\tau_2 \int_{-\infty}^{\tau_2} d\tau_1 A_4(\tau_1)A_4(\tau_2)\int \frac{d^3\mathbf{q}}{(2\pi)^3}\int \frac{d^3\mathbf{k}}{(2\pi)^3}F(\tau_1,\tau_2;k,q),$$

in which the function $F(\tau_1,\tau_2;k,q)$ has the following nine contributions:

$$\begin{aligned}
&(a_0) : (-4)(2)^2\mathrm{Im}\big[\mathcal{R}_p(\tau_0)^2\mathcal{R}_p^*(\tau_2)^2\big]\mathrm{Im}\big[\mathcal{R}_k{}'(\tau_1)^2\mathcal{R}_k'^*(\tau_2)^2\big]|\mathcal{R}_q(\tau_1)|^2\\
&(b_0) : (-4)(2)^2\mathrm{Im}\big[\mathcal{R}_p(\tau_0)^2\mathcal{R}_p^*(\tau_2)^2\big]\mathrm{Im}\big[\mathcal{R}_k(\tau_1)^2\mathcal{R}_k'^*(\tau_2)^2\big]|\mathcal{R}'_q(\tau_1)|^2\\
&(c_0) : (-4)(2)^3\mathrm{Im}\big[\mathcal{R}_p(\tau_0)^2\mathcal{R}_p^*(\tau_2)^2\big]\mathrm{Im}\big[\mathcal{R}_k(\tau_1)\mathcal{R}_k{}'(\tau_1)\mathcal{R}_k'^*(\tau_2)^2\big]\mathrm{Re}\big[\mathcal{R}_q(\tau_1)\mathcal{R}'_q(\tau_1)^*\big],
\end{aligned}\tag{A3}$$

and

$$\begin{aligned}
&(d_0) : (-4)(2)^4\mathrm{Im}\big[\mathcal{R}_p(\tau_0)^2\mathcal{R}_p^*(\tau_2)\mathcal{R}_p'^*(\tau_2)\big]\mathrm{Im}\big[\mathcal{R}_k{}'(\tau_1)^2\mathcal{R}_k'^*(\tau_2)\mathcal{R}_k^*(\tau_2)\big]|\mathcal{R}_q(\tau_1)|^2\\
&(e_0) : (-4)(2)^4\mathrm{Im}\big[\mathcal{R}_p(\tau_0)^2\mathcal{R}_p^*(\tau_2)\mathcal{R}_p'^*(\tau_2)\big]\mathrm{Im}\big[\mathcal{R}_k(\tau_1)^2\mathcal{R}_k'^*(\tau_2)\mathcal{R}_k^*(\tau_2)\big]|\mathcal{R}'_q(\tau_1)|^2\\
&(f_0) : (-4)(2)^5\mathrm{Im}\big[\mathcal{R}_p(\tau_0)^2\mathcal{R}_p^*(\tau_2)\mathcal{R}_p'^*(\tau_2)\big]\mathrm{Im}\big[\mathcal{R}_k(\tau_1)\mathcal{R}_k{}'(\tau_1)\mathcal{R}_k'^*(\tau_2)\mathcal{R}_k^*(\tau_2)\big]\mathrm{Re}\big[\mathcal{R}_q(\tau_1)\mathcal{R}'_q(\tau_1)^*\big],
\end{aligned}\tag{A4}$$

and

$$\begin{aligned}
&(m) : (-4)(2)^2\mathrm{Im}\big[\mathcal{R}_p(\tau_0)^2\mathcal{R}_p'^*(\tau_2)^2\big]\mathrm{Im}\big[\mathcal{R}_k{}'(\tau_1)^2\mathcal{R}_k^*(\tau_2)^2\big]|\mathcal{R}_q(\tau_1)|^2,\\
&(n) : (-4)(2)^2\mathrm{Im}\big[\mathcal{R}_p(\tau_0)^2\mathcal{R}_p'^*(\tau_2)^2\big]\mathrm{Im}\big[\mathcal{R}_k(\tau_1)^2\mathcal{R}_k^*(\tau_2)^2\big]|\mathcal{R}'_q(\tau_1)|^2,\\
&(r) : (-4)(2)^3\mathrm{Im}\big[\mathcal{R}_p(\tau_0)^2\mathcal{R}_p'^*(\tau_2)^2\big]\mathrm{Im}\big[\mathcal{R}_k(\tau_1)\mathcal{R}_k{}'(\tau_1)\mathcal{R}_k^*(\tau_2)^2\big]\mathrm{Re}\big[\mathcal{R}_q(\tau_1)\mathcal{R}'_q(\tau_1)^*\big].
\end{aligned}\tag{A5}$$

Now, we consider the contribution $\langle\hat{O}\rangle_{A_4B_4}$, yielding

$$\begin{aligned}
\langle\hat{O}\rangle_{A_4B_4} = &\int_{-\infty}^{\tau_0} d\tau_2\int_{-\infty}^{\tau_2} d\tau_1 A_4(\tau_1)B_4(\tau_2)\Big[\prod_i^4\int\frac{d^3\mathbf{q}_i}{(2\pi)^3}(2\pi)^3\delta^3(\sum_i\mathbf{q}_i)\Big]\Big[\prod_j^4\int\frac{d^3\mathbf{k}_i}{(2\pi)^3}(2\pi)^3\delta^3(\sum_i\mathbf{k}_i)\Big]\\
&\times\Big\langle\Big[(\hat{\mathcal{R}}_{\mathbf{q}_1}\hat{\mathcal{R}}_{\mathbf{q}_2}\hat{\mathcal{R}}'_{\mathbf{q}_3}\hat{\mathcal{R}}'_{\mathbf{q}_4})(\tau_1),\Big[(\hat{\mathcal{R}}_{\mathbf{p}_1}\hat{\mathcal{R}}_{\mathbf{p}_2})(\tau_0),(\hat{\mathcal{R}}_{\mathbf{k}_1}\hat{\mathcal{R}}_{\mathbf{k}_2}\hat{\mathcal{R}}_{\mathbf{k}_3}\hat{\mathcal{R}}_{\mathbf{k}_4})(\tau_2)\Big]\Big]\Big\rangle i^2\mathbf{k}_3\cdot\mathbf{k}_4.
\end{aligned}\tag{A6}$$

The leading contributions are those in which $\hat{\mathcal{R}}_{\mathbf{p}}(\tau_0)$ does not contract with $\hat{\mathcal{R}}_{\mathbf{k}_3}(\tau_2)$ and $\hat{\mathcal{R}}_{\mathbf{k}_4}(\tau_2)$. Correspondingly, the leading terms will be similar to terms (a_0), (b_0), and (c_0) in the analysis of $\langle\hat{O}\rangle_{A_4A_4}$. More specifically, the leading terms are

$$\begin{aligned}
&(a_1) : (-4)(2)^2\mathrm{Im}\big[\mathcal{R}_p(\tau_0)^2\mathcal{R}_p^*(\tau_2)^2\big]\mathrm{Im}\big[\mathcal{R}_k{}'(\tau_1)^2\mathcal{R}_k^*(\tau_2)^2\big]|\mathcal{R}_q(\tau_1)|^2(-i^2k^2)\\
&(b_1) : (-4)(2)^2\mathrm{Im}\big[\mathcal{R}_p(\tau_0)^2\mathcal{R}_p^*(\tau_2)^2\big]\mathrm{Im}\big[\mathcal{R}_k(\tau_1)^2\mathcal{R}_k^*(\tau_2)^2\big]|\mathcal{R}'_q(\tau_1)|^2(-i^2k^2)\\
&(c_1) : (-4)(2)^3\mathrm{Im}\big[\mathcal{R}_p(\tau_0)^2\mathcal{R}_p^*(\tau_2)^2\big]\mathrm{Im}\big[\mathcal{R}_k(\tau_1)\mathcal{R}_k{}'(\tau_1)\mathcal{R}_k^*(\tau_2)^2\big]\mathrm{Re}\big[\mathcal{R}_q(\tau_1)\mathcal{R}'_q(\tau_1)^*\big](-i^2k^2).
\end{aligned}\tag{A7}$$

Now consider $\langle\hat{O}\rangle_{B_4A_4}$, yielding,

$$\begin{aligned}
\langle\hat{O}\rangle_{B_4A_4} = &\int_{-\infty}^{\tau_0} d\tau_2\int_{-\infty}^{\tau_2} d\tau_1 B_4(\tau_1)A_4(\tau_2)\Big[\prod_i^4\int\frac{d^3\mathbf{q}_i}{(2\pi)^3}(2\pi)^3\delta^3(\sum_i\mathbf{q}_i)\Big]\Big[\prod_j^4\int\frac{d^3\mathbf{k}_i}{(2\pi)^3}(2\pi)^3\delta^3(\sum_i\mathbf{k}_i)\Big]\\
&\times\Big\langle\Big[(\hat{\mathcal{R}}_{\mathbf{q}_1}\hat{\mathcal{R}}_{\mathbf{q}_2}\hat{\mathcal{R}}_{\mathbf{q}_3}\hat{\mathcal{R}}_{\mathbf{q}_4})(\tau_1),\Big[(\hat{\mathcal{R}}_{\mathbf{p}_1}\hat{\mathcal{R}}_{\mathbf{p}_2})(\tau_0),(\hat{\mathcal{R}}_{\mathbf{k}_1}\hat{\mathcal{R}}_{\mathbf{k}_2}\hat{\mathcal{R}}'_{\mathbf{k}_3}\hat{\mathcal{R}}'_{\mathbf{k}_4})(\tau_2)\Big]\Big]\Big\rangle i^2\mathbf{q}_3\cdot\mathbf{q}_4.
\end{aligned}\tag{A8}$$

The leading contributions are $(a_0), (b_0), (c_0), (d_0), (e_0)$, and (f_0) in $\langle\hat{O}\rangle_{A_4A_4}$, yielding

$$\begin{aligned}
&(a_2) : (-4)(2)^2\text{Im}\big[\mathcal{R}_p(\tau_0)^2\mathcal{R}_p^*(\tau_2)^2\big]\text{Im}\big[\mathcal{R}_k(\tau_1)^2\mathcal{R}_k'^*(\tau_2)^2\big]|\mathcal{R}_q(\tau_1)|^2(-i^2k^2)\\
&(b_2) : (-4)(2)^2\text{Im}\big[\mathcal{R}_p(\tau_0)^2\mathcal{R}_p^*(\tau_2)^2\big]\text{Im}\big[\mathcal{R}_k(\tau_1)^2\mathcal{R}_k'^*(\tau_2)^2\big]|\mathcal{R}_q(\tau_1)|^2(-i^2q^2)\\
&(c_2) : (-4)(2)^4\text{Im}\big[\mathcal{R}_p(\tau_0)^2\mathcal{R}_p^*(\tau_2)^2\big]\text{Im}\big[\mathcal{R}_k(\tau_1)^2\mathcal{R}_k'^*(\tau_2)^2\big]|\mathcal{R}_q(\tau_1)|^2(i^2\mathbf{k}\cdot\mathbf{q})
\end{aligned} \tag{A9}$$

and

$$\begin{aligned}
&(d_2) : (-4)(2)^4\text{Im}\big[\mathcal{R}_p(\tau_0)^2\mathcal{R}_p^*(\tau_2)\mathcal{R}_p'^*(\tau_2)\big]\text{Im}\big[\mathcal{R}_k(\tau_1)^2\mathcal{R}_k'^*(\tau_2)\mathcal{R}_k^*(\tau_2)\big]|\mathcal{R}_q(\tau_1)|^2(-i^2k^2)\\
&(e_2) : (-4)(2)^4\text{Im}\big[\mathcal{R}_p(\tau_0)^2\mathcal{R}_p^*(\tau_2)\mathcal{R}_p'^*(\tau_2)\big]\text{Im}\big[\mathcal{R}_k(\tau_1)^2\mathcal{R}_k'^*(\tau_2)\mathcal{R}_k^*(\tau_2)\big]|\mathcal{R}_q(\tau_1)|^2(-i^2q^2)\\
&(f_2) : (-4)(2)^6\text{Im}\big[\mathcal{R}_p(\tau_0)^2\mathcal{R}_p^*(\tau_2)\mathcal{R}_p'^*(\tau_2)\big]\text{Im}\big[\mathcal{R}_k(\tau_1)^2\mathcal{R}_k'^*(\tau_2)\mathcal{R}_k^*(\tau_2)\big]\mathcal{R}_q(\tau_1)|^2(i^2\mathbf{k}\cdot\mathbf{q})
\end{aligned} \tag{A10}$$

From the six terms above, one can check that the terms (c_2) and (f_2) make zero contributions after performing the double momentum integrals of the form $\int d^3\mathbf{q}\, d^3\mathbf{k}\,(\mathbf{q}\cdot\mathbf{k})\mathcal{F}(\tau_1,\tau_2;k,q)$, which vanishes.

Finally, considering $\langle\hat{O}\rangle_{B_4B_4}$, we have

$$\begin{aligned}
\langle\hat{O}\rangle_{B_4B_4} = &\int_{-\infty}^{\tau_0} d\tau_2\int_{-\infty}^{\tau_2} d\tau_1 B_4(\tau_1)B_4(\tau_2)\Big[\prod_i^4\int\frac{d^3\mathbf{q}_i}{(2\pi)^3}(2\pi)^3\delta^3(\sum_i\mathbf{q}_i)\Big]\Big[\prod_j^4\int\frac{d^3\mathbf{k}_i}{(2\pi)^3}(2\pi)^3\delta^3(\sum_i\mathbf{k}_i)\Big]\\
&\times\Big\langle\Big[(\hat{\mathcal{R}}_{\mathbf{q}_1}\hat{\mathcal{R}}_{\mathbf{q}_2}\hat{\mathcal{R}}_{\mathbf{q}_3}\hat{\mathcal{R}}_{\mathbf{q}_4})(\tau_1),\Big[(\hat{\mathcal{R}}_{\mathbf{p}_1}\hat{\mathcal{R}}_{\mathbf{p}_2})(\tau_0),(\hat{\mathcal{R}}_{\mathbf{k}_1}\hat{\mathcal{R}}_{\mathbf{k}_2}\hat{\mathcal{R}}_{\mathbf{k}_3}\hat{\mathcal{R}}_{\mathbf{k}_4})(\tau_2)\Big]\Big]\Big\rangle(\mathbf{q}_3\cdot\mathbf{q}_4)(\mathbf{k}_3\cdot\mathbf{k}_4).
\end{aligned} \tag{A11}$$

The leading contributions are the (a_0) and (b_0) terms, yielding

$$\begin{aligned}
&(a_3) : (-4)(2)^2\text{Im}\big[\mathcal{R}_p(\tau_0)^2\mathcal{R}_p^*(\tau_2)^2\big]\text{Im}\big[\mathcal{R}_k(\tau_1)^2\mathcal{R}_k^*(\tau_2)^2\big]|\mathcal{R}_q(\tau_1)|^2(-i^2k^2)^2\\
&(b_3) : (-4)(2)^2\text{Im}\big[\mathcal{R}_p(\tau_0)^2\mathcal{R}_p^*(\tau_2)^2\big]\text{Im}\big[\mathcal{R}_k(\tau_1)^2\mathcal{R}_k^*(\tau_2)^2\big]|\mathcal{R}_q(\tau_1)|^2(-i^2k^2)(-i^2q^2)
\end{aligned} \tag{A12}$$

So, in total, we have 15 leading terms (i.e., containing p^{-3}) given by $(a_0), (b_0), (c_0), (d_0), (e_0), (f_0), (a_1), (b_1), (c_1), (a_2), (b_2), (d_2), (e_2), (a_3)$ and (b_3) listed above. Out of these 15 contributions, only the terms $(b_0), (c_0), (d_0), (e_0), (f_0)$ scale as $\Delta N^2 e^{12\Delta N}$, while the remaining terms scale like $\Delta N e^{12\Delta N}$. In the limit of a large enough ΔN, we may neglect the latter contributions as the subleading corrections.

Calculating the leading terms from $(b_0), (c_0), (d_0), (e_0), (f_0)$, and neglecting the common factor $(2\pi)^3\delta^3(\mathbf{p}_1+\mathbf{p}_2)$ and $\frac{(4\pi)^2}{(2\pi)^6}$ from the double azimuthal integrals over momentum, we obtain

$$(b_0) \quad : \frac{27H^6}{512M_P^6\epsilon_i^3hp^3}(h^2+28h-384)\Delta N^2e^{12\Delta N}+\mathcal{O}(\Delta N), \tag{A13}$$

$$(c_0) \quad : \frac{-27H^6}{64M_P^6\epsilon_i^3hp^3}(h^2+8h+96)\Delta N^2e^{12\Delta N}+\mathcal{O}(\Delta N), \tag{A14}$$

$$(d_0) \quad : \frac{9H^6}{256M_P^6\epsilon_i^3hp^3}(h^2+8h+48)\Delta N^2e^{12\Delta N}+\mathcal{O}(\Delta N), \tag{A15}$$

$$(e_0) \quad : \frac{-9H^6}{256M_P^6\epsilon_i^3hp^3}(h^2+16h+96)\Delta N^2e^{12\Delta N}+\mathcal{O}(\Delta N), \tag{A16}$$

$$(f_0) \quad : \frac{-9H^6}{16M_P^6\epsilon_i^3hp^3}(h+6)\Delta N^2e^{12\Delta N}+\mathcal{O}(\Delta N). \tag{A17}$$

In comparison, we also present the remaining ten subleading terms $(a_0), (a_1), \ldots, (b_3)$ as well:

$$(a_0) \quad : \frac{3H^6\Delta N e^{12\Delta N}}{5600 M_P^6 \epsilon_i^3 h p^3}(346h - 2997) + \mathcal{O}(\Delta N^0), \tag{A18}$$

$$(a_1) \quad : \frac{H^6\Delta N e^{12\Delta N}}{206976000 M_P^6 \epsilon_i^3 h p^3}(808500h^2 + 51180091h - 44241120) + \mathcal{O}(\Delta N^0), \tag{A19}$$

$$(b_1) \quad : \frac{3H^6\Delta N e^{12\Delta N}}{71680 M_P^6 \epsilon_i^3 h p^3}(1544h - 33024) + \mathcal{O}(\Delta N^0), \tag{A20}$$

$$(c_1) \quad : \frac{3H^6\Delta N e^{12\Delta N}}{179200 M_P^6 \epsilon_i^3 h p^3}(2100h^2 + 82832h - 344832) + \mathcal{O}(\Delta N^0), \tag{A21}$$

$$(a_2) \quad : \frac{H^6\Delta N e^{12\Delta N}}{5017600 M_P^6 \epsilon_i^3 h p^3}(78400h^2 + 1628972h - 21780864) + \mathcal{O}(\Delta N^0), \tag{A22}$$

$$(b2) \quad : \frac{9H^6\Delta N e^{12\Delta N}}{17920 M_P^6 \epsilon_i^3 h p^3}(35h^2 - 246h - 10368) + \mathcal{O}(\Delta N^0), \tag{A23}$$

$$(d2) \quad : \frac{3H^6\Delta N e^{12\Delta N}}{5017600 M_P^6 \epsilon_i^3 p^3}(4900h^2 - 162656h - 1374912) + \mathcal{O}(\Delta N^0), \tag{A24}$$

$$(e2) \quad : \frac{-3H^6\Delta N e^{12\Delta N}}{20070400 M_P^6 \epsilon_i^3 p^3}(4900h^2 - 178080h + 5806080) + \mathcal{O}(\Delta N^0), \tag{A25}$$

$$(a3) \quad : \frac{H^6\Delta N e^{12\Delta N}}{3763200 M_P^6 \epsilon_i^3 p^3}(478062h - 3175200) + \mathcal{O}(\Delta N^0), \tag{A26}$$

$$(b3) \quad : \frac{H^6\Delta N e^{12\Delta N}}{313600 M_P^6 \epsilon_i^3 p^3}(58637h - 423360) + \mathcal{O}(\Delta N^0). \tag{A27}$$

Combining the five leading contributions $(b_0), (c_0), (d_0), (e_0), (f_0)$, and including all numerical factors, we obtain our final result:

$$\langle \mathcal{R}_{\mathbf{p}_1}\mathcal{R}_{\mathbf{p}_2}\rangle|_{2-\text{loops}} = (2\pi)^3\delta^3(\mathbf{p}_1+\mathbf{p}_2)\frac{-27H^6(4\pi)^2}{512M_P^6\epsilon_i^3 h p^3 (2\pi)^6}(23h^2+132h+1152)\Delta N^2 e^{12\Delta N} + \mathcal{O}(\Delta N).$$

Now, multiplying by the factor $\frac{p^3}{2\pi^2}$ to construct the dimensionless power spectrum, we end up with our fractional two-loop correction as follows:

$$\frac{\Delta\mathcal{P}^{(2-\text{loop})}}{\mathcal{P}_{\text{CMB}}} \simeq -\frac{27(23h^2+132h+1152)}{8h} e^{12\Delta N} N^2 \mathcal{P}_{\text{CMB}}^2 + \mathcal{O}(\Delta N), \tag{A28}$$

where $\mathcal{P}_{\text{CMB}}$ is the tree-level power spectrum for the CMB scale modes given in Equation (44).

References

1. Kristiano, J.; Yokoyama, J. Constraining Primordial Black Hole Formation from Single-Field Inflation. *Phys. Rev. Lett.* **2024**, *132*, 221003. [CrossRef] [PubMed]
2. Kristiano, J.; Yokoyama, J. Note on the bispectrum and one-loop corrections in single-field inflation with primordial black hole formation. *Phys. Rev. D* **2024**, *109*, 103541. [CrossRef]
3. Riotto, A. The Primordial Black Hole Formation from Single-Field Inflation is Not Ruled Out. *arXiv* **2023**, arXiv:2301.00599.
4. Riotto, A. The Primordial Black Hole Formation from Single-Field Inflation is Still Not Ruled Out. *arXiv* **2023**, arXiv:2303.01727.
5. Choudhury, S.; Gangopadhyay, M.R.; Sami, M. No-go for the formation of heavy mass Primordial Black Holes in Single Field Inflation. *Eur. Phys. J. C* **2024**, *84*, 884. [CrossRef]
6. Choudhury, S.; Panda, S.; Sami, M. PBH formation in EFT of single field inflation with sharp transition. *Phys. Lett. B* **2023**, *845*, 138123. [CrossRef]
7. Choudhury, S.; Panda, S.; Sami, M. Quantum loop effects on the power spectrum and constraints on primordial black holes. *J. Cosmol. Astropart. Phys.* **2023**, *11*, 066. [CrossRef]

8. Choudhury, S.; Panda, S.; Sami, M. Galileon inflation evades the no-go for PBH formation in the single-field framework. *J. Cosmol. Astropart. Phys.* **2023**, *8*, 078. [CrossRef]
9. Choudhury, S.; Karde, A.; Panda, S.; Sami, M. Realisation of the ultra-slow roll phase in Galileon inflation and PBH overproduction. *J. Cosmol. Astropart. Phys.* **2024**, *7*, 034. [CrossRef]
10. Choudhury, S.; Sami, M. Large fluctuations and Primordial Black Holes. *arXiv* **2024**, arXiv:2407.17006. [CrossRef]
11. Firouzjahi, H. One-loop corrections in power spectrum in single field inflation. *J. Cosmol. Astropart. Phys.* **2023**, *10*, 006. [CrossRef]
12. Motohashi, H.; Tada, Y. Squeezed bispectrum and one-loop corrections in transient constant-roll inflation. *J. Cosmol. Astropart. Phys.* **2023**, *8*, 069. [CrossRef]
13. Firouzjahi, H.; Riotto, A. Primordial Black Holes and loops in single-field inflation. *J. Cosmol. Astropart. Phys.* **2024**, *2*, 021. [CrossRef]
14. Tasinato, G. Large $|\eta|$ approach to single field inflation. *Phys. Rev. D* **2023**, *108*, 043526. [CrossRef]
15. Franciolini, G.; Iovino, A.J.; Taoso, M.; Urbano, A. Perturbativity in the presence of ultraslow-roll dynamics. *Phys. Rev. D* **2024**, *109*, 123550. [CrossRef]
16. Firouzjahi, H. Loop corrections in gravitational wave spectrum in single field inflation. *Phys. Rev. D* **2023**, *108*, 043532. [CrossRef]
17. Maity, S.; Ragavendra, H.V.; Sethi, S.K.; Sriramkumar, L. Loop contributions to the scalar power spectrum due to quartic order action in ultra slow roll inflation. *J. Cosmol. Astropart. Phys.* **2024**, *5*, 046. [CrossRef]
18. Cheng, S.L.; Lee, D.S.; Ng, K.W. Primordial perturbations from ultra-slow-roll single-field inflation with quantum loop effects. *J. Cosmol. Astropart. Phys.* **2024**, *3*, 008. [CrossRef]
19. Fumagalli, J.; Bhattacharya, S.; Peloso, M.; Renaux-Petel, S.; Witkowski, L.T. One-loop infrared rescattering by enhanced scalar fluctuations during inflation. *J. Cosmol. Astropart. Phys.* **2024**, *4*, 029. [CrossRef]
20. Nassiri-Rad, A.; Asadi, K. Induced gravitational waves from non-attractor inflation and NANOGrav data. *J. Cosmol. Astropart. Phys.* **2024**, *4*, 009. [CrossRef]
21. Meng, D.S.; Yuan, C.; Huang, Q.G. One-loop correction to the enhanced curvature perturbation with local-type non-Gaussianity for the formation of primordial black holes. *Phys. Rev. D* **2022**, *106*, 063508. [CrossRef]
22. Cheng, S.L.; Lee, D.S.; Ng, K.W. Power spectrum of primordial perturbations during ultra-slow-roll inflation with back reaction effects. *Phys. Lett. B* **2022**, *827*, 136956. [CrossRef]
23. Fumagalli, J. Absence of one-loop effects on large scales from small scales in non-slow-roll dynamics. *arXiv* **2023**, arXiv:2305.19263.
24. Tada, Y.; Terada, T.; Tokuda, J. Cancellation of quantum corrections on the soft curvature perturbations. *J. High Energy Phys.* **2024**, *1*, 105. [CrossRef]
25. Firouzjahi, H. Revisiting loop corrections in single field ultraslow-roll inflation. *Phys. Rev. D* **2024**, *109*, 043514. [CrossRef]
26. Iacconi, L.; Mulryne, D.J. Multi-field inflation with large scalar fluctuations: Non-Gaussianity and perturbativity. *J. Cosmol. Astropart. Phys.* **2023**, *9*, 033. [CrossRef]
27. Davies, M.W.; Iacconi, L.; Mulryne, D.J. Numerical 1-loop correction from a potential yielding ultra-slow-roll dynamics. *J. Cosmol. Astropart. Phys.* **2024**, *4*, 050. [CrossRef]
28. Iacconi, L.; Mulryne, D.; Seery, D. Loop corrections in the separate universe picture. *J. Cosmol. Astropart. Phys.* **2024**, *6*, 062. [CrossRef]
29. Kristiano, J.; Yokoyama, J. Comparing sharp and smooth transitions of the second slow-roll parameter in single-field inflation. *J. Cosmol. Astropart. Phys.* **2024**, *10*, 036. [CrossRef]
30. Kristiano, J.; Yokoyama, J. Generating large primordial fluctuations in single-field inflation for PBH formation. *arXiv* **2024**, arXiv:2405.12149.
31. Ballesteros, G.; Egea, J.G. One-loop power spectrum in ultra slow-roll inflation and implications for primordial black hole dark matter. *J. Cosmol. Astropart. Phys.* **2024**, *7*, 052. [CrossRef]
32. Kawaguchi, R.; Tsujikawa, S.; Yamada, Y. Roles of boundary and equation-of-motion terms in cosmological correlation functions. *Phys. Lett. B* **2024**, *856*, 138962. [CrossRef]
33. Braglia, M.; Pinol, L. No time to derive: Unraveling total time derivatives in in-in perturbation theory. *J. High Energy Phys.* **2024**, *8*, 068. [CrossRef]
34. Firouzjahi, H. Loop corrections in the bispectrum in ultraslow-roll inflation with PBHs formation. *Phys. Rev. D* **2024**, *110*, 043519. [CrossRef]
35. Caravano, A.; Franciolini, G.; Renaux-Petel, S. Ultra-Slow-Roll Inflation on the Lattice: Backreaction and Nonlinear Effects. *arXiv* **2024**, arXiv:2410.23942.
36. Caravano, A.; Inomata, K.; Renaux-Petel, S. Inflationary Butterfly Effect: Nonperturbative Dynamics from Small-Scale Features. *Phys. Rev. Lett.* **2024**, *133*, 15. [CrossRef]
37. Saburov, S.; Ketov, S.V. Quantum Loop Corrections in the Modified Gravity Model of Starobinsky Inflation with Primordial Black Hole Production. *Universe* **2024**, *10*, 354. [CrossRef]
38. Seery, D. One-loop corrections to the curvature perturbation from inflation. *J. Cosmol. Astropart. Phys.* **2008**, *2*, 006. [CrossRef]
39. Seery, D. One-loop corrections to a scalar field during inflation. *J. Cosmol. Astropart. Phys.* **2007**, *11*, 025. [CrossRef]
40. Senatore, L.; Zaldarriaga, M. On Loops in Inflation. *J. High Energy Phys.* **2010**, *12*, 008. [CrossRef]
41. Pimentel, G.L.; Senatore, L.; Zaldarriaga, M. On Loops in Inflation III: Time Independence of zeta in Single Clock Inflation. *J. High Energy Phys.* **2012**, *7*, 166. [CrossRef]

42. Inomata, K.; Braglia, M.; Chen, X.; Renaux-Petel, S. Questions on calculation of primordial power spectrum with large spikes: The resonance model case. *J. Cosmol. Astropart. Phys.* **2023**, *4*, 011; Erratum in *J. Cosmol. Astropart. Phys.* **2023**, *9*, E01. [CrossRef]
43. Ivanov, P.; Naselsky, P.; Novikov, I. Inflation and primordial black holes as dark matter. *Phys. Rev. D* **1994**, *50*, 7173–7178. [CrossRef] [PubMed]
44. Garcia-Bellido, J.; Morales, E.R. Primordial black holes from single field models of inflation. *Phys. Dark Univ.* **2017**, *18*, 47–54. [CrossRef]
45. Germani, C.; Prokopec, T. On primordial black holes from an inflection point. *Phys. Dark Univ.* **2017**, *18*, 6–10. [CrossRef]
46. Biagetti, M.; Franciolini, G.; Kehagias, A.; Riotto, A. Primordial Black Holes from Inflation and Quantum Diffusion. *J. Cosmol. Astropart. Phys.* **2018**, *7*, 032. [CrossRef]
47. Khlopov, M.Y. Primordial Black Holes. *Res. Astron. Astrophys.* **2010**, *10*, 495–528. [CrossRef]
48. Özsoy, O.; Tasinato, G. Inflation and Primordial Black Holes. *Universe* **2023**, *9*, 203. [CrossRef]
49. Byrnes, C.T.; Cole, P.S. Lecture notes on inflation and primordial black holes. *arXiv* **2021**, arXiv:2112.05716.
50. Escrivà, A.; Kuhnel, F.; Tada, Y. Primordial Black Holes. *arXiv* **2022**, arXiv:2211.05767. [CrossRef]
51. Pi, S. Non-Gaussianities in primordial black hole formation and induced gravitational waves. *arXiv* **2024**, arXiv:2404.06151.
52. Maldacena, J.M. Non-Gaussian features of primordial fluctuations in single field inflationary models. *J. High Energy Phys.* **2003**, *305*, 013. [CrossRef]
53. Jarnhus, P.R.; Sloth, M.S. de Sitter limit of inflation and nonlinear perturbation theory. *J. Cosmol. Astropart. Phys.* **2008**, *2*, 013. [CrossRef]
54. Arroja, F.; Koyama, K. Non-gaussianity from the trispectrum in general single field inflation. *Phys. Rev. D* **2008**, *77*, 083517. [CrossRef]
55. Inomata, K. Superhorizon Curvature Perturbations Are Protected against One-Loop Corrections. *Phys. Rev. Lett.* **2024**, *133*, 141001. [CrossRef] [PubMed]
56. Kawaguchi, R.; Tsujikawa, S.; Yamada, Y. Proving the absence of large one-loop corrections to the power spectrum of curvature perturbations in transient ultra-slow-roll inflation within the path-integral approach. *arXiv* **2024**, arXiv:2407.19742. [CrossRef]
57. Fumagalli, J. Absence of one-loop effects on large scales from small scales in non-slow-roll dynamics II: Quartic interactions and consistency relations. *arXiv* **2024**, arXiv:2408.08296.
58. Kinney, W.H. Horizon crossing and inflation with large eta. *Phys. Rev. D* **2005**, *72*, 023515. [CrossRef]
59. Namjoo, M.H.; Firouzjahi, H.; Sasaki, M. Violation of non-Gaussianity consistency relation in a single field inflationary model. *Europhys. Lett.* **2013**, *101*, 39001. [CrossRef]
60. Martin, J.; Motohashi, H.; Suyama, T. Ultra Slow-Roll Inflation and the non-Gaussianity Consistency Relation. *Phys. Rev. D* **2013**, *87*, 023514. [CrossRef]
61. Chen, X.; Firouzjahi, H.; Namjoo, M.H.; Sasaki, M. A Single Field Inflation Model with Large Local Non-Gaussianity. *Europhys. Lett.* **2013**, *102*, 59001. [CrossRef]
62. Morse, M.J.P.; Kinney, W.H. Large-η constant-roll inflation is never an attractor. *Phys. Rev. D* **2018**, *97*, 123519. [CrossRef]
63. Lin, W.C.; Morse, M.J.P.; Kinney, W.H. Dynamical Analysis of Attractor Behavior in Constant Roll Inflation. *J. Cosmol. Astropart. Phys.* **2019**, *9*, 063. [CrossRef]
64. Dimopoulos, K. Ultra slow-roll inflation demystified. *Phys. Lett. B* **2017**, *775*, 262–265. [CrossRef]
65. Chen, X.; Firouzjahi, H.; Komatsu, E.; Namjoo, M.H.; Sasaki, M. In-in and δN calculations of the bispectrum from non-attractor single-field inflation. *J. Cosmol. Astropart. Phys.* **2013**, *1312*, 039. [CrossRef]
66. Akhshik, M.; Firouzjahi, H.; Jazayeri, S. Effective Field Theory of non-Attractor Inflation. *J. Cosmol. Astropart. Phys.* **2015**, *7*, 048. [CrossRef]
67. Akhshik, M.; Firouzjahi, H.; Jazayeri, S. Cosmological Perturbations and the Weinberg Theorem. *J. Cosmol. Astropart. Phys.* **2015**, *12*, 027. [CrossRef]
68. Mooij, S.; Palma, G.A. Consistently violating the non-Gaussian consistency relation. *J. Cosmol. Astropart. Phys.* **2015**, *11*, 025. [CrossRef]
69. Bravo, R.; Mooij, S.; Palma, G.A.; Pradenas, B. A generalized non-Gaussian consistency relation for single field inflation. *J. Cosmol. Astropart. Phys.* **2018**, *5*, 024. [CrossRef]
70. Finelli, B.; Goon, G.; Pajer, E.; Santoni, L. Soft Theorems For Shift-Symmetric Cosmologies. *Phys. Rev. D* **2018**, *97*, 063531. [CrossRef]
71. Passaglia, S.; Hu, W.; Motohashi, H. Primordial black holes and local non-Gaussianity in canonical inflation. *Phys. Rev. D* **2019**, *99*, 043536. [CrossRef]
72. Pi, S.; Sasaki, M. Logarithmic Duality of the Curvature Perturbation. *Phys. Rev. Lett.* **2023**, *131*, 011002. [CrossRef]
73. Özsoy, O.; Tasinato, G. Consistency conditions and primordial black holes in single field inflation. *Phys. Rev. D* **2022**, *105*, 023524. [CrossRef]
74. Firouzjahi, H.; Riotto, A. Sign of non-Gaussianity and the primordial black holes abundance. *Phys. Rev. D* **2023**, *108*, 123504. [CrossRef]
75. Namjoo, M.H. One consistency relation for all single-field inflationary models. *J. Cosmol. Astropart. Phys.* **2024**, *5*, 041. [CrossRef]
76. Namjoo, M.H.; Nikbakht, B. Non-Gaussianity consistency relations and their consequences for the peaks. *J. Cosmol. Astropart. Phys.* **2024**, *8*, 005. [CrossRef]

77. Cai, Y.F.; Chen, X.; Namjoo, M.H.; Sasaki, M.; Wang, D.G.; Wang, Z. Revisiting non-Gaussianity from non-attractor inflation models. *J. Cosmol. Astropart. Phys.* **2018**, *5*, 012. [CrossRef]
78. Weinberg, S. *The Quantum Theory of Fields;* Volume 1: Foundations; Cambridge University Press: Cambridge, UK, 2005. [CrossRef]
79. Cheung, C.; Creminelli, P.; Fitzpatrick, A.L.; Kaplan, J.; Senatore, L. The Effective Field Theory of Inflation. *J. High Energy Phys.* **2008**, *803*, 014. [CrossRef]
80. Cheung, C.; Fitzpatrick, A.L.; Kaplan, J.; Senatore, L. On the consistency relation of the 3-point function in single field inflation. *J. Cosmol. Astropart. Phys.* **2008**, *802*, 021. [CrossRef]
81. Hassan Firouzjahi and Bahar Nikbakht, *in progress.*
82. Chen, X.; Huang, M.X.; Shiu, G. The Inflationary Trispectrum for Models with Large Non-Gaussianities. *Phys. Rev. D* **2006**, *74*, 121301. [CrossRef]
83. Chen, X.; Hu, B.; Huang, M.X.; Shiu, G.; Wang, Y. Large Primordial Trispectra in General Single Field Inflation. *J. Cosmol. Astropart. Phys.* **2009**, *8*, 008. [CrossRef]
84. Weinberg, S. Quantum contributions to cosmological correlations. *Phys. Rev. D* **2005**, *72*, 043514. [CrossRef]

universe

MDPI

Review

Primordial Black Holes: Formation, Spin and Type II

Tomohiro Harada

Department of Physics, Rikkyo University, Toshima, Tokyo 171-8501, Japan; harada@rikkyo.ac.jp

Abstract: Primordial black holes (PBHs) may have formed through the gravitational collapse of cosmological perturbations that were generated and stretched during the inflationary era, later entering the cosmological horizon during the decelerating phase, if their amplitudes were sufficiently large. In this review paper, we will briefly introduce the basic concept of PBHs and review the formation dynamics through this mechanism, the estimation of the initial spins of PBHs and the time evolution of type II fluctuations, with a focus on the radiation-dominated and (early) matter-dominated phases.

Keywords: black holes; early universe; general relativity

1. Introduction

Black holes may have formed in the early Universe. This possibility was first considered by Zel'dovich and Novikov (1966) [1] and Hawking (1971) [2]. These black holes have been termed primordial black holes (PBHs). The observational relevance of PBHs to cosmology has been established by the subsequent work of Carr and Hawking (1974) [3] and Carr (1975) [4]. See Refs. [5,6] for the recent brief reviews of the history and the future of PBHs.

Although our Universe can be well approximated by a homogeneous and isotropic cosmological model—the Friedmann–Lemaître–Robertson–Walker (FLRW) solution—on large scales, cosmological structures such as galaxy clusters, galaxies, planetary systems, and stars must still form. These structures arise from primordial fluctuations. Observations of anisotropies in the cosmic microwave background (CMB) radiation have revealed that the amplitude of these primordial fluctuations is generally very small. One of the most promising mechanisms to generate such fluctuations is inflation. In this scenario, quantum fluctuations provide the origin of the perturbations, strongly suggesting that the probability of generating large-amplitude perturbations would be small but finite. During inflation, these perturbations expand along with the inflationary expansion of the Universe. After inflation, when the Universe is reheated, sufficiently large-amplitude perturbations could eventually overcome cosmic expansion and collapse into black holes due to their self-gravity. Unlike the formation of typical astrophysical black holes, the formation of PBHs does not require stellar evolution.

The formation of PBHs through primordial fluctuations was initially proposed by Hawking (1971) [2] and Carr and Hawking (1974) [3] and further developed by Carr (1975) [4]. This formation mechanism has since been confirmed through numerical simulations based on numerical relativity, pioneered by Nadezhin, Novikov and Polnarev (1978) [7]. In this review paper, we will explore the recent developments in PBH formation studies and their implications for cosmology.

The motivation for PBH research is multi-faceted. Since PBHs can, in principle, be observed, they provide valuable information about the early Universe. In this sense, PBHs can be regarded as the fossils of the early Universe. Furthermore, they are also one of the most promising candidates for dark matter. Moreover, Hawking evaporation plays an important role in the study of PBHs. Hawking recognised PBHs as a unique laboratory for studying black hole evaporation when he discovered this phenomenon in 1974 [8,9].

Hawking evaporation is key to understanding quantum gravity, where the information loss problem has been extensively discussed. The details of this evaporation may depend on quantum gravity and high-energy physics. Recently, several gravitational wave observatories have become available as observational tools to study strongly gravitating but otherwise dark objects, following LIGO's first direct detection of gravitational waves in 2015 [10]. It has been realized that gravitational waves provide a unique means of probing PBHs [11]. This suggests that PBHs lie at the intersection of various growing fields of modern physics, such as cosmology, general relativity, gravitational waves, quantum gravity and high-energy physics.

As observational data accumulate, gravitational waves are becoming an increasingly important means of observing our Universe. More than 100 events have been observed by the LIGO–Virgo–KAGRA (LVK) collaboration, and many of these have been identified as binary black holes. Several groups proposed the possibility that these binary black holes may be of cosmological origin, as many of them are approximately $30M_{\odot}$, which is more massive than was typically expected based on the standard theory of stellar evolution [12–14]. As more observational data accumulate, information about not only the masses but also the spins of binary black holes is obtained for some events [15], which may provide a new clue to the origin of these binary black holes. From the obtained mass function of binary black holes, a search for the population of PBHs was carried out [16]. It has also been argued that the existence of subsolar candidates in LIGO data would be a smoking gun for PBHs because no astrophysical scenario has yet been established to form subsolar-mass black holes [17]. Most recently, the NANOGrav collaboration reported evidence for nanohertz gravitational waves [18], which might be consistent with gravitational waves induced by scalar perturbations that could have produced the amount of PBHs responsible for the binary black holes observed by the LVK collaboration (e.g., [19]). The recent observations with the James Web Space Telescope have identified several very massive galaxies at high redshifts, which might indicate a tension with predictions in the standard ΛCDM model but can be explained if we include PBH clusters in the formation scenario [20–22].

I would also comment on the possibility of PBHs constituting all or a considerable fraction of dark matter. We usually discuss this in terms of the fraction of PBHs to cold dark matter (CDM), $f(M) = \Omega_{\rm PBH}(M)/\Omega_{\rm CDM}$, as a function of the mass of PBHs, M. This fraction is observationally constrained in some mass ranges but not in others. See Figure 10 of Carr, Kohri, Sendouda and Yokoyama (2021) [23] for an overview of the constraints. Recent observational constraints indicate two intriguing windows for dark matter. One is $M \sim 10^{17}$–10^{23} g, where $f(M)$ is virtually unconstrained. This implies that PBHs could account for all the CDM for this mass range. The other mass range is $M \sim 1$–$10^3 M_{\odot}$, where $f(M) \lesssim 0.1$, which is of interest in the context of terrestrial gravitational wave observations. Although there is currently a mass window in which all the CDM might be explained by PBHs, a stricter constraint could be placed on this mass window in the near future. Even if this turns out to be the case, it does not imply that studies of PBHs are without value. In this context, I would like to quote a profound statement by Bernard Carr: "Indeed their study may place interesting constraints on the physics relevant to these areas even if they never formed" [24].

The aim of this article is to briefly review the theoretical studies of PBH formation from primordial fluctuations, the recent estimation of the spins of PBHs and the formation of PBHs generated from the so-called type II configurations. In Section 2, we present the basic concept of PBHs. We discuss their masses, Hawking evaporation, the fraction of their contribution to the whole dark matter and the formation probability. In Section 3, we discuss the formation of PBHs. We focus on the formation mechanism from fluctuations generated by inflation and present the key ideas of this scenario. In Section 4, we discuss the spins of PBHs. The spins of PBHs have often been regarded just as negligible until recently. We quantitatively discuss the spins of PBHs formed not only in a radiation-dominated era but also in an (early) matter-dominated era. In Section 5 we discuss the initial configurations featured with their spatial geometry with throat structure, which is called type II, and the formation of PBHs from them based on the results of the recently conducted numerical simulations. In Section 6, we conclude the paper.

2. Basic Concept of Primordial Black Holes

2.1. Mass

The striking feature of PBHs is that they can have a large range of possible mass scales from $\sim 10^{-5}$ g to $\sim 10^{56}$ g depending on the formation scenarios, at least in principle. Although the mass of the PBH may depend on the scenario, we usually assume that it can be approximately given by the mass enclosed within the cosmological horizon or Hubble horizon at the formation time t_f from the big bang as

$$M \simeq M_H(t_f) \simeq \frac{c^3}{G} t_f \simeq 1 M_\odot \left(\frac{t_f}{10^{-5}\ \text{s}} \right), \tag{1}$$

where c and G are the speed of light and the gravitational constant, respectively, for which the gravitational radius is given by

$$R_g \simeq 1\ \text{km} \left(\frac{M}{M_\odot} \right). \tag{2}$$

So, we can say "the smaller, the older". Table 1 shows the relation between the formation time and the initial PBH mass. Nevertheless, we should keep a caveat in mind that the mass of PBHs could be much smaller than the horizon mass in certain formation scenarios.

Table 1. The cosmological time from Big Bang and the initial mass of PBHs if they are formed then.

Cosmological Time	Mass of PBHs
$\sim 10^{-43}$ s [Planck time]	$\sim 10^{-5}$ g [Planck mass]
$\sim 10^{-23}$ s	$\sim 10^{15}$ g [Critical Mass]
$\sim 10^{-5}$ s [QCD crossover]	$\sim 10^{33}$ g [Solar mass]
$\sim 10^{12}$ s [Matter-radiation equality]	$\sim 10^{50}$ g
$\sim 10^{19}$ s [Present epoch]	$\sim 10^{56}$ g [Mass of the observable Universe]

The mass of PBHs may have changed in time after they formed. As for the mass accretion, earlier works suggest that it does not so significantly affect the mass, at least in radiation domination [3]. This was confirmed by numerical simulations (e.g., [25]). On the other hand, Hawking evaporation is considered to decrease the mass of PBHs. The mass loss is considered to be virtually negligible if the mass of the PBH is much more massive than the critical mass $\sim 10^{15}$ g, which will be discussed below.

2.2. Evaporation

Based on quantum field theory in curved spacetimes, Hawking (1974) [8,9] found that black holes emit black body radiation. The temperature T_H of the black body, which is called the Hawking temperature, is proportional to the surface gravity of the horizon and is given for the Schwarzschild black hole by

$$T_H = \frac{\hbar c^3}{8\pi G M k} \simeq 100\ \text{MeV} \left(\frac{M}{10^{15} \text{g}} \right)^{-1}, \tag{3}$$

where $\hbar$ and k are the reduced Planck constant and the Boltzmann constant, respectively. This is called the Hawking evaporation. If we assume that the black hole loses its mass due to this radiation of quantum fields according to the Stefan–Boltzmann law, which is the so-called semi-classical approximation, we obtain

$$\frac{dM}{dt} = -\frac{g_{\rm eff}\hbar c^4}{15360\pi G^2 M^2}, \tag{4}$$

where $g_{\rm eff} \sim 100$ is the effective degrees of freedom and the grey-body factor is neglected. This implies that the evaporation timescale $t_{\rm ev}$ is given by

$$t_{\rm ev} \simeq \frac{G^2 M^3}{g_{\rm eff}\hbar c^4} \simeq 10\ {\rm Gyr}\left(\frac{M}{10^{15}{\rm g}}\right)^3, \tag{5}$$

after which the black hole of mass M loses almost all of its mass. As the black hole decreases its mass, the temperature becomes higher and higher and the evaporation timescale becomes shorter and shorter. Although the black hole decreases its mass very slowly in most of its life, it rapidly loses its remaining mass at its final moment, which is called a black hole explosion. When it becomes as light as the Planck mass $m_{\rm Pl} \sim 10^{-5}$ g, the semi-classical approximation necessarily breaks down and the subsequent evolution of the black hole should greatly depend on quantum gravity.

From Equation (5), the critical mass is approximately given by 10^{15} g for which $t_{\rm ev} = t_0$ is satisfied, where t_0 is the age of the Universe. So, if $M \lesssim 10^{15}$ g, PBHs have dried up after the explosion until now, whether they leave Planck mass relics or not. If $M \sim 10^{15} - 10^{17}$ g, PBHs are currently emitting X-rays and γ rays and can be observed through those emissions. If $M \gtrsim 10^{17}$ g, the evaporation is mostly negligible and the mass of the PBH remains almost constant until now. It should also be noted that very recently there has been a debate on a specific hypothetical scenario that assumes that the Hawking evaporation is strongly affected due to the back reaction effect even if the black hole is macroscopic [26].

2.3. Probability

To discuss the formation, the fraction $\beta(M)$ of the Universe which goes into PBHs when the mass contained within the cosmological horizon is M, is often used. This can also be regarded as the formation probability of PBHs. If we consider PBHs formed in the radiation-dominated phase, PBHs act as nonrelativistic particles surrounded by relativistic particles. Therefore, the energy density of PBHs, $\rho_{\rm PBH}$, decays as a^{-3}, while the energy density of relativistic particles, $\rho_{\rm rad}$, decays as a^{-4}, where a is the scale factor of the Universe. This implies $\rho_{\rm PBH}/\rho_{\rm rad} \propto a$, that is, PBHs are condensed during the radiation-dominated era in proportion to the scale factor. Taking this concentration effect into account, $\beta(M)$ is related to the the current fraction of PBHs of mass M to all the CDM

$$f(M) = \left.\frac{\Omega_{\rm PBH}}{\Omega_{\rm CDM}}\right|_{t=t_0} \tag{6}$$

through

$$\beta(M) \simeq 2\times 10^{-18}\left(\frac{M}{10^{15}\ {\rm g}}\right)^{1/2} f(M) \tag{7}$$

for $M \gtrsim 10^{15}$ g [4,23]. This formula can derived from the expansion law $a(t) \propto t^{1/2}$ for the flat Friedmann–Lemaître–Robertson–Walker (FLRW) solution in radiation domination, the mass within the horizon at the matter-radiation equality $M_{\rm eq} \sim 10^{50}$ g and Equation (1). This equation is important because it connects the theory and the observation of PBHs. Thus, for example, only a tiny probability $\beta \simeq 2\times 10^{-17}$ is enough to explain all of the dark matter if $M = 10^{17}$ g, while $\beta \simeq 2\times 10^{-8}$ is necessary for that if $M = 30 M_\odot$.

For $M \gtrsim 10^{15}$ g, Equation (7) combined with the observational constraint on $f(M)$ gives the observational constraint on $\beta(M)$. As for $M \lesssim 10^{15}$ g, PBHs have evaporated away until now. However, the evaporation of PBHs may spoil big bang nucleosynthesis for $M \sim 10^{10}$–10^{13} g if $\beta(M)$ is too large. Such a consideration gives the observational constraint on $\beta(M)$ for $M \gtrsim 10^{-5}$ g, where the lower limit is usually assumed to be the Planck mass. See Figure 18 of Ref. [23] for more details.

3. Formation

3.1. Overview

The serious study of PBH formation dates back to Carr (1975) [4]. One of its aims is to theoretically predict $\beta(M)$ and, therefore, $f(M)$ and other observables of PBHs for a given cosmological scenario. We can obtain the information of the early Universe from observational data on PBHs only through these studies. Apart from such an observational motivation, through the study of formation, we can understand physics in PBH formation and investigate new phenomena and/or new physics in highly nonlinear general relativistic dynamics and high-energy physics. There have been a lot of scenarios proposed for PBH formation. One of the most standard ones is the direct collapse of a large amplitude of primordial perturbations generated by inflation [27,28]. See also Ref. [11] for a recent review of this scenario. This can be considered as 'inevitable' as inflationary cosmology has been regarded as an essential part of standard cosmology. Among the other alternative scenarios, domain wall collapse, bubble nucleation, collapse of string networks and phase transitions have attracted more attention than others. Hereafter, we will focus on the formation scenario from fluctuations generated by inflation. Studies on this scenario have developed more than on other scenarios. The key ideas to this scenario are the following: fluctuations generated by inflation, long-wavelength solutions, formation threshold, black hole critical behavior, dependence on the equation of state (EOS) and statistics on the abundance estimate. We will briefly view these topics below. We will hereafter use the units in which $c = G = 1$ over this paper unless they are explicitly given.

3.2. Fluctuations Generated by Inflation

The most striking feature of inflation is that it can not only solve the flatness problem and the horizon problem but also provide the mechanism to generate fluctuations through quantum effects. Those fluctuations seed structure formation of different scales in our Universe and provide anisotropies in CMB currently observed with high accuracy.

The evolution of fluctuations in the inflationary cosmology is schematically illustrated in Figure 1. The quantum fluctuations generated by inflation are stretched to length scales much larger than the Hubble horizon scale cH^{-1} because the fluctuation scale, which is proportional to a, expands faster than the Hubble horizon scale, which is proportional to t, due to the accelerated expansion. This process is also considered to remove quantum coherence from the fluctuations. The generated classical fluctuations are further stretched away compared to the Hubble length as long as the inflationary phase continues. After the inflation ends, the expansion of the Universe begins to decelerate. There is a possibility that the Universe might experience the early matter-dominated phase due to the harmonic oscillation of an inflation field. In any case, the Universe eventually is dominated by the radiation field, tightly coupled relativistic particles in an almost complete thermal equilibrium state. This process is called reheating. If we consider the scale of the fluctuation to be much larger than the Hubble horizon in the decelerated Universe, the scale of the fluctuation, which is proportional to a, expands slower than the Hubble horizon scale, which is proportional to t. The time when the fluctuation scale is as large as the Hubble scale is called the horizon entry of the perturbation. Although inflationary cosmology has been becoming an essential part of standard cosmology, there has been no standard inflation model until now. There are lots of inflation models, each of which gives the power spectrum $P_\zeta(k)$ and the other statistics of the curvature perturbations ζ. See Ref. [11] for details of different inflation models in the context of PBH formation.

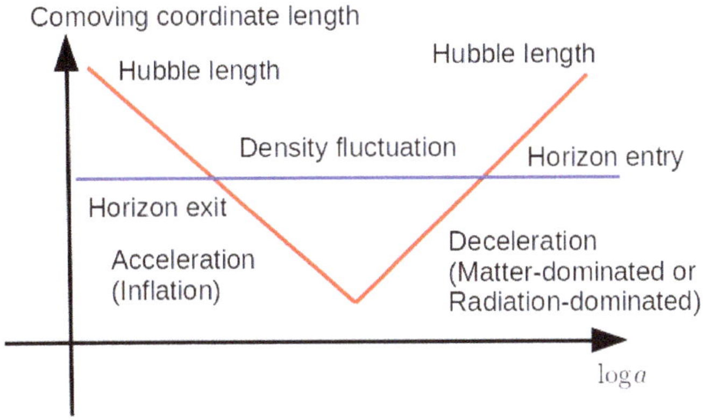

Figure 1. The schematic figure for the evolution of the scales of fluctuations generated by inflation. The quantum fluctuations generated by inflation are stretched to length scales that are much larger than the Hubble horizon scale cH^{-1} and are classical. In the decelerated phase after inflation, the scale of the fluctuation expands slower than the Hubble horizon scale. The time when the fluctuation scale becomes as large as the Hubble scale is called the horizon entry of the perturbation. After the horizon entry, the perturbation can collapse into a black hole leading to PBH formation if its amplitude is sufficiently large.

3.3. Large-Amplitude Long-Wavelength Solutions

As discussed above, fluctuations are generated and stretched to super-horizon scales by inflation. According to the standard cosmological scenario, the Universe must become radiation-dominated and decelerate after inflation. Then, the scale of the fluctuations become as large as the horizon scale, or in other words, the fluctuations re-enter the horizon. Then, if the amplitude of perturbation is nonlinearly large, the fluctuation can collapse into a black hole in the radiation-dominated era. This scenario has been conventionally studied for many years. In some scenarios, there can be an early matter-dominated era before the standard late one. In this case, we also need to think about the collapse of the fluctuation in the early matter-dominated era.

Based on the above scenarios, we discuss the fluctuations of super-horizon scales in the decelerated expansion after inflation. This situation is schematically illustrated in Figure 2. As we have already seen, in order to discuss PBH formation in radiation domination, it is necessary to deal with a nonlinearly large amplitude of perturbation of super-horizon scales. To obtain such perturbations, we solve the Einstein equation by a series expansion assuming that the spatial derivative is regarded as much smaller than the time derivative. This scheme is called gradient expansion and the solutions obtained by this scheme are called long-wavelength solutions.

First, we make the standard $3+1$ decomposition of the spacetime. We foliate four-dimensional spacetimes with constant t spacelike hypersurfaces Σ_t. Let x^i and $\gamma_{ij}(t,x^k)$ be spatial coordinates on Σ_t and the metric tensor on Σ_t, respectively, where $i,\ j,\ k,\ \ldots$ run over 1, 2 and 3. Then, the line element in four dimensions can be written by

$$ds^2 = -\alpha^2 dt^2 + \gamma_{ij}(dx^i + \beta^i dt)(dx^j + \beta^j dt), \tag{8}$$

where α and β^i are functions of t and x^i and called the lapse function and the shift vector, respectively. The functions α and β^i specify the time slicing, i.e., how the spacetime is foliated with constant t spacelike hypersurfaces, and the coordinate threading, i.e., how the constant x^i world lines cross the constant t hypersurfaces, respectively. We can choose the lapse and the shift according to the purposes so that we can fix four degrees of the gauge freedom in the Einstein equation. Using this decomposition, we can reformulate the Einstein equation in the form of the second-order time development, i.e., the Cauchy problem. See, e.g., Refs. [29,30] for more details.

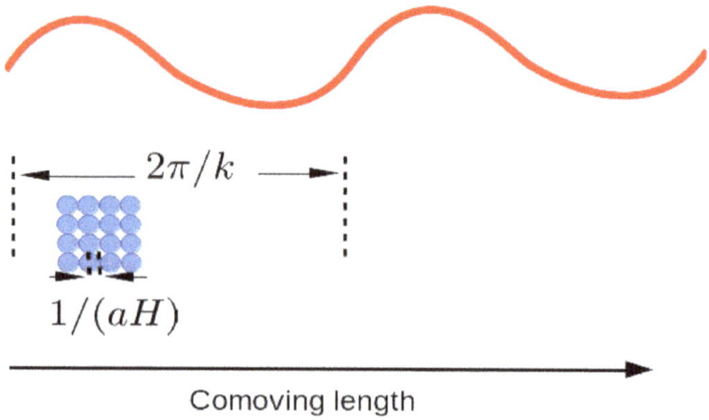

Figure 2. Schematic figure of long-wavelength solutions. The long-wavelength solutions are obtained under the assumption that the comoving length scale $2\pi/k$ of the perturbation in consideration is much longer than the Hubble horizon length $1/(aH)$, which gives the size of Hubble patches denoted with small blue disks.

In the next step towards the long-wavelength solutions, we take the following procedures [31–35]. In our situation, it is more useful to further decompose γ_{ij} in the form

$$\gamma_{ij} = e^{2\zeta}a^2(t)\tilde{\gamma}_{ij}, \tag{9}$$

where $\tilde{\gamma}_{ij}$ and ζ are both the functions of t and x^i so that $\det(\tilde{\gamma}_{ij}) = \det(\eta_{ij})$ should hold, where η_{ij} is the time-independent metric of the flat 3-space, while $a(t)$ is the scale factor of the reference flat FLRW spacetime. Thus, we have the line element in general spacetimes in four dimensions in the following $3+1$ form:

$$ds^2 = -\alpha^2 dt^2 + e^{2\zeta}a^2(t)\tilde{\gamma}_{ij}(dx^i + \beta^i dt)(dx^j + \beta^j dt). \tag{10}$$

We call this the cosmological $3+1$ conformal decomposition.

Then, we assume that the scale of the perturbation is much larger than the Hubble horizon scale, i.e., that

$$\epsilon := \frac{k}{aH} \tag{11}$$

is much smaller than unity, where k denotes the comoving wave number corresponding to the comoving scale of the perturbation. Applying the gradient expansion in powers of ϵ for the Einstein equation, we can obtain growing-mode solutions with an assumption that the zeroth-order solutions in powers of ϵ take the following form:

$$ds^2 = -dt^2 + e^{2\zeta_0(x)}a^2(t)\eta_{ij}dx^i dx^j, \tag{12}$$

where $\zeta_0(x)$ is the zeroth-order part of ζ, which is a function solely of $x = (x^1, x^2, x^3)$ and identified with primordial curvature perturbation. Note that this assumption is compatible with the comoving slice, the constant-mean-curvature (CMC) slice and the uniform-density slice but not with the conformal Newtonian gauge condition, which is often used in cosmology. This is because, as is well known, in the conformal Newtonian gauge, the density perturbation in the linear order approaches a time-independent function of x in the super-horizon limit, while it approaches zero in the same limit in the cosmological long-wavelength solutions formulated in Refs. [32–34], although we do not exclude the existence of other formulations that are compatible with the conformal Newtonian gauge. The higher-order terms of the solutions are obtained in terms of the function $\zeta_0(x)$ if we impose the appropriate gauge conditions. In other words, $\zeta_0(x)$, which is not suppressed due to the long-wavelength scheme, generates the long-wavelength solutions. For example, the density perturbation δ is $O(\epsilon^2)$ and is given by

$$\delta_{\rm CMC} \simeq -\frac{4}{3a^2H^2}e^{-\frac{5\zeta_0}{2}}\Delta_{\rm flat}e^{\frac{\zeta_0}{2}} \tag{13}$$

in the CMC slice.

If we apply this formulation to the spherically symmetric spacetime, where the zeroth order metric can be written in the following form

$$ds^2 = -dt^2 + e^{2\zeta_0(r)}a^2(t)[dr^2 + r^2(d\theta^2 + \sin^2\theta d\phi^2)], \tag{14}$$

in the conformally flat coordinates, we can show that this is equivalent to the asymptotically quasihomogeneous solutions developed in Ref. [34] in the Misner–Sharp formulation, where the comoving slice and the comoving thread are adopted, and the explicit transformation between them is given in Ref. [35].

3.4. Formation Threshold in Radiation Domination

If the amplitude of the perturbation, which is generated by inflation and enters the horizon in the radiation-dominated era, is sufficiently large, it will directly collapse into a black hole. Since the linear perturbation in the radiation-dominated era does not significantly grow, such a perturbation must be nonlinearly large. This prevents us from accessing the full dynamics of PBH formation with analytical methods. Only full numerical relativity simulations can accurately describe the general relativistic dynamics of PBH formation, which has been pioneered by Nadezhin, Novikov and Polnarev (1978) [7]. In fact, it has been established by several analytical and numerical works such as Refs. [4,7] that PBHs really form in this scenario.

This scenario naturally implies that there exists a threshold for PBH formation. Carr (1975) derived the threshold $\delta_{\rm th} \sim 1/3$ for radiation domination according to the Jeans scale argument in terms of δ_H, the density perturbation at the horizon entry of the perturbation. In numerical relativity, the threshold value is obtained as $\delta_{\rm th} \sim 0.45$ in terms of the density perturbation averaged in the comoving slice over $0 < r < r_0$, where r_0 is the radius of the overdense region, at the horizon entry of the overdense region, where δ_H is regarded as that in the nontrivial lowest order of the long-wavelength expansion scheme [35–38]. Harada, Yoo and Kohri (2013) [39] refined Carr's argument from a general relativistic point of view and analytically derived the threshold $\delta_{\rm th} \simeq 0.41$ in the comoving slice. Although the averaged density perturbation δ_H might look straightforward to interpret, it has difficulty in its interpretation. This is because if we calculate the averaged density perturbation at the horizon entry using the nontrivial lowest order of the solutions, it cannot be a real physical value, as the latter can only be obtained after the full numerical simulation and is not necessarily useful.

Shibata and Sasaki (1999) [32] defined a compaction function in the CMC slice. Although it was intended to equal the ratio of the excess in the Misner–Sharp mass, or equivalently the Kodama mass, to the areal radius, it is not equal to that but

$$\mathcal{C}_{\rm SS} \approx \frac{1}{2}\left[1 - \left(1 + r\zeta'\right)^2\right], \tag{15}$$

where the weak equality denotes the equality in the long-wavelength limit and we have omitted the subscript 0 in ζ according to the convention [35,40]. This has been shown to have a geometrical origin as a compactness function in the static spacetime obtained by removing the scale factor in the long-wavelength limit [41]. It is proportional to the ratio of the Misner–Sharp mass excess to the areal radius in the comoving slice as

$$\mathcal{C}_{SS}(r) \approx \frac{3}{4} C_{com}(r), \tag{16}$$

where

$$C_{com} := \frac{2\delta M_{com}}{R} \tag{17}$$

with δM_{com} being the Misner–Sharp mass excess in the comoving slice. Note the factor of 2 on the right-hand side of Equation (17). The threshold value is $\sim$0.4 in terms of the maximum value of the Shibata–Sasaki compaction function $\mathcal{C}_{SS}(r)$ in the long-wavelength limit [32,35,38]. This is more straightforward than the averaged density perturbation because, for the compaction function description, we only have to take the long-wavelength limit of the solutions, where all the higher-order contributions naturally disappear. This is probably why people have favored using the compaction function. On the other hand, this threshold can be transformed to $\sim$0.5 in terms of the density perturbation in the comoving slice averaged over the ball $0 < r < r_m$ at its horizon entry, where r_m is the radius at which $\mathcal{C}_{SS}$ takes the maximum [42] again with a caveat that this value is only correct in the nontrivial lowest order of the long-wavelength expansion. The averaging over $0 < r < r_m$ is more consistent and applicable than over the overdense ball $0 < r < r_0$ as the definition of r_0 is problematic for density perturbations without an underdense region.

In the last decade, great effort has been paid to reveal the profile dependence of the threshold [42–44]. In particular, Escrivà, Sheth and Germani (2020) [44] found a universal threshold

$$\bar{\mathcal{C}}_{com} \simeq \frac{2}{5}$$

in terms of $\bar{\mathcal{C}}_{com}$, the spatial average of C_{com} over $0 < r < r_m$, where $C_{com}(r)$ takes a maximum at $r = r_m$. This holds within 2% accuracy over different profiles they surveyed. This gives a new threshold condition that uses the maximum $C_{com}(r_m)$ and its second-order derivative $C''_{com}(r_m)$. See Ref. [44] for details. More recently, Ianniccari, Iovino, Kehagias, Perrone and Riotto [45] reported some numerical coincidence between this threshold and that for the existence of a circular photon orbit.

There is a possibility that isocurvature perturbations may be produced depending on the cosmological scenario. Yoo, Harada, Hirano, Okawa and Sasaki (2021) [46], through numerical relativity simulations by introducing a massless scalar field perturbation into the radiation-dominated Universe, demonstrated that isocurvature perturbations can also produce PBHs in the radiation-dominated era. As isocurvature perturbations can play important roles in various intriguing inflationary scenarios, further studies are anticipated to achieve a comprehensive understanding of the formation threshold from isocurvature.

3.5. Softer Equation of State

There remain a lot of theoretical possibilities for the thermal history of the Universe. In addition to the standard phases of radiation dominance and subsequent matter dominance, there may be an early matter-dominated phase and/or phase transition or crossover phase in which the equation of the state of the dominant component of the Universe is significantly softer than the radiation fluid. Although PBH formation has been conventionally studied in the radiation-dominated phase, the scenarios in other phases with such significantly soft equations of state will also be important. This is because as we will see below, $\beta(M)$ strongly depends on the threshold, which is smaller for the softer equation of state and, hence, PBH formation can be strongly enhanced so that PBHs formed in this phase may dominate those formed in the radiation-dominated era, provided that the power spectrum density perturbation is nearly scale-invariant.

This scenario was already suggested by Carr (1975) [4], where the threshold density perturbation was estimated to $\delta_H \sim w$ for the linear EOS $p = w\rho$ using the simple application of the Jeans criterion in Newtonian gravity. Harada, Yoo and Kohri (2013) [39] refined it by comparing the free-fall time and the sound-crossing time in a simplified toy model of the fully general relativistic spacetime considering gauge difference to find the threshold value

$$\delta_H \simeq \frac{3(1+w)}{5+3w}\sin^2\left(\frac{\pi\sqrt{w}}{1+3w}\right) \tag{18}$$

in the comoving slice. This analytical expression showed a good agreement with the results of numerical relativity simulation [38,47] for $0 < w \leq 1/3$ as seen in Figure 3. These studies showed that the threshold for the EOS $p = w\rho$ is an increasing function of w for $0 < w \leq 1/3$ and approaches 0 as $w \to 0$. For example, for the QCD crossover, w drops to $\sim$0.23 from 1/3 [48], this suggests enhanced generation of PBHs. Although w is time-dependent in this case, the threshold has also been calculated based on the Jeans criterion [49]. Subsequently, full numerical relativity simulations have revealed that the PBH formation will be enhanced by a factor of the order of 1000 [50,51]. See Ref. [52] for a recent review on this subject.

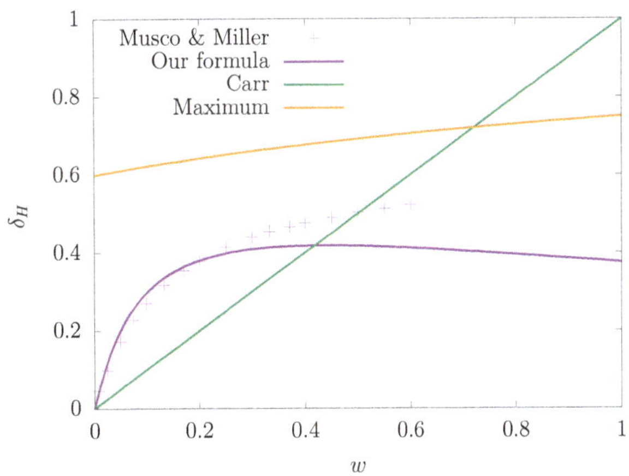

Figure 3. EOS dependence of the PBH formation threshold. The horizontal axis is the parameter w for the EOS $p = w\rho$, while the vertical axis is the threshold value in terms of δ_H, the density perturbation over the overdense region, $0 < r < r_0$, at its horizon entry in the comoving slice. The crosses show the numerical result obtained in Ref. [38]. The purple curve is the plot of Equation (18), which shows an agreement with the numerical result within $\sim$20%, while the green curve shows Carr's formula $\delta_H \simeq w$. See Ref. [39] for more details.

Musco and Papanikolaou (2022) [53] studied the effect of anisotropic stress, which is the deviation from the perfect fluid description, on PBH formation in radiation domination and found that if this is large enough, it could lead to a significant variation in the abundance of PBHs.

3.6. Matter Domination

If the argument in Section 3.5 for the EOS $p = w\rho$ applied to $w \approx 0$ for the matter-dominated phase, one might consider that PBHs could be overproduced. However, this cannot be correct because the realistic physical system is highly nonspherical and the deviation from spherical symmetry will grow during the collapse in matter domination. Not only after the standard matter-radiation equality time but also in a possible early matter-dominated phase, which may naturally occur in the preheating process or in the strong phase transition in the Universe, PBH formation will be enhanced. This is one of the interesting epochs to study in the context of PBH formation. Since the condition for PBH formation is not solely determined by the pressure gradient force, we would need a totally different treatment for this phase from that for the radiation-dominated phase.

The PBH formation in matter domination has been pioneered by Khlopov and Polnarev (1980) [54–57], where the effects of anisotropy and inhomogeneity are studied as an obstruction of the PBH formation process. Harada, Yoo, Kohri, Nakao and Jhingan (2016) [58] revisited the anisotropic effects by combining the picture of the pancake collapse of dark matter and the hoop conjecture by Thorne, which claims that black holes with horizons form when and only when a mass M gets compactified into a region whose circumference in every direction is $\mathscr{C} \lesssim 4\pi M$ [59,60]. This suppression due to the anisotropic effect is schematically shown in Figure 4. They not only qualitatively reproduced the result of Refs. [54,56] but also updated the coefficient as

$$\beta_{\rm aniso}(M) \simeq 0.05556\sigma_H^5(M), \tag{19}$$

where $\sigma_H(M)$ is the standard deviation of δ_H in the mass scale of M and the Gaussian distribution for density perturbation is assumed.

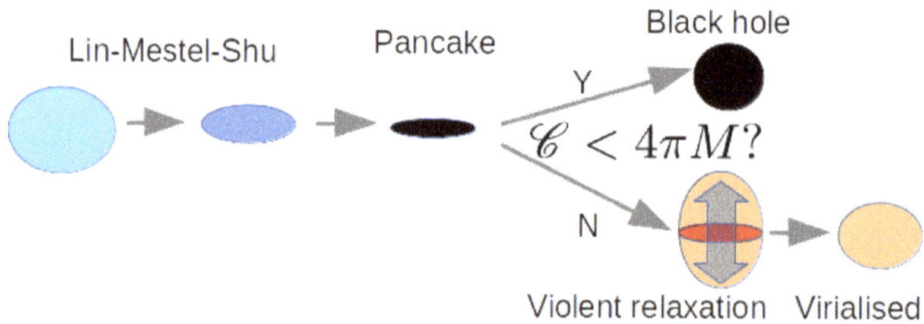

Figure 4. Schematic illustration of the scenario on the anisotropic effect in the formation of PBHs in matter domination. $\mathscr{C}$ denotes the hoop of the surface of the gravitating object of mass M. The scenario is the following. The collapse of a dust ball is unstable against nonspherical perturbations towards the pancake. Whether or not a horizon forms around the pancake is subject to the hoop conjecture. If the condition $\mathscr{C} \lesssim 4\pi M$ is satisfied, a black hole forms. Otherwise, the collapse leads to virialization by acquiring velocity dispersion through violent relaxation. See Ref. [58] for details.

Kopp, Hofmann and Weller (2010) [61] modeled the spherical formation of PBHs in matter domination with the Lemaître–Tolman–Bondi (LTB) solution and Harada and Jhingan (2015) [62] extended it to a nonspherical formation with the Szekeres quasispherical solutions. Kokubu, Kyutoku, Kohri and Harada (2018) [63] revisited the inhomogeneity effects by utilizing the LTB solution and not only qualitatively reproduced the result of Refs. [54,56] but also updated the coefficient so that the additional suppression factor is given by

$$\beta_{\rm inhom}(M) \simeq 3.70\sigma_H^{3/2}(M) \tag{20}$$

with the caveat that its physical effect largely depends on the assumption that black hole formation is prevented by the appearance of an extremely high-density region before the black hole horizon formation surrounding it.

Harada, Kohri, Sasaki, Terada and Yoo (2022) [64] showed that the effects of velocity dispersion that may have been generated in the possible nonlinear growth of perturbation in the earlier phase can suppress PBH formation. The effects of the angular momentum can play important roles and suppress PBH formation for smaller σ_H [65]. We will later discuss the angular momentum in PBH formation in the matter-dominated era in the context of the initial spins of PBHs.

There is another intriguing approach to PBH formation during the early matter-dominated era. Such an era is thought to be naturally realized during the preheating phase following inflation when an oscillating massive scalar field dominates the Universe. It is possible to gain further insights into PBH formation during this phase by directly studying the perturbations in the Universe dominated by a massive scalar field. This approach was explicitly proposed by Padilla, Hidalgo and Malik (2022) [66–68], and numerical relativity simulations have been conducted based on this method, involving a mixture of a massive scalar field and a massless scalar field [69–71]. This approach may involve highly fruitful physics, and further studies will be necessary for a comprehensive understanding of this scenario.

3.7. Critical Behavior

One of the great achievements of numerical relativity is the discovery of critical behavior in gravitational collapse, which is also called black hole critical behavior. This was first discovered by Choptuik (1993) [72] in the spherically symmetric system of a massless scalar field and followed by Evans and Coleman (1994) [73] in the spherical system of a radiation fluid. The phenomena were theoretically understood in terms of the renormalization group analysis by Koike, Hara and Adachi (1995) [74]. These early studies were all on asymptotically flat spacetimes and whether they apply to PBH formation was not so trivial because of the different boundary conditions and the existence of the characteristic scale, the Hubble horizon length. The critical behavior in the PBH formation was discovered by Niemeyer and Jedamzik (1999) [75] but subsequently questioned [76]. Finally, Musco and Miller (2013) [38] beautifully confirmed the critical behavior in PBH formation. The essence of the critical behavior in the context of PBH formation is the following. Let us consider a one-parameter family of initial data of the Cauchy problem, for which we can choose the averaged density perturbation δ_H as the parameter of the family. The features of the critical behavior do not depend on the choice of the one-parameter family of the initial data. As we have already seen, there is a threshold value $\delta_{\rm th}$, beyond which a PBH forms. Then, the evolution of the initial data which has the critical value $\delta_H = \delta_{\rm th}$ approaches a particular member of continuously self-similar solutions, which is called a critical solution. If δ_H is slightly above the critical value, we have PBH formation and the scaling law for the mass of the formed PBH, M, holds as follows:

$$M \approx K M_H (\delta_H - \delta_{\rm th})^{\gamma}, \tag{21}$$

where $\gamma \simeq 0.36$ is called a critical exponent and K is a nondimensional positive constant that depends on the shape of the initial perturbation, varying between 3 and 30 with a typical value $\sim$5. In fact, the critical behavior, such as the critical solution and the critical exponent, does not depend on the choice of the one-parameter family of initial data, which is called universality. Note, however, that the critical solution and the critical exponent do depend on the matter field and the equation of state even if it is a perfect fluid [77].

In the context of PBH formation, if we assume that δ_H obeys some reasonable statistical distribution, the critical behavior implies that a tiny fraction of perturbations of mass scale M_H can produce PBHs that are much smaller than M_H, while a large fraction still produce those of the order of M_H. This becomes important especially if we consider $\beta(M)$ for $M \sim 10^{16}$–10^{17} g. This is because the PBHs of the critical mass $\sim 10^{15}$ g are severely constrained by observation through its X-ray or gamma-ray emission, while those of $\sim 10^{16}$–10^{17} g are not. For example, the formation probability $\beta(M)$ with the horizon mass $M = 10^{16}$ g is severely constrained by such a tiny fraction of PBHs of the critical mass $\sim 10^{15}$ g considerably smaller than 10^{16} g [78,79].

3.8. Abundance Estimation and Statistics

Even if we can identify the threshold condition and the classical dynamics of PBH formation and subsequent evolution, it is not sufficient to determine $\beta(M)$. Clearly, we also need statistical properties of the perturbations. In inflationary cosmology, the statistical properties of fluctuations generated by inflation, such as the power spectrum $P_\zeta(k)$ and other statistical properties can be predicted, at least in principle, if we fix the inflation model.

Carr (1975) [4,39] simply assumed that δ_H obeys a Gaussian distribution and obtained the following formula

$$\beta(M) \simeq 2\frac{1}{\sqrt{2\pi}\sigma_H(M)}\int_{\delta_{\rm th}}^{\delta_{\rm max}} d\delta e^{-\frac{\delta^2}{2\sigma_H^2(M)}} \simeq \sqrt{\frac{2}{\pi}}\frac{\sigma_H(M)}{\delta_{\rm th}}e^{-\frac{\delta_{\rm th}^2}{2\sigma_H^2(M)}}, \tag{22}$$

where $\delta_{\rm max} = 2/3$ and $\sigma_H(M)$ are the possible maximum value and the standard deviation of δ_H in the mass scale of M and in the last approximation $\delta_{\rm th} \gg \sigma_H(M)$ is assumed. The factor of two in the above equation comes from the recommendation for structure formation by Press and Schechter (1975) [80,81], whereas whether or not this factor applies to PBH formation is highly nontrivial. Since we need at least $\beta(M) \gtrsim 10^{-18}$ for PBHs to contribute to a considerable fraction of dark matter, we can conclude that only the Gaussian tail beyond $\sim 8\sigma_H$ is responsible for PBHs. This implies that introducing small non-Gaussianities can significantly enhance the formation of PBHs [82]. See Ref. [83] for a recent review on non-Gaussianities in PBH formation and their link to induced gravitational waves. If we consider PBHs formed in the radiation-dominated era and assume $\delta_{\rm th} \sim 0.45$ or ~ 0.5, we need at least $\sigma_H(M) \gtrsim 0.05$ or $P_\zeta(k) \gtrsim 0.01$. This is much larger than the observed value $P_\zeta(k) \simeq 10^{-10}$ for the CMB anisotropies, although it does not immediately exclude the considerable formation of PBHs because the scales of the PBHs and the CMB anisotropies are usually very different from each other.

Although Carr's formula (22), which is also called the Press–Schechter approximation, is very useful, this is considered as a very rough approximation. One of the reasons is that even if the curvature perturbation obeys a Gaussian distribution, the averaged density perturbation δ_H in the comoving slice cannot because it must be within a finite interval between the minimum -1 and the maximum $2/3$ [61]. The other is that since PBHs will form only at very rare peaks of the perturbation, peak theory should apply, which is known to give a physically reasonable prediction for galaxy formation [84]. The prediction of peak theory may be significantly different from that of Carr's formula for the estimation of PBH abundance. Currently, there are a few variations in the application of peak theory to the estimation of the PBH abundance, which comes from the theoretical ambiguity caused by an incomplete understanding of nonlinear, nonspherical and multi-scale general relativistic dynamics [85–89]. See also Ref. [90] for a recent review on this topic.

4. Initial Spins

If the black hole no hair conjecture holds in astrophysics, stationary black holes in vacuum should be well approximated by Kerr black holes which are characterized only by two parameters, the mass and the angular momentum. As we have already discussed, the mass of PBHs is approximately equal to that within the Hubble horizon at the time of formation. So, what determines the spin of PBHs? If black holes of masses from several to several tens of solar masses are observed, it would be very difficult to distinguish between PBHs and astrophysical black holes. Astrophysical black holes form in the final stage of the evolution of massive stars. If we observe an isolated black hole, the only information of its own other than its mass is its spin. So, if the spins of PBHs are expected to be very different from those of astrophysical black holes, the observation of spin can be potentially decisive information to distinguish between the two populations. Here, we focus on the initial spins of PBHs. However, it should be noted that the accretion and merger history after their formation could also greatly affect their spins depending on the scenarios [91] and, hence, the theoretically estimated values for the initial spins should be compared with the observation with great care.

4.1. Spins of Primordial Black Holes Formed in Radiation Domination

It was discussed that the formation process of PBHs in radiation domination is well approximated by spherically symmetric dynamics since Carr (1975) [4]. Recent studies have revealed that this early argument is basically correct. Chiba and Yokoyama (2017) [92] gave an upper bound $\sim$0.4 on the root mean square of the nondimensional Kerr parameter $a_* = a/M$, where a and M are the Kerr parameter and the mass of the black hole. He and Suyama (2019) [93] discussed the effect of angular momentum on the threshold.

Peak theory features more accurate quantitative analyses. For PBHs formed in radiation domination, the threshold density perturbation $\delta_{\rm th}$ is of the order of the unity, which is considerably larger than its standard deviation. In other words, PBHs can form only at very rare peaks. Peak theory predicts that there is only a very small deviation from spherical symmetry for such rare peaks. This is very important to not only justify the spherical symmetry assumption but also estimate the angular momentum of PBHs formed in radiation domination. Based on the perturbative analysis based on peak theory, De Luca, Desjacques, Franciolini, Malhotra and Riotto (2019) [94] concluded that the root mean square of the nondimensional Kerr parameter a_* is of the order of 10^{-2}, while Harada, Yoo, Kohri, Koga and Monobe (2021) estimated the root mean square of a_* to be in the order of 10^{-3} and showed that a small fraction of PBHs of masses much smaller than the mass enclosed within the Hubble horizon, $M \ll M_H$, as a result of critical phenomena, can have much larger spins [95].

Based on this analysis combined with peak theory, Koga, Harada, Tada, Yokoyama and Yoo (2022) [96] calculated the distribution of the effective spin parameter $\chi_{\rm eff}$ of binary black holes, which is a well-determined observable of the spins obtained from gravitational waveforms from inspiralling binaries. Figure 5 shows the distribution functions of the parameters a_* and M of isolated PBHs on the top panel and of the mass ratio $q := M_2/M_1$, the Chirp mass $\mathcal{M} := (M_1M_2)^{3/5}/(M_1+M_2)^{1/5}$ and $\chi_{\rm eff} := (a_{*1}\cos\theta_1 + qa_{*2}\cos\theta_2)/(1+q)$ of binary PBHs on the bottom panels, where M_i, a_{*i} and θ_i ($i = 1$, 2) are individual masses, individual nondimensional Kerr parameters and the angles of individual spins with respect to the orbital angular momentum, respectively, with $M_1 \geq M_2$ being assumed. Although there is a tendency for larger spins with smaller masses because of the critical phenomena, the probabilities for large values of the spin parameters are strongly suppressed. See Ref. [96] for details.

In the above analyses, it was assumed that the power spectrum of curvature perturbation is nearly monochromatic. Recently, it has been revealed that if this assumption is relaxed, the root mean square of a_* can be slightly larger for some set of broad power spectra but should still be bounded by the value of the order of 10^{-3} [97].

4.2. Spins of Primordial Black Holes Formed with a Soft Equation of State

It is interesting to ask how much the results in Section 4.1 depend on the properties of the matter fields in the cosmological phase when the PBHs formed. In particular, it was shown that the PBH production is significantly enhanced in the QCD crossover, where the effective value of w drops from 1/3 to $\sim$0.23 [50,51]. Saito, Harada, Koga and Yoo (2023) [98] showed that for a soft EOS parameterized by $p = w\rho$, the root mean square of a_* is a decreasing function of w and can be well fitted by the power law $\propto w^{-0.49}$. However, since the dependence is weak for $w \simeq 0.2$–$1/3$, the initial spins are only modestly enhanced for the QCD crossover such as to $\sim$0.003 from $\sim$0.002 for radiation. It also suggests that a_* can be very large if $w \ll 1$, although the analysis in Ref. [98] is not well justified in the limit $w \to 0$, where the treatment for matter domination should apply.

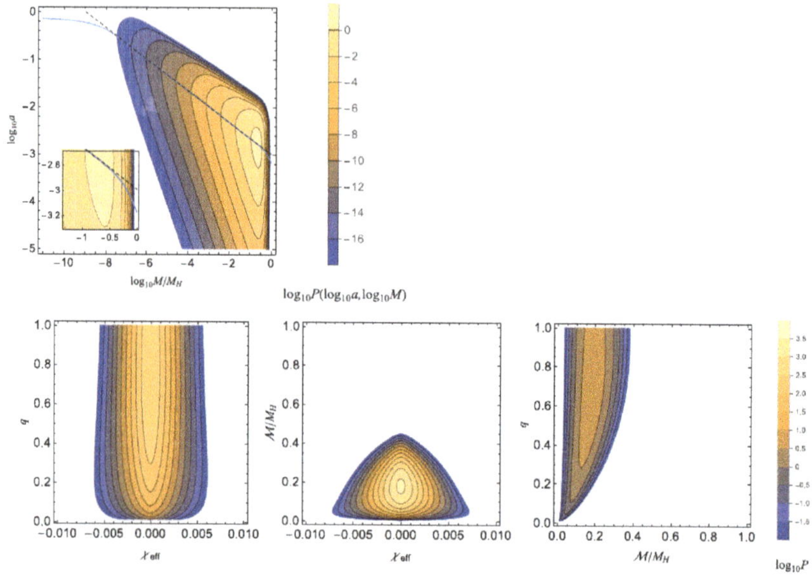

Figure 5. Contour maps of the distribution functions of (M, a_*) for isolated PBHs on the top panel, M and a_* are the mass and the nondimensional Kerr parameter, respectively, and $(\chi_{\rm eff}, q)$, $(\chi_{\rm eff}, \mathcal{M})$ and $(\mathcal{M}, q)$ for binary PBHs from left to right on the bottom panels, where $\chi_{\rm eff}$, $\mathcal{M}$ and q are the effective spin parameter, the Chirp mass and the mass ratio of the binary PBHs, respectively. The definitions of q, $\mathcal{M}$ and $\chi_{\rm eff}$ are $q := M_2/M_1$, $\mathcal{M} := (M_1 M_2)^{3/5}/(M_1 + M_2)^{1/5}$ and $\chi_{\rm eff} := (a_{*1}\cos\theta_1 + q a_{*2}\cos\theta_2)/(1+q)$, respectively, where M_i, a_{*i} and θ_i $(i = 1, 2)$ are individual masses, individual nondimensional Kerr parameters and the angles of individual spins with respect to the orbital angular momentum, respectively, with $M_1 \geq M_2$ being assumed. Taken from Ref. [96].

4.3. Spins of Primordial Black Holes Formed in (Early) Matter Domination

Since PBH formation will be enhanced in the (early) matter-dominated phase, it is very important to predict their spins. As we have seen before, nonspherical effects may become important. In fact, Harada, Yoo, Kohri and Nakao (2017) [65] investigated the effects of angular momentum for PBH formation in this phase. Both the first-order and second-order effects can potentially play important roles. The first-order effect generates angular momentum through the nonsphericity of the region to collapse into a black hole, which is generally misaligned with a mode wave number vector of the velocity perturbation. This is schematically illustrated in Figure 6, where the ellipsoidal region that will collapse is misaligned with the wave number of the velocity perturbation denoted by the red arrows and, hence, it carries nonvanishing angular momentum. The secondary effect also gives angular momentum through the coupling of two independent modes of linear perturbation, which is illustrated in Figure 1 of Ref. [65].

Although the dynamics of PBH formation in matter domination are expected to be very complicated, a perturbative calculation under certain working assumptions gives

$$\langle a_*^2 \rangle \sim \sigma_H^{-1/2}, \tag{23}$$

where a_* is the nondimensional spin parameter of the region to collapse and σ_H is the standard deviation of δ_H at the horizon entry. Although the nondimensional numerical factor of the order of the unity on the right-hand side should be determined, this implies that most PBHs have spins $a_* = O(1)$ if $\sigma_H \sim 0.1$. The angular momentum effects will strongly suppress PBH formation if σ_H is even much smaller because of the Kerr bound $|a_*| \leq 1$. These results have recently been updated based on peak theory [99].

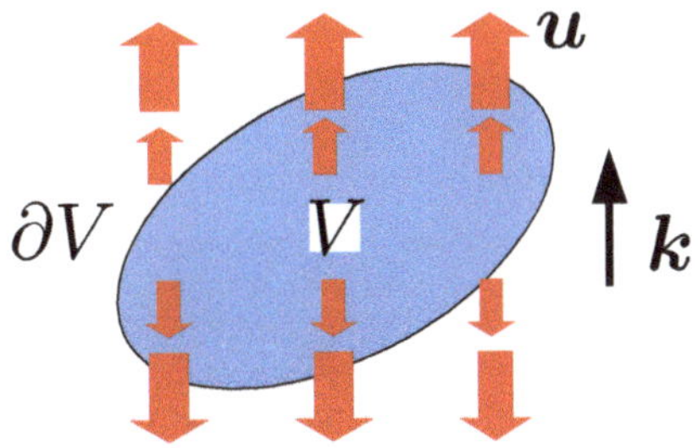

Figure 6. Schematic illustration of the first-order effect to generate angular momentum. The blue region V denotes the region that will collapse, which is approximated by an ellipsoid. The red arrows denote the velocity perturbation $\boldsymbol{u}$ with a wave number vector being parallel to the vertical direction. Since the ellipsoidal region V is misaligned with the wave number $\boldsymbol{k}$ of the velocity perturbation, it carries nonvanishing angular momentum. The angular momentum can increase in time as the perturbation grows. This can also be understood through the effect of torque exerted on the boundary ∂V. See Ref. [65] for details.

4.4. Nonspherical Simulation of PBH Formation

Although we have discussed the initial spins of PBHs, we have neglected nonspherical nonlinear general relativistic dynamics in the formation of black holes for simplicity. It is clearly important to numerically simulate the nonspherical formation of PBHs and investigate how much the initially nonspherical initial data will affect the formation threshold and the initial spins of the produced PBHs. This is complementary to the perturbative analysis as it can check the validity of assumptions made.

Yoo, Harada and Okawa (2020) [100] conducted a 3D numerical simulation of nonspherical PBH formation in radiation domination based on numerical relativity for the first time. They prepared the long-wavelength solutions as initial data, for which initial nonsphericity is expected to be typically very small according to peak theory. However, to make the numerical results clearer, they put $\sim$10% nonsphericity in the initial data, which is much larger than the values expected from peak theory for PBH formation. They found that even such large nonsphericity changed the threshold of PBH formation only by $\sim$1%. See Figure 7 which is taken from Ref. [100] for the initial density perturbation on the left panel and the apparent horizon formation in the course of gravitational collapse on the right panel. We can see that in the simulation with the near-threshold value, a very small apparent horizon forms at the very central region. This implies that we need a very high resolution near the center. This problem was attacked with a rescaled radial coordinate [100].

Yoo (2024) [101] conducted further numerical relativity simulations for the EOS $p = w\rho$ with $w = 0.2$ and $1/3$ with much higher resolution and accuracy. For this purpose, not only the rescaled radial coordinate but also the multi-level mesh refinement scheme was adopted. In such simulations, the estimation of the spin is not straightforward. In this work, it was implemented as follows: For the Kerr black hole, we can write the spin parameter a_* using the event horizon configuration through the following relation

$$a_* = \frac{\sqrt{4\pi A(d^2 - \pi A)}}{d^2}, \tag{24}$$

where d and A are the equatorial circumference and the area, respectively. Yoo (2024) [101] estimated the spin of the PBH as $|a_*| \ll 0.1$ assuming Equation (24) for the numerically found apparent horizon. This is consistent with the perturbative estimates discussed in Sections 4.1 and 4.2.

As far as I am aware, no 3D numerical relativity simulations for nonspherical PBH formation in the (early) matter-dominated phase have been carried out using a fluid or dust description of the matter fields. In the scalar field approach to the early matter-dominated phase, De Jong, Aurrekoetxea, Lim and França (2023) [70] studied nonspherical PBH formation through 3D numerical relativity simulations and reported that the spin in the final stage of formation would be negligible. Although this result is highly suggestive, further systematic studies are anticipated in the near future.

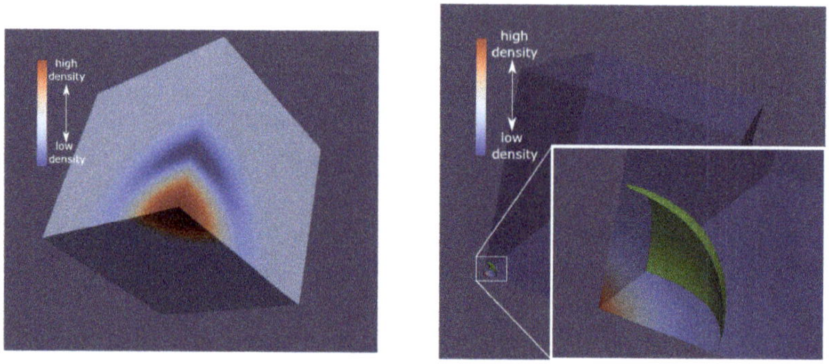

Figure 7. The initial density perturbation is on the left panel and the apparent horizon formation in the course of gravitational collapse is on the right panel. We can see that an apparent horizon forms in the central region, which is very small compared to the size of the whole computational domain. It implies that to resolve the apparent horizon for the near-critical collapse requires very high resolution at least near the center. Both were taken from Ref. [100] with permission.

5. Type II Perturbation and Type B PBH

5.1. Introduction to Type II Perturbation

In the FLRW solution, the geometry of the constant cosmological time t spacelike hypersurface is given by a constant curvature space with a scale factor $a(t)$. The metric can be written in the following familiar form:

$$ds^2 = -dt^2 + a^2(t)\left[\frac{1}{1-K\tilde{r}^2}d\tilde{r}^2 + \tilde{r}^2(d\theta^2 + \sin^2\theta d\phi^2)\right]. \tag{25}$$

The spatial curvature K can be zero, positive, or negative. For $K \neq 0$, the geometry has a finite curvature radius $1/\sqrt{|K|}$ up to the scale factor. The positive curvature solution generally has particular time evolution in $a(t)$, which begins to expand from the Big Bang, and reaches maximum expansion and collapses to a big crunch, while for the zero and negative cases it just continues expanding from the Big Bang to infinity. As is well known, for the positive-curvature FLRW spacetime, there is a coordinate singularity at $\tilde{r} = 1/\sqrt{K}$ in the line element (25). We can go beyond this singularity by introducing the radial coordinate χ in place of $\tilde{r}$ so that the line element can be written in the following form:

$$ds^2 = -dt^2 + a^2(t)[d\chi^2 + \sin^2\chi(d\theta^2 + \sin^2\theta d\phi^2)], \tag{26}$$

where K is normalized to 1 using the freedom in rescaling $a(t)$. We can see that the coordinate singularity which was at $r = 1$ in the (t, r) coordinates is now resolved at $\chi = \pi/2$ in the (t, χ) coordinates. Roughly speaking, a type II perturbation is the perturbation that covers the surface $\chi = \pi/2$, while a type I does not.

In the perturbation theory in the flat FLRW solution, there are growing and decaying modes. We can usually neglect decaying modes in the early Universe if we are interested in subsequent structure formation including PBH formation. Then, we can expect that the regions with positive and negative density perturbations can be described by locally positive and negative curvature FLRW solutions, respectively. In fact, this is really true in the long-wavelength solutions because Equation (14), which is the metric in the long-wavelength limit, can be rewritten in the following form [34,35]:

$$ds^2 = -dt^2 + a^2(t)\left[\frac{1}{1-K(\tilde{r})\tilde{r}^2}d\tilde{r}^2 + \tilde{r}^2(d\theta^2 + \sin^2\theta d\phi^2)\right] \quad (27)$$

in a new radial coordinate $\tilde{r}$. Since type II perturbations need larger amplitudes of curvature perturbation, they are much rarer than type I in a standard probability distribution function of the curvature perturbation, whereas type II can be dominant in a particular inflationary scenario with large non-Gaussianities [102].

5.2. Positive Curvature Region, Type II Perturbation and Separate Universe Condition

As we have discussed, the region with a positive density perturbation can locally be well described by the positive-curvature FLRW solution at least in the long-wavelength limit. To understand the geometry and the dynamics of such a system, we introduce a toy model, which consists of a positive-curvature FLRW region surrounded by a flat FLRW region [39,61]. We call this model the three-zone model. Figure 8 gives its schematic figure. The line elements in regions I and III are written in the following forms, respectively,

$$ds^2 = -dt^2 + a_{\rm I}^2(t)\left[d\chi^2 + \sin^2\chi(d\theta^2 + \sin^2\theta d\phi^2)\right] \quad (28)$$

for $0 < \chi < \chi_a$ and

$$ds^2 = -dt^2 + a_{\rm III}^2(t)\left[dr^2 + r^2(d\theta^2 + \sin^2\theta d\phi^2)\right] \quad (29)$$

for $r_b < r$, where $a_{\rm I}(t)$ and $a_{\rm III}(t)$ are the scale factors of the positive-curvature and flat FLRW solutions; in region II it cannot be written in a simple form. Although this toy model is not only unrealistic but also difficult to justify as the whole evolution of the spacetime unless we make strong assumptions about region II, it is still valid as the long-wavelength solution and useful to understand the concept of nonlinear perturbation. In this model, we have two independent physical length scales of region I, the curvature radius, which is normalized to 1 in the above metric, and the size of the region, which is given by χ_a. As χ_a is increased from 0 to $\pi/2$, the comoving areal radius of region I, which is given by $\sin\chi_a$, monotonically increases from 0 to 1. However, as χ_a is further increased from $\pi/2$ to π, the comoving areal radius turns to decrease from 1 to 0. This consideration implies the separate universe configuration for which $\chi_a = \pi$, which is notified in Ref. [3]. Kopp, Hofmann and Weller (2010) [61] classified the configurations with $0 < \chi_a < \pi/2$ and with $\pi/2 < \chi_a < \pi$ into types I and II, respectively, while the marginal case is given by a three-hemisphere $\chi_a = \pi/2$, where we neglect the contribution of region II.

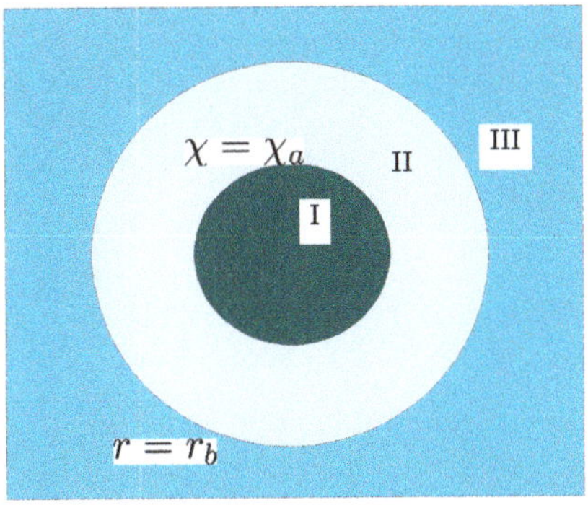

Figure 8. The three-zone model of the positive density perturbation. Regions I and III are described by the closed and flat FLRW solutions, while region II is an underdense matching layer. The spheres of $\chi = \chi_a$ and $r = r_b$ give the outer edge of region I and the inner edge of region III, respectively. See Refs. [39,61] for more details.

Apart from this toy model, it is still true that there should be two independent physical length scales to characterize the spatial geometry, the curvature radius and the size of the curved region as schematically illustrated in Figure 9. It should be noted that this figure does not show the sequence of time evolution but each configuration gives a set of initial data for each time development of the perturbation. It is physically interesting to think of a nearly separate universe configuration, while a totally separate universe has nothing to do with our observable Universe.

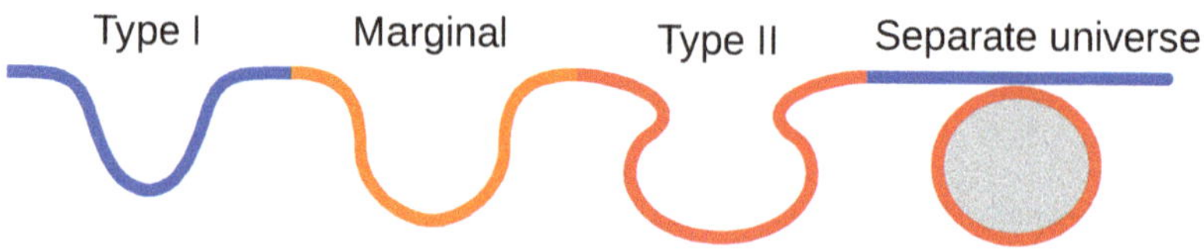

Figure 9. Classification of spatial configurations of overdense perturbations. The type II configuration is with the throat structure in the spatial geometry, while the type I is not. Note that the sequence does not correspond to the time evolution but to different initial curvature perturbations.

5.3. Type II Perturbation and Its Time Development

Here, we review the sets of initial data in terms of types I and II based on the recent work by Uehara, Escrivà, Harada, Saito and Yoo (2024) [103] for radiation domination. Here, the system is assumed to be spherically symmetric. They first constructed the long-wavelength solutions by choosing a function $\zeta(r)$. One of the functional forms they tried for $\zeta(r)$ is given by

$$\zeta(r) = \mu e^{-(1/2)k^2 r^2} W(r), \tag{30}$$

where the constant μ gives the amplitude of perturbation and $W(r)$ is an appropriate window function chosen so that the Gaussian tail can be eliminated and ζ can be smoothly matched to 0 at the outer boundary of the numerical domain. The areal radius R and the Shibata–Sasaki compaction function C_{SS} are plotted as functions of r in Figures 1b and 2a of [103], respectively, where we can see that for $\mu \gtrsim 1.4$, R is no longer a monotonic function of the radial coordinate r, which implies that there is a throat and that $C_{SS}(r)$ has two peaks with the value of $1/2$ and a minimum in between. See Refs. [41,103] for the account for this peculiar behavior of the compaction function. These features are essentially the same in matter domination as discussed in Ref. [61] using the LTB solution.

Then, the time development of those initial data was constructed with a standard numerical relativity scheme based on the Baumgarte–Shapiro–Shibata–Nakamura formalism but adjusted for the spherical formation of PBHs. Figure 3 of [103] summarizes the evolution of the spacetime for the long-wavelength solutions generated by the curvature perturbation given by Equation (30). For $\mu = 0.5$, the amplitude of perturbation is so small that it cannot collapse but disperses away. For $\mu = 1.2$ and 1.8, the amplitude of perturbation is so large that it can collapse into a black hole. This suggests that the critical value of μ for the black hole formation is between 0.5 and 1.2.

5.4. Type B Horizon Structure

In the asymptotically flat spacetimes, black holes are conventionally defined by event horizons. However, in a cosmological setting, event horizons are not necessarily useful because the definition of an event horizon is teleological in the sense that it is only based on the infinite future and because the structure of infinities in cosmological spacetimes can be very different from asymptotically flat ones. Furthermore, in the expanding Universe, there may be another type of horizon, i.e., a cosmological horizon. As we will see below, we encounter the interplay between these two types of horizons for some classes of PBH formation from the type II configuration. This implies that we need to distinguish between these two types of horizons. For these reasons, we adopt the notion of trapping horizons, a quasilocal formulation of horizons [104,105]. This is akin to the notion of apparent horizons in the present setting. Here, we define horizons using the pair of null expansions θ_+ and θ_- along outgoing and ingoing radial null coordinates ξ_+ and ξ_-, respectively. Black hole horizons can be defined with one vanishing and one negative null expansion so that any radial light ray there cannot expand its cross-section. Similarly, cosmological horizons can be defined with one vanishing and one positive null expansion so that any radial light ray there cannot contract its cross-section. The black hole horizons and cosmological horizons are called future trapping horizons and past trapping horizons, respectively. In principle, we may also have a horizon with both null expansions vanishing, which is called a bifurcating trapping horizon [106]. In Appendix A, we will introduce trapping horizons in a more mathematical manner, whereas the intuitive understanding discussed above is still useful.

With the above terminology, we can now describe the structure of trapping horizons. We assign ξ_+ and ξ_- as the null coordinates being the standard ones in the far region which is asymptotic to the flat FLRW solution, which we assume to exist. There is a big difference in the structure of trapping horizons and trapped regions between $\mu = 1.2$ and $\mu = 1.8$. We schematically plot the horizon structures inferred by numerical simulations in Figure 10. For $\mu = 1.2$, as shown on the top panel, there is a past trapped region and a future-trapped region disconnected from each other, with the associated past and future trapping horizons being separate. We call this horizon structure type A. For $\mu = 1.8$, as shown on the bottom panel, the past and future trapped regions have contacts at two points. The two points of intersection, which are two spheres, correspond to bifurcating trapping horizons. There appears to be an untrapped region enclosed by the future and past trapping horizons and the two bifurcating trapping horizons, where $\theta_+ < 0$ and $\theta_- > 0$. We call this horizon structure type B.

This numerical result for the perturbation indicates that there is a peculiar structure for trapping horizons and trapped regions, if μ is greater than some critical value. This was also true if we took another choice of $\zeta(r)$, so the above features are at least general to some extent. So, we can regard this structure as common for a sufficiently large amplitude of curvature perturbation of type II. We infer the existence of the marginal horizon structure between types A and B, which is shown on the middle panel of Figure 10. We can say that a type II perturbation does not always result in a type B structure for radiation. This is due to the effect of pressure, because it is known that for the dust case a type II perturbation necessarily entails the horizon structure of type B. So, we can conclude that there are at least PBHs with structures of types I-A, II-A and II-B in radiation domination.

It should be noted that the type B horizon structure is not known in the gravitational collapse in asymptotically flat spacetimes and is, therefore, very unique to PBH formation as long as the author is aware. For this reason, the result of the type II-B PBH formation is very intriguing in the context of the variety of black hole formation in general relativity. On the other hand, their relevance to observational cosmology is yet unclear.

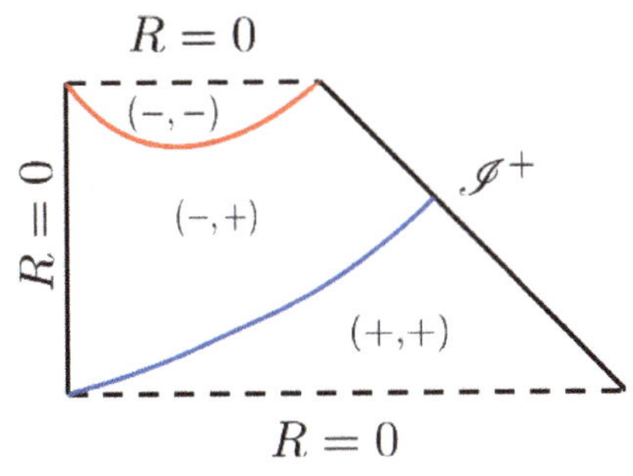

Figure 10. *Cont.*

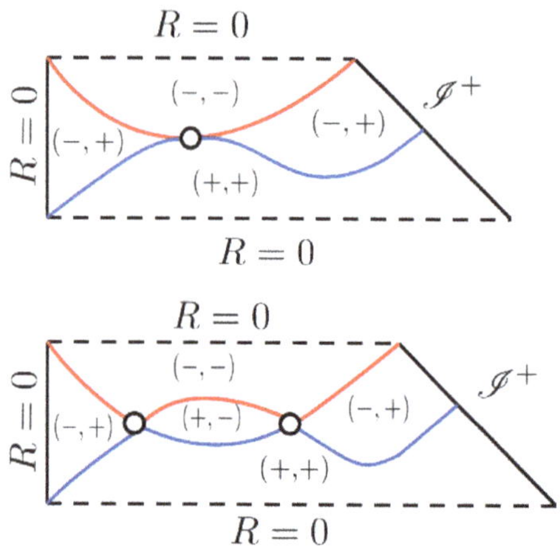

Figure 10. Structures of trapping horizons and trapped regions inferred by numerical simulations. The structures indicated on the top, middle and bottom panels are called type A, marginal and type B, respectively. The signs of the radial null expansions $(\text{sign}(\theta_-), \text{sign}(\theta_+))$ are shown for each region. The regions are divided by trapping horizons, where $\theta_+\theta_- = 0$. The future and past trapping horizons are denoted by the red and blue curves, respectively. The bifurcating trapping horizons are denoted by open circles. The black dashed lines denote spacetime singularities, while the black solid lines denote the regular centers and the null infinities. This figure is comparable to Figure 7 of Ref. [103]. See text for more details.

6. Conclusions

In the recent development of research on this subject, it has been revealed that PBHs lie at the intersection of various developing branches of modern physics. In this article, we review their basic concept, formation, spins and link to the type II configuration.

In Section 2, we present the basic concept of PBHs. The mass of PBHs is usually considered to be the mass scale within the cosmological horizon at the time of their formation. Mass accretion may significantly increase the PBH mass depending on the evolution scenario, whereas it has been shown to be negligible during the evolution in radiation domination. The mass of PBHs can be significantly reduced by Hawking evaporation, and they would have evaporated completely by now if they were lighter than the critical mass of approximately 10^{15} g, although the details of the evaporation process are still somewhat under debate. We can obtain constraints on the fraction $f(M)$ of PBHs of mass M in relation to all dark matter through different observations. The fraction $f(M)$ can be transformed into the formation probability $\beta(M)$ depending on the cosmological evolutionary scenario. The standard cosmic history implies that a very small value of $\beta(M)$, as small as $\sim 10^{-17}$ for $M \sim 10^{17}$ g, can yield $f(M) = O(1)$, i.e., can explain all dark matter, because the PBHs' contribution to the energy of the Universe increases in proportion to the scale factor during the radiation-dominated phase.

In Section 3, we discuss PBH formation. A detailed and precise understanding of PBH formation physics has become increasingly important. The basic question in this study is how to predict $\beta(M)$ and other observationally significant quantities from a given cosmological scenario. Focusing on PBH formation from fluctuations generated by inflation, the key terms are inflation models, long-wavelength solutions, thresholds, softer EOS, matter domination, critical behaviors and statistics.

In Section 4, we discuss the initial spins of PBHs. PBHs formed during radiation domination are unlikely to have large spins. Perturbative studies show that the nondimensional Kerr parameter of these PBHs is typically in the order of 10^{-3}. In contrast, PBHs formed during matter domination can acquire large spins, at least initially. The effect of mass accretion after formation on the nondimensional Kerr parameter needs to be studied carefully. It is evident that numerical simulations based on numerical relativity should shed light on this problem.

In Section 5, we review the recently conducted numerical relativity simulations of the time evolution of type II perturbations during radiation domination. This is particularly relevant to a very rare perturbation peak or a specific inflationary scenario. The resulting structures of trapping horizons can be classified into two types: one is standard for PBHs, and the other is very unique, featured with the crossing of trapping horizons as bifurcating trapping horizons. We refer to these as types A and B, respectively, while we can also discuss the marginal structure. The numerical simulations suggest that the evolution of type II perturbations can be classified into type A and type B based on their horizon structure. Thus, we may call them types II-A and II-B.

Finally, I must acknowledge that there are many interesting issues concerning the formation of PBHs that cannot even be mentioned in this article. The study of PBH formation not only requires a deep understanding of physical phenomena within known standard physics but also offers the opportunity to explore unknown new physics through PBHs. Both of these aspects will play important roles in the future of PBH formation studies.

Funding: This research was funded by JSPS KAKENHI Grant Numbers JP20H05853 and JP24K07027.

Data Availability Statement: No data are associated with the manuscript.

Acknowledgments: The main part of this article is based on exciting collaborations with Bernard Carr, Albert Escrivà, Jaume Garriga, Shin-Ichi Hirano, Hayami Iizuka, Sanjay Jhingan, Yasutaka Koga, Kazunori Kohri, Takafumi Kokubu, Koutaro Kyutoku, Hideki Maeda, Takeru Monobe, Tomohiro Nakama, Ken-Ichi Nakao, Hirotada Okawa, A. G. Polnarev, Daiki Saito, Misao Sasaki, Yuichiro Tada, Takahiro Terada, Koichiro Uehara, Jun'ichi Yokoyama, Shuichiro Yokoyama and Chul-Moon Yoo. The author thanks Cristiano Germani, Mohammad Ali Gorji and the other participants of the conference "Barcelona Black Holes (BBH) I: Primordial Black Holes", where this work was first presented, for helpful discussions and comments. The author is also very grateful to Ilia Musco and Theodoros Papanikolaou for their very helpful comments. The author is grateful to CENTRA, Departamento de Física, Instituto Superior Técnico—IST at Universidade de Lisboa, and Niels Bohr International Academy at Niels Bohr Institute for their hospitality during the writing of this manuscript.

Conflicts of Interest: The author declares no conflicts of interest.

Appendix A. Trapping Horizons

For a spherically symmetric spacetime, we can generally introduce radial null coordinates $\xi_\pm$ such that the line element can be written in the following double-null form:

$$ds^2 = -2e^{-f(\xi_+,\xi_-)}(\xi_+,\xi_-)d\xi_+d\xi_- + R^2(\xi_+,\xi_-)(d\theta^2 + \sin^2\theta d\phi^2). \tag{A1}$$

We introduce the future-directed radial null vectors $l^a_\pm \propto (\partial_\pm)^a$ such that $g_{ab}l^a_+ l^b_- = -1$, where $\partial_\pm = \partial/\partial\xi_\pm$. Then, we define $\theta_\pm$ as

$$\theta_\pm := l^a_\pm \partial_a \ln(R^2). \tag{A2}$$

We call a two-sphere specified with (ξ_+,ξ_-) a future (past) trapped sphere if $\theta_+\theta_- > 0$ and $\theta_+ + \theta_- < (>)0$. We call a two-sphere specified with (ξ_+,ξ_-) a future (past) marginal sphere if $\theta_+\theta_- = 0$ and $\theta_+ + \theta_- < (>)0$. We call a two-sphere specified with (ξ_+,ξ_-) a bifurcating marginal sphere if $\theta_+ = \theta_- = 0$. We call a two-sphere specified with (ξ_+,ξ_-) an untrapped sphere if $\theta_+\theta_- < 0$. We call a spacetime region a future (past) trapped region if any two-sphere given by (ξ_+,ξ_-) in the region is a future (past) trapped sphere. We call a spacetime region an untrapped region if any two-sphere given by (ξ_+,ξ_-) in the region is an untrapped sphere. We call a hypersurface foliated by future (past) marginal spheres a future (past) trapping horizon. We call a hypersurface foliated by bifurcating marginal spheres a bifurcating trapping horizon.

See Refs. [104–106] for more complete rigorous discussions and proofs of the basic properties of trapping horizons.

References

1. Zel'dovich, Y.B.; Novikov, I.D. The Hypothesis of Cores Retarded during Expansion and the Hot Cosmological Model. *Sovie Astron. AJ (Engl. Transl.)* **1967**, *10*, 602.
2. Hawking, S. Gravitationally Collapsed Objects of Very Low Mass. *Mon. Not. R. Astron. Soc.* **1971**, *152*, 75–78. [CrossRef]
3. Carr, B.J.; Hawking, S.W. Black holes in the early Universe. *Mon. Not. R. Astron. Soc.* **1974**, *168*, 399–415. [CrossRef]
4. Carr, B.J. The Primordial black hole mass spectrum. *Astrophys. J.* **1975**, *201*, 1–19. [CrossRef]
5. Carr, B.J.; Green, A.M. The History of Primordial Black Holes. *arXiv* **2024**, arXiv:2406.05736.
6. Riotto, A.; Silk, J. The Future of Primordial Black Holes: Open Questions and Roadmap. *arXiv* **2024**, arXiv:2403.02907.
7. Nadezhin, D.K.; Novikov, I.D.; Polnarev, A.G. The hydrodynamics of primordial black hole formation. *Sov. Astron.* **1978** *22*, 129–138.
8. Hawking, S.W. Black hole explosions. *Nature* **1974**, *248*, 30–31. [CrossRef]
9. Hawking, S.W. Particle Creation by Black Holes. *Commun. Math. Phys.* **1975**, *43*, 199–220; Erratum in *Commun. Math. Phys.* **1976** *46*, 206. [CrossRef]
10. Abbott, B.P.; Abbott, R.; Abbott, T.; Abernathy, M.R.; Acernese, F.; Ackley, K.; Aggarwal, N.; Aguiar, O.D.; Aiello, L.; Cavalieri, R. et al. Observation of Gravitational Waves from a Binary Black Hole Merger. *Phys. Rev. Lett.* **2016**, *116*, 061102. [CrossRef]
11. Sasaki, M.; Suyama, T.; Tanaka, T.; Yokoyama, S. Primordial black holes—Perspectives in gravitational wave astronomy. *Class Quant. Grav.* **2018**, *35*, 063001. [CrossRef]
12. Sasaki, M.; Suyama, T.; Tanaka, T.; Yokoyama, S. Primordial Black Hole Scenario for the Gravitational-Wave Event GW150914 *Phys. Rev. Lett.* **2016**, *117*, 061101; Erratum in *Phys. Rev. Lett.* **2018**, *121*, 059901. [CrossRef] [PubMed]
13. Bird, S.; Cholis, I.; Muñoz, J.B.; Ali-Haïmoud, Y.; Kamionkowski, M.; Kovetz, E.D.; Raccanelli, A.; Riess, A.G. Did LIGO detect dark matter? *Phys. Rev. Lett.* **2016**, *116*, 201301. [CrossRef] [PubMed]
14. Clesse, S.; García-Bellido, J. The clustering of massive Primordial Black Holes as Dark Matter: Measuring their mass distribution with Advanced LIGO. *Phys. Dark Univ.* **2017**, *15*, 142–147. [CrossRef]
15. Abbott, B.P.; Abbott, R.; Abbott, T.D.; Acernese, F.; Ackley, K.; Adams, C.; Adams, T.; Addesso, P.; Adhikari, R.X.; Cao, J.; et al GW170104: Observation of a 50-Solar-Mass Binary Black Hole Coalescence at Redshift 0.2. *Phys. Rev. Lett.* **2017**, *118*, 221101, Erratum in *Phys. Rev. Lett.* **2018**, *121*, 129901. [CrossRef]
16. Franciolini, G.; Baibhav, V.; De Luca, V.; Ng, K.K.Y.; Wong, K.W.K.; Berti, E.; Pani, P.; Riotto, A.; Vitale, S. Searching for a subpopulation of primordial black holes in LIGO-Virgo gravitational-wave data. *Phys. Rev. D* **2022**, *105*, 083526. [CrossRef]
17. Carr, B.; Clesse, S.; Garcia-Bellido, J.; Hawkins, M.; Kuhnel, F. Observational evidence for primordial black holes: A positivist perspective. *Phys. Rept.* **2024**, *1054*, 1–68. [CrossRef]
18. Agazie, G.; Anumarlapudi, A.; Archibald, A.M.; Arzoumanian, Z.; Baker, P.T.; Bécsy, B.; Blecha, L.; Brazier, A.; Brook, P.R.; NANOGrav Collaboration; et al. The NANOGrav 15 yr Data Set: Evidence for a Gravitational-wave Background. *Astrophys. J. Lett.* **2023**, *951*, L8. [CrossRef]
19. Afzal, A.; Agazie, G.; Anumarlapudi, A.; Archibald, A.M.; Arzoumanian, Z.; Baker, P.T.; Bécsy, B.; Blanco-Pillado, J.J.; Blecha, L.; NANOGrav Collaboration; et al. The NANOGrav 15 yr Data Set: Search for Signals from New Physics. *Astrophys. J. Lett.* **2023**, *951*, L11; Erratum in *Astrophys. J. Lett.* **2024**, *971*, L27. [CrossRef]
20. Liu, B.; Bromm, V. Accelerating Early Massive Galaxy Formation with Primordial Black Holes. *Astrophys. J. Lett.* **2022**, *937*, L30. [CrossRef]
21. Hütsi, G.; Raidal, M.; Urrutia, J.; Vaskonen, V.; Veermäe, H. Did JWST observe imprints of axion miniclusters or primordial black holes? *Phys. Rev. D* **2023**, *107*, 043502. [CrossRef]
22. Yuan, G.W.; Lei, L.; Wang, Y.Z.; Wang, B.; Wang, Y.Y.; Chen, C.; Shen, Z.Q.; Cai, Y.F.; Fan, Y.Z. Rapidly growing primordial black holes as seeds of the massive high-redshift JWST Galaxies. *Sci. China Phys. Mech. Astron.* **2024**, *67*, 109512. [CrossRef]
23. Carr, B.; Kohri, K.; Sendouda, Y.; Yokoyama, J. Constraints on primordial black holes. *Rept. Prog. Phys.* **2021**, *84*, 116902. [CrossRef] [PubMed]
24. Carr, B.J. Primordial black holes as a probe of cosmology and high energy physics. In *Quantum Gravity*; Lecture Notes in Physics; Springer: Berlin/Heidelberg, Germany, 2003; Volume 631, pp. 301–321. [CrossRef]
25. Escrivà, A. Simulation of primordial black hole formation using pseudo-spectral methods. *Phys. Dark Univ.* **2020**, *27*, 100466. [CrossRef]
26. Dvali, G.; Eisemann, L.; Michel, M.; Zell, S. Black hole metamorphosis and stabilization by memory burden. *Phys. Rev. D* **2020**, *102*, 103523. [CrossRef]
27. Carr, B.J.; Lidsey, J.E. Primordial black holes and generalized constraints on chaotic inflation. *Phys. Rev. D* **1993**, *48*, 543–553. [CrossRef]
28. Carr, B.J.; Gilbert, J.H.; Lidsey, J.E. Black hole relics and inflation: Limits on blue perturbation spectra. *Phys. Rev. D* **1994**, *50*, 4853–4867. [CrossRef]
29. Wald, R.M. *General Relativity*; Chicago University Press: Chicago, IL, USA, 1984. [CrossRef]

30. Poisson, E. *A Relativist's Toolkit: The Mathematics of Black-Hole Mechanics*; Cambridge University Press: Cambridge, UK, 2009. [CrossRef]
31. Tomita, K. Primordial Irregularities in the Early Universe. *Prog. Theor. Phys.* **1972**, *48*, 1503–1516. [CrossRef]
32. Shibata, M.; Sasaki, M. Black hole formation in the Friedmann universe: Formulation and computation in numerical relativity. *Phys. Rev. D* **1999**, *60*, 084002. [CrossRef]
33. Lyth, D.H.; Malik, K.A.; Sasaki, M. A General proof of the conservation of the curvature perturbation. *J. Cosmol. Astropart. Phys.* **2005**, *5*, 004. [CrossRef]
34. Polnarev, A.G.; Musco, I. Curvature profiles as initial conditions for primordial black hole formation. *Class. Quant. Grav.* **2007**, *24*, 1405–1432. [CrossRef]
35. Harada, T.; Yoo, C.M.; Nakama, T.; Koga, Y. Cosmological long-wavelength solutions and primordial black hole formation. *Phys. Rev. D* **2015**, *91*, 084057. [CrossRef]
36. Musco, I.; Miller, J.C.; Rezzolla, L. Computations of primordial black hole formation. *Class. Quant. Grav.* **2005**, *22*, 1405–1424. [CrossRef]
37. Musco, I.; Miller, J.C.; Polnarev, A.G. Primordial black hole formation in the radiative era: Investigation of the critical nature of the collapse. *Class. Quant. Grav.* **2009**, *26*, 235001. [CrossRef]
38. Musco, I.; Miller, J.C. Primordial black hole formation in the early universe: Critical behavior and self-similarity. *Class. Quant. Grav.* **2013**, *30*, 145009. [CrossRef]
39. Harada, T.; Yoo, C.M.; Kohri, K. Threshold of primordial black hole formation. *Phys. Rev. D* **2013**, *88*, 084051; Erratum in *Phys. Rev. D* **2014**, *89*, 029903. [CrossRef]
40. Harada, T.; Yoo, C.M.; Koga, Y. Revisiting compaction functions for primordial black hole formation. *Phys. Rev. D* **2023**, *108*, 043515. [CrossRef]
41. Harada, T.; Iizuka, H.; Koga, Y.; Yoo, C.M. Geometrical origin for the compaction function for primordial black hole formation. *arXiv* **2024**, arXiv:2409.05544.
42. Musco, I. Threshold for primordial black holes: Dependence on the shape of the cosmological perturbations. *Phys. Rev. D* **2019**, *100*, 123524. [CrossRef]
43. Nakama, T.; Harada, T.; Polnarev, A.G.; Yokoyama, J. Identifying the most crucial parameters of the initial curvature profile for primordial black hole formation. *J. Cosmol. Astropart. Phys.* **2014**, *1*, 037. [CrossRef]
44. Escrivà, A.; Germani, C.; Sheth, R.K. Universal threshold for primordial black hole formation. *Phys. Rev. D* **2020**, *101*, 044022. [CrossRef]
45. Ianniccari, A.; Iovino, A.J.; Kehagias, A.; Perrone, D.; Riotto, A. The Black Hole Formation—Null Geodesic Correspondence. *Phys. Rev. Lett.* **2024**, *133*, 081401. [CrossRef] [PubMed]
46. Yoo, C.M.; Harada, T.; Hirano, S.; Okawa, H.; Sasaki, M. Primordial black hole formation from massless scalar isocurvature. *Phys. Rev. D* **2022**, *105*, 103538. [CrossRef]
47. Escrivà, A.; Germani, C.; Sheth, R.K. Analytical thresholds for black hole formation in general cosmological backgrounds. *J. Cosmol. Astropart. Phys.* **2021**, *1*, 030. [CrossRef]
48. Byrnes, C.T.; Hindmarsh, M.; Young, S.; Hawkins, M.R.S. Primordial black holes with an accurate QCD equation of state. *J. Cosmol. Astropart. Phys.* **2018**, *8*, 041. [CrossRef]
49. Papanikolaou, T. Toward the primordial black hole formation threshold in a time-dependent equation-of-state background. *Phys. Rev. D* **2022**, *105*, 124055. [CrossRef]
50. Escrivà, A.; Bagui, E.; Clesse, S. Simulations of PBH formation at the QCD epoch and comparison with the GWTC-3 catalog. *J. Cosmol. Astropart. Phys.* **2023**, *5*, 004. [CrossRef]
51. Musco, I.; Jedamzik, K.; Young, S. Primordial black hole formation during the QCD phase transition: Threshold, mass distribution, and abundance. *Phys. Rev. D* **2024**, *109*, 083506. [CrossRef]
52. Jedamzik, K. Primordial black hole formation during cosmic phase transitions. *arXiv* **2024**, arXiv:2406.11417.
53. Musco, I.; Papanikolaou, T. Primordial black hole formation for an anisotropic perfect fluid: Initial conditions and estimation of the threshold. *Phys. Rev. D* **2022**, *106*, 083017. [CrossRef]
54. Khlopov, M.Y.; Polnarev, A.G. Primordial black holes as a cosmological test of grand unification. *Phys. Lett. B* **1980**, *97*, 383–387. [CrossRef]
55. Polnarev, A.G.; Khlopov, M.Y. Primordial Black Holes and the ERA of Superheavy Particle Dominance in the Early Universe. *Sov. Astron.* **1981**, *25*, 406.
56. Polnarev, A.G.; Khlopov, M.Y. Dustlike Stages in the Early Universe and Constraints on the Primordial Black-Hole Spectrum. *Sov. Astron.* **1982**, *26*, 391–395.
57. Polnarev, A.G.; Khlopov, M.Y. Cosmology, primordial black holes, and supermassive particles. *Sov. Phys. Uspekhi* **1985**, *28*, 213–232. [CrossRef]
58. Harada, T.; Yoo, C.M.; Kohri, K.; Nakao, K.I.; Jhingan, S. Primordial black hole formation in the matter-dominated phase of the Universe. *Astrophys. J.* **2016**, *833*, 61. [CrossRef]
59. Klauder, J.R. (Ed.) *Magic Without Magic—John Archibald Wheeler. A Collection of Essays in Honor of His 60th Birthday*; Freeman: San Francisco, CA, USA, 1972.
60. Misner, C.W.; Thorne, K.S.; Wheeler, J.A. *Gravitation*; W. H. Freeman: San Francisco, CA, USA, 1973.

61. Kopp, M.; Hofmann, S.; Weller, J. Separate Universes Do Not Constrain Primordial Black Hole Formation. *Phys. Rev. D* **2011**, *83*, 124025. [CrossRef]
62. Harada, T.; Jhingan, S. Spherical and nonspherical models of primordial black hole formation: Exact solutions. *Prog. Theor. Exp. Phys.* **2016**, *2016*, 093E04. [CrossRef]
63. Kokubu, T.; Kyutoku, K.; Kohri, K.; Harada, T. Effect of Inhomogeneity on Primordial Black Hole Formation in the Matter Dominated Era. *Phys. Rev. D* **2018**, *98*, 123024. [CrossRef]
64. Harada, T.; Kohri, K.; Sasaki, M.; Terada, T.; Yoo, C.M. Threshold of primordial black hole formation against velocity dispersion in matter-dominated era. *J. Cosmol. Astropart. Phys.* **2023**, *2*, 038. [CrossRef]
65. Harada, T.; Yoo, C.M.; Kohri, K.; Nakao, K.I. Spins of primordial black holes formed in the matter-dominated phase of the Universe. *Phys. Rev. D* **2017**, *96*, 083517; Erratum in *Phys. Rev. D* **2019**, *99*, 069904. [CrossRef]
66. Padilla, L.E.; Hidalgo, J.C.; Malik, K.A. New mechanism for primordial black hole formation during reheating. *Phys. Rev. D* **2022**, *106*, 023519. [CrossRef]
67. Padilla, L.E.; Hidalgo, J.C.; Núñez, D. Long-wavelength nonlinear perturbations of a complex scalar field. *Phys. Rev. D* **2021**, *104*, 083513. [CrossRef]
68. Hidalgo, J.C.; Padilla, L.E.; German, G. Production of PBHs from inflaton structures. *Phys. Rev. D* **2023**, *107*, 063519. [CrossRef]
69. de Jong, E.; Aurrekoetxea, J.C.; Lim, E.A. Primordial black hole formation with full numerical relativity. *J. Cosmol. Astropart. Phys.* **2022**, *3*, 029. [CrossRef]
70. de Jong, E.; Aurrekoetxea, J.C.; Lim, E.A.; França, T. Spinning primordial black holes formed during a matter-dominated era. *J. Cosmol. Astropart. Phys.* **2023**, *10*, 067. [CrossRef]
71. Aurrekoetxea, J.C.; Clough, K.; Muia, F. Oscillon formation during inflationary preheating with general relativity. *Phys. Rev. D* **2023**, *108*, 023501. [CrossRef]
72. Choptuik, M.W. Universality and scaling in gravitational collapse of a massless scalar field. *Phys. Rev. Lett.* **1993**, *70*, 9–12. [CrossRef]
73. Evans, C.R.; Coleman, J.S. Observation of critical phenomena and selfsimilarity in the gravitational collapse of radiation fluid. *Phys. Rev. Lett.* **1994**, *72*, 1782–1785. [CrossRef]
74. Koike, T.; Hara, T.; Adachi, S. Critical behavior in gravitational collapse of radiation fluid: A Renormalization group (linear perturbation) analysis. *Phys. Rev. Lett.* **1995**, *74*, 5170–5173. [CrossRef]
75. Niemeyer, J.C.; Jedamzik, K. Dynamics of primordial black hole formation. *Phys. Rev. D* **1999**, *59*, 124013. [CrossRef]
76. Hawke, I.; Stewart, J.M. The dynamics of primordial black hole formation. *Class. Quant. Grav.* **2002**, *19*, 3687–3707. [CrossRef]
77. Maison, D. Nonuniversality of critical behavior in spherically symmetric gravitational collapse. *Phys. Lett. B* **1996**, *366*, 82–84. [CrossRef]
78. Niemeyer, J.C.; Jedamzik, K. Near-critical gravitational collapse and the initial mass function of primordial black holes. *Phys. Rev. Lett.* **1998**, *80*, 5481–5484. [CrossRef]
79. Yokoyama, J. Cosmological constraints on primordial black holes produced in the near critical gravitational collapse. *Phys. Rev. D* **1998**, *58*, 107502. [CrossRef]
80. Press, W.H.; Schechter, P. Formation of galaxies and clusters of galaxies by selfsimilar gravitational condensation. *Astrophys. J.* **1974**, *187*, 425–438. [CrossRef]
81. Peebles, P.J.E. *Principles of Physical Cosmology*; Princeton University Press: Princeton, NJ, USA, 2020.
82. Young, S.; Byrnes, C.T. Primordial black holes in non-Gaussian regimes. *J. Cosmol. Astropart. Phys.* **2013**, *8*, 052. [CrossRef]
83. Pi, S. Non-Gaussianities in primordial black hole formation and induced gravitational waves. *arXiv* **2024**, arXiv:2404.06151.
84. Bardeen, J.M.; Bond, J.R.; Kaiser, N.; Szalay, A.S. The Statistics of Peaks of Gaussian Random Fields. *Astrophys. J.* **1986**, *304*, 15–61. [CrossRef]
85. Yoo, C.M.; Harada, T.; Garriga, J.; Kohri, K. Primordial black hole abundance from random Gaussian curvature perturbations and a local density threshold. *Prog. Theor. Exp. Phys.* **2018**, *2018*, 123E01; Erratum in *Prog. Theor. Exp. Phys.* **2024**, *2024*, 049202. [CrossRef]
86. Yoo, C.M.; Harada, T.; Hirano, S.; Kohri, K. Abundance of Primordial Black Holes in Peak Theory for an Arbitrary Power Spectrum. *Prog. Theor. Exp. Phys.* **2021**, *2021*, 013E02; Erratum in *Prog. Theor. Exp. Phys.* **2024**, *2024*, 049203. [CrossRef]
87. Germani, C.; Sheth, R.K. Nonlinear statistics of primordial black holes from Gaussian curvature perturbations. *Phys. Rev. D* **2020**, *101*, 063520. [CrossRef]
88. Kitajima, N.; Tada, Y.; Yokoyama, S.; Yoo, C.M. Primordial black holes in peak theory with a non-Gaussian tail. *J. Cosmol. Astropart. Phys.* **2021**, *10*, 053. [CrossRef]
89. Germani, C.; Sheth, R.K. The Statistics of Primordial Black Holes in a Radiation-Dominated Universe: Recent and New Results. *Universe* **2023**, *9*, 421. [CrossRef]
90. Young, S. Computing the abundance of primordial black holes. *arXiv* **2024**, arXiv:2405.13259.
91. De Luca, V.; Franciolini, G.; Pani, P.; Riotto, A. The evolution of primordial black holes and their final observable spins. *J. Cosmol. Astropart. Phys.* **2020**, *4*, 052. [CrossRef]
92. Chiba, T.; Yokoyama, S. Spin Distribution of Primordial Black Holes. *Prog. Theor. Exp. Phys.* **2017**, *2017*, 083E01. [CrossRef]
93. He, M.; Suyama, T. Formation threshold of rotating primordial black holes. *Phys. Rev. D* **2019**, *100*, 063520. [CrossRef]

94. De Luca, V.; Desjacques, V.; Franciolini, G.; Malhotra, A.; Riotto, A. The initial spin probability distribution of primordial black holes. *J. Cosmol. Astropart. Phys.* **2019**, *5*, 018. [CrossRef]
95. Harada, T.; Yoo, C.M.; Kohri, K.; Koga, Y.; Monobe, T. Spins of primordial black holes formed in the radiation-dominated phase of the universe: First-order effect. *Astrophys. J.* **2021**, *908*, 140. [CrossRef]
96. Koga, Y.; Harada, T.; Tada, Y.; Yokoyama, S.; Yoo, C.M. Effective Inspiral Spin Distribution of Primordial Black Hole Binaries. *Astrophys. J.* **2022**, *939*, 65. [CrossRef]
97. Banerjee, I.K.; Harada, T. Spin of Primordial Black Holes from Broad Power Spectrum: Radiation Dominated Universe. *arXiv* **2024**, arXiv:2409.06494.
98. Saito, D.; Harada, T.; Koga, Y.; Yoo, C.M. Spins of primordial black holes formed with a soft equation of state. *J. Cosmol. Astropart. Phys.* **2023**, *7*, 030. [CrossRef]
99. Saito, D.; Harada, T.; Koga, Y.; Yoo, C.M. Revisiting spins of primordial black holes in a matter-dominated era based on peak theory. *arXiv* **2024**, arXiv:2409.00435. [CrossRef]
100. Yoo, C.M.; Harada, T.; Okawa, H. Threshold of Primordial Black Hole Formation in Nonspherical Collapse. *Phys. Rev. D* **2020**, *102*, 043526; Erratum in *Phys. Rev. D* **2023**, *107*, 049901. [CrossRef]
101. Yoo, C.M. Primordial black hole formation from a nonspherical density profile with a misaligned deformation tensor. *arXiv* **2024**, arXiv:2403.11147. [CrossRef]
102. Escrivà, A.; Atal, V.; Garriga, J. Formation of trapped vacuum bubbles during inflation, and consequences for PBH scenarios. *J. Cosmol. Astropart. Phys.* **2023**, *10*, 035. [CrossRef]
103. Uehara, K.; Escrivà, A.; Harada, T.; Saito, D.; Yoo, C.M. Numerical simulation of type II primordial black hole formation. *arXiv* **2024**, arXiv:2401.06329.
104. Hayward, S.A. General laws of black hole dynamics. *Phys. Rev. D* **1994**, *49*, 6467–6474. [CrossRef]
105. Hayward, S.A. Gravitational energy in spherical symmetry. *Phys. Rev. D* **1996**, *53*, 1938–1949. [CrossRef]
106. Maeda, H.; Harada, T.; Carr, B.J. Cosmological wormholes. *Phys. Rev. D* **2009**, *79*, 044034. [CrossRef]

universe MDPI

Article

Quantum Loop Corrections in the Modified Gravity Model of Starobinsky Inflation with Primordial Black Hole Production

Sultan Saburov [1] **and Sergei V. Ketov** [1,2,3,*]

[1] Interdisciplinary Research Laboratory, Tomsk State University, Tomsk 634050, Russia; saburov@mail.tsu.ru
[2] Department of Physics, Tokyo Metropolitan University, 1-1 Minami-ohsawa, Hachioji 192-0397, Tokyo, Japan
[3] Kavli Institute for the Physics and Mathematics of the Universe (WPI), The University of Tokyo Institutes for Advanced Study, Kashiwa 277-8583, Chiba, Japan
* Correspondence: ketov@tmu.ac.jp

Abstract: A modified gravity model of Starobinsky inflation and primordial black hole production is proposed in good (within 1σ) agreement with current measurements of the cosmic microwave background radiation. The model is an extension of the singularity-free Appleby–Battye–Starobinsky model by the R^4 term with different values of the parameters whose fine-tuning leads to the efficient production of primordial black holes on smaller scales with the asteroid-size masses between 10^{16} g and 10^{20} g. Those primordial black holes may be part (or the whole) of the current dark matter, while the proposed model can be confirmed or falsified by the detection or absence of the induced gravitational waves with the frequencies in the 10^{-2} Hz range. The relative size of quantum (loop) corrections to the power spectrum of scalar perturbations in the model is found to be of the order of 10^{-3} or less, so that the model is not ruled out by the quantum corrections.

Keywords: inflation; primordial black holes; dark matter; gravitational waves

1. Introduction

The Starobinsky model of inflation [1] as the modified $(R+\alpha R^2)$ gravity is theoretically well motivated (see, e.g., Refs. [2,3] for a recent review), being in excellent agreement with the current cosmic microwave background (CMB) radiation measurements [4–6]. The Starobinsky model can be extended within modified $F(R)$ gravity in order to describe double slow-roll (SR) inflation with an ultra-slow-roll (USR) phase by engineering the function $F(R)$ leading to a near-inflection point in the inflaton potential below the scale of inflation [7–9]. It results in large density perturbations whose gravitational collapse leads to the production of primordial black holes (PBHs).

Adding the near-inflection point and the USR phase requires fine-tuning the model parameters [10,11], which often lowers the value of the CMB tilt n_s of scalar perturbations and thus leads to a tension with CMB measurements [12], while large perturbations may imply significant non-Gaussianity and quantum (loop) corrections that may invalidate classical single-field models of inflation and PBH production [13–17].

In this paper, we derive the peak amplitude and the frequency of the PBH-production-induced stochastic gravitational waves (GWs) and estimate quantum loop corrections in the model of Ref. [8] by using the δN formalism [18,19]. We use the natural units with $c=\hbar=M_P=1$ where M_P is the reduced Planck mass in our equations, while restoring them for the values of dimensional physical observables.

2. The Model

The phenomenological model [8] of inflation and PBH production has the $F(R)$ gravity action

$$S=\frac{1}{2}\int d^4x\sqrt{-g}\,F(R), \tag{1}$$

whose F-function of the spacetime scalar curvature R reads

$$F(R) = (1 + g \tanh b)R + gE_{AB} \ln\left[\frac{\cosh\left(\frac{R}{E_{AB}} - b\right)}{\cosh(b)}\right] + \frac{R^2}{6M^2} - \delta\frac{R^4}{48M^6}, \tag{2}$$

where the first three terms are known in the literature as the Appleby–Battye–Starobinsky (ABS) model [20] with the Starobinsky mass $M \approx 1.3 \times 10^{-5}$ defining the scale of the first SR phase of inflation. The ABS parameter

$$E_{AB} = \frac{R_0}{2g \ln(1 + e^{2b})} \tag{3}$$

has the new scale R_0 defining the second SR phase of inflation below the Starobinsky scale.[1] The other parameters g and b define the shape of the inflaton potential and have to be fine-tuned in order to obtain a near-inflection point. The last term in Equation (2) may be considered as a quantum gravity correction that was employed in Ref. [8] in order to obtain good (within 1σ) agreement with the measured CMB value of n_s. The function (2) obeys the no-ghost (stability) conditions, $F'(R) > 0$ and $F''(R) > 0$, for the relevant values of R, avoids singularities, obeys the Newtonian limit, and describes double inflation with three phases (SR-USR-SR) after fine-tuning the parameters [8].

To produce PBHs, one needs a large enhancement of the power spectrum of scalar perturbations by seven orders of magnitude against the CMB spectrum. Then, as was shown in Ref. [8], the parameters (R_0, g, b) should be fine-tuned as

$$R_0 \approx 3.00M^2, \quad g \approx 2.25 \quad \text{and} \quad b \approx 2.89. \tag{4}$$

It leads to the production of PBHs with asteroid-size masses in the range between 10^{16} g and 10^{20} g [8], exceeding the Hawking (black hole) evaporation limit of 10^{15} g, so that those PBHs may form part (or the whole) of dark matter (DM) in the current universe [21].

A modified $F(R)$-gravity is known to be equivalent to the quintessence (scalar–tensor gravity) in terms of the canonical inflaton field ϕ with the scalar potential $V(R(\phi))$ in the parametric form [22],

$$V(R) = \frac{F'R - F}{2(F')^2}, \quad \phi(R) = \sqrt{\frac{3}{2}} \ln F', \tag{5}$$

where the primes denote the derivatives with respect to R. The (numerically obtained) profile of the inflaton potential $V(\phi)$ for some values of R_0 and δ is given in Figure 1.

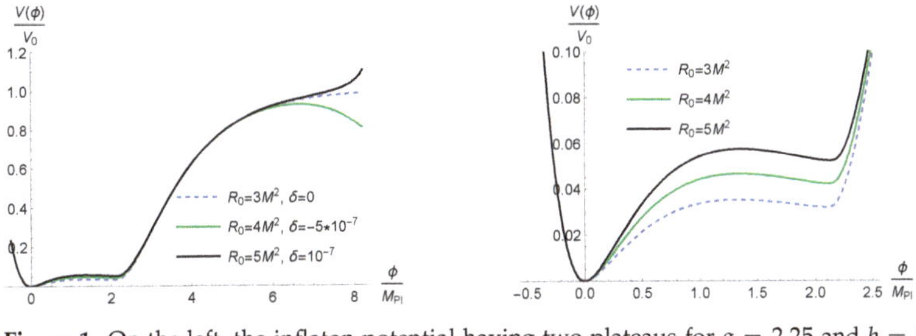

Figure 1. On the left, the inflaton potential having two plateaus for $g = 2.25$ and $b = 2.89$ with $V_0 = \frac{3}{4}M^2$. On the right, zoom of the potential for lower values of ϕ with a near-inflection point. The potential is unstable for negative values of δ and has the infinite plateau for $\delta = 0$ describing the Starobinsky inflation.

The standard SR conditions are given by $\epsilon_{\rm sr} \ll 1$ and $|\eta_{\rm sr}| \ll 1$, where

$$\epsilon_{\rm sr}(\phi) = \frac{1}{2}\left(\frac{V'(\phi)}{V(\phi)}\right)^2 \quad \text{and} \quad \eta_{\rm sr}(\phi) = \frac{V''(\phi)}{V(\phi)}, \tag{6}$$

while the time clock is conveniently defined by the number N of e-folds, $N(t) = \int_t H(\tilde{t})d\tilde{t}$, where $H(t)$ is the Hubble function. The CMB tilt n_s of scalar perturbations and the tensor-to-scalar ratio r are related to the values of the SR parameters at the horizon exit with the standard pivot scale $k_* = 0.05\ \text{Mpc}^{-1}$. In the model [8], the tensor-to-scalar ratio r is well inside the current observational bound, $r < 0.032$, and the tilt n_s agrees within 1σ with the current CMB measurements [4–6], $n_s = 0.9649 \pm 0.0042$, with $\delta \sim 10^{-7}$.

The primordial spectrum $P_\zeta(k)$ of 3-dimensional scalar (density) perturbations $\zeta(x)$ in a flat Friedman universe is defined by the 2-point correlation function as

$$\left\langle \frac{\delta\zeta(x)}{\zeta} \cdot \frac{\delta\zeta(y)}{\zeta} \right\rangle = \int \frac{d^3k}{k^3} e^{ik\cdot(x-y)} \frac{P_\zeta(k)}{P_0}, \tag{7}$$

where $k = 2\pi/\lambda$ is the co-moving number. The scale k is simply related to the e-folds number N via $N = -\int^k d\tilde{k}/\tilde{k}$. Though the USR phase has dynamics different from the SR one, the dimensionless power spectrum of scalar perturbations in the SR approximation

$$P_\zeta = \frac{H^2}{8\pi^2\epsilon_{\rm sr}} \tag{8}$$

appears to be a good approximation in the USR phase as well. The power spectrum is given in Figure 2 in the best case given by the 3rd row of Table 1 in Ref. [8]. Accordingly, the SR parameter $\epsilon_{\rm sr}$ drops to very low values, indicating the USR phase.

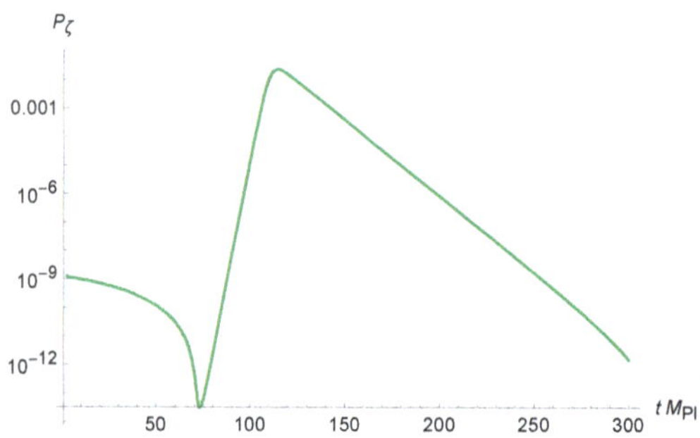

Figure 2. The primordial power spectrum $P_\zeta(t)$ of scalar perturbations from Equation (8). A derivation of the spectrum from the Mukhanov–Sasaki equation leads to a very similar result.

The power spectrum-related observables in our model best case are given by [8]

$$n_s \approx 0.965, \quad r \approx 0.0095, \quad M_{\rm PBH} \approx 1.0 \cdot 10^{20}\ \text{g}. \tag{9}$$

The corresponding peak in the power spectrum of Figure 2 can be roughly approximated by the lognormal fit [23]

$$P_\zeta^{\rm peak}(k) \approx \frac{A_\zeta}{\sqrt{2\pi}\Delta} \exp\left[\frac{-\ln^2(k/k_p)}{2\Delta^2}\right] \tag{10}$$

with the amplitude $A_\zeta \approx 0.06$ and the width $\Delta \approx 1.5$, where $k_p \approx 4.5 \cdot 10^{12}\ \text{Mpc}^{-1}$ is the location of the peak; see Figure 3.

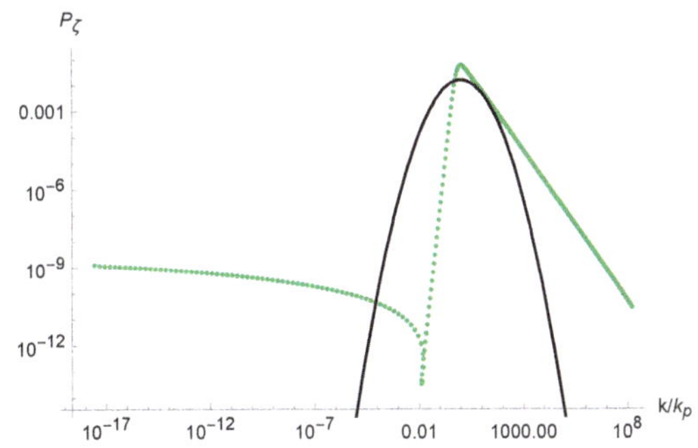

Figure 3. The primordial power spectrum $P_\zeta(k)$ of scalar perturbations (in green) and the lognormal fit (in black). The green lines (both dotted and solid) represent the results of our numerical calculations of the spectrum. The solid green line against the dotted green line highlights a slightly better agreement with Equation (8).

The value of $\eta_{\rm sr} \approx -0.025$ in the USR phase practically does not depend upon the parameters R_0 and δ. The more illuminating functions are given by the Hubble flow parameters

$$\epsilon_H = -\frac{\dot{H}}{H^2}, \quad \eta_H = \epsilon_H - \frac{\dot{\epsilon}_H}{2\epsilon_H H}. \tag{11}$$

Though $\epsilon_{\rm sr}$ and ϵ_H can be identified, the evolution of $\eta_{\rm sr}$ and η_H is different during the USR phase; see Figure 4.

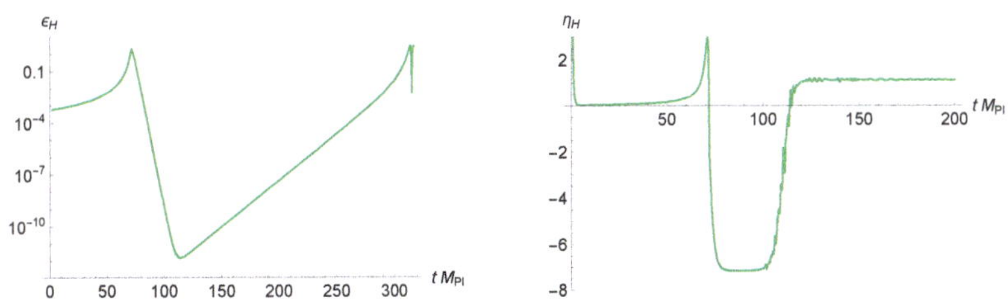

Figure 4. The evolution of $\epsilon_H(t)$ and $\eta_H(t)$ with the initial conditions $\phi_{in} = 7.01$ and $\dot{\phi}_{in} = 0$, and the parameters $\delta = 2.7 \times 10^{-8}$ and $R_0 = 3.0\ M^2$. In the USR phase, the value of $\eta_H(t)$ is close to -6.

3. PBH-Induced GW and PBH-DM Density Fraction

PBH production in the early universe leads to stochastic gravitational waves (GWs) different from primordial GWs caused by inflation. The current energy density fraction of those PBH-induced GWs can be computed in the second order with respect to perturbations as [24,25]

$$\begin{aligned}\Omega_{\rm GW}(k) =& \frac{c_g \Omega_{r,0}}{36} \int_0^{\frac{1}{\sqrt{3}}} {\rm d}d \int_{\frac{1}{\sqrt{3}}}^{\infty} {\rm d}s \left[\frac{(d^2 - 1/3)(s^2 - 1/3)}{s^2 - d^2}\right]^2 \times \\ & \mathcal{P}_\zeta\left(\frac{k\sqrt{3}}{2}(s+d)\right) \mathcal{P}_\zeta\left(\frac{k\sqrt{3}}{2}(s-d)\right) \left[\mathcal{I}_c(d,s)^2 + \mathcal{I}_s(d,s)^2\right],\end{aligned} \tag{12}$$

where the functions $\mathcal{I}_c(d,s)$ and $\mathcal{I}_s(d,s)$ are

$$\begin{aligned}\mathcal{I}_c(d,s) &= -36\pi \frac{(s^2+d^2-2)^2}{(s^2-d^2)^3}\theta(s-1), \\ \mathcal{I}_s(d,s) &= -36\frac{s^2+d^2-2}{(s^2-d^2)^2}\left[\frac{s^2+d^2-2}{(s^2-d^2)}\ln\left|\frac{d^2-1}{s^2-1}\right| + 2\right],\end{aligned} \tag{13}$$

with $\Omega_{r,0} \sim 8.6 \times 10^{-5}$ being the current energy density fraction of radiation, $\theta(s-1)$ is the step function, and $c_g \approx 0.4$.

A numerical calculation of Equation (12) with the power spectrum in Figure 2 yields the result given in Figure 5 in green. The numerical plot can be well approximated by the lognormal fit given in Figure 5 in black, with the analytic formula

$$_{\rm GW}(k) = \frac{A_{\rm GW}}{\sqrt{2\pi}\sigma_{\rm GW}} \exp\left[-\frac{\ln^2(k/k_p)}{2\sigma_{\rm GW}^2}\right], \tag{14}$$

the amplitude $A_{\rm GW} \approx 5.6 \cdot 10^{-8}$, and the width $\sigma_{\rm GW} \approx \Delta/\sqrt{2} \approx 1.06$, where Δ is the width of the power spectrum in Figure 2. The $\Omega_{\rm GW}^{\rm peak}(k)$ near the peak is roughly given by $10^{-6}\, \mathcal{P}_\zeta^2(k)$, in agreement with the estimates in Refs. [26,27].

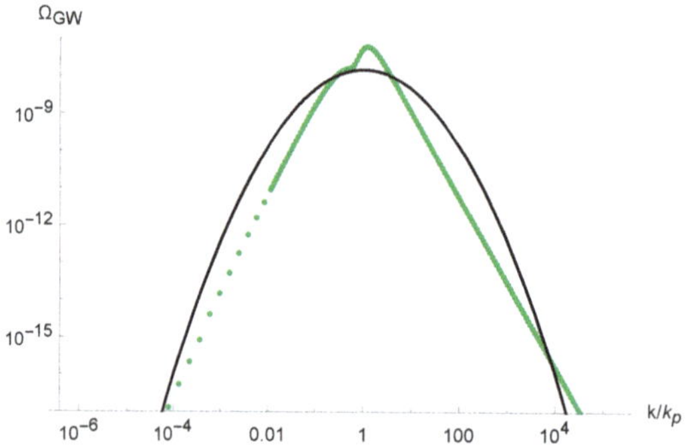

Figure 5. The PBH-induced stochastic GW density fraction (in green) with the lognormal fit (in black) in Equation (14). The dotted and solid green lines mean that the log-normal fit (14) applies only to the top of the peak.

The induced GW frequencies f_p are related to the PBH masses as [28]

$$f_p \approx 5.7\left(\frac{M_\odot}{M_{\rm PBH}}\right)^{1/2} 10^{-9}\ {\rm Hz}, \tag{15}$$

where the Sun mass is given by $M_\odot \approx 2 \cdot 10^{33}$ g. Given the PBH masses of 10^{20} g, as in our model, it results in the GW frequency $f_p \approx 0.0255$ Hz. It is higher than the GW frequencies between 3 and 400 nHz detected by NANOGrav [29]. A detection of the GW peak with that frequency in the 10^{-2} Hz range would provide observational support to our model provided the peak is due to generation of the secondary GWs. A more specific comparison of our predictions with future GW observations is possible by plotting the GW spectrum in our model against the expected sensitivity curves in future space-based gravitational interferometers such as LISA [30,31], TianQin [32], Taiji [33,34], and DECIGO [35]; see Figure 6, where we have used Refs. [36–38] for the colored curves. The space-based experiments are expected to be sensitive to stochastic GWs in the frequencies between 10^{-3} and 10^{-1} Hz, while the predicted black curve in Figure 6 in our model belongs to that frequency range.

The PBH-in-DM density fraction $f(k)$ on scale k can be estimated in the Press–Schechter formalism [39] as

$$f(k) = \frac{\Omega_{\rm PBH}(k)}{\Omega_{\rm DM}} \approx \frac{1.2 \cdot 10^{24} \beta_f(k)}{\sqrt{M_{\rm PBH}(k) {\rm g}^{-1}}} \approx 17.1 (k \cdot {\rm Mpc}) \beta_f(k), \tag{16}$$

where [40,41]

$$\beta_f(k) \simeq \frac{\sigma(k)}{\sqrt{2\pi}\delta_c} e^{-\frac{\delta_c^2}{2\sigma^2(k)}}, \quad \sigma^2(k) = \frac{16}{81}\int \frac{dq}{q}\left(\frac{q}{k}\right)^4 e^{-q^2/k^2} P_\zeta(q), \tag{17}$$

with the constant δ_c depending upon the shape of the PBH peak in the power spectrum and representing the density threshold for PBH formation. The integral in Equation (17) can be estimated as

$$\sigma^2 \approx \frac{3.84}{81} P_\zeta^{\text{peak}}, \tag{18}$$

where P_ζ^{peak} is the value of the power spectrum at the PBH peak. Then, Equations (16) and (17) imply

$$f(k) \sim \frac{10^{23}}{\sqrt{M_{\text{PBH}}(k)\text{g}^{-1}}} \left(\frac{\sqrt{P_\zeta^{\text{peak}}}}{\delta_c}\right) e^{-9.3\,\delta_c^2/P_\zeta^{\text{peak}}}. \tag{19}$$

This equation demonstrates the high sensitivity of the PBH-in-DM fraction upon the value of $\delta_c^2/P_\zeta^{\text{peak}}$. In the case of the power spectrum in Figure 2, we have $\delta_c \approx 0.45$ and $P_\zeta^{\text{peak}} \approx 0.05$. Given the range of the model parameters in Ref. [8], Equation (19) gives the PBH fraction in DM between 0.1% and 100%. In addition, the Press–Schechter formalism itself should be considered with a grain of salt because it was found to be unreliable [42–44]. It is also worth mentioning that even a small PBH fraction could have an important role in cosmology [21].

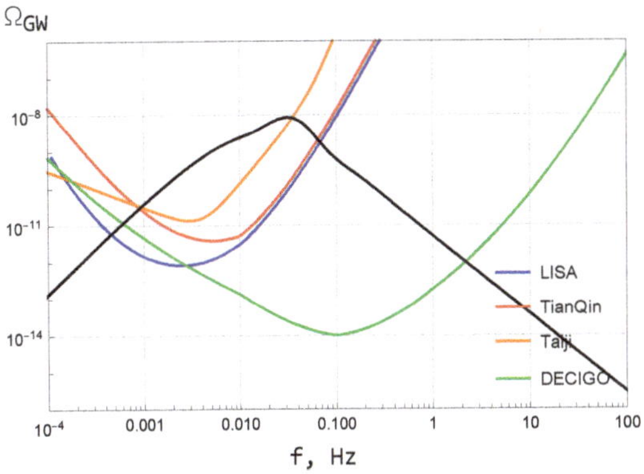

Figure 6. The GW density induced by the power spectrum of scalar perturbations in our model (in black) against the expected sensitivity curves for future space-based GW experiments (in color).

4. Loop Corrections

In the δN formalism [19] for single-field inflation, a scalar (comoving curvature) perturbation $\delta N \equiv \zeta$ is a function of variation $\delta\phi$ of inflaton ϕ_i at its initial value,

$$\delta N = N'(\phi)\delta\phi + \frac{N''(\phi)}{2!}\delta\phi^2 + \frac{N'''(\phi)}{3!}\delta\phi^3 + \ldots, \tag{20}$$

where perturbations are not assumed to be small. The power spectrum of scalar perturbations is defined by a two-point function of Fourier components $\zeta_{\mathbf{p}}$ as

$$\langle \zeta_{\mathbf{p_1}} \zeta_{\mathbf{p_2}} \rangle = (2\pi)^3 \delta^3(\mathbf{p_1} + \mathbf{p_2}) P(p), \quad P_\zeta(p) = \frac{k^3}{2\pi^2} P(p), \tag{21}$$

where

$$\zeta_{\mathbf{p}} = N'\delta\phi_{\mathbf{p}} + \frac{N''}{2!}\int \frac{d^3q}{(2\pi)^3}\delta\phi_{\mathbf{q}}\delta\phi_{\mathbf{p}-\mathbf{q}} + \frac{N'''}{3!}\int \frac{d^3q_1}{(2\pi)^3}\frac{d^3q_2}{(2\pi)^3}\delta\phi_{\mathbf{q}_1}\delta\phi_{\mathbf{q}_2}\delta\phi_{\mathbf{p}-\mathbf{q}_1-\mathbf{q}_2} + \dots \tag{22}$$

in terms of external 3D momenta $\mathbf{p}$ and loop momenta $\mathbf{q}$. Substituting Equation (22) into Equation (21) yields the loop expansion of the power spectrum $P_\zeta(p)$. In order to apply that to a particular model, one has to know the function $N(\phi)$ explicitly. It was derived in Ref. [16],

$$N_{\rm total}(\phi_i) \simeq \frac{1}{\eta_{\rm sr}}\ln\left[1+\frac{\eta_{\rm sr}}{\sqrt{2\epsilon_{\rm sr}}}(\phi_i-\phi_s)\right] - \frac{1}{3}\ln\left[1+\frac{3(\phi_s-\phi_e)}{\pi_s}\right] + \frac{1}{\eta_{\rm V}}\ln[-2\eta_{\rm V}\pi_e - 6\sqrt{2\epsilon_{\rm V}}], \tag{23}$$

where the first term refers to the SR(I) phase with the initial value ϕ_i and the end value ϕ_s, the second term refers to the USR phase with the initial momentum value π_s and the end momentum value π_e, and the third terms refers to the SR(II) phase with the slow-roll parameters $\epsilon_{\rm V}$ and $\eta_{\rm V}$. The subscripts (s,e) refer to values of any quantity at the start and end of the USR phase, respectively. The leading contribution comes from the first term in Equation (23).

To compute loop corrections, one has to calculate the derivatives $N^{(n)}(\phi)$. The first three derivatives can be estimated as follows:

$$N' \approx \frac{1}{\sqrt{2\epsilon_s}} \approx \frac{e^{-3\Delta N_{\rm USR}}}{\sqrt{2\epsilon_e}}, \quad N'' \approx -\frac{\eta_{\rm sr}}{2\epsilon_s}, \quad N''' \approx \frac{2\eta_{\rm sr}^2}{(2\epsilon_s)^{3/2}}, \tag{24}$$

where $\Delta N_{\rm USR}$ is the duration of the USR phase and $\eta_{\rm sr} \equiv \eta_{\rm sr}(\phi_s)$.

To compute loop corrections to the amplitude of the power spectrum, we considered the effective action up to the third order with respect to $\delta\phi(x,t)$ on the background $\phi(t)$,

$$S_{\delta\phi} = S^{(2)}_{\delta\phi} + S^{(3)}_{\delta\phi} = \tfrac{1}{2}\int d^4x\sqrt{-g}\left(g_{\mu\nu}\partial^\mu\delta\phi\partial^\nu\delta\phi - V_{,\phi\phi}\delta\phi^2\right) + \tfrac{1}{2}\int d^4x\sqrt{-g}\left(-\tfrac{1}{3}V_{,\phi\phi\phi}\delta\phi^3\right). \tag{25}$$

A comparable contribution of the quartic coupling $V_{,\phi\phi\phi\phi}$ was investigated in Ref. [45]. When using the FLRW background, the effective action reads

$$S_{\delta\phi} = \tfrac{1}{2}\int d^4x\, a^3(t)\left[\dot{\delta\phi}^2 - a^{-2}(t)(\partial\delta\phi)^2 - V_{,\phi\phi}\delta\phi^2\right] + \int d^4x\left[-\tfrac{a^3(t)}{3!}V_{,\phi\phi\phi}\delta\phi^3\right], \tag{26}$$

where $\partial \equiv \sum_i \partial_i$ is a sum of spatial derivatives. The mode functions arising in the solutions to classical equations of motion from the action (26) with Bunch–Davies initial conditions,

$$u_k(\tau) = \frac{H}{\sqrt{2k^3}}(1+ik\tau)e^{-ik\tau}, \tag{27}$$

are written down in terms of the conformal time $d\tau = \frac{dt}{a}$, where $-\infty < \tau \le 0$. The CMB modes that left the horizon during inflation are given by $u_k(0) = \frac{H}{\sqrt{2k^3}}$.

Canonical quantization implies a decomposition into positive and negative parts, as well as the commutation relations (in the interaction picture)

$$\begin{aligned} \delta\phi^I_{\mathbf{k}}(\tau) &= \delta\phi^+_{\mathbf{k}}(\tau) + \delta\phi^-_{\mathbf{k}}(\tau) = u_k(\tau)a_{\mathbf{k}} + u^*_k(\tau)a^\dagger_{-\mathbf{k}}, \\ [a_{\mathbf{p}}, a^\dagger_{\mathbf{q}}] &= (2\pi)^3\delta^3(\boldsymbol{p}-\boldsymbol{q}), \ [a_{\mathbf{p}}, a_{\mathbf{q}}] = [a^\dagger_{\mathbf{p}}, a^\dagger_{\mathbf{q}}] = 0. \end{aligned} \tag{28}$$

To obtain the one-loop correction according to Equations (21) and (22), one has to evaluate the three-point correlation function. For this purpose, we applied the in-in formalism that gives

$$\langle\delta\phi_{\mathbf{p}}\delta\phi_{\mathbf{q}}\delta\phi_{-\mathbf{p}-\mathbf{q}}\rangle = \langle\overline{T}e^{i\int_{t_0}^t d\tilde{t}H_{\rm int}}\,\delta\phi^I_{\mathbf{p}}\delta\phi^I_{\mathbf{q}}\delta\phi^I_{-\mathbf{p}-\mathbf{q}}\,Te^{-i\int_{t_0}^t d\tilde{t}H_{\rm int}}\rangle, \tag{29}$$

where T and $\overline{T}$ stand for the time ordering and anti-time ordering, respectively, t_0 and t are the times associated with the subhorizon and superhorizon scales, respectively, and $H_{\rm int}(\tilde{t})$ is the interaction Hamiltonian in the third order,

$$H_{\rm int}(\tilde{t}) = \frac{1}{3!}a^3(\tilde{t})V_{,\phi\phi\phi}\int d^3x(\delta\phi^I(x,\tilde{t}))^3 = \frac{1}{3!(2\pi)^6}a^3(\tilde{t})V_{,\phi\phi\phi}\int d^3kd^3\lambda\delta\phi^I_{\mathbf{k}}(\tilde{t})\delta\phi^I_{\lambda}(\tilde{t})\delta\phi^I_{-\mathbf{k}-\lambda}(\tilde{t}). \quad (30)$$

The standard (Friedmann and Klein–Gordon) equations of motion yield the following asymptotic approximation for the third derivative of the potential in terms of the Hubble flow parameters (11):

$$V_{,\phi\phi\phi} \simeq -\frac{3H\dot{\eta}_H}{2\sqrt{2\epsilon_H}}. \quad (31)$$

Expanding the T-exponentials in Equation (29) to the first order with respect to $H_{\rm int}$, we find

$$\begin{aligned}&\langle\delta\phi_{\mathbf{p}}(t)\delta\phi_{\mathbf{q}}(t)\delta\phi_{-\mathbf{p}-\mathbf{q}}(t)\rangle \approx -i\int_{t_0}^{t} d\tilde{t}\langle[\delta\phi^I_{\mathbf{p}}(t)\delta\phi^I_{\mathbf{q}}(t)\delta\phi^I_{-\mathbf{p}-\mathbf{q}}(t), H^{int}(\tilde{t})]\rangle \\ &= 2{\rm Im}\left(\int_{t_0}^{t} d\tilde{t}\langle\delta\phi^I_{\mathbf{p}}(t)\delta\phi^I_{\mathbf{q}}(t)\delta\phi^I_{-\mathbf{p}-\mathbf{q}}(t)H^{int}(\tilde{t})\rangle\right).\end{aligned} \quad (32)$$

After substituting Equations (27), (28), (30), and (31) into Equation (32) and using Wick's theorem, we derived the three-point correlator as follows:

$$\begin{aligned}&\langle\delta\phi_{\mathbf{p}}\delta\phi_{\mathbf{q}}\delta\phi_{-\mathbf{p}-\mathbf{q}}\rangle = \\ &= -\frac{1}{2(2\pi)^6}{\rm Im}\left(\int_{-\infty}^{0} d\tilde{\tau}a^3(\tilde{\tau})\frac{H\eta'(\tilde{\tau})}{\sqrt{2\epsilon}}\int d^3kd^3\lambda\langle\delta\phi^+_{\mathbf{p}}\delta\phi^+_{\mathbf{q}}\delta\phi^+_{-\mathbf{p}-\mathbf{q}}\delta\phi^-_{\mathbf{k}}(\tilde{\tau})\delta\phi^-_{\lambda}(\tilde{\tau})\delta\phi^-_{-\mathbf{k}-\lambda}(\tilde{\tau})\rangle\right) \\ &= -\frac{\Delta\eta}{2(2\pi)^6}{\rm Im}\left(\frac{H_e a_e^3}{\sqrt{2\epsilon_e}}\int d^3kd^3\lambda\langle\delta\phi^+_{\mathbf{p}}(\tau)\delta\phi^+_{\mathbf{q}}(\tau)\delta\phi^+_{-\mathbf{p}-\mathbf{q}}(\tau)\delta\phi^-_{\mathbf{k}}(\tau_e)\delta\phi^-_{\lambda}(\tau_e)\delta\phi^-_{-\mathbf{k}-\lambda}(\tau_e)\rangle\right) \\ &\approx \frac{3!(2\pi)^3|\Delta\eta|}{2}{\rm Im}\left(\frac{H_e^4 a_e^3}{\sqrt{2\epsilon_e}}\frac{H_0^3(1-iq\tau_e)^2}{(2q^3)^2 2p^3}e^{2iq\tau_e}\right),\end{aligned} \quad (33)$$

where the $H(0) = H_0$ denotes the Hubble value during the SR(I), and the reference time was chosen at $\tau = 0$ because we were only interested in the power spectrum on superhorizon scales relevant to CMB, and $p \ll q$. To avoid divergences, the vacuum expectation value was normalized by the volume of the entire system.

The dynamics of the parameter η_H implies it is essentially constant everywhere except for the moments of a decrease or an increase (corresponding to τ_s and τ_e, respectively), while the moment of the increase is particularly significant (see Figure 4). The approximate solution (23) to the equations of motion in our model is smooth as well as the corresponding $\eta_H(t)$ function defined by the second Equation (11). To simplify our calculations, we employed the derivative of η_H with respect to the conformal time as the (Dirac) delta function, $\eta'_H(\tau) \sim \delta(\tau - \tau_e)\Delta\eta$, where $\Delta\eta \approx -6$ is the depth of the pit, inside integrations, which corresponds to a sharp transition. Via integration, the delta function fixes the entire integrand at the time τ_e corresponding to the end of the USR stage.

Equations (20)–(22) lead to a recovery of the tree-level contribution (8) as the leading term in the loop expansion of the power spectrum, as well as the first (one-loop) contribution as follows:

$$P_\zeta^{1-{\rm loop}}(p) \equiv \frac{N'N''}{2}\int\frac{d^3q}{(2\pi)^3}\langle\delta\phi_{\mathbf{p}}\delta\phi_{\mathbf{q}}\delta\phi_{-\mathbf{p}-\mathbf{q}}\rangle. \quad (34)$$

After substituting Equations (24) and (33) into Equation (34), we found

$$\begin{aligned}&P_\zeta^{1-{\rm loop}}(p) \approx \frac{\eta_{\rm sr}e^{-3\Delta N_{\rm USR}}}{2\sqrt{2\epsilon_e}2\epsilon_s}\int_{\rm USR} dq(4\pi q^2)\frac{3!|\Delta\eta|}{2}{\rm Im}\left(\frac{H_e^4 a_e^3}{\sqrt{2\epsilon_e}}\frac{H_0^3(1-iq\tau_e)^2}{(2q^3)^2 2p^3}e^{2iq\tau_e}\right) \\ &\approx \eta_{\rm sr}e^{-3\Delta N_{\rm USR}}P_\zeta(p)P_\zeta^{\rm PBH}\frac{(2\pi)^3|\Delta\eta|}{4}\left(\frac{H_0}{H_e}\right),\end{aligned} \quad (35)$$

where $P_\zeta^{\rm PBH} \sim 10^{-2}$ is the fixed amplitude of the power spectrum on the small scales associated with the short-wavelength PBH modes exiting the horizon during the USR phase of inflation. The value of $H_0/H_e \approx 5$ defines the ratio of the inflation and PBH scales in our model.

It is evident from Equation (35) that the dependence of the one-loop correction upon the slow-roll parameter η_{sr} comes from the second derivative N'', the exponential factor depending upon $\Delta N_{\rm USR}$ arises from the first derivative N', and the dependence upon $P_\zeta^{\rm PBH}$ and $\Delta\eta$ comes from the third derivative $V_{,\phi\phi\phi}$, namely, from $\eta'(\tau)$.

A detailed calculation of the higher-loop corrections is highly involved and is not given here. However, it is possible to obtain a rough estimate of the two-loop correction by using the approximative formula given in Ref. [16],

$$P_\zeta^{2-{\rm loop}}(p) \approx \frac{N'N'''}{3!}|\delta\phi_{\mathbf{p}}|^2 \int \frac{d^3q}{(2\pi)^3}|\delta\phi_{\mathbf{q}}|^2 \sim \eta_{\rm sr}^2 \Delta N_{\rm USR} P_\zeta(p) P_\zeta^{\rm PBH}. \tag{36}$$

In our model, according to the plot on the right-hand side of Figure 4, we have $\Delta N_{\rm USR} \approx 3.1$.

Therefore, the relative size of the one-loop and two-loop corrections from PBH production to the power spectrum at the CMB pivot scale $k^* = 0.05$ are

$$\frac{P_\zeta^{1-{\rm loop}}(k^*=0.05)}{P_\zeta(k^*=0.05)} \approx \eta_{\rm sr} e^{-3\Delta N_{\rm USR}} P_\zeta^{\rm PBH} \frac{(2\pi)^3|\Delta\eta|}{4}\left(\frac{H_0}{H_e}\right) \approx 10^{-3} \tag{37}$$

and

$$\frac{P_\zeta^{2-{\rm loop}}(k^*=0.05)}{P_\zeta(k^*=0.05)} \sim \eta_{\rm sr}^2 \Delta N_{\rm USR} P_\zeta^{\rm PBH} \approx 10^{-5}, \tag{38}$$

where we used $P_\zeta(k^*=0.05) \approx 2\cdot 10^{-9}$ for the CMB power spectrum. Therefore, the one-loop contribution is suppressed by the factor $\eta_{\rm sr} e^{-3\Delta N_{\rm USR}}$, whereas the two-loop contribution is suppressed by $\eta_{\rm sr}^2 \Delta N_{\rm USR}$ (we recall that $\eta_{\rm sr} \approx -0.025$ in our model). As regards the higher n-loop corrections, their structure includes the suppression factor $\eta_{\rm sr}^n P_\zeta^{\rm PBH} \sim 10^{-2n-2}$ so that they are expected to be negligible too.

The relative smallness of loop corrections in our model is in agreement with the considerations of Refs. [16,17,46–48] but in disagreement with the results of Refs. [13,14]. Our calculations were based on the δN formalism, also used in Ref. [16], whereas the calculations performed in Refs. [13,14] were based on the in-in formalism. It is beyond the scope of our investigation to compare the two formalisms. [2]

The amplitude of the power spectrum during USR was fixed in our approach, while we effectively assumed a sharp transition in part of our analytic calculations. The sharpness of transitions can be quantitatively estimated by the parameter h defined by [18]

$$h = 6\frac{\sqrt{2\epsilon_V}}{\dot{\phi}\,(t_e)} = -6\frac{\sqrt{2\epsilon_V}}{\pi_e} = -6\sqrt{\frac{\epsilon_V}{\epsilon_s}}e^{3\Delta N_{\rm USR}}, \tag{39}$$

where π_e is the inflaton momentum at the end of the USR inflation, $\epsilon_s \approx 1$ is the SR parameter at the end of the SR(I) or at the beginning of USR, and ϵ_V is the SR parameter at the end of USR. In our model, by using Figure 4, we obtained $h \approx -0.66$, which implies a sharp (though rather mild) transition because, on the one hand, $|h|$ is not much less than one but on the other hand, it is still away from a truly sharp transition with the "standard" value $h = -6$ used in Ref. [13]. As was demonstrated in Ref. [45], the lower value of h also justifies ignoring the quartic coupling in our analysis. As a result, the one-loop correction in our model appeared to be small against the tree-level contribution, as in Refs. [17,49].

5. Conclusions

The main new results of this paper were given by Figure 6, Equations (19) and (35). It follows that the modified gravity model [8] of Starobinsky inflation with PBH production may generate a significant part (or the whole) of dark matter from PBHs, while it was not ruled out by quantum loop corrections because the latter were relatively small by a factor of 10^{-3} against the tree-level (classical) contribution. The key role in the last conclusion was played by the derivatives of the function $N(\phi)$ during the USR phase, describing superhorizon curvature perturbations in the δN formalism, which led to the suppression of loop contributions. It is worth mentioning that our results only apply to the particular phenomenological model of PBH production related to Starobinsky inflation.

The predicted frequency $f_p \approx 2.55 \cdot 10^{-2}$ Hz of the PBH production-induced stochastic GWs is in the range between 10^{-3} Hz and 10^{-1} Hz of the frequencies that are expected to be sensitive to the future space-based gravitational interferometers.

As was recently pointed out in the literature [50–53], the standard result for the primordial black hole survival at present, based on the Hawking semiclassical evaporation formula, may be relaxed below 10^{15} g, when going beyond the semiclassical approximation. Should this be the case, fine-tuning the parameters in our model for efficient PBH production (needed for DM) may be significantly relaxed.

Author Contributions: S.S. and S.V.K. have contributed equally. All authors have read and agreed to the published version of the manuscript.

Funding: S.S. and S.V.K. were supported by Tomsk State University. S.V.K. was also supported by Tokyo Metropolitan University, the Japanese Society for Promotion of Science under the grant no. 22K03624, and the World Premier International Research Center Initiative (MEXT, Japan).

Data Availability Statement: No data are associated with the manuscript.

Acknowledgments: We acknowledge discussion and correspondence with Matthew Davies, Gia Dvali, Guillem Domenech, Andrew Gow, Jason Kristiano, Peter Kazinsky, Sayantan Choudhury, Kin-Wang Ng, and Alexei Starobinsky. This paper is devoted to memory of late Alexei Starobinsky.

Conflicts of Interest: The authors declare no conflict of interest.

Notes

1 It differs from Ref. [20] where R_0 was related to the dark energy scale.

2 See, however, Ref. [45] for a partial comparison.

References

1. Starobinsky, A.A. A new type of isotropic cosmological models without singularity. *Phys. Lett. B* **1980**, *91*, 99–102. [CrossRef]
2. Ketov, S.V. On the equivalence of Starobinsky and Higgs inflationary models in gravity and supergravity. *J. Phys. A* **2020**, *53*, 084001. [CrossRef]
3. Ivanov, V.R.; Ketov, S.V.; Pozdeeva, E.O.; Vernov, S.Y. Analytic extensions of Starobinsky model of inflation. *J. Cosmol. Astropart. Phys.* **2022**, *3*, 058. [CrossRef]
4. Akrami, Y.; Arroja, F.; Ashdown, M.; Aumont, J.; Baccigalupi, C.; Ballardini, M.; Banday, A.J.; Barreiro, R.B.; Bartolo, N.; Basak, S.; et al. Planck 2018 results. X. Constraints on inflation. *Astron. Astrophys.* **2020**, *641*, A10. [CrossRef]
5. Ade, P.A. et al. [BICEP, Keck Collaboration] Improved Constraints on Primordial Gravitational Waves using Planck, WMAP, and BICEP/Keck Observations through the 2018 Observing Season. *Phys. Rev. Lett.* **2021**, *127*, 151301. [CrossRef]
6. Tristram, M.; Banday, A.J.; Górski, K.M.; Keskitalo, R.; Lawrence, C.R.; Andersen, K.J.; Barreiro, R.B.; Borrill, J.; Colombo, L.P.L.; Eriksen, H.K.; et al. Improved limits on the tensor-to-scalar ratio using BICEP and Planck data. *Phys. Rev. D* **2022**, *105*, 083524. [CrossRef]
7. Frolovsky, D.; Ketov, S.V.; Saburov, S. Formation of primordial black holes after Starobinsky inflation. *Mod. Phys. Lett. A* **2022**, *37*, 2250135. [CrossRef]
8. Saburov, S.; Ketov, S.V. Improved Model of Primordial Black Hole Formation after Starobinsky Inflation. *Universe* **2023**, *9*, 323. [CrossRef]
9. Kamenshchik, A.Y.; Pozdeeva, E.O.; Tribolet, A.; Tronconi, A.; Venturi, G.; Vernov, S.Y. The Superpotential Method and the Amplification of Inflationary Perturbations. *arXiv* **2024**. [CrossRef]

10. Geller, S.R.; Qin, W.; McDonough, E.; Kaiser, D.I. Primordial black holes from multifield inflation with nonminimal couplings. *Phys. Rev. D* **2022**, *106*, 063535. [CrossRef]
11. Cole, P.S.; Gow, A.D.; Byrnes, C.T.; Patil, S.P. Primordial black holes from single-field inflation: A fine-tuning audit. *arXiv* **2023**. [CrossRef]
12. Karam, A.; Koivunen, N.; Tomberg, E.; Vaskonen, V.; Veermäe, H. Anatomy of single-field inflationary models for primordial black holes. *J. Cosmol. Astropart. Phys.* **2023**, *3*, 013. [CrossRef]
13. Kristiano, J.; Yokoyama, J. Ruling Out Primordial Black Hole Formation From Single-Field Inflation. *arXiv* **2024**. [CrossRef]
14. Choudhury, S.; Panda, S.; Sami, M. Quantum loop effects on the power spectrum and constraints on primordial black holes. *J. Cosmol. Astropart. Phys.* **2023**, *11*, 066. [CrossRef]
15. Cheng, S.-L.; Lee, D.-S.; Ng, K.-W. Primordial perturbations from ultra-slow-roll single-field inflation with quantum loop effects. *arXiv* **2024**. [CrossRef]
16. Firouzjahi, H.; Riotto, A. Primordial Black Holes and Loops in Single-Field Inflation. *arXiv* **2023**. [CrossRef]
17. Davies, M.W.; Iacconi, L.; Mulryne, D.J. Numerical 1-loop correction from a potential yielding ultra-slow-roll dynamics. *arXiv* **2024**. [CrossRef]
18. Cai, Y.-F.; Chen, X.; Namjoo, M.H.; Sasaki, M.; Wang, D.-G.; Wang, Z. Revisiting non-Gaussianity from non-attractor inflation models. *J. Cosmol. Astropart. Phys.* **2018**, *5*, 12. [CrossRef]
19. Abolhasani, A.A.; Firouzjahi, H.; Naruko, A.; Sasaki, M. *Delta N Formalism in Cosmological Perturbation Theory*; WSP: London, UK, 2019. [CrossRef]
20. Appleby, S.A.; Battye, R.A.; Starobinsky, A.A. Curing singularities in cosmological evolution of F(R) gravity. *J. Cosmol. Astropart. Phys.* **2010**, *6*, 5. [CrossRef]
21. Carr, B.; Kohri, K.; Sendouda, Y.; Yokoyama, J. Constraints on primordial black holes. *Rept. Prog. Phys.* **2021**, *84*, 116902. [CrossRef]
22. Maeda, K.-I. Towards the Einstein-Hilbert Action via Conformal Transformation. *Phys. Rev. D* **2021**, *39*, 3159. [CrossRef]
23. Frolovsky, D.; Ketov, S.V. Fitting power spectrum of scalar perturbations for primordial black hole production during inflation. *Astronomy* **2023**, *2*, 47–57. [CrossRef]
24. Espinosa, J.R.; Racco, D.; Riotto, A. A Cosmological Signature of the SM Higgs Instability: Gravitational Waves. *J. Cosmol. Astropart. Phys.* **2018**, *9*, 012. [CrossRef]
25. Kohri, K.; Terada, T. Semianalytic calculation of gravitational wave spectrum nonlinearly induced from primordial curvature perturbations. *Phys. Rev. D* **2018**, *97*, 123532. [CrossRef]
26. Pi, S.; Sasaki, M. Gravitational Waves Induced by Scalar Perturbations with a Lognormal Peak. *J. Cosmol. Astropart. Phys.* **2020**, *9*, 037. [CrossRef]
27. Domènech, G. Scalar Induced Gravitational Waves Review. *Universe* **2021**, *7*, 398. [CrossRef]
28. Luca, V.D.; Franciolini, G.; Riotto, A. NANOGrav Data Hints at Primordial Black Holes as Dark Matter. *Phys. Rev. Lett.* **2021**, *126*, 041303. [CrossRef] [PubMed]
29. Agazie, G. et al. [NANOGrav Collaboration] The NANOGrav 15 yr Data Set: Evidence for a Gravitational-wave Background. *Astrophys. J. Lett.* **2023**, *951*, L8. [CrossRef]
30. Amaro-Seoane, P. et al. [LISA Collaboration] Laser Interferometer Space Antenna. *arXiv* **2017**. [CrossRef]
31. Smith, T.L.; Caldwell, R. LISA for Cosmologists: Calculating the Signal-to-Noise Ratio for Stochastic and Deterministic Sources. *Phys. Rev. D* **2019**, *100*, 104055. [CrossRef]
32. Luo, J. et al. [TianQin Collaboration] TianQin: A space-borne gravitational wave detector. *Class. Quant. Grav.* **2016**, *33*, 035010. [CrossRef]
33. Gong, X.; Lau, Y.K.; Xu, S.; Amaro-Seoane, P.; Bai, S.; Bian, X.; Cao, Z.; Chen, G.; Chen, X.; Ding, Y.; et al. Descope of the ALIA mission. *J. Phys. Conf. Ser.* **2015**, *610*, 012011. [CrossRef]
34. Ruan, W.-H.; Guo, Z.-K.; Cai, R.-G.; Zhang, Y.-Z. Taiji program: Gravitational-wave sources. *Int. J. Mod. Phys. A* **2020**, *35*, 2050075. [CrossRef]
35. Kudoh, H.; Taruya, A.; Hiramatsu, T.; Himemoto, Y. Detecting a gravitational-wave background with next-generation space interferometers. *Phys. Rev. D* **2006**, *73*, 064006. [CrossRef]
36. Thrane, E.; Romano, J.D. Sensitivity curves for searches for gravitational-wave backgrounds. *Phys. Rev. D* **2013**, *88*, 124032. [CrossRef]
37. Schmitz, K. New Sensitivity Curves for Gravitational-Wave Signals from Cosmological Phase Transitions. *J. High Energy Phys.* **2021**, *1*, 97. [CrossRef]
38. Aldabergenov, Y.; Addazi, A.; Ketov, S.V. Testing Primordial Black Holes as Dark Matter in Supergravity from Gravitational Waves. *Phys. Lett. B* **2021**, *814*, 136069. [CrossRef]
39. Press, W.H.; Schechter, P. Formation of galaxies and clusters of galaxies by selfsimilar gravitational condensation. *Astrophys. J.* **1974**, *187*, 425–438. [CrossRef]
40. Inomata, K.; Kawasaki, M.; Mukaida, K.; Tada, Y.; Yanagida, T.T. Inflationary Primordial Black Holes as All Dark Matter. *Phys. Rev. D* **2017**, *96*, 043504. [CrossRef]
41. Aldabergenov, Y.; Addazi, A.; Ketov, S.V. Primordial black holes from modified supergravity. *Eur. Phys. J. C* **2020**, *80*, 917. [CrossRef]

42. Franciolini, G.; Kehagias, A.; Matarrese, S.; Riotto, A. Primordial Black Holes from Inflation and non-Gaussianity. *J. Cosmol. Astropart. Phys.* **2018**, *3*, 16. [CrossRef]
43. Germani, C.; Sheth, R.K. The Statistics of Primordial Black Holes in a Radiation-Dominated Universe: Recent and New Results. *Universe* **2023**, *9*, 421. [CrossRef]
44. Figueroa, D.G.; Raatikainen, S.; Rasanen, S.; Tomberg, E. Implications of stochastic effects for primordial black hole production in ultra-slow-roll inflation. *J. Cosmol. Astropart. Phys.* **2022**, *5*, 27. [CrossRef]
45. Firouzjahi, H. One-loop corrections in power spectrum in single field inflation. *J. Cosmol. Astropart. Phys.* **2023**, *10*, 6. [CrossRef]
46. Inomata, K. Curvature Perturbations Protected Against One Loop. *arXiv* **2024**. [CrossRef]
47. Kristiano, J.; Yokoyama, J. Generating large primordial fluctuations in single-field inflation for PBH formation. *arXiv* **2024**. [CrossRef]
48. Kristiano, J.; Yokoyama, J. Comparing sharp and smooth transitions of the second slow-roll parameter in single-field inflation. *arXiv* **2024**. [CrossRef]
49. Tada, Y.; Terada, T.; Tokuda, J. Cancellation of quantum corrections on the soft curvature perturbations. *J. High Energy Phys.* **2024**, *1*, 105. [CrossRef]
50. Dvali, G.; Eisemann, L.; Michel, M.; Zell, S. Black hole metamorphosis and stabilization by memory burden. *Phys. Rev. D* **2020**, *102*, 103523. [CrossRef]
51. Michel, M.; Zell, S. The Timescales of Quantum Breaking. *Fortsch. Phys.* **2023**, *71*, 2300163. [CrossRef]
52. Alexandre, A.; Dvali, G.; Koutsangelas, E. New Mass Window for Primordial Black Holes as Dark Matter from Memory Burden Effect. *arXiv* **2024**. [CrossRef]
53. Thoss, V.; Burkert, A.; Kohri, K. Breakdown of Hawking Evaporation opens new Mass Window for Primordial Black Holes as Dark Matter Candidate. *arXiv* **2024**. [CrossRef]

universe

MDPI

Review

The Statistics of Primordial Black Holes in a Radiation-Dominated Universe: Recent and New Results

Cristiano Germani [1,2,*,†] and Ravi K. Sheth [3,†]

1 Institut de Ciencies del Cosmos (ICCUB), Universitat de Barcelona, Martí i Franquès 1, E08028 Barcelona, Spain
2 Departement de Física Quàntica i Astrofisica, Universitat de Barcelona, Martí i Franquès 1, E08028 Barcelona, Spain
3 Department of Physics and Astronomy and Center for Particle Cosmology, University of Pennsylvania, 209 S. 33rd Street, Philadelphia, PA 19104, USA
* Correspondence: germani@icc.ub.edu
† These authors contributed equally to this work.

Abstract: We review the nonlinear statistics of Primordial Black Holes that form from the collapse of over-densities in a radiation-dominated Universe. We focus on the scenario in which large over-densities are generated by rare and Gaussian curvature perturbations during inflation. As new results, we show that the mass spectrum follows a power law determined by the critical exponent of the self-similar collapse up to a power spectrum dependent cutoff, and that the abundance related to very narrow power spectra is exponentially suppressed. Related to this, we discuss and explicitly show that both the Press–Schechter approximation and the statistics of mean profiles lead to wrong conclusions for the abundance and mass spectrum. Finally, we clarify that the transfer function in the statistics of initial conditions for Primordial Black Holes formation (the abundance) does not play a significant role.

Keywords: Primordial Black Holes; Dark Matter; statistics

1. Setup

Primordial Black Holes (PBHs), if they exist, are arguably the most economical form of Dark Matter. A PBH is the final state of a large over-density collapse that happened long before matter–radiation equality. A black hole is surely of a primordial origin if its mass is lower than the solar one. In the realm in which PBHs account for all of Dark Matter, this is precisely where their masses need to be [1].

Assuming only the standard model of particle physics, before matter–radiation equality and after inflation the universe is in a state dominated by radiation. Because black holes only interact gravitationally, they behave as dust particles. In other words, in a homogeneous and isotropic universe (i.e., a Friedman–Robertson–Walker (FRW) universe) with metric element

$$ds^2 = -dt^2 + a(t)^2 d\vec{x} \cdot d\vec{x}\,, \tag{1}$$

the energy density of radiation scales as a^{-4} and that of Primordial Black Holes scales as a^{-3}. In the minimal scenario in which the perturbations in the mean FRW universe were seeded by inflation, large over-densities might only have been generated at scales much smaller than those related to the cosmic microwave background radiation observed today. Thus, for a high reheating temperature PBHs will have had a long period to increase their energy density relatively to radiation. In this scenario, the initial density of PBHs is tiny; therefore, the formation of a PBH represents a rare event. This has a particular resonance with the theory of inflation, as inflation mainly generates perturbative over-densities. However, because perturbations are quantum in nature, their amplitudes are statistically distributed, meaning that non-perturbative over-densities can be generated as well, albeit rarely.

Thus far, there is no evidence of (perturbative) deviations from the simple Gaussian statistics; therefore, for the purposes of discussion, we assume it here in order to investigate the statistics of PBH formation from a given Gaussian-distributed initial seed of curvature perturbations. Our procedure can be generalized to other distributions as well.

Before moving on, it is necessary to quickly review the conditions for PBH formation in an expanding universe. Interested readers may consult the review of [2] for a more technical perspective; here, we provide a more intuitive overview.

In the hoop conjecture of Thorne [3], a black hole forms if a portion of fluid with "mass" M in an asymptotically flat spacetime can be enclosed within a hoop of perimeter $2\pi R_s$, where $R_s \equiv 2M$ ($G_N = \hbar = c = 1$) is the Schwarzschild radius of the portion of that fluid. For a fluid of density ρ, we can define an instantaneous mass ($dt = 0$) as

$$M\Big|_{dt=0} \equiv 4\pi \int_0^R \rho \, \bar{R}^2 d\bar{R}\Big|_{dt=0}, \tag{2}$$

where we have assumed that the energy density of the fluid is isotropic and R is the areal radius. In an expanding universe, this mass is known as the Misner–Sharp mass (MS) [4]. The scenario we are interested in here is a localized large over-density (defined in a specific gauge that we clarify later on) evolving in a homogeneous and isotropic radiation-dominated universe. Because of the assumption of asymptotic flatness in the hoop conjecture, we could attempt to obtain similar conditions in an expanding universe by subtracting the infinitely long wavelength mode (the background energy density) from the MS mass. Thus, we define the instantaneous over-mass as

$$\delta M(R)\Big|_{dt=0} \equiv 4\pi \int_0^R \delta\rho \, \bar{R}^2 d\bar{R}\Big|_{dt=0}, \tag{3}$$

where $\delta\rho \equiv \rho - \rho_b$ is the over-density with respect to the background ρ_b.

The extension of the hoop conjecture to an expanding universe would then be that a collapse into a black hole starts at position R_{BH} whenever

$$\mathcal{C}\Big|_{dt=0} \equiv \frac{2\delta M(R_{BH})\Big|_{dt=0}}{R_{BH}}\Bigg|_{dt=0} \sim 1\,. \tag{4}$$

We can now remove the fixed time and define

$$\mathcal{C}(r,t) \equiv \frac{8\pi \int_0^r \delta\rho \, R(\bar{r},t)^2 \partial_{\bar{r}} R(\bar{r},t)\, d\bar{r}}{R(r,t)}, \tag{5}$$

where r is the radial coordinate of the general isotropic metric

$$ds^2 = -A(r,t)^2 dt^2 + \frac{\partial_r R(r,t)^2 dr^2}{\Gamma(r,t)^2} + R(r,t)^2 d\Omega_2^2\,. \tag{6}$$

The metric function Γ is the "boost factor" of the fluid (being 1 for vanishing fluid velocity and gravitational mass [4]); thus, the Misner–Sharp mass is the equivalent of the "rest mass" of the fluid. The function $\mathcal{C}$, defined in Equation (5), was called the *compaction function* in [5], in which the compaction factor was defined without the factor 2. The factor 2 was introduced in [6] to resemble the Schwarzschild potential; see [7] for the latest interpretations of the earlier work of [5]. In the same paper, it was shown that, in agreement with the hoop conjecture, a black hole would inexorably form whenever $\mathcal{C} \sim 1$.

Let us now return to our primordial black hole scenario. We have already mentioned that whenever PBHs are generated by large inflationary perturbations during the radiation epoch they are generically rare. Assuming a Gaussian distribution of perturbations, this implies an approximate spherical symmetry around the peak of such rare large perturbations [8]. Thus, in this regime the condition for collapse into black holes can be provided in terms of $\mathcal{C}$:

The first trapped horizon is formed at a radius equaling the maximum of the compaction function [6]; in other words, a black hole will form at a scale $r = r_{\rm BH}(t_{\rm BH})$ solution of $\partial_r \mathcal{C} = 0$ whenever $\mathcal{C}(r_{\rm BH}, t_{\rm BH}) \sim 1$. The threshold for PBH formation can then be provided in terms of a critical value $\mathcal{C}_c(r_m(t_i), t_i)$, where $t_i \neq t_{\rm BH}$ is some initial time [6,9].

Initial Conditions and Threshold

The question we answer in this section is:

Under which initial conditions for $\mathcal{C}$ will a BH form at some later time $t = t_{\rm BH}$?

Under the inflationary paradigm, perturbations set in at scales larger than the cosmological horizon. On such scales and at the leading order in gradient expansion, the metric (6) can always be recast (see, e.g., [10] for a review) into a local FRW metric:

$$ds_l^2 \simeq -dt_l^2 + a_l(t_l)^2 dr_l^2 + r_l^2 a_l(t_l)^2 d\Omega_2^2 \tag{7}$$

where the subscript l refers to local coordinates and functions and $a_l(t)$ to the spatial rescaling due to a long-wavelength perturbation, i.e., a perturbation with a wavelength larger than the cosmological horizon of the background. Up to decaying terms, we find that $a_l(t_l) \simeq a(t_l)$, where a is the background scale factor in local time [11].

The idea is then to find appropriate initial conditions for $\mathcal{C}$ at super-horizon scales such that a future black hole would form. At the leading order in gradient expansion, however, the compaction function vanishes, making it necessary to go beyond this concept.

If the typical co-moving scale of the perturbation is r_0, then we can define the parameter

$$\sigma(t) \equiv \frac{R_H(t)^2}{R_p(t)^2} = \frac{a_i^2}{H^2 a(t)^2 r_0^2} = \frac{a_i^2}{\dot{a}(t)^2 r_0^2} \ll 1\,, \tag{8}$$

where a_i is the scale factor at some initial time t_i. Thus, $R(t_i, r) = r a(t_i)$.

If we specialize to the case of radiation,

$$\sigma(t) = \frac{3\,\frac{a(t)^2}{a_i^2}}{(l_p^4 \rho_0)\,(r_0/l_p)^2}\,, \tag{9}$$

where ρ_0 is the background energy-density at $t = t_i$, then we have now replaced units by using l_p as the Planck scale. Because the universe is expanding, a necessary condition is then to choose an initial time such that

$$l_p^4 \rho_0 \gg \frac{l_p^2}{r_0^2 a_i^2}\,. \tag{10}$$

As already anticipated, a generic spherically symmetric metric can be recast in the following diagonal form:

$$ds^2 = -A(r,t)^2 dt^2 + \frac{\partial_r R(r,t)^2}{1 - K(r,t) r^2} dr^2 + R(r,t)^2 d\Omega_2^2\,. \tag{11}$$

At super-horizon scales, we can then expand each function in (11) in powers of $\sigma(t)$ [12,13]. However, it turns out that the zeroth order in σ is ambiguous, and we might instead write [14]

$$1 - K(r,t) r^2 = (1 - K_i(r) r^2)\tilde{K}(r,t)\,, \tag{12}$$

where $K_i(r)$ is the curvature K at an initial time $t = t_i$ where the perturbation is at super-horizon scales and $\tilde{K}(r,t) = 1 + \sum_{n=1}^{\infty} \tilde{K}_n(r)\sigma(t)^n$. For consistency, K should then be the next-to-leading order in gradient expansion; and indeed, we find that $K_i \sim \mathcal{O}(l_p^2/r_0^2)$ [14] when scales smaller than the horizon are cut off.

With this, at the leading order in σ (though fully nonlinearly) we finds that for the case of radiation we have [6,15]

$$\mathcal{C} \simeq \frac{2}{3} K_i(r) r^2 \,. \tag{13}$$

As can be seen, at super-horizon scales the compaction function is approximately constant. Thus, a black hole can only be formed when the typical scale of the perturbation re-enters the horizon. This typical scale, as we explain, is related to the size r_m in which the compaction function is maximal. Moreover, as it is found numerically [6], the co-moving location of this maximum does not change much up to the moment in which the hoop conjecture conditions are met. Thus, we shall henceforth approximate it as constant.

Our prescription for the black hole formation is then that a black hole eventually forms whenever $\mathcal{C}$ at its maximum is larger than a certain critical value $\mathcal{C}_c$, which we specify later on. The constant(s) $\mathcal{C}_c$ are called the "thresholds".

The curvature K_i is related to the initial co-moving over-density as follows [6]:

$$a_i^2 \delta\rho_i(r) \propto (K_i + \frac{r}{3} K_i') \tag{14}$$

where $f' \equiv \partial_r f$ for a given function f. At the maximum of the compaction function $r = r_m$, where we want to define a threshold for the black hole formation, we have

$$\mathcal{C}(r_m) \propto \delta\rho_i(r_m)\, r_m^2 \,. \tag{15}$$

Therefore, the threshold on the compaction function greatly differs from other previous prescriptions, which considered a "threshold" of the over-density amplitude at its peak $r \sim 0$. Here, we mean a radius small enough to be closer to the highest point of the over-density while being large enough to retain the super-horizon approximation.. In the presence of an over-density it is clear that $r_m \neq 0$, as $\mathcal{C} \geq 0$; thus, the relationship between $\delta\rho(0)$ and $\mathcal{C}(r_m)$ depends on the full profile of the over-density, and any statistics based on the distribution of $\delta\rho_i(0)$ are limited to very specific statistical realizations. We return to this point later on.

The initial spatial metric at time $t = t_i$ and at super-horizon scales is then

$$ds^2\Big|_{t=t_i} \simeq \frac{a_i^2\, dr^2}{1 - K_i(r) r^2} + a_i^2\, r^2\, d\Omega_2^2 \tag{16}$$

recalling that $R(t_i, r) = a_i\, r$. However, inflationary initial conditions are usually phrased using a different conformal form

$$ds^2\Big|_{t=t_i} \simeq a_i^2\, e^{2\zeta(\hat{r})} \left(d\hat{r}^2 + \hat{r}^2 d\Omega_2^2 \right) , \tag{17}$$

where ζ is the curvature perturbation in the synchronous gauge. For both expressions to be the same, it must be the case that both

$$r = e^{\zeta(\hat{r})} \hat{r} \quad \text{and} \quad \frac{dr}{\sqrt{1 - K_i(r) r^2}} = e^{\zeta(\hat{r})}\, d\hat{r}. \tag{18}$$

However, the expression on the left implies that $dr/d\hat{r} = e^{\zeta(\hat{r})}[1 + \hat{r}\,\partial_{\hat{r}}\zeta(\hat{r})]$; along with the expression on the right, this implies that

$$K_i(r) r^2 = 1 - \left[1 + \hat{r}\,\partial_{\hat{r}}\zeta(\hat{r})\right]^2 . \tag{19}$$

In the $\hat{r}$ coordinates, the maximum of the compaction function is at that $\hat{r} = \hat{r}_m$ where $\partial_{\hat{r}}(\hat{r}\partial_{\hat{r}}\zeta) = 0$; due to the regularity condition $K_i(r)r^2 < 1$, the amplitude of this maximum cannot exceed 2/3.

As for the over-density, it is obvious that the threshold $\mathcal{C}_c$ cannot be generically written in terms of the value of ζ at its central peak unless strong assumptions are made concerning the statistical realizations of ζ. Moreover, because the theory is invariant under shifts of ζ at super-horizon scales, its peak value only has a meaning by fixing it at certain scales. In contrast, the value of the compaction function at its maximum is an observable. Previous attempts at statistical descriptions of PBH formations have assumed that the spread of the curvature profile around the mean is negligible [16–18]. Following [19], we instead scan all possible compaction function shapes, defining the corresponding threshold for each realization.

2. Analytic Formula for the Threshold

As we have discussed earlier, a gravitational collapse occurs at positions where the maximum of the compaction function is over threshold. The threshold is that value of the compaction function for which a black hole of mass zero is eventually formed. Far away from the maximum the fluid is dispersed away, whereas close to the center the regularity of the curvature $K(r)$ requires the compaction function to decay as r^2. Therefore, what matters most for the threshold is the form of the compaction function around its maximum. Moreover, because the conditions to form a black hole are tied to the local gradient pressures, the threshold only approximately depends on the local shape of the compaction function. At the maximum ($r = r_m$), we can then fit the compaction function using its value $\mathcal{C}(r_m)$ and its normalized curvature $q \equiv \frac{-r_m^2 \mathcal{C}''(r_m)}{4\mathcal{C}(r_m)}$ [20]; see [21] for other equations of state.

In [20], the following fitting function was used:

$$\mathcal{C}_{fit}(r) = \mathcal{C}(r_m)\,\frac{r^2}{r_m^2} e^{\frac{1}{q}\left(1-\left[\frac{r}{r_m}\right]^{2q}\right)}\,, \tag{20}$$

where $\mathcal{C}(r_m)$, r_m, and q are calculated from the compaction function of the perturbation.

The average of the fitting function

$$\bar{\mathcal{C}}(r) = \frac{3}{r_m^3}\theta(r - r_m)\int_0^{r_m} \mathcal{C}_{fit}(r')r'^2\, dr' \tag{21}$$

would then be a fictitious top-hat compaction function which has a threshold for radiation equal to $\bar{\mathcal{C}}_c = \frac{2}{5}$ [9]. With this, by inverting (21) we can obtain the threshold for $\mathcal{C}(r_m)$ [20]:

$$\mathcal{C}_c(q) = \frac{4}{15} e^{-\frac{1}{q}} \frac{q^{1-\frac{5}{2q}}}{\Gamma(\frac{5}{2q}) - \Gamma(\frac{5}{2q}, \frac{1}{q})}\,. \tag{22}$$

Note that $\mathcal{C}_c \geq \frac{2}{5}$; this lower bound has been confirmed by numerical studies. Numerical simulations show that the critical value (22) and its dependence on q are accurate to within 2% [20].

If (22) is provided in terms of derivatives with respect to $\hat{r}$ rather than r, e.g., in [22], then

$$q = \frac{-\hat{r}_m^2 \partial_{\hat{r}}^2 \mathcal{C}(\hat{r}_m)}{4\mathcal{C}(\hat{r}_m)(1 - \frac{3}{2}\mathcal{C}(\hat{r}_m))}\,. \tag{23}$$

With this analytical formula for the threshold of PBH formation, we are now in a position to calculate the statistical abundance of PBHs in our universe.

3. The Statistics of Compaction Function

A scalar field (in particular, the inflaton) does not have a preferred direction; therefore, we expect isolated perturbations to be spherical. This symmetry is broken only by the interference of two or more nearby perturbations [8]. Thus, focusing on rare events, as in the case of PBHs formed during the radiation era, it is the case that spherical symmetry around a rare peak of a Gaussianly distributed amplitude is well preserved.

The observed cosmological curvature perturbations at the cosmic microwave background (CMB) scale are extremely close to (multivariate) Gaussian [23]. We assume that this is true on all scales. Nevertheless, it should be noted by the reader that for most of the inflationary evolution related to PBH formation, e.g., [24–26], this assumption may be too strong, e.g., [27–29].

In the Gaussian case, then, as discussed above, we have

$$\mathcal{C}(\hat{r},\vec{x}_0) = \frac{2}{3}\left[1 - \left(1 + \hat{r}\,\partial_{\hat{r}}\zeta(\hat{r},\vec{x}_0)\right)^2\right] , \tag{24}$$

where x_0 defines the *center* of the spherical distribution and $\hat{r} = \|\vec{x} - \vec{x}_0\|$ in the coordinates (17). Although the combination $\hat{r}\,\partial_{\hat{r}}\zeta(\hat{r},\vec{x}_0)$ is independent on the choice of radial units, $\hat{r}$ is not.Therefore, we fix the scale factor at matter–radiation equality $a_{eq} = 1$ such that $\hat{r}$ represents physical distances.

Before developing the full statistics in detail, it is interesting to note that even if ζ (and its derivatives) are multi-Gaussian distributed, $\mathcal{C}$ follows a non-centrally peaked multi-χ^2 distribution. Thus, rare configurations of $\mathcal{C}$ generically do not coincide with rare configurations of ζ.

PBHs are then distributed according to the multi-χ^2-statistics of $\mathcal{C}$ constrained under the following conditions:

There is a center: $\mathcal{C}(\hat{r},\vec{x})$ has a peak, and there exists a position $\vec{x}_0$ such that

$$\vec{\nabla}\mathcal{C}(\hat{r},\vec{x})\Big|_{\vec{x}=\vec{x}_0} = 0\,,\ \text{and}\,,\ \nabla^2\mathcal{C}(\hat{r},\vec{x})\Big|_{\vec{x}=\vec{x}_0} < 0\,. \tag{25}$$

This condition is studied in peaks theory [8]. At this level, we can already see that the statistics we are looking for are very different from the Press–Schechter/excursion set approach from [30,31], which is often used in the literature. Whereas peaks theory seeks to describe the point process which describes the positions around which collapse occurs, the Press–Schechter calculation does *not* describe a point process; it only aims at a statistical description of the mass fraction in bound objects. It returns biased (incorrect) estimates of this mass fraction, as it assumes that this can be done by considering the statistics of all positions in space [32] rather than the special subset of positions around which collapse occurs [33]. This point has been discussed extensively in the literature on halo formation during matter domination [32–34]; because essentially all of that discussion remains relevant during radiation domination, we do not repeat it here.)

$\mathcal{C}(\hat{r},\vec{x}_0)$ **has a maximum at** $\hat{r} = \hat{r}_m$: this condition is verified for

$$\partial_{\hat{r}}\mathcal{C}(\hat{r},\vec{x}_0)\Big|_{\hat{r}=\hat{r}_m} = 0\,,\text{and}\,,\ \partial^2_{\hat{r}}\mathcal{C}(\hat{r},\vec{x}_0)\Big|_{\hat{r}=\hat{r}_m} < 0\,. \tag{26}$$

This important condition appeared for the first time in the nonlinear statistics of [19], and is based on the excursion set peaks formalism from [34]. In our statistics, we calculate the probability that the compaction function has a maximum for a given scale r, thereby scanning all possible realizations of $\mathcal{C}$. Other proposed statistics [16–18] have instead considered only the maximum associated with the averaged profile. This difference is crucial. Using our procedure, we are able to associate a threshold for *each* statistical realization of the perturbation's shapes (or the above q variable) instead of considering only the statistics of $\mathcal{C}_c$ for a *fixed* (presumably average!) q.

$\mathcal{C}(\hat{r}_m, \vec{x}_0)$ **is over the threshold:** the condition in this case is

$$\mathcal{C}(\hat{r}_m, \vec{x}_0) \geq \mathcal{C}_c(q) \,. \tag{27}$$

Again, this condition crucially differs from those employed in [16–18], with respectively considered thresholds for the co-moving curvature and for the over-density at the peak (i.e., at $r = 0$).

3.1. The Statistical Variables and the Role of the Transfer Function

As we have already commented, a direct connection to inflation is obtained by considering $\zeta(\vec{x})$ as the main statistical variable. Strictly speaking, this is the curvature perturbation calculated in the past infinity as if radiation was always dominating the universe's evolution [6] (in the language of [6], in the limit of $t \to 0$). Practically, however, numerical simulations start at a finite time from the cosmological singularity. Thus, one would be tempted to consider the statistics of PBHs at the time t_{BH}, at which the maximum of the compaction function crosses the cosmological horizon (e.g., [22,35,36]), where nonlinear effects start to play a crucial role. The statistics would then be developed by hoping that the evolution of ζ, via the perturbative transfer function, would retain its invariant Gaussian nature up to the point of crossing the horizon. However, it is expected that the time evolution of curvature perturbations in the nonlinear regime would badly break our assumption of Gaussianity. For example, in the context of stochastic inflation, the nonlinear regime is no longer Markovian [11].

Even accepting the assumption of Gaussianity, because the perturbation at $t = t_{BH}$ is outside the regime of validity of gradient expansion the use of $\zeta(r_m, t_{BH})$ is *not consistent* with the quasi-homogeneous initial conditions of the Misner–Sharp system [14], which really use $\zeta(\vec{x}) \equiv \zeta(\vec{x}, t \to 0)$. Thus, using $\zeta(\vec{x}, t_{BH})$ would mean setting a different numerical problem from the one studied in the literature [6,9]. In turn, this choice would lead to a background-dependent threshold which cannot be obtained as a simple generalization of the one employed here (the threshold suggested in [22] tries to generalize the one in [20] using the transfer function in the definition of the compaction function maximum; however, as we have discussed in the text this is inconsistent).

Thus, we accept the small error caused by not being able to run the simulation from the infinite past, and consider the initial conditions for black hole formation at the leading order in gradient expansion. This error is smaller than the error we already accepted by using our analytical formula for the threshold. One can appreciate it by looking at the simulated Hamiltonian constraint deviation caused by the use of quasi-homogeneous conditions at a finite time [9].

Having defined the relevant variable, we are now equipped to study its statistical proprieties. In Fourier space, we have

$$\zeta(\vec{x}) = \int \frac{d^3k}{(2\pi)^3} \, e^{i\vec{k}\cdot\vec{x}} \zeta_k \,, \tag{28}$$

where the ζ_ks modes, assumed to be Gaussian, coincide with the Fourier modes of the curvature perturbations in synchronous gauge and at the leading order in gradient expansion. More precisely, the amplitudes follow a Rayleigh distribution and the phases a uniform distribution.

While the curvature perturbation is a function of $\vec{x}$, $\mathcal{C}$ is a function of $\hat{r}$ and $\vec{x}_0$. Thus, we need to specify the meaning of $\hat{r}\partial_{\hat{r}}\zeta(\hat{r}, \vec{x}_0)$; we can integrate the Laplacian of ζ over a ball B centered on $\vec{x}_0$, obtaining (henceforth, we drop the hats)

$$\int_B d^3x \, \nabla^2\zeta = 4\pi r^2 \partial_r \zeta(r, \vec{x}_0) \,, \tag{29}$$

where we have assumed spherical symmetry around $\vec{x}_0$ (having in mind a rare peak from inflation). The same relation can be written as

$$r\partial_r\zeta(r,\vec{x}_0) = \frac{1}{4\pi r}\int d^3x\nabla^2\zeta\,\theta(r - \|\vec{x}-\vec{x}_0\|)\,. \tag{30}$$

Using the Fourier decomposition (28), we obtain

$$r\partial_r\zeta(r,\vec{x}_0) = -\frac{1}{4\pi r}\int \frac{d^3k}{(2\pi)^3}\,k^2\zeta_k \int d^3x\, e^{i\vec{k}\cdot\vec{x}}\theta(r - \|\vec{x}-\vec{x}_0\|)\,. \tag{31}$$

For the second integral, we can use polar coordinates:

$$\int d^3x\, e^{i\vec{k}\cdot\vec{x}}\theta(r - \|\vec{x}-\vec{x}_0\|) = 2\pi e^{i\vec{k}\cdot\vec{x}_0}\int d\tilde{r}d\phi\, \tilde{r}^2 e^{ik\tilde{r}\cos\phi}\theta(r - \|\vec{x}-\vec{x}_0\|) = e^{i\vec{k}\cdot\vec{x}_0}\frac{4\pi}{k^3}(-kr\cos(kr)+\sin(kr))$$

allowing us to use the suggestive form

$$r\partial_r\zeta = \frac{1}{3}\int \frac{d^3k}{(2\pi)^3}e^{i\vec{k}\cdot\vec{x}_0}(kr)^2\zeta_k W_{\rm TH}(kr)\,, \tag{32}$$

where we have used the definition of the Fourier-transformed top-hat window function

$$W_{\rm TH}(kr) = 3\,\frac{\sin(kr) - kr\cos(kr)}{(kr)^3}\,. \tag{33}$$

We stress here that the top-hat window function *was not added by hand*, and is simply encoded in the variable $r\partial_r\zeta$. However, this appearance of $W_{\rm TH}$ ensures that sub-horizon modes of $r\partial_r\zeta$ (for r larger than the cosmological horizon size) are cut away, a necessary condition for the initial conditions of the Misner–Sharp equations. Evidently, then, in order to construct the compaction function we ought to be interested in the Laplacian of the curvature perturbation.

This Laplacian is related to the linear density perturbation used in previous statistics [16]. Hence, for greater intuitiveness we can define the new variable as

$$g_r(\vec{x}_0) \equiv -\frac{4}{3}r\partial_r\zeta = -\frac{4}{9}\int \frac{d^3k}{(2\pi)^3}e^{i\vec{k}\cdot\vec{x}_0}(kr)^2\zeta_k W_{\rm TH}(kr)\,. \tag{34}$$

Note that g_r is a linear combination of ζ_k, meaning that it similarly follows a Gaussian statistics.

Although the compaction function is *fully nonlinear*, for greater intuitiveness it should be noted that $g_r(\vec{x}_0)$ is the would-be smoothed *linear* over-density on a ball of radius r at super-horizon scales. This results from the Poisson equation of linear perturbations in an FRW Universe. It should be stressed here that this linear over-density is only an auxiliary (i.e., non-physical) function in the regime that we are interested in.

With $g_r(\vec{x}_0)$, we can now write the compaction function as

$$\mathcal{C}(r,\vec{x}_0) = g_r(\vec{x}_0)\Big(1 - \frac{3}{8}g_r(\vec{x}_0)\Big). \tag{35}$$

Finally, note that when we impose that the condition that $\vec{x}_0$ must be a maximum, the coordinates can be shifted to locally fix $\vec{x}_0 = \vec{0}$.

3.2. Statistical Conditions I: Existence of Isolated Peaks

In what follows, we search for a special center $\vec{x}_0 = \vec{x}_p$ which is a peak for the compaction function. Indeed, as already discussed, the first constraint to form a PBH is the existence of an isolated peak in the compaction function which defines the spatial center of the approximately spherically symmetric initial over-density (again, we are assuming rare peaks). Then, we need to restrict the statistical realizations of the compaction function to those that fulfill the constraints

$$\vec{\nabla}_{\vec{x}_0}\mathcal{C}(r,\vec{x}_0)=0 \qquad \text{and} \qquad \nabla^2_{\vec{x}_0}\mathcal{C}(r,\vec{x}_0)<0\,. \tag{36}$$

The first condition identifies positions which are extremes, while the second ensures that these extremes are local maxima, i.e., peaks.

The condition for an extreme of $\mathcal{C}$ is satisfied either where $\vec{\nabla}_{\vec{x}}g_r(\vec{x}_0)=0$ or where $g_r(\vec{x}_0)=4/3$; however, as the latter has probability zero and would produce a PBH of negligible mass, we discard it here. It can now be seen that the condition for the peak of the compaction function is the same condition for an extreme of the linear over-density, as considered earlier in [16]. This extreme is a maximum whenever $0<g_r(\vec{x}_0)<4/3$ and a minimum for $4/3<g_r(\vec{x}_0)<8/3$. The first region corresponds to the so-called type I collapse and the second to type II [37]. Although type II has yet to be thoroughly explored, it is believed that such peaks collapse rapidly [38]. However, because high peaks are extremely rare and type IIs are even rarer than type Is, we only consider type Is in what follows (i.e., $g_r(\vec{x}_0)\leq 4/3$). Moreover, because g_r must exceed 0.49 for PBH formation (corresponding to the limit $\mathcal{C}_c\geq 0.4$ [20]), the conditions for the existence of an (isolated) center of the compaction function are

$$0.49\lesssim g_r(\vec{x}_p)\leq\frac{4}{3}, \qquad \vec{\nabla}_{\vec{x}_p}g_r(\vec{x}_p)=0\,, \qquad \text{and} \qquad \nabla^2_{\vec{x}_p}g_r(\vec{x}_p)<0\,. \tag{37}$$

In a discretized sense (see [8]), the total number density of peaks is

$$n_{peaks}=\sum_{\vec{x}_p}\delta^{(3)}(\vec{x}-\vec{x}_p)\,; \tag{38}$$

the idea is to use the statistical variable g_r and its derivatives instead of the peak(s) position(s).

We can expand g_r around the maximum(s) ($\vec{x}_p$):

$$g_r(\vec{x})\simeq g_r(\vec{x}_p)+\frac{1}{2}\nabla_i\nabla_j g_r(\vec{x}_p)(x^i-x^i_p)(x^j-x^j_p)\,. \tag{39}$$

Defining $\vec{\eta}(\vec{x})\equiv\vec{\nabla}g_r(\vec{x})$ and $\chi_{ij}\equiv\nabla_i\nabla_j g_r(\vec{x_p})$, we have

$$n_{peaks}=\sum_{\vec{x}_p}\left|\det\chi_{ij}\right|\delta^{(3)}(\vec{\eta})\,, \tag{40}$$

where we have used the position around the peak $\eta_i(\vec{x})\simeq\sum_j\chi_{ij}(\vec{x}-\vec{x}_p)^j$.

The sum over peaks can now be replaced by the probability of having a peak in a position $\vec{x}_p$. As already discussed, we assume that ζ, and consequently g_r, follow multi-Gaussian distributions. By this we mean that ζ is the anti-Fourier transformation of the Gaussian random variables ζ_k. While the probability distribution we look for is a Gaussian on $g_r(\vec{x})$, as we have discussed, concerning the peaks it is enough to use second-order expansion g_r around the peak value and consider $p(g_r(\vec{x}))\to p(g_r(\vec{x}_p),\chi_{ij},\vec{\eta}(\vec{x}_p))$.

Statistically, vector, scalar, and tensor quantities can decouple; thus, $p(g_r,\chi_{ij},\vec{\eta})=p(g_r)p(\chi^T_{ij})p(\chi_r|g_r)p(\vec{\eta})$, where χ^T is the traceless part of the Hessian matrix of g_r. We additionally define $\chi_r\equiv -r^2\nabla^2 g_r(\vec{x}_p)$, which is the trace part of the Hessian. The latter is a scalar, and as such correlates with g_r. All of these probabilities obviously remain multi-Gaussians.

Then, the mean number of peaks is

$$n_{peaks}\to\int d\chi_{ij}d\vec{\eta}\;dg_r\left|\det\chi_{ij}\right|\delta^{(3)}(\vec{\eta})\theta(\chi_r)\;p(g_r)p(\chi^T_{ij})p(\chi_r|g_r)p(\vec{\eta})\,, \tag{41}$$

where the Dirac delta and the Heaviside theta define $\vec{x}_p$ as an extreme that is a maximum.

The Dirac delta constraint is easy to integrate:

$$\int d\vec{\eta}\, \delta^{(3)}(\vec{\eta})\, p(\vec{\eta}) \to \frac{1}{(2\pi)^{3/2}\sigma_\eta^3}, \tag{42}$$

where the third power is due to the fact that the distribution is three-dimensional and $\sigma_\eta \equiv \sqrt{\det(\langle \eta_i \eta_j \rangle)}$.

The integral in χ_{ij}^T is more involved. However, assuming an approximate rotation invariance (we consider high and rare peaks), the determinant of the Hessian simply provides the normalized trace value to the cube, i.e., $\left|\det \chi_{ij}\right| \sim \frac{\chi_r^3}{r^3}$, while the integral over $d\chi^T$ only provides the variance. All in all, then, we have

$$\int d\chi_{ij} \left|\det\chi_{ij}\right|\, p(\chi_{ij})\, \theta(\chi_r) \sim \int_0^\infty d\chi_r\, \frac{\sigma_\chi^3 \chi_r^3}{r^3}\, p(\chi_r|g_r). \tag{43}$$

The exact integration in χ^T can actually be done exactly, leading to

$$n_{peaks} = \int dg_r \int_0^\infty d\chi_r\, \frac{f(\chi_r/\sigma_\chi)}{(2\pi r_*^2)^{3/2}}\, p(g_r)\, p(\chi_r|g_r)\,, \tag{44}$$

where the explicit calculation to find $f(x)$ (from the integration in χ^T) can be found in appendix A of [8] The function is

$$f(x) = \frac{x^3 - 3x}{2}\left[\operatorname{erf}\left(\sqrt{\frac{5}{2}}x\right) + \operatorname{erf}\left(\sqrt{\frac{5}{2}}\frac{x}{2}\right)\right] + \sqrt{\frac{2}{5\pi}}\left[\left(\frac{31x^2}{4} + \frac{8}{5}\right)e^{-5x^2/8} + \left(\frac{x^2}{2} - \frac{8}{5}\right)e^{-5x^2/2}\right], \tag{45}$$

and, as anticipated, $f(x) \to x^3$ for $x \to \infty$ (i.e., large isolated peaks); finally, $r_* \equiv r\sigma_\eta/\sigma_\chi$.

We pause here for a moment. The integral in g_r has been intentionally left indefinite, for the reason that only a subset of peaks that are the maximum of the compaction function and over-threshold are related to PBH formation. Because we are looking for rare configurations, we assume the existence of only one maximum per peak, thereby discarding the cloud-in-cloud possibility. In this respect, a local maximum would be a global one. A new condition, that $r = r_m$ must be a maximum, should then be supplemented in the statistics of peaks, which is what we do in the next section.

3.3. Statistical Conditions II: Maximum of Compaction Function

The second constraint we must impose is the existence of a maximum for the compaction function. Here, note that 'maximum' refers to the variation of r, rather than the variation of the spatial position $\vec{x}$, meaning that when the spatial peak position $\vec{x_p}$ has been found the maximum can be related to the behavior of the derivatives of C with respect to r. Thus, what we really want to count is

$$\frac{dn_{peaks \cap max}}{dr} = \sum_{\vec{x}_p, r_m} \delta(\vec{x} - \vec{x}_p)\delta(r - r_m)\,, \tag{46}$$

where r_m denotes the compaction function maximum for each $\vec{x}_p$, where we have assumed that a peak and a maximum of the compaction function do not happen at more than one smoothing scale.

As before, for type I black holes the maxima of C coincide with those of $g_r(\vec{x}_0)$; hence, we can consider the following expansion around r_m:

$$\frac{dg_r}{dr} = \left.\frac{dg_r}{dr}\right|_{r_m} + \left.\frac{d^2 g_r}{dr^2}\right|_{r_m}(r - r_m) + \ldots = \left.\frac{d^2 g_r}{dr^2}\right|_{r_m}(r - r_m) + \ldots. \tag{47}$$

In what follows, it is useful to define the following dimensionless quantities related to expansion in r:

$$v_r \equiv r\,\frac{dg_r}{dr}, \qquad w_r \equiv -r^2\,\frac{d^2 g_r}{dr^2}, \qquad \text{and} \qquad \chi_r \equiv -r^2\nabla^2 g_r. \tag{48}$$

The curvature of the compaction function is actually related to the Hessian around the peak; indeed, we have

$$2g_r + w_r = -\frac{4}{9}\int \frac{d^3k}{(2\pi)^3} e^{i\vec{k}\cdot\vec{x}_0}(kr)^4 \zeta_k W_{\rm TH}(kr) \ = \chi_r. \tag{49}$$

This shows that for the variables associated with the 'infinite' past (i.e., χ_r, the (dimensionless) Laplacian of g_r, and w_r), the second derivative with respect to scale the r differs by $2g_r$, i.e., the curvatures with respect to position and scale (χ_r and w_r, respectively) differ by $2g_r$.

With the above definitions, we have

$$v_r = -w_{r_m}\,\frac{(r - r_m)}{r_m} + \ldots, \tag{50}$$

thus, we have

$$\delta(r - r_m) = \frac{w}{r_m}\delta(v_r) \sim \frac{w}{r_m}\delta(v_{r_m}) \tag{51}$$

whenever the Dirac delta is imposed. The density of these states is then obtained by replacing the following in (44):

$$p(g_r)\,dg_r \to p(g_{r_m}, v_{r_m}, w)\,dv_{r_m}\,dw\,dg_{r_m}\,. \tag{52}$$

By changing the sum in r_m into an integral, we finally obtain the number of peaks per unit volume having a maximum at some r (we now remove for simplicity all the indices r):

$$n_{peaks\,\cap\,max} = \int_{r_{min}}^{r_{max}} \frac{dr}{r}\int_0^\infty d\chi \int_0^\infty dw\,w \int dg\,\frac{f(\chi/\sigma_\chi)}{(2\pi r_*^2)^{3/2}}\,p(g, v=0, w)\,p(\chi|g, v=0, w)\,\theta(\chi) \tag{53}$$

where the integral over only positive values of w specifies that r_m is a maximum.

Because $\chi = 2g + w$, i.e., χ is completely determined by g and w, $p(\chi|g, v=0, w) \to \delta(\chi - (2g + w))$; thus, we finally obtain

$$n_{peaks\,\cap\,max} = \int_{r_{min}}^{r_{max}} \frac{dr}{r}\int_0^\infty dw\,w \int dg\,\frac{f(\chi/\sigma_\chi)}{(2\pi r_*^2)^{3/2}}\,p(g, v=0, w)\,, \tag{54}$$

where one should read $\chi = 2g + w$. Note that because both w and g are positive we do not need to add an extra constraint for the maximum in $\vec{x}_p$ (i.e., $\theta(\chi_r)$).

The integral in r requires further explanation. First of all, the minimal radius should be larger than the Hubble radius at the initial time in order for the gradient expansion to be valid. Second, because of Hawking evaporation, black holes less massive than $\approx 10^{15}\,gr$ will have completely evaporated by now, and should not be counted [1]. Thus (replacing Planck units $M_p \simeq 2\times 10^{-5} gr$), because $M_\bullet \sim M_p^2 H_r^{-1} = ra(t_r)M_p^2$, we fix

$$r_{min} = \mathrm{MAX}\left[\left(\frac{10^{12}\,gr}{M_p^2}\right)^{3/4} H_{eq}^{-1/4}, (H(t_i)a(t_i))^{-1}\right], \tag{55}$$

where we have used $a_{eq} = 1$ and H_{eq} is the Hubble scale at matter equality. Whether the Hubble radius at initial conditions or the Hawking evaporation limit ought to be considered as the minimal radius depends on the specific inflationary model under investigation.

Similarly, the maximal scale we are interested in here is the Horizon size at matter–radiation equality:

$$r_{max} = H_{eq}^{-1} \,. \tag{56}$$

Finally, note that the integral in r is really the integral of all configurations that have a maximum for the compaction function for a given smoothing scale r; hence, we remove the sub-index m from now on.

3.4. Statistics Condition III: Being over the Threshold

We are now finally in a position to implement the over-threshold condition for the integral in g_r. Converse to the case of finding a local maximum for the compaction function, in general the threshold value is nonlocal, i.e., it depends on the profile realization of g_r at all smoothing scales. Thus, the number of peaks we are looking for is the subset of $n_{peaks \cap max}$ such that for any possible smoothing scale configuration *with the same peak position and same* r_m the corresponding g_{r_m} is over-threshold. Obviously, this would lead to an untreatable computational problem.

To bypass this issue, previous approaches have considered only the mean g_r profile [16] (here, we remind the reader that g_r happens to equal the would be linear over-density [19]), and have consequently associated r_m with the location of the compaction function maximum when the mean profile is used. However, unless all possible realizations of g_r for any smoothing scale r have negligible spread from $\langle g_r \rangle$, this peak counting will be grossly wrong (similar arguments have been made for the ζ statistics in [17,18]). Here, we instead adopt the more refined argument already outlined in the introduction.

Although it is true that the threshold at the maximum of the compaction function depends upon the full g_r profile [6], to a very good approximation the threshold only depends on the curvature of the compaction function around a maximum r_m [20]. Thus, we can simply consider the ensemble of all possible compaction function curvatures for any given smoothing scale $r = r_m$, then associate a threshold with each of these.

The realization of this condition requires some algebra; in the region $g \leq \frac{4}{3}$ and at the maximum, we have

$$-r^2\partial_r^2 C\Big|_{r=r_m} = -r^2\partial_r^2 g_r\left(1-\frac{3}{4}g_r\right)\Big|_{r=r_m} = w\sqrt{1-\frac{3}{2}C_c} \,. \tag{57}$$

leading to

$$q = \frac{w}{4C_c(q)\sqrt{1-\frac{3}{2}C_c(q)}} \,, \tag{58}$$

and implying that $C_c(q) = C_c(q(w)) = C_c(w)$. On the other hand, the threshold in terms of g is

$$g_c(w) = \frac{4}{3}\left(1-\sqrt{1-\frac{3}{2}C_c(w)}\right) \tag{59}$$

Thus, the number density of peaks that would eventually collapse into Primordial Black Holes is

$$n_\bullet = \int_{r_{min}}^{r_{max}} \frac{dr}{r} \int_0^\infty dw\, w \int_{g_c(w)}^{\frac{4}{3}} dg \frac{f(\chi/\sigma_\chi)}{(2\pi r_*^2)^{3/2}}\, p(g, v=0, w) \,. \tag{60}$$

3.5. Energy Density in PBHs

Super-horizon over-threshold perturbations that enter the horizon at the time $ra(t_r) = H(t_r)^{-1}$ quickly collapse into black holes (here, we assume instantaneous formation) [6].

Up to a deviation of about $\Delta\mathcal{C} \equiv \mathcal{C} - \mathcal{C}_c \sim 10^{-2}$ from the threshold value [9], the mass is distributed according to the following scaling law [39]:

$$M_\bullet = \frac{\mathcal{K}}{2H_r}\left[\mathcal{C}(g) - \mathcal{C}_c(w)\right]^{0.36}, \tag{61}$$

where $H_r \equiv H(t_r)$. The function $\mathcal{K}$ is of $\mathcal{O}(1)$, and depends on the specific curvature profile chosen; thus, it depends slightly on r. Nevertheless, we keep it constant in our estimation of abundance, accepting the related error $\mathcal{O}(1)$. To fix its value, we take the typical value $\mathcal{K} \simeq 6$ from [9].

For $\Delta\mathcal{C} \geq 10^{-2}$, the mass distribution starts to deviate from the scaling law with a maximum deviation of $\sim 15\%$ [40]. Because the error on using (61) is on a similar order as that emerging from fixing $\mathcal{K}$, we keep it and retain the scaling law throughout.

Because we are interested in the energy density of PBHs at matter–radiation equality, we need to be careful about the evolution of the number density. Thus far, we have calculated the number density per co-moving volume. When black holes are formed, as we have already mentioned in the introduction, their energy density simply dilutes as dust, i.e., as the inverse of the volume expansion. At super-horizon scales and the leading order in gradient expansion, the power spectrum is constant; thus, peak positions in co-moving volumes do not change, i.e., at the leading order there are no intrinsic velocities between peaks. Obviously, at the next-to-leading order in gradient expansion this would not exactly be the case. While we do not consider this subtlety further here, we do mention that it would be incorrect to try to capture the intrinsic velocities between peaks by considering a linear transfer function in the statistics, for similar reasons as those outlined before in the case of threshold definition. More specifically, the transfer function capturing the time evolution of the power spectrum at linear order (even assuming that Gaussianity is not broken at next-to-leading order in gradients), is *next* to the leading order in gradient expansion. Therefore, considering a transfer function in the statistical correlators would imply the necessity of considering both new nonlinear thresholds and new nonlinear initial conditions. What should be done instead is to calculate the initial number density in co-moving volumes as done here, then simulate the velocity dispersion of the initial peaks at later times. We are not aware of any of such numerical simulation to date; thus, we accept the small related error in order to consider only the leading order in gradient expansion.

Having made these remarks, the total energy density of PBHs at matter–radiation equality is

$$\frac{d^2\rho_\bullet}{dr dM_\bullet} \sim \sum_{\vec{x}_0, r_m, M_\bullet, \mathcal{C}(r_m) > \mathcal{C}_c} \delta(\vec{x} - \vec{x}_0)\delta(r - r_m)\, \delta\left(M_\bullet - \frac{\mathcal{K}}{2H_r}\left[\mathcal{C}(g) - \mathcal{C}_c(w)\right]^{0.36}\right) M_\bullet \tag{62}$$

or

$$\rho_\bullet = \int_{r_{min}}^{r_{max}} \frac{dr}{r} \int_0^\infty dw\, w \int_{g_c(w)}^{\frac{4}{3}} dg \frac{f(\chi/\sigma_\chi)}{(2\pi r_*^2)^{3/2}} p(g, v = 0, w) \frac{\mathcal{K}}{2H_r}\left[\mathcal{C}(g) - \mathcal{C}_c(w)\right]^{0.36}, \tag{63}$$

where we have again set $a_{eq} = 1$. We can now define the fractional density of PBHs at equality, $\beta_\bullet \equiv \frac{\rho_\bullet}{\rho_{eq}}$, as follows:

$$\begin{aligned}
\beta_\bullet &= \frac{8\pi}{3H_{eq}^2} \int_{r_{min}}^{r_{max}} \frac{dr}{r} \int_0^\infty dw\, w \int_{g_c(w)}^{\frac{4}{3}} dg \frac{f(\chi/\sigma_\chi)}{(2\pi r_*^2)^{3/2}} p(g, v = 0, w) \frac{\mathcal{K}}{2H_r}\left[\mathcal{C}(g) - \mathcal{C}_c(w)\right]^{0.36} \\
&= \frac{4\pi\mathcal{K}}{3(H_{eq} r_{eq})^3} \int_{r_{min}}^{r_{max}} \frac{dr}{r} \frac{(r/r_{eq})^2}{(r/r_{eq})^3} \int_0^\infty dw\, w \int_{g_c(w)}^{\frac{4}{3}} dg \frac{f(\chi/\sigma_\chi)}{[2\pi(r_*/r)^2]^{3/2}} p(g, v = 0, w) \left[\mathcal{C}(g) - \mathcal{C}_c(w)\right]^{0.36}.
\end{aligned} \tag{64}$$

recalling that $H_{eq} r_{eq} = a_{eq}^{-1}$ and $a_{eq} = 1$ in our units.

The final expression makes it easy to see that the abundance of PBHs with respect to radiation grows with the scale factor from formation. If we ignore the scaling solution and only focus on order of magnitudes, i.e., we consider that the PBH mass is provided simply by the mass within the horizon at formation, then we can ignore the integrals over w and g in the mass while retaining the dependence on r. Then, because $\mathcal{K}/(2H_r) \propto r^2$ and $(2\pi r_*^2)^{3/2} \propto r^3$, we find that we obtain $r_{\rm eq}/r = a_{eq}/a(t_r)$ for each r. This is the scaling we should expect; the integrals over w and g serve to return the actual fraction of patches which produce PBHs, and the fact that they depend on r shows how the energy density that is stored in PBHs builds up over time, i.e., as r increases.

3.6. Mass Spectrum of PBHs

The mass distribution per proper volume (known as the 'mass function') is

$$\frac{dn_\bullet(M_\bullet)}{dM_\bullet} \sim \sum_{\vec{x}_0, r_m, \mathcal{C}(r_m)>\mathcal{C}_c} \delta(\vec{x}-\vec{x}_0)\delta(r-r_m)\,\delta\left(M_\bullet - \frac{\mathcal{K}}{2H_r}\left[\mathcal{C}(g)-\mathcal{C}_c(w)\right]^{0.36}\right), \tag{65}$$

leading to

$$\frac{dn(M_\bullet)}{dM_\bullet} = \int_{r_{\rm min}}^{r_{\rm max}} \frac{dr}{r} \int_0^\infty dw\, w \int_{g_c(w)}^{4/3} dg\, \frac{p(g,w,v=0)\, f(x)}{(2\pi r_*^2)^{3/2}}\, \delta_{\rm D}\left(M_\bullet - \mathcal{K}M_{\rm H}[C(g)-C_c(w)]^{0.36}\right), \tag{66}$$

where $g_c(w) = (4/3)\,[1-\sqrt{1-3C_c(w)/2}]$ and we use $x \equiv \frac{\mathcal{X}}{\sigma_\mathcal{X}}$.

We can write the previous expression in a different form:

$$\frac{dn(M_\bullet)}{dM_\bullet} = \int \frac{dr}{r} \int_0^\infty \frac{dw}{w}\, \frac{d^3n(M_\bullet, r, w, v=0)}{d\ln r\, d\ln w\, dM_\bullet}, \tag{67}$$

where

$$\frac{d^3n(M_\bullet, r, w, v=0)}{d\ln r\, d\ln w\, dM_\bullet} \equiv \int_{g_c(w)}^{4/3} dg\, \frac{w\, p(g,w,v=0)\, f(x)}{(2\pi r_*^2)^{3/2}}\, \delta_{\rm D}\left(M_\bullet - \mathcal{K}M_{\rm H}[C(g)-C_c(w)]^{0.36}\right). \tag{68}$$

The delta function suggests that we define

$$M_g \equiv \mathcal{K}M_{\rm H}[C(g)-C_c(w)]^{0.36} = \mathcal{K}M_{\rm H}(2/3)^{0.36}[1-\tilde{C}_c(w)-(1-\tilde{g})^2]^{0.36}, \tag{69}$$

where we have additionally defined $\tilde{g} \equiv g/(4/3)$ and $\tilde{C} \equiv C/(2/3)$ (i.e., both quantities are defined in units of their maximum possible value) and $M_H \equiv 1/(2H_r)$. Note that $M_g = 0$ when $g = g_c(w)$ and $M_g = [1-\tilde{C}_c(w)]^{0.36}$ when $g = 4/3$. Inverting, we can write g as a function of M_g, which yields

$$g = \frac{4}{3}\left[1 - \sqrt{1-\tilde{C}_c(w) - \frac{3}{2}\left(\frac{M_g}{\mathcal{K}M_{\rm H}}\right)^{1/0.36}}\right], \tag{70}$$

thus,

$$\left.\frac{dg}{dM_g}\right|_{w\ \rm fixed} = \frac{2}{3}\frac{(3/2)(M_g/\mathcal{K}M_{\rm H})^{1/0.36}/0.36M_g}{\sqrt{1-\tilde{C}_c(w)-\frac{3}{2}\left(\frac{M_g}{\mathcal{K}M_{\rm H}}\right)^{1/0.36}}} = \frac{(M_g/\mathcal{K}M_{\rm H})^{1/0.36}/0.36M_g}{1-\tilde{g}}. \tag{71}$$

As a result, instead of the original integral in dg at fixed w in (68), we can consider an integral in dM_g. By defining $g_\bullet$ such that $M_g = M_\bullet$, we obtain

$$\frac{d^3n(M_\bullet, r, w, v=0)}{d\ln r\, d\ln w\, d\ln M_\bullet} = \frac{(M_\bullet/\mathcal{K}M_{\rm H})^{1/0.36}/0.36}{\sqrt{1-\frac{3C_c(w)}{2}-\frac{3}{2}\left(\frac{M_\bullet}{\mathcal{K}M_{\rm H}}\right)^{1/0.36}}}\,\frac{w\, p(g_\bullet, w, v=0)\, f(x_\bullet)}{(2\pi r_*^2)^{3/2}} \tag{72}$$

recalling that for the type I black holes the term in the square root is always positive.

The r-dependence in this expression arises from the r-dependence of the various correlators (which we describe below) and from $M_{\rm H} = M_{\rm eq}\,(r/r_{\rm eq})^2$. This shows that the power law $(M_\bullet/\mathcal{K}M_{\rm H})^{1/0.36}$, which derives from the scaling solution, is a generic prediction of our approach (see a similar result for the case of bubbles formation [41], where the power law in the mass spectrum similarly results from a Jacobian). The question, then, is whether the other terms and the subsequent integrations over w and r modify this power law. Before we answer this question, we should stress again that our approach in Equation (66) explicitly integrates over PBH profile shapes (parametrized by w) and PBH formation times (parametrized by r); it does *not* assume that either of these distributions are sharply peaked.

For notational convenience, it is convenient to define

$$N_{rw}(M_\bullet) \equiv \frac{d^3n(M_\bullet, r, w, v=0)}{d\ln r\, d\ln w\, d\ln M_\bullet}, \qquad N_r(M_\bullet) \equiv \frac{d^2n(M_\bullet, r, v=0)}{d\ln r\, d\ln M_\bullet} \qquad \text{and} \qquad N(M_\bullet) \equiv \frac{dn(M_\bullet, v=0)}{d\ln M_\bullet}, \tag{73}$$

where the second expression is the result of integrating the first over w, the final expression is the result of additionally integrating over r, and it is understood that we always have $v = 0$. Thus,

$$N(M_\bullet) = \int_{r_{\rm min}}^{r_{\rm max}} \frac{dr}{r}\, N_r(M_\bullet) = \int_{r_{\rm min}}^{r_{\rm max}} \frac{dr}{r} \int_0^\infty \frac{dw}{w}\, N_{rw}(M_\bullet). \tag{74}$$

If we think of $N(M_\bullet)$ as the final PBH 'mass function' (i.e., at equality), then it is the sum of all PBHs formed at earlier times as indexed by r, $N_r(M_\bullet)$; thus, at any fixed time we can have a range of compaction function 'shapes' indexed by w, with $N_{rw}(M_\bullet)$ being the mass function at fixed r and w.

3.7. The Probability Distribution

Now that we have implemented the constraints, we need to specify the probability distribution $p(g, v=0, w)$. For this, it is useful to use the identities

$$p(g, v=0, w) = p(v=0)\, p(g|v=0)\, p(w|g, v=0) = p(v=0)\, p(w|v=0)\, p(g|w, v=0), \tag{75}$$

where, e.g., $p(w|g, v=0)$ is the conditional probability of having w given g and $v=0$. Because the ζ_k are Gaussian by assumption, all the other variables are as well; thus, we only need to consider two-point correlators. Defining

$$\sigma_j^2(r) = \frac{16}{81}\int \frac{dk}{k}\,(kr)^4\, \mathcal{P}_\zeta(k)\, W_{\rm TH}^2(kr)\,(kr)^{2j} \qquad \text{where} \qquad \mathcal{P}_\zeta(k) = \frac{k^3\,\langle\zeta_k\zeta_{k'}\rangle}{2\pi^2}\,\delta^{(3)}(k+k'), \tag{76}$$

we have

$$\begin{aligned}
&\sigma_g^2 \equiv \langle g^2\rangle = \sigma_0^2, \qquad \langle\eta_i\eta_j\rangle = \frac{\sigma_1^2}{3}\delta_{ij}, \qquad \sigma_\chi^2 \equiv \langle\chi^2\rangle = \sigma_2^2\,, \qquad \sigma_w^2 \equiv \langle w^2\rangle = \sigma_2^2 - 4\sigma_1^2 + 4\sigma_0^2, \\
&\sigma_v^2 \equiv \langle v^2\rangle = \frac{d\langle gv\rangle}{d\ln r} - \langle gv\rangle + \sigma_1^2 - 2\sigma_0^2, \qquad \text{where} \qquad \langle gv\rangle = \frac{1}{2}\frac{d\sigma_0^2}{d\ln r}
\end{aligned} \tag{77}$$

and

$$\begin{aligned}
&\langle g\chi\rangle = \sigma_1^2\,, \qquad \langle gw\rangle = \sigma_1^2 - 2\sigma_0^2\,, \qquad \langle w\chi\rangle = \sigma_2^2 - 2\sigma_1^2\,, \\
&\langle v\chi\rangle = \frac{1}{2}\frac{d\sigma_1^2}{d\ln r} - \sigma_1^2\,, \qquad \langle vw\rangle = \langle v\chi\rangle - 2\langle vg\rangle.
\end{aligned} \tag{78}$$

With these correlators in hand, we can now write

$$p(w, v = 0) = \frac{e^{-\frac{w^2}{2\sigma_w^2(1-\gamma_{wv}^2)}}}{\sqrt{4\pi^2\sigma_v^2\sigma_w^2(1-\gamma_{wv}^2)}} \quad \text{and} \quad p(g|w, v = 0) = \frac{e^{\frac{-(g-\langle g|v=0,w\rangle)^2}{2\Sigma^2_{g|vw}}}}{\sqrt{2\pi\Sigma^2_{g|vw}}}, \tag{79}$$

where we have used the fact that $p(v = 0) = 1/\sqrt{2\pi\sigma_v^2}$ and we have

$$\langle g|v = 0, w\rangle = \sigma_g \frac{\gamma_{wg} - \gamma_{wv}\gamma_{vg}}{1 - \gamma_{wv}^2} \frac{w}{\sigma_w} \tag{80}$$

and

$$\Sigma^2_{g|wv} = \sigma_g^2 \frac{1 - \gamma_{gv}^2 - \gamma_{wv}^2 - \gamma_{wg}^2 + 2\gamma_{gv}\gamma_{wv}\gamma_{wg}}{1 - \gamma_{wv}^2}. \tag{81}$$

Note that the probability $p(g|w, v = 0)$ is *not* centered on $g = 0$ due to its conditional nature; furthermore, we have used the normalized (Pearson) correlation coefficient $\gamma_{ab} \equiv \langle ab\rangle/\sigma_a\sigma_b$. Unless $\mathcal{P}_\zeta$ is a power law, these additionally depend on r.

Before we consider explicit examples, because $p(g|w, v = 0)$ is a Gaussian we expect the mass function $N_{rw}(M_\bullet)$ to be a power law in $M_\bullet$ times a Gaussian cutoff. Note that this cutoff is not simply $\exp[-(M_\bullet/M_f)^{1/0.36}]$, as suggested in [42] by the use of Press–Schechter formalism, both because $g_\bullet$ is a more complicated function of $M_\bullet$ (see Equation (70)) and because $p(g_\bullet|w, v = 0)$ is not centered on zero. This shows once more that the Press–Schechter formalism is not adequate for PBHs.

4. Illustrative Examples

We now consider a number of examples that demonstrate the implications of our approach. It is useful in the following to note that $\sigma_j \propto r^j$ for sufficiently large r; this is a consequence of the built-in top-hat filter along with the fact that we do not employ an additional transfer function when computing statistics; for further discussion, see [19].

4.1. Sharp Feature

Previous work assumes that

$$\mathcal{P}_{\text{sharp}}(k) = A_s\, k_p\, \delta(k - k_p) \tag{82}$$

produces a PBH with a well defined mass. In this case, the idea is to set k_p such as to produce approximately asteroid-mass objects while setting A_s by requiring that their abundance accounts for all the Dark Matter.

However, upon setting $\kappa \equiv k_p r$ we have

$$\sigma_j^2(r) = A_s\,\kappa^4\, W^2(\kappa)\,\kappa^{2j} \qquad \sigma_w^2(r) = (\kappa^2 - 2)^2\,\sigma_0^2 \qquad \sigma_v^2(r) = A_s\,\kappa^4\left(3j_0(\kappa) - W(\kappa)\right)^2 \tag{83}$$

and

$$\langle gw\rangle = \sigma_0\sigma_w \qquad \langle wv\rangle = \sigma_w\sigma_v \qquad \langle gv\rangle = \sigma_0\sigma_v. \tag{84}$$

As a result, $\gamma_{gv} = \gamma_{wv} = \gamma_{gw} = 1$, suggesting that g, v, and w only differ from one another by multiplicative factors. Hence, as we are interested in $v = 0$, the others are peaked to zero; however, $g = 0$ results in no PBHs at all!

While this power spectrum relates the linear and nonlinear statistics (see [19]), in the nonlinear case the contribution of PBHs from a very peaked power spectrum is exponentially suppressed. This does not happen for statistics using the mean profiles [16–18], as the constraint of having a maximum in the compaction function ($v = 0$) does not enter into the statistics.

4.2. Broad Feature

Next, we consider models that have power over a broad range of scales before cutting off exponentially at $kr_p \gg 1$:

$$\mathcal{P}_{\rm broad}(k) = A_s\,(k\,r_p)^p\,\exp(-k\,r_p). \tag{85}$$

On physical grounds, we expect $p \leq 4$ [43,44]. Here, A_s and r_p are a characteristic amplitude and scale, respectively.

The appearance of r_p means that it is more convenient to express our results in terms of dimensionless quantities, e.g., because of the factor of r_{eq}/r inside the integral in the second part of Equation (64), it is better to work with

$$\tilde{\beta} \equiv (r_p/\mathcal{K}r_{eq})\,\beta_\bullet \tag{86}$$

rather than $\beta_\bullet$ (i.e., the mass fraction at equality) itself. Likewise, it is better to work with scaled number densities

$$\tilde{N}(M_\bullet) \equiv \frac{4\pi\,r_p^3}{3}\,N(M_\bullet) = \frac{N(M_\bullet)}{\rho_{eq}/\mathcal{K}M_p}\,\frac{r_p}{\mathcal{K}r_{eq}}, \tag{87}$$

where, the first equality is dimensionless (a volume times a number density) and the final expression shows that $\tilde{N}$ scales the number density by $(\rho_{eq}/\mathcal{K}m_p)$, meaning that the remaining factors are the same as those which arise when defining $\tilde{\beta}_\bullet$.

4.2.1. Mass Function

Figure 1 shows the dimensionless scaled mass function $\tilde{N}$ when $p = 0$. The two panels show different values of A_s for which $\sigma_0 = 0.33$ (left) and 0.6 (right). In each panel, the thick red curve shows $\tilde{N}(M_\bullet)$; this is a power law at low masses (the magenta curve shows a power law of the expected slope, 1/0.36) which is truncated exponentially at larger masses. Moreover, $\tilde{N}(M_\bullet)$ is built up over time by summing over many $\tilde{N}_r(M_\bullet)$; the solid curves show these for a few choices of r (with a larger r extending to larger $M_\bullet$).

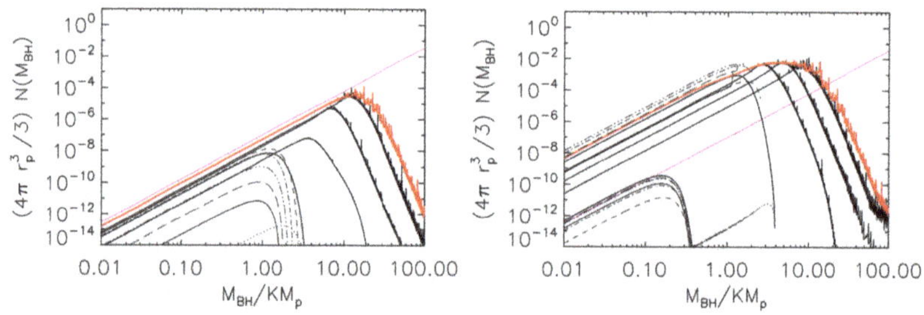

Figure 1. Distribution of $M_\bullet$ when $\mathcal{P}(k)$ has slope $p = 0$ and amplitude A_s such that $\sigma_0 = 0.33$ (**left**) and $\sigma_0 = 0.6$ (**right**). The thick red curve in each panel shows the scaled mass function $\tilde{N}(M_\bullet)$ (Equation (87)), which results from summing over $\tilde{N}_r(M_\bullet)$ distributions, a selection of which are shown as thick black curves (with a larger r extending to larger $M_\bullet$). Each of these results from summing over different $\tilde{N}_{rw}(M_\bullet)$ (cf. Equation (74)), which we show for a few representative values of w (dotted, short-dashed, dot-dashed, dot-dot-dot-dashed, long dashed). At small r, increasing w decreases the maximum mass, while at larger r the mass function is a power law with a small divergence at the largest allowed masses; the amplitude of this power law depends on w and qualitatively traces the distribution of w, i.e., it is small at both small and large values of w. The magenta line shows a power law of slope 1/0.36, which we argue in the text to be a good approximation at small $M_\bullet$.

At any r, the shape of N_r is a power law at low masses with an exponential truncation at larger masses. The amplitude of the power law part is clearly not monotonic in r, as neither a very small nor a very large r contributes to the final $N(M_\bullet)$. Because r is related to the time of PBH formation, this explicitly shows that PBHs form over a range of times and that at any given time they can form with a wide spectrum of masses. Moreover, while there is a preferred formation time for N_r as r changes, e.g., at the r for which the amplitude of the power law is greatest, the peak mass at this time is not the same as the peak mass of the red curve ($N(M_\bullet)$). Thus, we cannot assume either a delta function in PBH formation times or an equivalent dominance of the mean profile for the compaction function.

Each $N_r(M_\bullet)$ distribution, i.e., the mass function of black holes formed at a given time, is built from summing over different $N_{rw}(M_\bullet)$ distributions, i.e., the mass function of black holes formed at a given time and with a given compaction function curvature. The thin curves of different styles in Figure 1 show $N_{rw}(M_\bullet)$ for a few choices of r and w. The cutoff properties of $N_{rw}(M_\bullet)$ can be understood as arising from the interplay between the pole in $dg_\bullet/d\ln M_\bullet$ (Equation (72)) and the exponential suppression from $p(g_\bullet)$, with both acting to modify what would otherwise be a power law of slope $1/0.36 = 2.78$. In particular, here it can be seen explicitly that the exponential cutoff arises from the fact that for a given r and w there is a maximum possible mass which is set by requiring $g_\bullet \leq 4/3$; note that this is slightly more stringent than just saying that the mass cannot exceed that within the horizon. For a given A_s, this maximum is larger if r is larger and smaller if w is larger, which is because $C_c(w)$ increases as w increases. Finally, just as we cannot assume a delta function in r, we cannot assume a delta function in w either. This once again shows that the use of a mean profile would provide an incorrect result.

To illustrate how the shape rather than the amplitude of the power spectrum affects the predictions, Figure 2 shows a similar analysis of the case in which $p = 2$. The mass spectrum clearly peaks at a lower $M_\bullet/M_p$ than before, and the range of r that contributes significantly is smaller than before as well; however, the qualitative trends remain: a power law of slope $1/0.36$ at lower masses is truncated at larger masses, at first gradually and then exponentially.

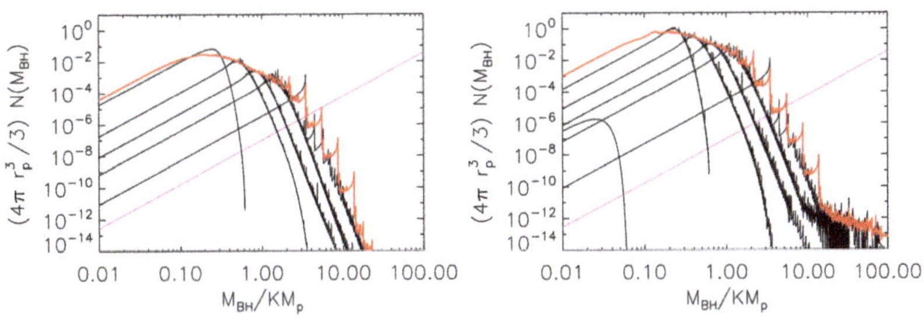

Figure 2. Same as in the previous figure, except now with $\mathcal{P}(k)$ having a slope $p = 2$ and without any N_{rw} curves. The right-hand panel has $\sigma_0 = 0.5$, while the left hand panel has the same σ_0 as in Figure 1; however, the value of A_s is similar to that in the right hand panel of Figure 1. For the same A_s, the mass spectrum peaks at smaller $M_\bullet$ compared to $p = 0$, while the power law slope at low masses (magenta) is the same.

4.2.2. Dependence of the Abundance on the Power Spectrum

Figure 3 shows how the predicted abundance (Equation (64)) for a given maximal scale r_{max} depends on $\mathcal{P}(k)$. The two panels are for two different $\mathcal{P}(k)$, and the different curves in each panel are for different amplitudes A_s. Because the horizontal axis is a proxy for the formation time, the flattening of the curves at $r \gg r_p$ is another indication that PBHs do not form at very late times. The steepness of these curves, i.e., how rapidly they rise to their asymptotic value, is a measure of the narrowness of the distribution of PBH formation times. Evidently, in the panel on the right this distribution is rather narrow, whereas in the panel on the left it is broader. In both panels, a smaller amplitude of $\mathcal{P}(k)$ results in a broader range of formation times.

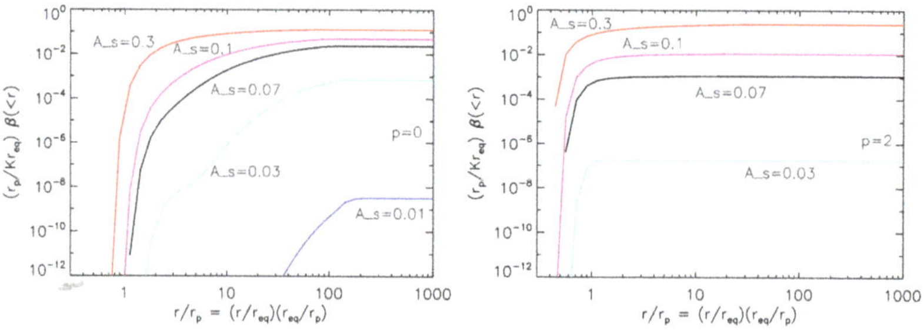

Figure 3. Dependence of scaled mass fraction at equality (Equation (86)), that is, in PBHs which formed before r; the panels show results for $\mathcal{P}(k)$ having $p = 0$ (**left**) and $p = 2$ (**right**); the different curves in each panel are for different amplitudes A_s (as labeled). In both cases, this quantity is smaller if A_s is smaller. For $p = 2$, most PBHs form when r is slightly smaller than r_p; for $n = 0$ the range of r/r_p is broader, especially for small amplitudes.

As expected, it can be observed that the abundances related to very small power spectrum amplitudes are exponentially suppressed. This is due to the fact that essentially all variances decrease with A_s. Thus, for a very small power spectrum, in order to have $\beta_\bullet \sim 1$ we would need r_p to be much smaller than r_{eq}. On the contrary, a relatively large amplitude is related to a power spectrum peak position closer to the scale of matter–radiation equality.

Finally, another power spectrum shape that has appeared in the literature is the log-normal one, which sets

$$\mathcal{P}_{\rm LN}(k) = A_s\, {\rm e}^{-[\ln(k/k_p)]^2/2\sigma_{\rm LN}^2}. \tag{88}$$

Here, $\sigma_{\rm LN}$ determines the width of the feature, which is centered on k_p. For small $\sigma_{\rm LN}$, σ_j are not monotonic functions of r (see Figure 4); this can be understood by considering the delta function limit discussed in the sharp spectrum section and noting that $W(kr)$ oscillates (and strongly!). At larger r, the σ_j in this model is asymptotic with respect to the $\sigma_j \propto r^j$ scaling noted earlier. In this respect, the model does not provide any new insights compared to what we have already studied in this section, and we do not consider it further.

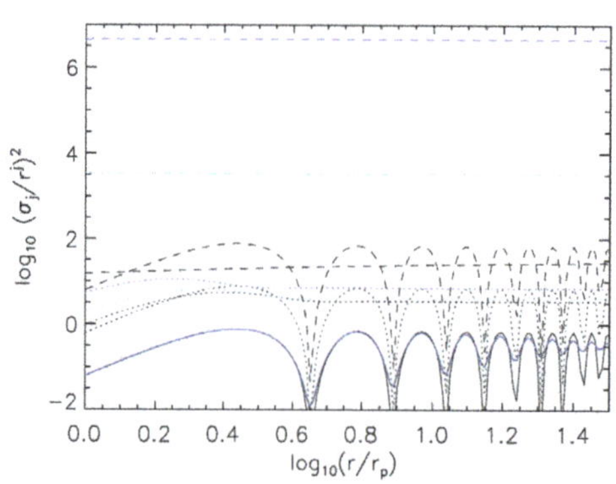

Figure 4. Value of $(\sigma_j/r^j)^2$ in the Log-normal model for $\sigma_{\rm LN} = 0.03$ (solid), 0.3 (dotted), and 3 (dashed) for $j = 0$ (black), $j = 1$ (cyan), and $j = 2$ (blue) with $A_s = 1$. Curves which drop to zero show the delta function limit ($\sigma_{\rm LN} \to 0$) of σ_0. Larger $\sigma_{\rm LN}$ are simply smeared versions of this limit, and at large r/r_p, $\sigma_j/r_j \to$ is constant.

5. Conclusions

In this paper, we have reviewed a framework for nonlinear estimation of the abundances of PBHs formed at times when radiation dominated the energy density of the universe. Our method explicitly differs from others in the literature in that it accounts for the fact that PBHs form at positions and scales where all possible statistical realizations of compaction functions (Equation (24)) are maximized. These positions and scales are intimately related to the formation time of black holes.

Apart from predicting the PBH abundance for any given inflationary power spectrum, we have studied their mass distribution, that is, the mass function. We show that the latter is generically a power law at low masses (Figures 1 and 2). The slope of this power law depends on the critical scaling law for PBH formation (Equation (72)), and is independent of the shape or amplitude of the underlying power spectrum of fluctuations. At larger masses, there is a cutoff which depends on the shape and amplitude of the power spectrum (Figures 1 and 2).

Our analysis shows that smaller amplitudes generically result in PBH formation that extends over a larger range in r (compare the top and bottom curves in left panel of Figure 3), i.e., over a longer range of times. In this regard, models which arbitrarily assume a single epoch of PBH formation related to a peak scale of the power spectrum (r_p) should be treated with skepticism. On the other hand, PBHs considerably lighter than $M_{\rm eq}$ (the mass of the cosmological horizon at matter–radiation equality) require $r_p \ll M_{\rm eq}^{-1}$ and a small A_s. Exactly how small depends on the shape of $\mathcal{P}(k)$, and is the subject of ongoing work. The fact that we can treat the whole of Dark Matter as being contained in PBHs, even with a small power spectrum, is related to the fact that the mass spectrum is broader when the amplitude of the power spectrum is smaller, as can be seen Figures 1 and 2. As a byproduct, this cautions against analyses which assume that all PBHs have the same mass.

Finally, we would like to remark that all of our results in this paper depend critically on *not* including a transfer function when integrating over a power spectrum to compute our statistics. We have provided a detailed discussion of why a transfer function should not be used in our setup, in particular because of the way in which the compaction function is defined.

Funding: CG is funded by the Proyecto Ministerial PID2019-105614GB-C22 and the 2021 SGR 00872 project of the Generalitat de Catalunya.

Data Availability Statement: No data have been used.

Acknowledgments: CG thanks Misao Sasaki and Shi Pi for discussions and comments on the first version of this paper, the participants of the molecule workshop "Revisiting cosmological nonlinearities in the era of precision surveys" YITP-T-23-03, where this work was presented, the Yukawa Institute for Theoretical Physics (Kyoto, Japan), and finally the Institute of Basic Science (Daejeon, South Korea) for hospitality during the writing of this paper. RKS thanks the ICTP for its hospitality during the summer of 2023. Both CG and RKS thank the Institute of Cosmos Sciences for supporting the organization of the first school on Primordial Black Holes, where this work was tpresented.

Conflicts of Interest: There is no conflict of interest.

References

1. Carr, B.; Kohri, K.; Sendouda, Y.; Yokoyama, J. Constraints on Primordial Black Holes. *Rep. Prog. Phys.* **2021**, *84*, 116902. [CrossRef] [PubMed]
2. Escrivà, A. PBH Formation from Spherically Symmetric Hydrodynamical Perturbations: A Review. *Universe* **2022**, *8*, 66. [CrossRef]
3. Thorne, K.S. Nonspherical Gravitational Collapse: A Short Review. In *Magic without Magic*; Klauder, J., Ed.; Freeman: San Francisco, CA, USA, 1972.
4. Misner, C.W.; Sharp, D.H. Relativistic equations for adiabatic, spherically symmetric gravitational collapse. *Phys. Rev.* **1964**, *136*, B571–B576. [CrossRef]
5. Shibata, M.; Sasaki, M. Black hole formation in the Friedmann universe: Formulation and computation in numerical relativity. *Phys. Rev. D* **1999**, *60*, 084002. [CrossRef]

6. Musco, I. Threshold for Primordial Black Holes: Dependence on the shape of the cosmological perturbations. *Phys. Rev. D* **2019**, *100*, 123524. [CrossRef]
7. Harada, T.; Yoo, C.M.; Koga, Y. Revisiting compaction functions. *arXiv* **2023**, arXiv:2304.13284.
8. Bardeen, J.M.; Bond, J.R.; Kaiser, N.; Szalay, A.S. The Statistics of Peaks of Gaussian Random Fields. *Astrophys. J.* **1986**, *304*, 15–61. [CrossRef]
9. Escrivà, A. Simulation of primordial black hole formation using pseudo-spectral methods. *Phys. Dark Univ.* **2020**, *27*, 100466. [CrossRef]
10. Cruces, D. Review on Stochastic Approach to Inflation. *Universe* **2022**, *8*, 334. [CrossRef]
11. Cruces, D.; Germani, C. Stochastic inflation at all order in slow-roll parameters: Foundations. *Phys. Rev. D* **2022**, *105*, 023533. [CrossRef]
12. Nadezhin, D.K.; Novikov, I.D.; Polnarev, A.G. The hydrodynamics of primordial black hole formation. *Sov. Astron.* **1978**, *22*, 129.
13. Novikov, I.D.; Polnarev, A.G. The Hydrodynamics of Primordial Black Hole Formation-Dependence on the Equation of State. *Sov. Astron.* **1980**, *24*, 147.
14. Polnarev, A.G.; Musco, I. Curvature profiles as initial conditions for primordial black hole formation. *Class. Quant. Grav.* **2007**, *24*, 1405–1432. [CrossRef]
15. Harada, T.; Yoo, C.M.; Nakama, T.; Koga, Y. Cosmological long-wavelength solutions and primordial black hole formation. *Phys. Rev. D* **2015**, *91*, 084057. [CrossRef]
16. Germani, C.; Musco, I. Abundance of Primordial Black Holes Depends on the Shape of the Inflationary Power Spectrum. *Phys. Rev. Lett.* **2019**, *122*, 141302. [CrossRef]
17. Yoo, C.M.; Harada, T.; Garriga, J.; Kohri, K. Primordial black hole abundance from random Gaussian curvature perturbations and a local density threshold. *PTEP* **2018**, *2018*, 123E01. [CrossRef]
18. Yoo, C.M.; Harada, T.; Hirano, S.; Kohri, K. Abundance of Primordial Black Holes in peak theory for an arbitrary power spectrum. *PTEP* **2021**, *2021*, 013E02. [CrossRef]
19. Germani, C.; Sheth, R.K. Nonlinear statistics of Primordial Black Holes from Gaussian curvature perturbations. *Phys. Rev. D* **2020**, *101*, 063520. [CrossRef]
20. Escrivà, A.; Germani, C.; Sheth, R.K. Universal threshold for primordial black hole formation. *Phys. Rev. D* **2020**, *101*, 044022. [CrossRef]
21. Escrivà, A.; Germani, C.; Sheth, R.K. Analytical thresholds for black hole formation in general cosmological backgrounds. *JCAP* **2021**, *2021*, 030. [CrossRef]
22. Musco, I.; Luca, V.D.; Franciolini, G.; Riotto, A. Threshold for Primordial Black Holes. II. A simple analytic prescription. *Phys. Rev. D* **2021**, *103*, 063538. [CrossRef]
23. Akrami, Y.; Arroja, F.; Ashdown, M.; Aumont, J.; Baccigalupi, C.; Ballardini, M.; Banday, A.J.; Barreiro, R.B.; Bartolo, N.; Basak, S.; et al. Planck 2018 results. IX. Constraints on primordial non-Gaussianity. *Astron. Astrophys.* **2020**, *641*, A9.
24. Germani, C.; Prokopec, T. On Primordial Black Holes from an inflection point. *Phys. Dark Univ.* **2017**, *18*, 6–10. [CrossRef]
25. Motohashi, H.; Hu, W. Primordial Black Holes and Slow-Roll Violation. *Phys. Rev. D* **2017**, *96*, 063503. [CrossRef]
26. Özsoy, O.; Tasinato, G. Inflation and Primordial Black Holes. *Universe* **2023**, *9*, 203. [CrossRef]
27. Atal, V.; Germani, C. The role of non-gaussianities in Primordial Black Hole formation. *Phys. Dark Univ.* **2019**, *24*, 100275. [CrossRef]
28. Atal, V.; Garriga, J.; Marcos-Caballero, A. Primordial black hole formation with non-Gaussian curvature perturbations. *JCAP* **2019**, *2019*, 073. [CrossRef]
29. Pi, S.; Sasaki, M. Logarithmic Duality of the Curvature Perturbation. *Phys. Rev. Lett.* **2023**, *131*, 011002. [CrossRef]
30. Press, W.H.; Schechter, P. Formation of galaxies and clusters of galaxies by selfsimilar gravitational condensation. *Astrophys. J.* **1974**, *187*, 425–438. [CrossRef]
31. Bond, J.R.; Cole, S.; Efstathiou, G.; Kaiser, N. Excursion set mass functions for hierarchical Gaussian fluctuations. *Astrophys. J.* **1991**, *379*, 440. [CrossRef]
32. Sheth, R.K. Symmetry in stochasticity: Random walk models of large scale structure. *Pramana-J. Phys.* **2011**, *77*, 169–184. [CrossRef]
33. Paranjape, A.; Sheth, R.K. Peaks theory and the excursion set approach. *MNRAS* **2012**, *426*, 2789–2796. [CrossRef]
34. Paranjape, A.; Sheth, R.K.; Desjacques, V. Excursion Set Peaks: A self-consistent model of dark halo abundances. *MNRAS* **2013**, *431*, 1503–1512. [CrossRef]
35. Kalaja, A.; Bellomo, N.; Bartolo, N.; Bertacca, D.; Matarrese, S.; Musco, I.; Raccanelli, A.; Verde, L. From Primordial Black Holes Abundance to Primordial Curvature Power Spectrum (and back). *JCAP* **2019**, *10*, 031. [CrossRef]
36. Luca, V.D.; Kehagias, A.; Riotto, A. How Well Do We Know the Primordial Black Hole Abundance? The Crucial Role of Non-Linearities when Approaching the Horizon. *arXiv* **2023**, arXiv:2307.13633.
37. Kopp, M.; Hofmann, S.; Weller, J. Separate Universes Do Not Constrain Primordial Black Hole Formation. *Phys. Rev. D* **2011**, *83*, 124025. [CrossRef]
38. Harada, T.; Carr, B.J.; Igata, T. Complete conformal classification of the Friedmann–Lemaître–Robertson–Walker solutions with a linear equation of state. *Class. Quantum Grav.* **2018**, *35*, 105011. [CrossRef]
39. Niemeyer, J.C.; Jedamzik, K. Dynamics of Primordial Black Hole Formation. *Phys. Rev. D* **1999**, *59*, 124013. [CrossRef]

40. Escrivà, A.; Romano, A.E. Effects of the shape of curvature peaks on the size of Primordial Black Holes. *JCAP* **2021**, *05*, 066. [CrossRef]
41. Escrivà, A.; Atal, V.; Garriga, J. Formation of trapped vacuum bubbles during inflation, and consequences for PBH scenarios. *arXiv* **2023**, arXiv:2306.09990.
42. Yokoyama, J. Cosmological constraints on Primordial Black Holes produced in the near-critical gravitational collapse. *Phys. Rev. D* **1998**, *58*, 107502 [CrossRef]
43. Özsoy, O.; Tasinato, G. On the slope of the curvature power spectrum in non-attractor inflation. *JCAP* **2020**, *04*, 048. [CrossRef]
44. Cole, P.S.; Gow, A.D.; Byrnes, C.T.; Patil, S.P. Steepest growth re-examined: Repercussions for primordial black hole formation. *arXiv* **2023**, arXiv:2204.07573.

universe MDPI

Review

Inflation and Primordial Black Holes

Ogan Özsoy [1,2,*] and Gianmassimo Tasinato [3,4]

[1] CEICO, Institute of Physics of the Czech Academy of Sciences, Na Slovance 1999/2, 182 21 Prague, Czech Republic
[2] Instituto de Física Téorica UAM/CSIC, Calle Nicolás Cabrera 13-15, Cantoblanco, 28049 Madrid, Spain
[3] Dipartimento di Fisica e Astronomia, Università di Bologna, Via Irnerio 46, 40129 Bologna, Italy
[4] Department of Physics, Swansea University, Swansea SA2 8PP, UK
* Correspondence: ogan.ozsoy@csic.es

Abstract: We review conceptual aspects of inflationary scenarios able to produce primordial black holes by amplifying the size of curvature fluctuations to the level required to trigger black hole formation. We identify general mechanisms to do so, both for single- and multiple-field inflation. In single-field inflation, the spectrum of curvature fluctuations is enhanced by pronounced gradients of background quantities controlling the cosmological dynamics, which can induce brief phases of non-slow-roll inflationary evolution. In multiple-field inflation, the amplification occurs through appropriate couplings with additional sectors characterized by tachyonic instabilities that enhance the size of their fluctuations. As representative examples, we consider axion inflation and two-field models of inflation with rapid turns in field space. We develop our discussion in a pedagogical manner by including some of the most relevant calculations and by guiding the reader through the existing theoretical literature, emphasizing general themes common to several models.

Keywords: inflation; primordial black holes; dark matter; early universe

1. Introduction

Primordial black holes: history of the concept

Inflation, a short period of accelerated expansion in the very early moments of the universe, has become one of the main pillars of modern cosmology [1,2]. Leaving aside its success in addressing the puzzles of the standard hot Big Bang cosmology, inflation provides an explanation for the quantum mechanical origin of structures such as galaxies (including our own) and the anisotropies in the cosmic microwave background (CMB) radiation [3]. In the last two decades, advances in observational cosmology and, in particular, observations of the CMB and of the large-scale structure (LSS) of our universe have confirmed the predictions of inflation and arguably established its status as the main theoretical framework describing the very early universe [4,5]. These successes notwithstanding, CMB and LSS probes only provide us with information on the early universe at the largest cosmological scales ($10^{-4} \lesssim k\,[\mathrm{Mpc}^{-1}] \lesssim 10^{-1}$) corresponding to a small fraction of the early stages of inflationary dynamics. Hence, while inflation provides us with a consistent, testable framework to understand the initial conditions in the universe at the largest scales, we do not have direct access to most of the inflationary dynamics or to the evolution of the universe in the early post-inflationary era. Importantly, these stages could be host to a number of interesting phenomena, including the production of stable relics such as dark matter (see, e.g., [6] for a historical review of dark matter), which is essential in understanding the world we observe today, as well as for establishing new physics. Indeed, the existence of non-luminous, cold dark matter (CDM) that constitutes a quarter of the total energy budget in the universe [7] is one of the most glaring pieces of evidence beyond Standard Model physics [8]. The absence of signatures from collider experiments, along with unsuccessful direct and indirect detection searches, have all made the DM puzzle particularly compelling [9].

An intriguing and economical explanation that might account for DM density in our universe is a scenario wherein DM is made of compact objects such as primordial black holes (PBHs). Pioneered by the works of Y. Zel'dovic and I. Novikov [10] and S. Hawking [11], the initial ideas in this direction began with the realization that PBHs could form as a result of the gravitational collapse of over-dense inhomogeneities in the early universe. In the mid 1970s, it was realized in the works of B. Carr [12,13] and G. Chapline [14] that PBHs could contribute to DM density and provide the seeds for the supermassive BHs populating our universe [15]. Following these theoretical progressions, the interest of the scientific community in PBHs increased in the mid 1990s with the reported detection of microlensing events from the MACHO collaboration [16]. An immediate interpretation of these results suggested the possibility that a significant fraction of mass density in our galaxy could be composed of subsolar-mass PBHs. However, these considerations were later rendered invalid by the findings of the EROS [17] and OGLE [18,19] collaborations, concluding that only a small fraction of mass in the Milky Way could be in the form of PBHs.

Stimulated both by the absence of signals for well-motivated particle DM candidates and the first detection of gravitational waves (GWs) from merging BHs by the the LIGO/VIRGO collaboration [20], a second surge of interest in PBHs was ignited (see Figure 1). In particular, different groups suggested that merging PBHs could be responsible for the observed GW signals, constituting a significant fraction of DM density in our universe [21–23]. Since the first appearance of these articles, a significant amount of effort has been pushed forward by the community to search for and constrain the abundance of PBHs by utilizing their gravitational and electromagnetic effects on the environment at small scales. Various experiments set stringent constraints on PBH abundance for the solar- and subsolar-mass range, leaving a viable window for this scenario of tiny PBH masses ($10^{-17} \lesssim M_{\rm pbh}\,[M_\odot] \lesssim 10^{-12}$ ($M_\odot \simeq 1.98 \times 10^{33}$ gr)) as the totality of DM (see, e.g., [24–26]).

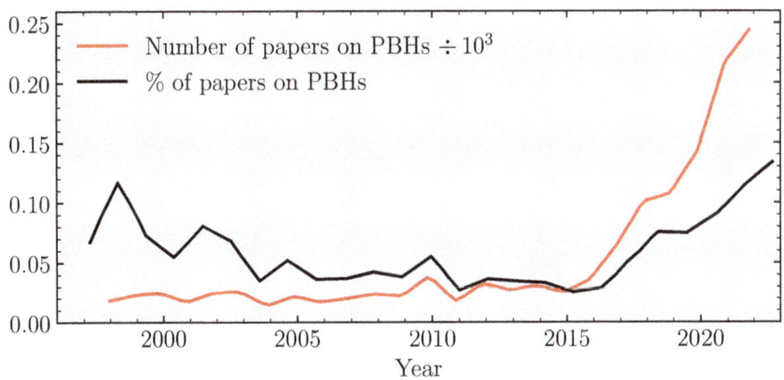

Figure 1. Total and relative number of manuscripts in the literature on arXiv from 1996 until today related to various aspects regarding primordial black holes. Spikes of activity in the literature, particularly after the mid 1990s due to claimed lensing events by the the MACHO collaboration and GW detection by LIGO in 2015, are clearly visible. Data were extracted from https://www.benty-fields.com/trending (accessed on 24 January 2023).

It should be noted that some of the constraints derived in the literature make specific assumptions about the formation process and the subsequent evolution of PBHs (such as monochromatic mass functions, clustering and accretion processes, etc.) and about other model-dependent specifics (such as non-Gaussianity) that could relax or tighten these constraints[1]. Since we mostly focus on the subject of inflationary model building, we will not review these issues and aforementioned constraints, but the interested reader can find more details in excellent reviews published recently, see e.g., [28–34].

PBHs are likely to form well before the end of the radiation-dominated era (i.e., before the so-called matter–radiation equality) and behave like cold and collision-less matter. Therefore, they constitute an interesting DM candidate if they are massive enough ($M_{\rm pbh} \gtrsim 10^{15}\,{\rm g} \simeq 10^{-18} M_\odot$) to ensure a lifetime comparable with the age of the universe [35][2]. In this context, a particularly appealing aspect of PBH dark matter is its economical and minimal structure in the sense that this scenario does not require any additional beyond-Standard-Model (BSM) physics (such as new particles and interactions), provided that one alters the not-so-well-constrained early universe at small scales by introducing a viable mechanism to account for the production of large density fluctuations required for PBH formation.

Similar to the generation of CMB anisotropies, a compelling and natural source of these perturbations in the early universe could be the quantum fluctuations that are stretched outside the horizon during inflation. However, in order to generate such over-dense regions that can collapse to form PBHs in the post-inflationary universe, one needs to devise a mechanism to enhance, by several orders of magnitude, the inflationary scalar perturbations at small scales ($k \gg k_{\rm cmb}$) (corresponding to late stages of inflation) far above the values required to match CMB observations. As the observed temperature anisotropies prefer a red-tilted power spectrum at CMB scales, this situation generally requires a blue-tilted power spectrum or some specific features at scales associated with PBH formation.

In the context of canonical single scalar field inflation, the first ideas in this direction appeared in works by P. Ivanov, P. Naselsky and I. Novikov [38] (see also [39]). In particular, these authors showed that if the inflation potential has a very flat plateau-like region for field ranges corresponding to the late stages of accelerated expansion, the inflationary dynamics enters a "non-attractor" regime called ultra-slow-roll (USR) inflation [40]. This leads to super-horizon growth of scalar perturbations [41–43] that can eventually trigger PBH formation in the post-inflationary universe[3]. Many explicit single-field inflationary models that exhibit similar local features were subsequently studied in the literature; for a partial list of popular works, see, e.g., [46–56] (see also [57–63] for earlier constructions). In the context of single scalar field inflation, another possibility to generate an enhancement in the scalar power spectrum is to invoke a variation of the sound speed of scalar fluctuations, for example, through a reduction in the speed of sound (c_s^2) [52,64,65] or through rapid oscillation (c_s^2), which triggers a resonant instability in the scalar sector [66,67].

From a top-down model-building perspective, a rich particle content during inflation is not just an interesting possibility, but appears to be a common outcome of many BSM theories [68]. Since the early days of research on PBHs, multifield inflationary scenarios have also attracted considerable attention as a natural way to realize enhancement in the scalar perturbations at small scales. For instance, large scalar perturbations may arise through instabilities arising in the scalar sector, e.g., during the waterfall phase of hybrid inflation [57,69,70] or due to turning trajectories in the multiscalar inflationary landscape, as reported recently in [71–78]. Another intriguing possibility in this context is to employ axion-gauge field dynamics during inflation [79–86]. In these models, particle production in the gauge field sector act as a source for the scalar fluctuations and can therefore be responsible for PBH formation.

A common feature of all inflationary scenarios capable of producing PBH populations is the inevitable production of a stochastic GW background (SGWB) induced through higher-order gravitational interactions between enhanced scalar and tensor fluctuations of the metric [87–89]. Interestingly, this signal may carry crucial information about the properties of its sources, including the amplitude, statistics and spectral shape of scalar perturbations (see, e.g., [90–98]) and could provide invaluable information about the underlying inflationary production mechanism. Furthermore, since the resulting GW background interacts very weakly with the intervening matter between the time of its formation and today, it leads to a rather clean probe of the underlying PBH formation scenario. This allows us to access inflationary dynamics on scales much smaller than those currently probed with CMB and LSS experiments through space- and ground-based GW interferometers including the Laser Interferometer Space Antenna (LISA) [99,100], Pulsar Timing Array (PTA) experiments [101,102] and DECIGO [103,104]. For a detailed review of induced SGWB and the dependence of its properties on the post-inflationary expansion history, see [105].

The structure of this review

If their origin is attributed to the large primordial fluctuations, PBHs may offer us a unique window to probe inflationary dynamics at sub-CMB scales. In this work, focusing mainly on the activity in the literature within the last few years, we aim to revisit and review different inflationary production mechanisms of PBHs[4] and their main predictions, in a heuristic and pedagogical manner. The audience we have in mind is graduate students or researchers in related fields who wish to learn about inflation and primordial black holes and to be guided through the large literature on the subject by emphasizing common conceptual themes behind many different realizations.

This review is organized as follows. In Section 2, we present a simplified, intuitive picture of PBH formation in the inflationary universe and provide an approximate estimate of the required conditions to produce PBHs from the perspective of inflationary dynamics. In Section 3, we discuss ideas to enhance the curvature power spectrum within single-field inflation, as required for PBH formation. These mechanisms exploit large gradients in background quantities that are converted into an amplification of fluctuations. Besides reviewing analytic findings, we also develop numerical analysis and provide a link to a code for reproducing our results (see Notes 21). In Section 4, we focus on multifield inflationary scenarios that can generate PBH populations including particle production during axion inflation or sudden turns in the multiscalar inflationary landscape. Finally, we end with a discussion on future directions in Section 5. We supplement this work with several technical appendices, where we provide useful formulas and calculations used in the main body of this work.

Conventions

Throughout this review, we work with natural units ($\hbar = c = 1$). We use the reduced Planck mass defined as $M_{\rm pl}^2 = (8\pi G)^{-1}$ and retain it in the equations unless otherwise stated. For time-dependent quantities, over dots and primes denote derivatives with respect to cosmological time (t) and conformal time ($d\tau \equiv dt/a(t)$), respectively, where $a(t)$ is the scale factor of the background FLRW metric ($g_{\mu\nu} = \mathrm{diag}(-1, a^2, a^2, a^2)$).

2. PBH Formation in the Early Universe

We start by providing a physical description of PBH formation in the early universe, as comprised of an early stage of inflation, followed by radiation and matter domination (for a mini review of background cosmology, see Appendix A). Our aim is to set the stage and relate basic properties of a PBH population—as their mass and abundance—with the features of primordial curvature fluctuations originating from inflation. For this purpose, in Section 2.1, we describe the mechanism of PBH formation in the post-inflationary universe, emphasizing its nature as a causal process controlled by inflationary quantum fluctuations. In Section 2.2, we discuss relevant concepts such as the threshold for collapse into black holes and the corresponding mass and collapse fraction of PBHs relevant for a computation of their abundance. Finally, in Section 2.3, we relate the PBH abundance to primordial physics during inflation with the aim of determining the amplitude of scalar fluctuations required to produce a population of PBHs with interesting consequences for cosmology. All the concepts we discuss form the basis and motivations for our analysis of inflationary mechanisms for PBH production, which we develop in Sections 3 and 4.

Main References: In compiling the materials for this section and to set the main framework for our discussion, we benefited from the ideas presented in the reviews by C. Byrnes and P. Cole [126] and M. Sasaki et al. [30], as well as the Ph.D. thesis of G. Franciolini [127].

2.1. PBH Formation as a Causal Process

An important concept in an expanding space–time is the horizon scale, which is crucial for understanding the causal properties of the dynamics of perturbations that are responsible for PBH formation. As an indicator of the rate of expansion of our universe, the Hubble rate ($H(t) \equiv \dot{a}(t)/a(t)$) has dimensions of inverse length (or time^{-1} in natural units). This makes the quantity H^{-1} (Hubble horizon) a natural candidate for a physical length scale in an expanding universe. Commonly referred to as the Hubble distance, the quantity $1/H$ (or c/H if one wishes to recover physical units) measures the distance that light travels within one unit of Hubble time. Therefore, it can be considered a good proxy for a (time-dependent) length scale controlling the size of a causal patch in our universe. Bearing in mind that we relate physical quantities to comoving quantities by the scale factor ($a(t)$), a useful quantity that guides us in this direction is the comoving Hubble horizon ($(aH)^{-1}$) and, in particular, its time evolution. When expressed in terms of the second derivative of the scale factor, the time derivative of the comoving horizon can be written as

$$\frac{\mathrm{d}}{\mathrm{d}t}\left(\frac{1}{aH}\right) = -\frac{\ddot{a}}{a^2H^2}\,. \tag{1}$$

Notice that during inflation, $\ddot{a} > 0$; hence, the comoving horizon scale is a decreasing function of time, whereas in a decelerating universe with $\ddot{a} < 0$ (i.e., in the post-inflationary universe before dark energy domination), this quantity is an increasing function of time. The property that the comoving horizon decreases during an accelerated expansion is perhaps the most important element to understand inflation as a solution of the horizon problem of hot Big Bang cosmology and a framework for the quantum mechanical origin of structures in our universe[5]. The time dependence of the comoving Hubble horizon is controlled by the value of the background equation of state (w) (EoS) as (see Appendix A)

$$(aH)^{-1} \propto a^{(1+3w)/2}. \tag{2}$$

Therefore, during inflation, $w \simeq -1$ and $(aH)^{-1} \propto a^{-1}$, while during the subsequent phases of the radiation-dominated (RDU) and matter-dominated universe (MDU), the comoving horizon evolves as $(aH)^{-1} \propto a^1$ and $(aH)^{-1} \propto a^{1/2}$, respectively. The evolution of the comoving horizon with respect to the logarithm-of-scale factor ($\ln(a)$) is illustrated in Figure 2.

When we study the statistical properties of fluctuations in Fourier space, we often label a given perturbation mode with a comoving length scale (k^{-1}) measured in units of megaparsecs (Mpc $= 3.26 \times 10^6$ light years $\simeq 3.1 \times 10^{19}$ km). Therefore, a crucial quantity to conceptualize the behavior of perturbations in the inflationary universe is the ratio of the wavelength of a given mode with respect to the size of the comoving Hubble horizon, i.e., $(aH)/k$. Fluctuations with wavelengths larger than the comoving horizon are referred to as superhorizon modes ($k < aH$), while subhorizon perturbations satisfy $k > aH$. Each mode crosses the horizon at $k = aH$. As shown in Figure 2, a typical fluctuation with a comoving size of k^{-1} (horizontal lines) begins its life deep inside the horizon (typically as a quantum fluctuation); then, it leaves the horizon to become a superhorizon mode, and, finally, it re-enters the comoving horizon in the post-inflationary universe. Large-scale modes (with smaller k) exit the horizon earlier than small-scale modes and re-enter the horizon at a later time in the post-inflationary era. For definiteness, in Figure 2, we represent the behavior of the comoving curvature perturbation ($\mathcal{R}$) [129,130] (see [128] for a textbook discussion), which plays an important role in our discussion.

In fact, apart from providing seeds for the observed cosmic microwave background (CMB) anisotropies at large scales, the dynamics of curvature fluctuations ($\mathcal{R}$) may also be at play for PBH formation, provided that cosmological fluctuations exhibit specific 'initial conditions' at small scales. For this purpose, denoting $k = k_{\rm pbh} \gg k_{\rm cmb}$ as the comoving momentum associated with PBH formation, we assume that the curvature power spectrum at these small scales acquires an amplification well above the level required to match CMB observations: $\mathcal{P}_{\mathcal{R}}(k_{\rm pbh}) \gg \mathcal{P}_{\mathcal{R}}(k_{\rm cmb}) \sim 10^{-9}$ [4] (more on this later). Soon after the end of inflation, i.e., after reheating[6], the modes associated with the enhancement (e.g., modes with a comoving size of $k_{\rm pbh}^{-1}$) become the seeds of density perturbations in the RDU:

$$\mathcal{P}_{\mathcal{R}}(k_{\rm pbh})^{1/2} \sim \frac{\delta\rho}{\rho}. \tag{3}$$

Since the comoving Hubble scale grows with respect to comoving scales in the RDU, the characteristic scale of perturbations eventually becomes comparable to the comoving horizon at $a = a_{\rm f}$, where $k_{\rm pbh} = (aH)_{\rm f}$ (the subscript f indicates PBH formation time). At this point, gravitational interactions can trigger the collapse of over-dense regions if the latter have a sufficient over-density above a certain collapse threshold (see the discussion below).

Notice that at subhorizon scales, the radiation pressure can overcome the gravitational collapse. Therefore, the production of PBHs effectively occurs around horizon re-entry. This implies that the concept of horizon re-entry is crucial for our understanding of PBH formation as a causal process. In fact, only when the physical wavelength of a perturbation becomes comparable to the causal distance ($1/H$) is gravity able to communicate the presence of an over-density and to initiate gravitational collapse. A schematic diagram that summarizes the discussion presented above is illustrated in Figure 2.

In what comes next, we briefly discuss relevant quantities such as the threshold for collapse and the mass and collapse fractions of PBHs, which are important for the computation of the current PBH abundance.

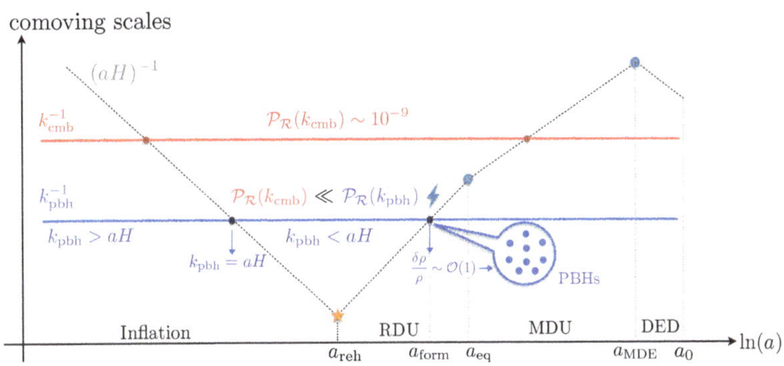

Figure 2. A sketch of the time evolution of curvature fluctuations ($\mathcal{R}$) (labeled by a comoving scale (k^{-1})) with respect to the comoving Hubble horizon (dotted lines) ($(aH)^{-1}$) in the early universe. In the post-inflationary universe, $a_{\rm reh}$ denotes the reheating time, $a_{\rm eq}$ refers to the time of matter–radiation equality, $a_{\rm MDE}$ refers to matter–dark energy equality and a_0 denotes the value of the scale factor today. The blue horizontal line indicates the comoving size of a representative small-scale perturbation responsible for PBH formation. If the power spectrum associated with these modes is enhanced during inflation, they can transfer their energy to density perturbations during radiation domination and ignite PBH formation upon horizon re-entry at $a = a_{\rm form} \equiv a_{\rm f}$.

2.2. *The Relevant Quantities for PBH Abundance*

The threshold for collapse

The first criterion of the collapse threshold—defined as $\delta_{\rm c}$ hereafter—for PBH formation was formulated by B. Carr in 1975 using a Jeans-type instability argument within Newtonian gravity [13]. In Carr's estimate, an over-density in RDU would collapse upon horizon re-entry if the fractional over-density of the perturbation is larger than the sound speed square of density perturbations (c_s^2), which is a measure of how fast a pressure wave caused by the over-density can travel from the center to the edge of a local fluctuation. In the RDU, the speed of sound of perturbations satisfies $c_s^2 = w = 1/3$ so that its square is directly related to EoS. This result implies that a perturbation can collapse to form PBHs if its over-density is larger than the pressure exerted by the radiation pressure[7]. Implementing general relativistic effects, the second analytical estimate was derived in [132], obtaining $\delta_c \simeq 0.4$ during the RDU. These results notwithstanding, further studies in this direction have found that the threshold depends on the initial profile of curvature perturbation [133–136][8].

At this point, we emphasize that a more precise criterion for the threshold of collapse is provided in terms of the so-called compaction function (which is a measure of excess mass within a given spatial volume; see, e.g., [135,139]) the maximum value of which results in δ_c. In this prescription, we note that contrary to the simplistic argument presented in Appendix B, there is no one-to-one correspondence between the threshold value and the peak over-density ((3) $\delta \equiv \delta\rho/\rho$), although they are related; see, e.g., [140]. Keeping this in mind, in the rest of this work, we will continue to refer to the threshold using the δ_c notation as commonly practiced in the literature.

An accurate characterization of the threshold requires a dedicated analysis of the evolution of perturbations in the non-linear regime after horizon re-entry, which can be done with the help of numerical simulations (see, e.g., [141] for a review). Utilizing the peak of the criterion of compaction function as the definition of the threshold (δ_c), simulations performed in the radiation fluid for different primordial perturbation profiles lead to a range of values ($\delta_c = 0.4 - 2/3$) depending on the shape of the density peak [135,142]. Although in general, threshold depends on the shape of the primordial perturbation, refs. [142,143] showed the existence of an approximate universal value for the threshold that depends only on the type of ambient fluid and geometry. For a detailed account of the threshold and its history, we refer the reader to a recent review [141].

The mass of PBHs

The characteristic mass of PBHs can be related to the mass contained within the Hubble horizon at the time of formation ($a = a_{\rm f}$) through an efficiency factor ($\gamma = 0.2$), as suggested by the analytical model developed in [13]:

$$M_{\rm pbh}^{(\rm f)} = \gamma\, M_H\Big|_{a=a_{\rm f}} = \gamma\, M_H^{(\rm eq)} \left(\frac{M_H^{(\rm f)}}{M_H^{(\rm eq)}}\right) = \left(\frac{a_{\rm f}}{a_{\rm eq}}\right)^2 \gamma\, M_H^{(\rm eq)}\,. \tag{4}$$

In this formula, $M_H(t) \equiv 4\pi\rho(t)/(3H(t)^3)$ is the time-dependent horizon mass, where the sub/superscripts "f" and "eq" denote quantities evaluated at the time of PBH formation and matter–radiation equality, respectively; we use the standard relations $H^2 \propto \rho \propto a^{-4}$ during the RDU. Noting that the horizon mass at the time of equality is given by $M_H^{(\rm eq)} \simeq 2.8 \times 10^{17}\, M_\odot$ [90], Equation (4) informs us that PBHs, contrary to astrophysical black holes, can, in principle, span a wide range of masses, depending on their formation time with respect to matter–radiation equality.

Making use of the time-dependent horizon mass as above, we can relate the PBH mass at formation to the characteristic size of the perturbations that leave the horizon during inflation and are responsible for PBH formation. For this purpose, we first rewrite the PBH mass at formation as

$$M_{\rm pbh}^{(\rm f)} = \left(\frac{\rho_{\rm f}}{\rho_{\rm eq}}\right)^{1/2} \left(\frac{H_{\rm eq}}{H_{\rm f}}\right)^2 \gamma\, M_H^{(\rm eq)}. \tag{5}$$

Using the property of entropy conservation ($g_s(T)\, T^3\, a^3 = $ constant) and the scaling property of the energy density with respect to the temperature of the plasma during the RDU ($\rho \propto g_*(T)\, T^4$), Equation (5) can be re-expressed as the following relation

$$\begin{aligned} M_{\rm pbh}^{(\rm f)}(k_{\rm pbh}) &= \left(\frac{g_*(T_{\rm f})}{g_*(T_{\rm eq})}\right)^{1/2} \left(\frac{g_s(T_{\rm eq})}{g_s(T_{\rm f})}\right)^{2/3} \left(\frac{k_{\rm eq}}{k_{\rm pbh}}\right)^2 \gamma\, M_H^{(\rm eq)}, \\ &\simeq \left(\frac{\gamma}{0.2}\right) \left(\frac{g_*(T_{\rm f})}{106.75}\right)^{-1/6} \left(\frac{k_{\rm pbh}}{3.2\times 10^5\, {\rm Mpc}^{-1}}\right)^{-2} 30 M_\odot\,. \end{aligned} \tag{6}$$

In the second line, we assume that the effective number of relativistic degrees of freedom in energy density and entropy are equal, i.e., we set $g_*(T) = g_s(T)$ and take $g_*(T_{eq}) \simeq 3.38$[9] with $k_{eq} \simeq 0.0104\,\text{Mpc}^{-1}$, accordingly with the latest Planck results [7]. Equation (6) indicates that for masses of PBHs that can be associated with recent LIGO observations ($M_{pbh} \simeq 30M_\odot$), the peak scale of perturbations responsible for PBH formation is much smaller compared to CMB scales ($k_{pbh} \gg k_{cmb}$). For subsolar-mass PBHs, the corresponding peak scale for PBH formation becomes progressively smaller. For example, considering the currently allowed sublunar range ($M_{moon} \simeq 3.7 \times 10^{-8}\,M_\odot$) of PBH masses ($10^{-17} \lesssim M_{pbh}[M_\odot] \lesssim 10^{-12}$), which are objects that can account for the totality of dark matter, the range of scales associated with PBH formation is quite small: $10^{12} \lesssim k_{pbh}\,[\text{Mpc}^{-1}] \lesssim 10^{15}$. See Table 1 for an easier-to-visualize summary of these considerations.

Elaborating on Equation (6), we can also derive a rough relation between the PBH mass at formation with the the number of e-folds (N_{pbh}) at which the PBH-forming modes leave the horizon during inflation. For this purpose, we first notice that $k_{pbh}/k_{cmb} = (aH)_{pbh}/(aH)_{cmb}$, where the values of the Hubble rate and scale factor should be evaluated at the scales of horizon exit during inflation (see Figure 2). Assuming a roughly constant slow-roll parameter ($\epsilon \equiv -\dot{H}/H^2 \ll 1$) between the horizon exit time of modes associated with CMB and PBH formation, respectively, we can relate the Hubble and the scale factor as $H_{pbh} = H_{cmb}\,e^{-\epsilon(N_{pbh}-N_{cmb})}$ and $a_{pbh} = a_{cmb}\,e^{N_{pbh}-N_{cmb}}$, where $N_{pbh} > N_{cmb}$ so that we count e-folds forward in time with respect to the horizon exit of the CMB mode[10]. Using the last two relations, we find $k_{pbh}/k_{cmb} \simeq e^{(N_{pbh}-N_{cmb})(1-\epsilon)}$; once plugged into Equation (6), assuming $k_{cmb} = 0.002\,\text{Mpc}^{-1}$, we find

$$M^{(f)}_{pbh}(N_{pbh}) \approx 7.7 \times 10^{17} M_\odot\; e^{-2(N_{pbh}-N_{cmb})(1-\epsilon)} \left(\frac{\gamma}{0.2}\right)\left(\frac{g_*(T_f)}{106.75}\right)^{-1/6}. \tag{7}$$

Modes that leave the horizon much later compared to CMB scales have $N_{pbh} - N_{cmb} \gg 1$; therefore, the exponential in Equation (7) can considerably reduce the overall large normalization, leading to small PBH masses.

Table 1. Range of PBH mass(es) vs. the corresponding wave number(s) (k_{pbh}) (see Equation (6)) of the primordial modes, together with the approximate horizon crossing time measured with respect to e-folding number of the pivot scale $k_{cmb} = 0.002\,\text{Mpc}^{-1}$ leaves the horizon during inflation, $\Delta N \equiv N_{pbh} - N_{cmb} > 0$ (see Equation (7)). The first row refers to the corresponding quantities for a typical supermassive black hole (SMBH), such as Sagittarius A* in the center of our galaxy [144]. The third row refers to asteroid-mass PBHs that can still account for a significant fraction (or all) of DM density today [26]. The corresponding mass of the PBHs in terms of the Earth's mass ($M_\odot \simeq 3.33 \times 10^5\,M_\oplus$) and the mass of Mount Everest ($M_{Everest} = 8.1 \times 10^{14}\,\text{kg} \simeq 4.1 \times 10^{-16}\,M_\odot$) are given in the last two columns on the right.

$M_{pbh}\,[M_\odot]$	ΔN	$k_{pbh}\,[\text{Mpc}^{-1}]$	$M_{pbh}\,[M_\oplus]$	$M_{pbh}\,[M_{Everest}]$
10^6	14	10^3	10^{11}	10^{21}
10^0–10^2	18–21	10^5–10^6	10^5–10^7	10^{15}–10^{17}
10^{-17}–10^{-12}	34–40	10^{12}–10^{15}	10^{-12}–10^{-7}	10^{-2}–10^3

PBH abundance

After discussing possible masses for PBHs and how they depend on the dynamics of inflation, we analyze the notion of abundance of PBHs relative to the energy density of other species. We can compute this quantity during two epochs: today and at PBH formation.

When considering the present-day fraction of PBH density, it is a common practice to relate the PBH abundance to present-day dark matter density, introducing the quantity

$$f_{\rm pbh} \equiv \frac{\Omega_{\rm pbh}}{\Omega_{\rm dm}}, \tag{8}$$

where for each species (i), we define $\Omega_i \equiv \rho_{i,0}/\rho_{c,0}$, with subscript "0" denoting quantities evaluated today and $\rho_{c,0} = 3H_0^2 M_{\rm pl}^2$ representing the critical density. Planck measurements provide the following value for the dark matter abundance [7],

$$\Omega_{\rm dm} h^2 = 0.120 \pm 0.001, \tag{9}$$

in terms of $h = 0.6736 \pm 0.0054$, which measures the Hubble rate (H_0) in units of $100\,{\rm km\,s^{-1}\,Mpc^{-1}}$.

We can then relate $f_{\rm pbh}$ today to the density fraction of PBHs at the epoch of their formation, denoting this quantity with β. In fact, since we assume that PBH formation takes place during the RDU and sinc after formation, the PBH density scales as $\rho_{\rm pbh} \propto a^{-3}$, we can write

$$\begin{aligned} \beta \equiv \left.\frac{\rho_{\rm pbh}}{\rho}\right|_{a=a_{\rm f}} = \frac{\rho_{\rm pbh,f}}{\rho_{\rm pbh,0}} \frac{\rho_{c,0}}{\rho_{\rm f}} \Omega_{\rm dm} f_{\rm pbh} &= \frac{\rho_{\rm pbh,f}}{\rho_{\rm pbh,0}} \frac{\rho_{c,0}}{\rho_{\rm eq}} \frac{\rho_{\rm eq}}{\rho_{\rm f}} \Omega_{\rm dm}\, f_{\rm pbh}, \\ &\simeq \frac{1}{2} \frac{a_{\rm f}}{a_{\rm eq}}\, \Omega_m^{-1}\, \Omega_{\rm dm}\, f_{\rm pbh} \end{aligned} \tag{10}$$

where Ω_m is the current matter density in the universe. In (10), we normalize the scale factor today as $a_0 = 1$ and use the fact that the total energy density evolves as $\rho \propto a^{-4}$ for $a_{\rm f} < a < a_{\rm eq}$ and $\rho(a_{\rm eq}) = 2\rho_{m,0}\, a_{\rm eq}^{-3}$. Using the conservation of total entropy ($g_s(T)\, T^3\, a^3 =$ constant), we can re-express the factor $a_{\rm f}/a_{\rm eq}$ appearing in Equation (10) as follows:

$$\begin{aligned} \frac{a_{\rm f}}{a_{\rm eq}} &= \frac{T_{\rm eq}}{T_{\rm f}} \left(\frac{g_s(T_{\rm eq})}{g_s(T_{\rm f})}\right)^{1/3}, \\ &\simeq 3.17 \times 10^{-9} \left(\frac{\gamma}{0.2}\right)^{-1/2} \left(\frac{g_*(T_{\rm f})}{106.75}\right)^{-1/12} \left(\frac{M_{\rm pbh}^{(\rm f)}}{M_\odot}\right)^{1/2}, \end{aligned} \tag{11}$$

where we make use of Equation (5) to relate $T_{\rm eq}/T_{\rm f}$ to the mass of PBHs at formation, and, as before, we assume $g_*(T) = g_s(T)$. Finally, plugging (11) into (10) and implementing Planck measurements on $\Omega_{\rm dm}$ (9) and Ω_m (see Equation (A18)), we directly relate the abundance of PBHs at formation (β) to their present-day fraction ($f_{\rm pbh}$) in terms of the PBH mass at formation:

$$\beta \simeq 1.33 \times 10^{-9} \left(\frac{\gamma}{0.2}\right)^{-1/2} \left(\frac{g_*(T_{\rm f})}{106.75}\right)^{-1/12} \left(\frac{M_{\rm pbh}^{(\rm f)}}{M_\odot}\right)^{1/2} f_{\rm pbh}\,. \tag{12}$$

Therefore, we learn that in the case in which PBH abundance accounts for the total DM density today ($f_{\rm pbh} \to 1$), the fraction of the total density in the form of PBHs (β) at the time of their formation takes extremely small values when considering an interesting range of masses ($M_{\rm pbh}^{(\rm f)}$). This situation reflects the fact that PBH formation in the early universe is a very rare event. There is also another way to parameterize β in terms of the (relative) number of collapsing regions to form PBHs. This approach is especially useful to relate the PBH abundance to the statistical properties of primordial fluctuations, as we discuss below.

Collapse fraction of PBHs at formation

The PBH abundance at formation can also be interpreted as the fraction (β) of local regions in the universe that has a density larger than a certain threshold. The standard treatment for estimating β is based on the so-called Press–Schechter model of gravitational collapse, which is widely used in the literature on large-scale structure formation [145][11],

$$\beta \equiv \left.\frac{\rho_{\rm pbh}}{\rho}\right|_{a=a_{\rm f}} \equiv \int_{\Delta_{\rm c}}^{\infty} P(\delta)\, {\rm d}\delta, \tag{13}$$

where $P(\delta)$ is the probability distribution function (PDF), which describes the likelihood that a given fluctuation has an over-density δ; we assume that a perturbation will collapse to form a BH if its amplitude is larger than a critical value (Δ_c). Notice that Δ_c is the critical density contrast and, in general, cannot be identified with the threshold $\Delta_{\rm c} \neq \delta_{\rm c}$, where the latter can be defined as the peak value of the compaction function and can be related to an averaged density contrast [141]. Note also that an alternative, more accurate approach to compute the PBH abundance requires the use of the PDF of the compaction function (see, e.g., [149]). A discussion of this approach and the actual relation between $\Delta_{\rm c}$ and $\delta_{\rm c}$ would bring us outside the pedagogical purposes of this review. For a detailed account of the use the PDF of the compaction function and the influence of resulting non-linearities on the PBH abundance, we refer the reader to [149] and references therein. For more details on the relation between $\Delta_{\rm c}$ and $\delta_{\rm c}$, see also the review [141]. Keeping an eye on the recent progress in the literature on these issues, we will continue to identify $\Delta_{\rm c} = \delta_{\rm c}$ for estimates of β in this and the next section and continue to work with a PDF on over-density δ.

Let us assume that the latter follows a Gaussian distribution,

$$P_G(\delta) = \frac{1}{\sqrt{2\pi}\sigma} e^{-(\delta-\mu)^2/2\sigma^2}, \tag{14}$$

where μ is the mean and σ^2 is the variance of the distribution. In Figure 3, we represent two Gaussian PDFs that have the same mean value and two different variances satisfying $\sigma_2^2 > \sigma_1^2$. As the second distribution is more "spread" with a larger variance compared to the first one, the probability of an over-density larger than the critical threshold ($\delta_{\rm c}$) is larger, as is PBH abundance β at formation, since the integral in Equation (13) has more support within the integration limits $\delta \in [\delta_{\rm c}, \infty]$. Hence, we can expect that the quantity (σ) plays an important role in estimating the PBH abundance.

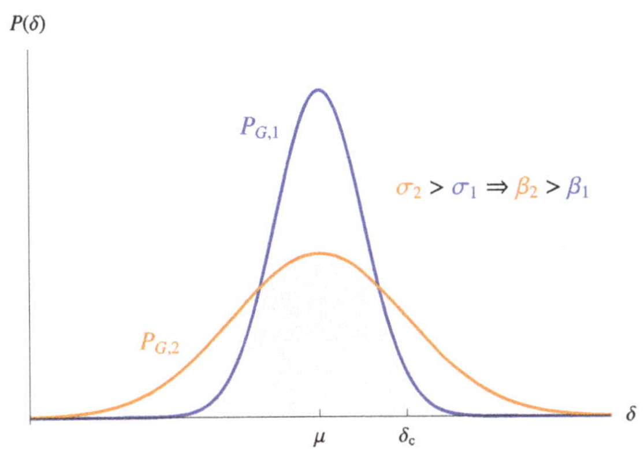

Figure 3. Two Gaussian PDFs of the over-density field (δ) with different variances ($\sigma_2^2 > \sigma_1^2$). Since the second distribution has a larger variance, the area under the curve above the critical threshold ($\delta_{\rm c} \leq \delta \leq \infty$) is larger, leading to larger PBH abundance (β) (13) at formation.

Using (12) and (13), we can estimate the required variance (σ^2) of (14) that can give rise to a large population of PBH today, as controlled by the quantity (β). Focusing on a distribution with zero mean $\mu = 0$ in (14) and integrating (13), we have

$$\beta = \int_{\delta_c}^{\infty} \frac{d\delta}{\sqrt{2\pi}\sigma} \exp\left(-\frac{\delta^2}{2\sigma^2}\right) = \frac{1}{2}\text{Erfc}\left(\frac{\delta_c}{\sqrt{2}\sigma}\right) \simeq \frac{\sigma}{\sqrt{2\pi}\delta_c} \exp\left(-\frac{\delta_c^2}{2\sigma^2}\right), \tag{15}$$

where $\text{Erfc}(x) = 1 - \text{Erf}(x)$ is the complementary error function, and in the last equality, we take $\delta_c \gg \sigma$. As we will learn shortly, this is a good approximation for all practical purposes. As concrete examples, substituting Equation (15) into Equation (12), we infer that a solar mass PBH population with $f_{\rm pbh} = 10^{-3}$ requires $\delta_c/\sigma \simeq 7$, whereas for a population with $M^{(f)}_{\rm pbh} = 10^{-12} M_\odot$ and $f_{\rm pbh} = 1$, we need $\delta_c/\sigma \simeq 7.9$. Assuming a threshold of $\delta_c = 0.4$, these results translate into $\sigma \simeq 0.06$ and $\sigma \simeq 0.05$, respectively. Notice also from (15) that the PBH abundance at the time of formation is exponentially sensitive to the variance of the distribution. We will dwell more on this dependence below. However, first, we discuss the implications of these findings in terms of the amplitude of scalar fluctuations generated during the phase of cosmic inflation.

2.3. Relating PBH Properties to Primordial Scalar Fluctuations

We now examine how to relate the notion of PBH abundance to the properties of the comoving curvature fluctuation ($\mathcal{R}$) [129,130], as produced in the early universe by cosmic inflation. $\mathcal{R}$ is conserved on super Hubble scales as the modes evolve from the inflationary phase to the RDU. We start by connecting the amplitude of $\mathcal{R}$ with the fractional over-density (δ), which triggers PBH formation, as we learned in our previous discussion. Working in Fourier space with Taylor expansion at the leading order in a gradient expansion (controlled by the small parameter k/aH) and at a linear order in $\mathcal{R}$, one finds [135]:

$$\delta(\vec{x}, t) \simeq \frac{2(1+w)}{(5+3w)} \frac{\nabla^2 \mathcal{R}(\vec{x})}{(aH)^2} + \ldots \quad \Longrightarrow \quad \delta_k \simeq -\frac{4}{9}\left(\frac{k}{aH}\right)^2 \mathcal{R}_k, \tag{16}$$

where $w = 1/3$ is used during the RDU, and $\ldots$ denotes terms of higher order ($\mathcal{O}(\mathcal{R}^2)$) in the curvature perturbation[12]. Defining the power spectrum of a Fourier variable (X_k) as

$$\langle X_k X_{k'} \rangle = \frac{2\pi^2}{k^3} \mathcal{P}_X(k)\, \delta^{(3)}(\vec{k} + \vec{k}'), \tag{17}$$

the relation between the power spectrum of over-density and curvature perturbation is then given by

$$\mathcal{P}_\delta(k) \simeq \frac{16}{81}\left(\frac{k}{aH}\right)^4 \mathcal{P}_\mathcal{R}(k). \tag{18}$$

In the computation of the density contrast, one should typically use a window function ($\mathcal{W}$) to smooth $\delta(\vec{x}, t)$ on a scale ($R \approx k^{-1} \approx (aH)^{-1}$) (e.g., on scales of size $k_{\rm pbh}^{-1}$ at horizon re-entry, as shown in Figure 2) relevant to PBH formation. Therefore, the variance of density contrast can be related to the primordial power spectrum as [30,152][13]

$$\sigma^2(R) \equiv \langle \delta^2 \rangle_R = \int_0^\infty d\ln q\, \mathcal{W}^2(q, R)\, \mathcal{P}_\delta(q) \simeq \frac{16}{81} \int_0^\infty d\ln q\, \mathcal{W}^2(q, R)\, (qR)^4\, \mathcal{P}_\mathcal{R}(q) \tag{19}$$

where $\mathcal{W}$ is the Fourier transform of a real space-window function. Popular choices for $\mathcal{W}$ include a volume-normalized Gaussian or a top-hat window function, the Fourier transforms of are respectively given by

$$\mathcal{W}(k, R) = \exp\left(-\frac{k^2 R^2}{2}\right), \qquad \mathcal{W}(k, R) = \frac{3\sin(kR) - 3kR\cos(kR)}{(kR)^3}. \tag{20}$$

When selecting a curvature power spectrum ($\mathcal{P}_\delta$) characterized by a narrow peak around the wave number $k_{\rm pbh}$, the integral in (19) can be approximated as $\sigma^2 \sim \mathcal{P}_\delta(k_{\rm pbh})$. Then, utilizing (18) at horizon entry ($k \simeq aH$) (i.e., at the time of PBH formation), since $81/16 \sim 5$, we can roughly relate the variance (σ) to the primordial curvature power spectrum as

$$\mathcal{P}_\mathcal{R}(k_{\rm pbh}) \sim 5\,\sigma^2 . \tag{21}$$

Finally, recall our considerations after Equation (15): a Gaussian PDF of δ requires $\sigma \simeq 0.06$ ($\sigma \simeq 0.05$) for $M^{\rm (f)}_{\rm pbh} = M_\odot$ ($M^{\rm (f)}_{\rm pbh} = 10^{-12}M_\odot$) to generate a population of $f_{\rm pbh} = 10^{-3}$ ($f_{\rm pbh} = 1$) today. Hence, we can estimate the amplitude of the scalar power spectrum needed at scales relevant for PBH formation:

$$\mathcal{P}_\mathcal{R}(k_{\rm pbh}) \sim 5\,\sigma^2 \sim 10^{-2} \qquad \text{for Gaussian fluctuations.} \tag{22}$$

This estimate holds for a wide range of subsolar PBH masses. This implies that we need a very large amplification of the curvature spectrum between large CMB and small PBH formation scales:

$$\Delta\mathcal{P}_\mathcal{R} \equiv \frac{\mathcal{P}_\mathcal{R}(k_{\rm pbh})}{\mathcal{P}_\mathcal{R}(k_{\rm cmb})} \sim 10^7 , \tag{23}$$

and the task is to produce such amplification in a controllable way by an appropriate inflationary mechanism.

It is worth pointing out that this estimate does not change much for even smaller-mass PBHs with $M^{\rm (f)}_{\rm pbh} < 10^{-12}M_\odot$ because the power spectrum has a logarithmic sensitivity to the PBH fraction (β). In order to see this, we can invert the expression (15) and use (22) to relate the primordial power spectrum of curvature perturbations to β as

$$\mathcal{P}_\mathcal{R}(k_{\rm pbh}) \sim 5\sigma^2 \sim \frac{5\,\delta_{\rm c}^2}{2\ln(1/\beta)} . \tag{24}$$

Now, as an extreme case, we can consider the smallest PBHs ($M^{\rm (f)}_{\rm pbh} \simeq 10^{-18}M_\odot$) that can survive until today (not yet eliminated by Hawking radiation), which have the tightest available observational constraints, restricting their current abundance to $f_{\rm pbh} \lesssim 10^{-9}$ [32]. Plugging these values into (12), PBH fraction at formation gives $\beta \lesssim 10^{-28}$, which, in turn, leads to the constraint $\mathcal{P}_\mathcal{R}(k_{\rm pbh}) \lesssim 6 \times 10^{-3}$ in (24) for a threshold of $\delta_{\rm c} = 0.4$. Therefore, we conclude that for Gaussian perturbations and for any PBH mass of interest, the amplitude of the scalar power spectrum relevant for PBH formation requires $\mathcal{P}_\mathcal{R} \sim 10^{-2}$ for any potentially observable PBH fraction ($f_{\rm PBH}$) today. The discussion presented above informs us that a small change in the amplitude of the power spectrum leads to a difference of many orders of magnitude in the fraction of regions collapsing into PBHs, as clearly implied by the exponential dependence of β on $\mathcal{P}_\mathcal{R}$ in (24). Similarly, a small change in the choice of threshold ($\delta_{\rm c}$) could lead to very different estimates in terms of β. For example, focusing on a fixed value of variance ($\sigma^2 \simeq 0.05$) as relevant for PBH formation, β (15) can change by various orders of magnitude if we reduce the threshold ($\delta_{\rm c}$) by about 20%. In fact,

$$\frac{\beta(\delta_{\rm c} = 0.4)}{\beta(\delta_{\rm c} = 1/3)} \simeq 10^{-5} , \tag{25}$$

demonstrating how tuned the conditions are for producing a cosmologically interesting population of PBHs.

Collapse fraction vs. curvature perturbation

While it is customary to use the smoothed density contrast at horizon crossing to estimate the number of collapsing regions, it is also possible to work directly with the comoving curvature perturbation to approximately compute the PBH fraction (β) at the time of formation. In this case, there is no need of to rely on the smoothing procedure of subhorizon fluctuations provided by the window functions [152,153]. Interestingly, as we will see later, this approach also provides a way to assess the effects of large primordial non-Gaussianity that might be present in some of the PBH-forming inflationary scenarios.

To understand the role of the primordial curvature fluctuation ($\mathcal{R}$), we approximate its variance with the power spectrum $\sigma_{\mathcal{R}}^2 \approx \mathcal{P}_{\mathcal{R}}$. Using the Press–Schechter approach with a Gaussian PDF for the curvature fluctuation spectrum, the fraction of collapsing regions at formation can be estimated as

$$\beta_G = \int_{\mathcal{R}_c}^{\infty} \mathrm{d}\mathcal{R}\, \frac{e^{-\mathcal{R}^2/(2\sigma_{\mathcal{R}}^2)}}{\sqrt{2\pi}\sigma_{\mathcal{R}}} \simeq \frac{1}{2}\mathrm{Erfc}\left(\frac{\mathcal{R}_c}{\sqrt{2\mathcal{P}_{\mathcal{R}}}}\right), \tag{26}$$

where $\mathcal{R}_c$ is the threshold. To roughly estimate $\mathcal{R}_c$, we can assume an almost scale-invariant power spectrum of $\mathcal{R}$ for a logarithmic range of wave numbers relevant to PBH formation. Making use of a Gaussian window function in (19) gives, in this case, $\sigma_{\mathcal{R}}^2 = \langle \mathcal{R}^2 \rangle \simeq 8\mathcal{P}_{\mathcal{R}}/81$. Finally, plugging the latter into (15), and comparing the resulting expression with (26), we obtain [152]:

$$\mathcal{R}_c \approx \frac{9}{2\sqrt{2}}\delta_c. \tag{27}$$

For a density threshold of $\delta_c = 0.4$, the relation presented above gives $\mathcal{R}_c \simeq 1.3$, which we set as the fiducial value for the estimates below. Following these considerations and using the formulas derived so far—in particular Equations (12) and (26)—we can then repeat the previous estimates and determine the approximate amplitude of the power spectrum required for PBH formation. The result is that for Gaussian primordial fluctuations, a sensible PBH population today requires $\mathcal{P}_{\mathcal{R}} \sim 10^{-2}$. These findings confirm our earlier results of Equation (22).

The case of non-Gaussian curvature fluctuations

So far, we have assumed that primordial fluctuations obey Gaussian statistics in order to estimate the amplitude of the power spectrum required for PBH formation. Since PBHs are expected to form through extremely rare large fluctuations (see Figure 3), any small deviations in the shape of the tail of the fluctuation distribution—which essentially depend on the amount of non-Gaussianity (i.e., skewness of the PDF)—can have a significant impact on the PBH abundance [153–161]. For the sake of obtaining a lower limit on the amplitude of the PBH-forming curvature power spectrum, we now consider scenarios with large primordial non-Gaussianity. A particularly interesting case of this type occurs if the main source of the curvature perturbation results from a higher-order interaction, where the distribution of $\mathcal{R}$ can be modeled as a χ^2 distribution (see, e.g., [79,153,162]):

$$\mathcal{R} = g^2 - \langle g^2 \rangle, \tag{28}$$

where g is a Gaussian random variable ($\langle g \rangle = 0$) with variance $\sigma_g^2 \equiv \langle g^2 \rangle$. The PDF of $\mathcal{R}$ in this case can be determined by making a change of variable $P_{\mathrm{NG}}(\mathcal{R}) = P_G(g)|\mathrm{d}g/\mathrm{d}\mathcal{R}|$, which takes the following form

$$P_{\mathrm{NG}}(\mathcal{R}) = \frac{e^{-(\mathcal{R}+\sigma_g^2)/2\sigma_g^2}}{2\sqrt{2\pi(\mathcal{R}+\sigma_g^2)}\,\sigma_g}. \tag{29}$$

Making a change of variable to a quantity t through the definition $\sigma_g^2 t = \mathcal{R} + \sigma_g^2 \longrightarrow \mathrm{d}\mathcal{R} = \sigma_g^2\,\mathrm{d}t$, the fraction of regions in the universe that can collapse to form PBHs can be estimated as

$$\beta_{\mathrm{NG}} = \int_{\mathcal{R}_c}^{\infty} \mathrm{d}\mathcal{R}\; P_{\mathrm{NG}}(\mathcal{R}) \simeq \frac{1}{2}\mathrm{Erfc}\left(\sqrt{\frac{1}{2} + \frac{\mathcal{R}_c}{\sqrt{2\mathcal{P}_{\mathcal{R}}}}}\right), \tag{30}$$

where in the last step, we approximate the variance as $\langle \mathcal{R}^2 \rangle = 2\langle g^2 \rangle^2 = 2\sigma_g^4 \approx \mathcal{P}_{\mathcal{R}}$, in order to express $\beta_{\rm NG}$ in terms of the curvature power spectrum[14]. We can now compute the amplitude of the curvature power spectrum required for PBH formation when the statistics of fluctuations are strongly non-Gaussian. Using (12), together with (30), a population of solar mass PBHs with $f_{\rm pbh} = 10^{-3}$ requires $\mathcal{P}_{\mathcal{R}} \simeq 1.5 \times 10^{-3}$, whereas for a population of PBHs with $M^{(\rm f)}_{\rm pbh} = 10^{-12} M_{\odot}$ and $f_{\rm pbh} = 1$, we find $\mathcal{P}_{\mathcal{R}} \simeq 9 \times 10^{-4} \simeq 10^{-3}$. Hence, we conclude that the required amplitude of the power spectrum is roughly given by

$$\mathcal{P}_{\mathcal{R}}(k_{\rm pbh}) \sim 10^{-3} \qquad \text{for non-Gaussian fluctuations.} \tag{31}$$

Compared to the case of Gaussian distributed curvature perturbation (see Equation (22)), we learn that the required amplitude of the power spectrum is reduced by about one order of magnitude. Therefore, a curvature perturbation with a smaller amplitude can produce the same amount of PBH abundance if non-Gaussianity is present (see Equation (28)). In particular, for $\mathcal{R}_{\rm c}^2 \gg \mathcal{P}_{\mathcal{R}}$, which is typically satisfied by a satisfactory approximation, we can disregard the $1/2$ factor in (30). Compared with Equation (26), the power spectrum required to generate the same collapse fraction of PBHs in both cases can be related as

$$\mathcal{P}_{\mathcal{R}_{\rm NG}} \simeq \frac{2}{\mathcal{R}_{\rm c}^2} \mathcal{P}^2_{\mathcal{R}_{\rm G}}. \tag{32}$$

To further illustrate these points, in Figure 4, we show the quantity (β) both for Gaussian and non-Gaussian cases, represented as a function of the curvature power spectrum. We learn from the figure that in the non-Gaussian case, within a phenomenologically relevant interval of β (see, e.g., (12)), a given value of the power spectrum leads to a much larger value of β. We emphasize that we focus on a specific type of non-Gaussian distribution (namely χ^2) to estimate the amplitude of the power spectrum required for PBH formation; therefore, the derived results could change for milder cases depending on the amplitude and sign of the non-Gaussianity (i.e., depending on whether the PDF in (26) is positively or negatively skewed) [153].

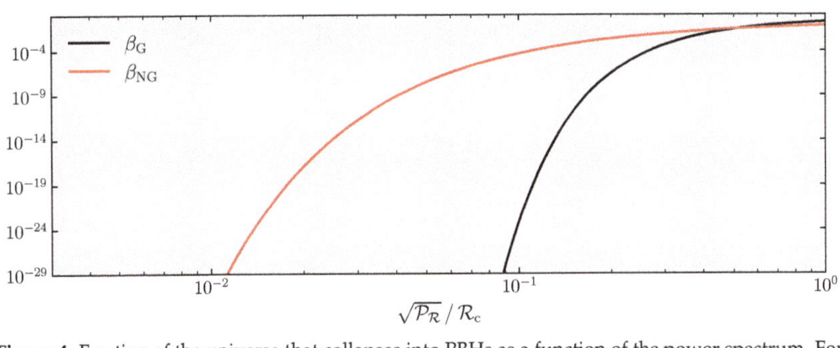

Figure 4. Fraction of the universe that collapses into PBHs as a function of the power spectrum. For a phenomenologically interesting interval of β (see, e.g., (12)) values, in the non-Gaussian case, we need a smaller amplitude of the power spectrum in order to generate the same amount of PBHs at horizon re-entry.

2.4. Brief Summary and the Path Ahead

Let us summarize the arguments that we have reviewed so far. We computed the required amplitude of the small-scale primordial power spectrum ($\mathcal{P}_{\mathcal{R}}(k_{\rm pbh})$) required to generate a sizable population of PBHs that can account for all or a fraction of DM density. The typical small scale of PBH formation ($k_{\rm pbh}$) is related to the BH mass through

Equation (6) (see Table 1 for examples). Compared with the power spectrum at large CMB scales, we need a

$$\Delta\mathcal{P}_{\mathcal{R}} \equiv \frac{\mathcal{P}_{\mathcal{R}}(k_{\rm pbh})}{\mathcal{P}_{\mathcal{R}}(k_{\rm cmb})} \sim 10^6 - 10^7 \tag{33}$$

enhancement in the spectrum amplitude between small and large scales, depending on the statistics obeyed by the primordial curvature perturbation (see Equations (22) and (31)).

We also learned that the PBH abundance is extremely sensitive to the amplitude of the primordial curvature spectrum. Notice that the results we reviewed are derived for the case of PBHs produced during the RDU; if early phase transitions or early phases of non-standard cosmology occur, the corresponding modified equations of state can also considerably influence the properties of the PBH population [143,163]. An interesting example is the QCD phase transition, which can lead to a high peak in the distribution of solar-mass PBHs several orders of magnitude larger than the corresponding values in the RDU [164,165].

There are various opportunities to improve and elaborate on these results. In our considerations, we assumed that PBHs form at a particular mass (Equations (4) and (6)), for example, by means of a sharply peaked primordial power spectrum; moreover, we ignored the effects of clustering [22,166–169] and mergers [170–172] on the PBH abundance and evolution. As shown in [140,173,174], assumptions about the shape of the primordial spectrum may alter the PBH abundance and the corresponding clustering properties. Another topic of debate concerns the use of over-density (δ) versus the curvature perturbation ($\mathcal{R}$) when computing the PBH abundance (see, e.g., [175] for a discussion on these issues). In light of the discussion presented above, we emphasize that the calculations we carried out in this section should be regarded as rough order-of-magnitude estimates, requiring more precise numerical analysis. Furthermore, in discussing the effects of non-Gaussianities on PBH formation, we stress that we computed the corresponding power spectrum only for an extreme example of χ^2 non-Gaussian statistics. For a detailed analysis of the impact of primordial non-Gaussianity on PBH formation and abundance, we refer the reader to the general discussion in [176]. An additional phenomenological consequence of PBH-forming inflationary scenarios is the inevitable production of a stochastic gravitational wave (GW) background [87,88]. In fact, an enhanced spectrum of curvature fluctuations, as needed to produce PBH, acts as a source of GW [89]. The characterization of this GW background (see, e.g., [93,94,96]), along with the properties of its scalar sources (see, e.g., [92,97,177]) and the corresponding forecasts for its detection, is an important avenue for the experimental probing of PBH-forming inflationary models. We refer the reader to [105] for a detailed recent review of scalar-induced GW backgrounds.

All the topics mentioned above are being actively developed by the PBH community. The arguments and results we reviewed in this section are sufficient for introducing our specific purpose, which is to review the theoretical foundation of inflationary scenarios leading to PBHs. In the following sections, we discuss different conceptual ideas and concrete inflationary mechanisms for obtaining the enhancement (33) of the curvature power spectrum, as needed to generate PBHs. We focus on the inflationary theory aspects only, without computing the resulting PBH abundance, as well as other phenomenological properties that are already covered in various recent complementary reviews [28–34].

3. Enhancement of Scalar Perturbations during Single-Field Inflation

We now focus our attention on inflationary scenarios able to lead to PBH formation. As we learned in the previous section, such scenarios are characterized by a significant enhancement in the curvature power spectrum at a scale ($k_{\rm pbh}$) (which depends on the

PBH mass) much smaller with respect to CMB scales ($k_{\rm cmb}$). The condition to satisfy is Equation (23), which we rewrite here as:

$$\frac{\mathcal{P}_{\mathcal{R}}(k_{\rm pbh})}{\mathcal{P}_{\mathcal{R}}(k_{\rm cmb})} \sim 10^7 \,. \tag{34}$$

We classify inflationary models as single-field (this section) or multifield type (next section) depending on whether the mechanism responsible for the enhancement in the scalar fluctuations relies on a single- or multifield scenario. In general, existing inflationary mechanisms amplify the spectrum of curvature fluctuations by means of significant gradients in the background evolution of fields responsible for inflation. In this section, we phrase our discussion as model-independently as possible, mostly focusing on conceptual aspects of the problem. We aim to discuss the dynamics and the general properties of curvature fluctuations in inflationary models leading to PBHs and refer to specific representative scenarios when necessary.

Main References: Our discussion in this section is based on [41,42,52,64].

3.1. The Dynamics of Curvature Perturbation

In order to analyze the behavior of the scalar power spectrum in single-field scenarios, we consider the second-order action of scalar perturbations around an inflationary phase of evolution. The background metric corresponds to a (quasi) de Sitter background with a nearly constant Hubble parameter (H). Cosmological inflation is controlled by a slow-roll parameter ($\epsilon \equiv \dot{H}/H^2$) satisfying $\epsilon \ll 1$, with $\epsilon = 1$ corresponding to the condition required to conclude the inflationary process. We work with conformal time ($\tau \leq 0$) during inflation (see, e.g., [178] for a classic survey of inflationary models).

The dynamics of scalar fluctuations can be formulated in terms of the comoving curvature perturbation ($\mathcal{R}$) [129,130], the second-order action (at the lowest order in derivatives)[15] of which takes the following form (see, e.g., [64])

$$S^{(2)}_{\mathcal{R}} = \frac{1}{2}\int \mathrm{d}\tau\, \mathrm{d}^3x\, \frac{2\,a^2(\tau) M^2(\tau)\, \epsilon(\tau)}{c_s^2(\tau)} \left[\mathcal{R}'^2 - c_s^2(\tau)(\vec{\nabla}\mathcal{R})^2\right]. \tag{35}$$

In this formula, c_s is the sound speed of the curvature perturbation, M is an effective time-dependent Planck mass and ϵ is the aforementioned slow-roll parameter.

First, we present a few initial words to contextualize single-field models aimed to produce PBHs, leading to a set of dynamics of curvature perturbation controlled by action (35). The simplest option to consider is PBH-forming models with unit sound speed and constant Planck mass characterized only by the shape of the potential ($V(\phi)$). As mentioned in the Introduction, models in this class require a potential characterized by a flat , plateau-like region (see, e.g., [46–53,55] for a choice of works studying this possibility; we will discuss its implications for the dynamics of curvature perturbations in the next subsection). PBH-forming potentials with the required characteristics can find explicit realizations, for example, in models of Higgs inflation [47,185–187],alpha attractors [188,189] and string inflation [51,52,190]. Considering more complex possibilities, PBH-generating models that exploit a time dependence for the sound speed are based on non-canonical kinetic terms for the inflation scalar, such as K-inflation [191,192] (see, e.g., [64–67,193–197] for concrete examples and Section 3.3 for some of their implications). Finally, scenarios with a time-dependent effective Planck mass can be generated by non-minimal couplings of the inflation scalar with gravity, as in the Horndeski action [198] and its cosmological applications to G-inflation scenarios [199]. Realizations of PBH-forming models in setups with non-minimal couplings belonging to the Horndeski sector include [200–203]. To the best of our knowledge, early universe models based on the more recent covariant DHOST actions [204–206] have not yet been explored in the context of PBH model building.

Interestingly, despite the many distinct concrete realizations, all single-field scenarios rely on a few common mechanisms to enhance the spectrum of curvature fluctuations, which exploit the behavior of background quantities. We are now going to discuss these mechanisms in a model-independent way. We treat M, c_s and ϵ as appearing in action (35) as time-dependent quantities controlled by the single scalar background profile that drives inflation. To start with, it is convenient to redefine the time variable in action (35) to adsorb the time-dependent c_s into a rescaled conformal time and impose an equal-scaling condition of time and space coordinates:

$$d\bar{\tau} = c_s d\tau \quad \Longrightarrow \quad S_{\mathcal{R}}^{(2)} = \frac{1}{2}\int d\bar{\tau}\, d^3x\, z^2(\bar{\tau})\left[\mathcal{R}'^2 - (\vec{\nabla}\mathcal{R})^2\right], \tag{36}$$

with a prime indicating a derivative with respect to $\bar{\tau}$, the rescaled conformal time. Importantly, we introduce a so-called time-dependent 'pump field' ($z(\bar{\tau})$) as

$$z^2(\bar{\tau}) = \frac{2\,a^2(\bar{\tau})\,M^2(\bar{\tau})\,\epsilon(\bar{\tau})}{c_s(\bar{\tau})}. \tag{37}$$

The dynamics of $\mathcal{R}$ are strongly tied to the time dependence of the pump field ($z(\bar{\tau})$) and, more generally, to the behavior of the background quantities (M, ϵ, c_s) that constitute it.

To analyze the evolution mode by mode, we work in Fourier space and write the Euler–Lagrange mode equation for curvature perturbation derived from the action (36):

$$\frac{1}{z^2(\bar{\tau})}\left[z^2(\bar{\tau})\mathcal{R}_k'(\bar{\tau})\right]' = -k^2\mathcal{R}_k(\bar{\tau}), \tag{38}$$

where $k \equiv |\vec{k}|$ is the magnitude of the wave number that labels a given mode. This is a differential equation involving derivatives along the time direction acting on the function $\mathcal{R}_k(\bar{\tau})$ depending on both time and momentum (k).

To express its solution, we implement a gradient expansion approach (see, e.g., [41,42,52]), starting from the solution in the limit of small $k/(aH)$ and including its momentum-dependent corrections, which solve (38) order by order in a $k/(aH)$ expansion. This approach is particularly suitable for our purpose of describing scenarios in which the size of small-scale curvature fluctuations ($k/(aH)$, large) differs considerably from that of large-scale curvature fluctuations ($k/(aH)$, small) (see condition (23)). Indeed, a gradient expansion allows us to better understand the physical origin of possible mechanisms that raise the curvature spectrum at small scales.

The most general solution of Equation (38), up to second order in powers of $k/(aH)$, is formally given by the following integral equation[16]

$$\mathcal{R}_k(\bar{\tau}) = \mathcal{R}_k^{(0)}\left[1 + \frac{\mathcal{R}_k^{(0)\prime}}{\mathcal{R}_k^{(0)}}\int_{\bar{\tau}_0}^{\bar{\tau}}\frac{d\bar{\tau}'}{\tilde{z}^2(\bar{\tau}')} - k^2\int_{\bar{\tau}_0}^{\bar{\tau}}\frac{d\bar{\tau}'}{\tilde{z}(\bar{\tau}')^2}\int_{\bar{\tau}_0}^{\bar{\tau}'}d\bar{\tau}''\,\tilde{z}^2(\bar{\tau}'')\,\frac{\mathcal{R}_k(\bar{\tau}'')}{\mathcal{R}_k^{(0)}}\right], \tag{39}$$

where the sub- and superscripts 0 denote a reference time, and a tilde over a time-dependent quantity indicates that it is normalized with respect to its value at $\bar{\tau} = \bar{\tau}_0$.

Typically, we are interested in relating the late time curvature perturbation at $\bar{\tau}$ to the same quantity computed at some earlier time (τ_0). For this purpose, it is convenient to identify $\bar{\tau}_0$ as the time coordinate evaluated soon after horizon crossing and $\mathcal{R}_k^{(0)}$ as the mode function computed at $\bar{\tau}_0$. In order to enhance the spectrum of curvature fluctuations at small scales (recall the PBH-forming condition of Equation (23)), we can envisage two possibilities. One option is to exploit the structure of Equation (39), making sure that its contributions within the square parentheses become increasingly important as time proceeds after modes leave the horizon. In this way, we generate a sizable scale dependence for $\mathcal{R}_k(\bar{\tau})$ after horizon crossing, with the possibility of amplifying the small-scale curvature spectrum. Alternatively, we can design methods that lead to significant scale dependence already at horizon crossing, i.e., for the quantity $\mathcal{R}_k^{(0)}$, which then maintains its value at superhorizon scales. We explore both these options in Sections 3.2 and 3.3, respectively.

To develop a quantitative discussion, it is convenient to introduce the so-called slow-roll parameters as

$$\eta \equiv \frac{\mathrm{d}\ln\epsilon}{\mathrm{d}N}, \; s \equiv \frac{\mathrm{d}\ln c_s}{\mathrm{d}N}, \; \mu \equiv \frac{\mathrm{d}\ln M^2}{\mathrm{d}N} \tag{40}$$

where, in our definition, we make use of the relation between e-foldings and the time coordinate ($\bar{\tau}$: $\mathrm{d}N = H\mathrm{d}t = (aH/c_s)\mathrm{d}\bar{\tau}$).

In standard models of inflation based on inflationary attractor dynamics, one imposes the so-called slow-roll conditions throughout the entire inflationary period, corresponding to the requirements

$$\eta,\, s,\, \mu,\, \frac{\mathrm{d}\eta}{\mathrm{d}N}, \frac{\mathrm{d}s}{\mathrm{d}N}, \frac{\mathrm{d}\mu}{\mathrm{d}N} \ll 1, \tag{41}$$

which imply that the pump field *always grows in time* $z^2 \propto (-\bar{\tau})^{-2}$ as $\bar{\tau} \to 0$ (see Equation (37)). As a consequence, the second and third terms in the general solution (39) decay as $(-\tau)^3$ and $(-\tau)^2$, respectively in the late time limit ($(-\tau) \to 0$). Hence, they can be identified as decaying modes[17] that rapidly cease to play any role in the dynamics of curvature perturbations. This is a regime of a *slow-roll attractor*, wherein soon after horizon crossing, the curvature perturbation settles into a nearly-constant configuration ($\mathcal{R}_k^{(0)}$), the spectrum of which is almost scale-invariant. In this case, the momentum-dependent terms in Equation (39) do not have the opportunity to raise the curvature spectrum at small scales.

Hence, in order to produce PBHs, we need to go beyond the slow-roll conditions of Equation (41), as first emphasized in [207]. Before discussing concrete ideas to do so, in view of numerical implementations, as well to improve our physical understanding, it is convenient to express the curvature perturbation (Equation (38)) in a way that makes more manifest the role of slow-roll parameters in controlling the mode evolution. We introduce a canonical variable (v_k) defined as

$$v_k(\bar{\tau}) = z(\bar{\tau})\mathcal{R}_k(\bar{\tau})\,. \tag{42}$$

Plugging this definition into Equation (38), we obtain the so-called Mukhanov–Sasaki equation, which reads

$$v_k''(\bar{\tau}) + \left(k^2 - \frac{z''}{z}\right)v_k(\bar{\tau}) = 0, \tag{43}$$

where

$$\frac{z''}{z} = 2\left(\frac{aH}{c_s}\right)^2 [1+\Theta]. \tag{44}$$

Expanding the derivatives of z (37) in terms of the slow-roll parameters of Equation (40), we define

$$\Theta \equiv -\frac{1}{2}(\epsilon+s) + \frac{3}{4}(\eta - s + \mu) + \frac{1}{8}(\eta - s + \mu)^2 - \frac{1}{4}(\epsilon+s)(\eta - s+\mu) + \frac{1}{4}\left(\frac{\mathrm{d}\eta}{\mathrm{d}N} - \frac{\mathrm{d}s}{\mathrm{d}N} + \frac{\mathrm{d}\mu}{\mathrm{d}N}\right). \tag{45}$$

Standard slow-roll attractor scenarios correspond to situations in which Θ is negligibly small; the quantity in Equation (44) then reads $2\,\bar{\tau}^{-2}$, leading to a scale-invariant curvature power spectrum. To break the scale invariance of curvature perturbation, we need to consider a sizable time-dependent Θ. We note that the expression (45) is exact and does not assume any slow-roll hierarchy as Equation (41). Hence, it can be used to study the system beyond slow roll, as we will do in the following section.

3.2. *Enhancement through the Resurrection of the Decaying Mode*

The idea

An interesting mechanism to enhance the curvature perturbation at superhorizon scales is suggested by the structure of the integrals within the square parentheses of Equation (39). Suppose that for a brief time interval, a given mode (k) experiences a background evolution during which the pump field (z) *rapidly decreases* after the horizon exit epoch ($\bar{\tau}_0$). Then, the 'decaying' mode can increase, and the integrals in the parentheses of Equation (39) can contaminate the nearly constant solution ($\mathcal{R}_k^{(0)}$), eventually leading to a late-time value ($\mathcal{R}_k(\bar{\tau}) \gg \mathcal{R}_k^{(0)}$) on superhorizon scales. This situation signals a significant departure from the attractor slow-roll regime discussed after Equation (41). In fact, in this case, the criterion for the enhancement of the curvature perturbation can be explicitly phrased in terms of the derivative of the pump-field transiently changing sign during a short time interval during inflation:

$$\frac{z'}{z} = \frac{aH}{c_s}\left[1 + \frac{\eta - s + \mu}{2}\right] < 0\,. \tag{46}$$

This condition implies that the combination of the slow-roll parameters ($\eta - s + \mu$) should be of order $\mathcal{O}(1)$ and negative during some e-folds during inflation, violating the slow-roll conditions (41). In particular, we require

$$\eta - s + \mu < -2\,. \tag{47}$$

If Equation (46) is satisfied, the slow-roll conditions (41) are not satisfied, and the contributions within parentheses in Equation (39) can increase. Strong time gradients of homogeneous background quantities, which lead to condition (47), can then be converted into a small-scale amplification of the curvature power spectrum. As discussed in [64], the expression (46), along with the considerations mentioned above, generalizes to a time-dependent sound speed and Planck mass according to the arguments first developed in [41,42].

Model building parameterization of the non-attractor phase

To illustrate a viable model that can generate a seven-order-of-magnitude enhancement required for PBH formation—see Equation (23)—we focus on canonical single-field models ($c_s \to 1$, $M \to M_{\rm pl}$ and $s \to 0$, $\mu \to 0$) in order to simplify our analysis. The background evolution for the single scalar field driving inflation is

$$\ddot{\phi} + 3H\dot{\phi} + V'(\phi) = 0\,, \tag{48}$$

where $V(\phi)$ is the scalar potential, and the time derivatives are carried on in coordinate time (t). The non-slow-roll dynamics are controlled by the properties of the potential (V), as we are going to discuss, and by its consequences for the behavior of the inflation velocity ($\dot{\phi}$).

Since in these scenarios, the pump field can be parameterized purely in terms of the slow-roll parameter (ϵ) as $z = a\sqrt{2\epsilon}\,M_{\rm pl}$ (see, e.g., Equation (37)), the linear dynamics of $\mathcal{R}_k$ (Equation (39)) are dictated by the first slow-roll parameter, the evolution of which is, in turn, determined by the sign and amplitude of the slow-roll parameter η. Hence, the criterion required to realize the desired growth in the spectrum can be simply parameterized as a condition of the second slow-roll parameter as $\eta < -2$ in (47).

From a concrete model-building perspective, scalar potentials ($V(\phi)$) that can induce this type of dynamics include a characteristic 'plateau' within a non-vanishing field range ($\Delta\phi \neq 0$) [38,39,208]. This property gives rise to phases of transient non-attractor dynamics of ultra slow-roll (USR) [40,43,44] or constant-roll (CR) [45,209,210] evolution, depending on the shape profile of the potential around the aforementioned feature. In particular, for USR evolution, the potential typically has a very flat plateau with $V' \simeq 0$, whereas for constant-roll evolution, $V' < 0$ so that the field climbs a hill by overshooting a local minimum[18]. As the scalar field, during its evolution, traverses such a flat region with negligible potential gradient, the acceleration term ($\ddot{\phi}$) is balanced by the Hubble damping term in the Klein–Gordon Equation (48), and the inflation speed is no longer controlled by the scalar potential. This phenomenon significantly changes the values of the inflation velocity ($\dot{\phi}$) during the transient non-attractor phase and inevitably leads to the violation of one of the slow-roll conditions:

$$\ddot{\phi} + 3H\dot{\phi} + V'(\phi) = 0 \implies \eta = -6 - \frac{2V'}{\dot{\phi}H} + 2\epsilon; \tag{49}$$

hence, $\eta \simeq -6$ ($\eta < -6$) for transient USR ($V' = 0$) and CR ($V' < 0$) phases [19]. We emphasize that since the non-slow-roll inflationary era is characterized by a large negative η for a brief interval of e-folds, the pump field, as well as the first slow-roll parameter (ϵ), quickly decays during this stage as required for the activation of the decaying modes. In fact,

$$\frac{\mathrm{d}\ln\epsilon}{\mathrm{d}N} \equiv \eta \quad \implies \quad z^2 \propto \epsilon \propto e^{-|\eta|\Delta N}, \tag{50}$$

where, for simplicity, we assume a constant η during the non-attractor phase. For explicit inflationary scenarios that can realize such transient phases in the context of PBH formation, see, e.g., [49,51,52,211]. Nevertheless, it is worth pointing out that, although possible, explicit constructions of suitable inflationary potentials involve a high degree of tuning to render the potentials extremely flat for a small region in the field range and ensure an appropriate transition for the scalar velocity among different epochs (see, e.g., the discussion in [50], as well as the comments at the end of this section).

After this general discussion of model building, in the analysis that follows, we do not need to work with an explicit form of potential ($V(\phi)$) to analyze the enhancement through the non-attractor dynamics. Instead, we exploit the general idea we are discussing in a model-independent way and model PBH-forming inflationary scenarios as a succession of distinct phases that connect smoothly with one another, each parameterized by a constant η (related approaches are developed in [212–216]). Our perspective encompasses the important features of scenarios based on the idea of transient resurrection of the decaying mode at superhorizon scales, satisfying Equations (46) and (47). In order to capture the transitions among phases, we multiply each phase by the smoothing function [217]:

$$\sigma(N, \Delta) = \frac{1}{2}\left[\tanh\left(\frac{N - N_i}{\Delta}\right) - \tanh\left(\frac{N - N_f}{\Delta}\right)\right], \tag{51}$$

where N denotes e-folds; N_i and N_f are the e-folding numbers at the beginning and end of the constant η phase, respectively; and Δ signifies the duration of the smoothing procedure. Keeping this smoothing prescription in mind, the inflationary evolution can be divided into three phases:

- **Phase I**. The initial phase of inflationary evolution is characterized by a standard slow-roll (SR) regime, where $\epsilon, \eta \ll 1$ and $\epsilon < \eta$ at the pivot scale ($k_{\rm cmb} = 0.05\ \mathrm{Mpc}^{-1}$) (assuming that modes at the pivot scale exit the horizon at the beginning of evolution ($N_{\rm b} = 0$)) in order to match Planck observations [4];
- **Phase II**. As the scalar field starts to traverse the flat plateau-like region in its potential, its dynamics eventually enter the non-attractor era, lasting a given number of e-folds of evolution. This phase is characterized by a large negative $\eta < -6$, during which the first slow-roll parameter (ϵ) decays exponentially:

$$\epsilon(N) \equiv \exp\left[\int_{60}^{N} \eta(N')\,\mathrm{d}N'\right]; \tag{52}$$

- **Phase III**. The final phase of evolution ensures a graceful exit from the non-attractor phase into a final slow-roll epoch, leading to the end of inflation. Since ϵ decays quickly in the non-attractor era, this final phase is characterized by a hierarchy between the slow-roll parameters:

$$\eta \gg \epsilon \tag{53}$$

where $\eta > 0$. We typically require a large positive η to bring back ϵ from its tiny values at the end of the non-attractor era, towards the value ($\epsilon = 1$) needed to conclude inflation. To capture this behavior accurately, we split the final phase of evolution into two parts, parameterizing η as

$$\eta(N) = \eta_{\rm III}^{(1)}\,\sigma_1(N,\Delta) + \eta_{\rm III}^{(2)}\,\sigma_2(N,\Delta)\,. \tag{54}$$

The relevant parameter choices to model the dynamics can be found in the third column of Table 2.

We note that our choice of η in the initial stage of the Phase III and in Phase II is not a coincidence; in most of the single-field modes, there exists a correspondence that relates the η values in Phase II and Phase III, i.e., $\eta_{\rm III} = -6 - \eta_{\rm II}$, which is a consequence of Wands' duality [218]. We will elaborate the consequence of this correspondence in the context of the power spectrum below, in particular for modes that exit the horizon as the background evolves from Phase II to Phase III.

Table 2. Parameter choices that characterize the background evolution of η smoothed by the function (51) in each phase. Note that the final phase of evolution is divided into two in orders to accommodate the end of inflation with $\epsilon = 1$ at $N_{\rm end} = 60$.

	Phase I	Phase II	Phase III	
η	0.02	−6.30	0.30	3.00
N_i	0.00	33.2	35.7	55.0
N_j	33.2	35.7	55.0	65.0
ν	1.00	0.50	1.00	2.00

In light of the discussion presented above, we can characterize the full background evolution using the Hubble hierarchy in (52) and $H(N) = H_{\rm end}\exp[-\int_{60}^{N}\epsilon(N')\mathrm{d}N']$, where $H_{\rm end}$ denotes the Hubble rate at the end of inflation, where $N_{\rm end} = 60$.

For a representative set of parameter choices (see Table 2), in Figure 5, we show an example of background evolution in which we plot ϵ, η and z'/z. The right panel of the figure makes manifest that the background evolution leads to $z'/z < 0$ for a short interval of e-folds ($N = 33$–35.7), as highlighted by the red region in the plot. In accordance with our discussion so far, this behavior is appropriate for triggering a significant enhancement in the power spectrum of curvature perturbation through the resurrection of the decaying mode.

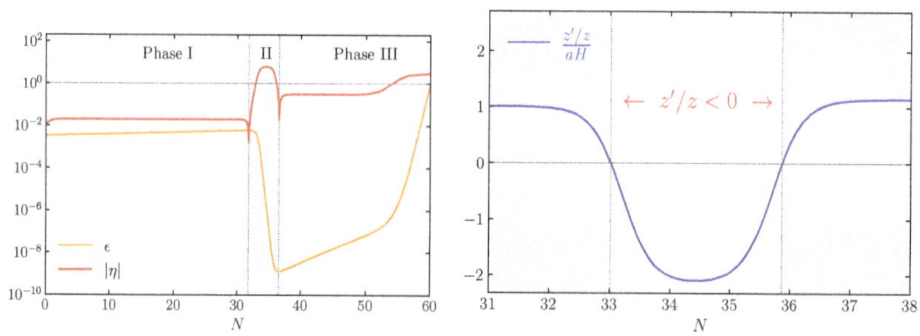

Figure 5. (**Left**) panel: evolution of ϵ and η in e-folds through the successive phases outlined in the main text. The green region indicates the range of e-fold numbers where $\eta < 0$, corresponding roughly to the beginning and end of the non-attractor phase. (**Right**) panel: the time evolution of $z'/z = aH(1 + \eta/2)$, with $z'/z < 0$ in the region highlighted in red.

Numerical analysis

Having obtained the background evolution, we are ready to describe mode evolution to obtain the power spectrum of curvature perturbation towards the end of inflation[20]:

$$\mathcal{P}_{\mathcal{R}}(k, N_{\rm end}) = \frac{k^3}{2\pi^2}\left|\frac{v_k(N_{\rm end})}{z(N_{\rm end})}\right|^2, \tag{55}$$

where to study the evolution of curvature perturbations, we make use of the canonical variable (v_k) and consider the Mukhanov–Sasaki system of Equations (43)–(45) after setting $s = \mu = 0$. In general, it is not possible to find full analytic solutions for this system of equations, and a numerical analysis is needed. However, as we will explain soon, interesting properties of the resulting curvature spectrum can be derived and understood analytically. We implement the numerical procedure explained in detail in the technical Appendix C, which solves the Mukhanov–Sasaki equation with Bunch–Davies initial conditions, and we provide a Python code that reproduces our numerical findings[21]. The resulting power spectrum is represented in Figure 6; it manifestly grows in amplitude towards small scales, exhibiting a peak at around $k_{\rm peak} \simeq 10^{12}$ Mpc$^{-1} \gg k_{\rm cmb}$. Notice that the spectrum grows as k^4 towards its peak and is characterized by a dip preceding the phase of steady growth [212]. We will have more to say soon about these features.

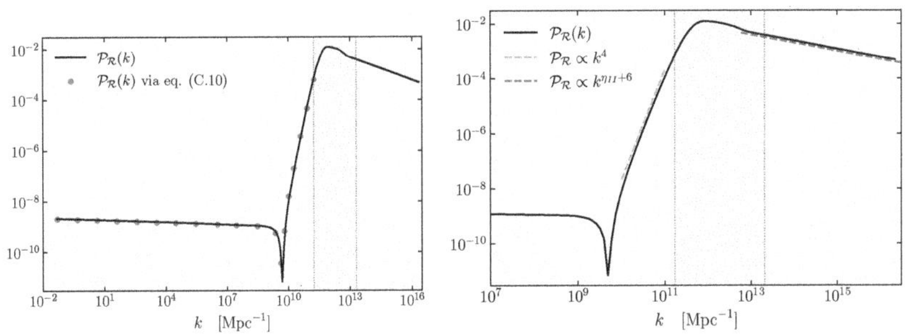

Figure 6. Power spectrum of curvature perturbation in the three-phase model described in the main text. The pale green region separated by the vertical lines denotes the range of modes that exit the horizon during the non-attractor era, whereas the light blue regions denote the range of modes that cross the horizon during the initial and final slow-roll era, respectively.

Interestingly, for the system under consideration, the bulk of the enhancement can be attributed to the active dynamics of the would-be 'decaying modes', i.e., the second and third terms of Equation (39). To show this explicitly, we study superhorizon solution of the curvature perturbation in Appendix C by applying the Formula (A38), which is a special case of Equation (39) applied to the canonical single-field scenario we discuss here. For a grid of wave numbers that exit the horizon during the initial slow-roll era, the amplitude of the power spectrum obtained in this way is shown by blue dots in Figure 6. The accuracy of these locations with respect to the full numerical result (black solid line) confirms our expectation that decaying modes in (39) play a crucial role for the enhancement of the curvature perturbation for this scenario. In the right panel of Figure 6, we zoom in on the growth and the subsequent decay of power spectrum following the peak.

The features of the spectrum: analytic considerations

Besides the numerical findings presented above, we can derive general analytic results for the spectrum of curvature fluctuations in scenarios activating the would-be decaying modes through a brief non-attractor era.

We start by noticing that for modes that leave the horizon during the initial slow-roll stage (leftmost light blue region in Figure 6), the spectrum shows characteristic features such as the presence of a dip, followed by an enhancement parameterized by a spectral index of $n_s - 1 = 4$ during the bulk of the growth [95,212,224–226]. The dip is physically caused by a disruptive interference between the 'constant' mode of curvature fluctuation at superhorizon scales and the 'decaying' mode that is becoming active and ready to contribute to the enhancement of the spectrum. The position and depth of the dip are analytically calculable in terms of other features of the spectrum, at least in a limit of short duration of the non-slow-roll epoch. It is found that the position of the dip in momentum space is proportional to the inverse fourth root of the enhancement of the spectrum, and the depth of the dip is proportional to the inverse square root of the enhancement of the spectrum [226]. These relations are valid for any single-field models that enhance the spectrum through a brief deviation from the standard attractor era, including cases with a time-varying sound speed and Planck mass. They are accompanied by consistency conditions on the squeezed limit of non-Gaussian higher-order point functions around the dip [227–229], as expected in single-field scenarios[22].

While in the considerations of the previous paragraph, we considered modes leaving the horizon during the first stage of slow-roll evolution, we can also derive analytic results for what happens during the non-attractor epoch. In fact, for modes that exit the horizon deep in the non-attractor era (light green region in the middle of Figure 6) and the following final slow-roll era, the spectrum behaves as expected in a standard slow-roll phase, with spectral index

$$n_s - 1 = -2\epsilon - \eta \simeq -\eta_{\rm III} = 6 + \eta_{\rm II} \,. \tag{56}$$

(Recall that the latin numbers II and III relate to the phases of evolution; see Equations (52) and (54).) This behavior is a manifestation of the duality invariance of perturbation spectra within distinct inflationary backgrounds, called Wands' duality (see, e.g., [218,231]). Wands' duality can be understood by noticing that the structure of the Mukhanov–Sasaki Equation (43) is unchanged by a redefinition of the pump field that leaves the combination z''/z invariant:

$$z(\tilde{\tau}) \quad\to\quad \tilde{z}(\tilde{\tau}) \equiv z(\tilde{\tau})\left(c_1 + c_2 \int^{\tilde{\tau}} \frac{d\tilde{\tau}}{z(\tilde{\tau})}\right) \qquad \Rightarrow \qquad \frac{\tilde{z}''}{\tilde{z}} = \frac{z''}{z} \tag{57}$$

where $c_{1,2}$ are arbitrary constants. If $z(\tilde{\tau})$ controls a phase of the slow-roll attractor, $\tilde{z} \propto 1/\tilde{\tau}$, a dual phase whose pump field ($z \propto \tilde{\tau}^2$) as given by Equation (57) describes a non-attractor era. Although the statistics of the canonical variable (v_k) are identical in the two regimes, the amplitude of the curvature perturbation spectrum ($\mathcal{R}_k$) increases in the non-attractor epoch. In scenarios in which the parameter η is considerably larger than the other slow-roll parameters, Wands' duality (57) analytically prescribes the relation (56), in agreement with the numerical findings plotted in Figure 6. Subtleties can arise in joining attractor and non-attractor phases, since consistency conditions can be violated [232] due to the effects of boundary conditions at the transitions between different epochs. All these considerations are relevant for our topic, given the sensitivity of PBH formation and properties on the shape of the spectrum near the peak.

For further detailed accounts of the characterization of the interesting features in the power spectrum of PBH-forming single-field scenarios, we refer the reader to [95,212,214,215,217,224–226,233].

Stochastic inflation and quantum diffusion

While, so far, we have focused on the predictions of the second-order action (35), non-linearities and non-Gaussian effects can play an important role in the production of PBHs, as we learned in Section 2.3. For the case of ultra-slow-roll (USR) models based on non-attractor phases of inflation, there are sources of non-Gaussianity associated with stochastic effects during inflation.

The stochastic approach to inflation pioneered by Starobinsky [234] constitutes a powerful formalism for describing the evolution of coarse-grained superhorizon fluctuations during inflation. It is based on a classical (but stochastic) Langevin equation, which reads in canonical single-field inflation (N is the number of e-folds, and we assume constant sound speed and Planck mass):

$$\frac{d\phi}{dN} = -\frac{V'}{3H^2} + \frac{H}{2\pi}\,\xi(N)\,. \tag{58}$$

Here, ϕ represents a coarse-grained version of superhorizon scalar fluctuations; V' is the derivative of the inflationary potential, which leads to a deterministic drift for the coarse-grained superhorizon mode; and ξ is a source of stochastic noise acting on long-wavelength fluctuations caused by the continuous kicks of modes that cross the cosmological horizon and pass from sub- to superhorizon scales during inflation.

Besides the physical insights that it offers, the inflationary stochastic formalism [234–242] offers the opportunity to obtain accurate results for the probability distribution function controlling coarse-grained superhorizon modes beyond any Gaussian approximation. As a classic example, by solving the Fokker–Planck equation associated with (58), the seminal work [238] analytically obtained the full non-Gaussian distribution functions for certain representative inflationary potentials, going beyond the reach of a perturbative treatment of the problem.

Returning to the discussion of a USR inflationary evolution for PBH scenarios, we can expect that stochastic effects can be very relevant in this context (see, e.g., [241,243–251]). In fact, since the amplitude of scalar fluctuations are amplified, the stochastic noise can increase relative to that under slow-roll inflation. Moreover, during USR, the derivative of the potential ($V' = 0$), the classical drift is absent, and the stochastic evolution is driven by stochastic effects only. Various researchers have studied this topic by solving the stochastic evolution equations [252–255] and found that non-Gaussian effects can change the predictions of PBH formation, depending on the duration of the USR phase. In fact, the stochastic noise modifies the tails for the curvature probability distribution function, which decays with an exponential (instead of a Gaussian) profile[23] and, consequently, tends to overproduce PBHs. Refs. [252–255] set constraints on the duration of the USR phase, which (depending on the scenario) can last, at most, a few e-folds before overproducing PBHs. There has been increased activity surrounding these subjects, and we refer the reader to the aforementioned literature for details on the state of the art with respect to this important topic.

3.3. *Growth in the Power Spectrum When the Decaying Modes Are Slacking*

Slow-roll violation without triggering decaying modes

We learned in the previous subsections that a possible way to enhance the spectrum of fluctuations at small scales with respect to its large-scale counterpart is to amplify the k-dependent corrections to the constant-mode solution ($\mathcal{R}_k^{(0)}$) within the parentheses of Equation (39).

However, as we anticipated in the paragraph following Equation (39), we can also design scenarios in which an enhanced time dependence of the slow-roll parameters leads to a scale-dependent curvature power spectrum at horizon crossing, even without exciting the decaying mode at superhorizon scales. The idea is to still make sure that the pump field ($z(\tau)$) increases with time—hence, conditions (46) and (47) are *not* satisfied, the decaying mode remains inactive and the terms within parentheses in Equation (39) can be neglected. However, at the same time, each individual slow-roll parameter changes considerably during a short time interval during inflation. The derivatives of slow-roll parameters can be large; they can contribute significantly to the quantity (Θ) controlling the Mukhanov–Sasaki equation, and they can influence the scale dependence of the curvature spectrum at horizon crossing (see Equations (43) and (45)).

We start this section by setting the stage and derive formulas to describe this possibility. We then present an explicit realization of this scenario. It is convenient to work with the canonical variable (v_k) defined through Equation (42), and solve the Mukhanov–Sasaki system in the form of the set of Equations (43)–(59). We assume that the pump field (z) is monotonic and always increasing with time, and we identify the sound horizon of fluctuations as $aH/c_s \simeq \sqrt{z''/(2z)}^{24}$. We can identify two asymptotic regimes for each mode k: (i) an early-time regime, when each mode is deep inside the horizon; and (ii) a late-time regime, when the modes are stretched to become a superhorizon. On the one hand, in the former regime, the modes satisfy $k^2 \gg z''/z$ and behave as the standard vacuum fluctuations in Minskowski space time

$$v_k(\bar{\tau}) = \frac{e^{-ik\bar{\tau}}}{\sqrt{2k}}, \qquad \text{for} \quad k^2 \gg \frac{z''}{z}. \tag{59}$$

On the other hand, later during inflation, the fluctuations are stretched outside the horizon, entering the second regime and eventually satisfying $k^2 \ll z''/z$, with a solution given by

$$v_k(\bar{\tau}) \simeq \mathcal{C}_{1,k}\, z + \mathcal{D}_{2,k}\, z \int \frac{\mathrm{d}\bar{\tau}'}{z^2(\bar{\tau}')} + \mathcal{O}(k^2), \tag{60}$$

where the finite k^2 corrections to this solution can be derived in a similar fashion as in (39). Recall that we are now interested in attractor background configurations; hence, we can neglect the last two terms in Equation (60) that rapidly decay. Shortly after horizon crossing, the canonical variable settles into the solution $v_k = z\, \mathcal{C}_{1,k}$. Using the field redefinition (42), we can identify the constant mode as the curvature perturbation at late times ($\mathcal{C}_{1,k} = \mathcal{R}_k = \mathcal{R}_k^{(0)}$). In order to determine its expression, we match the solutions sometime around horizon crossing ($\bar{\tau} = \bar{\tau}_0$), and we obtain

$$|\mathcal{C}_{1,k}|^2 = |\mathcal{R}_k^{(0)}|^2 = \frac{1}{2k}\frac{1}{z(\bar{\tau}_0)^2} = \frac{1}{2k}\frac{c_s}{a^2 M^2 2\epsilon}\bigg|_{\bar{\tau}=\bar{\tau}_0}. \tag{61}$$

The horizon-crossing time can be conveniently expressed as a leading contribution in a WKB approximation [64]:

$$k^2 = \frac{z''}{2z}\bigg|_{\bar{\tau}_0} = \left(\frac{aH}{c_s}\right)^2 (1+\Theta)\bigg|_{\bar{\tau}_0}, \tag{62}$$

with Θ given in Equation (45). Collecting there results, we can write the late-time power spectrum for curvature fluctuations as

$$\mathcal{P}_{\mathcal{R}}(k) \equiv \frac{k^3}{2\pi^2}\left|\frac{v_k(\bar{\tau})}{z}\right|^2 = \frac{k^3}{2\pi^2}\left|\mathcal{R}_k^{(0)}\right|^2 = \frac{H^2}{8\pi^2\, \epsilon\, c_s\, M^2}(1+\Theta)\bigg|_{\bar{\tau}_0}. \tag{63}$$

From (63), we observe that rapid changes in the background quantities (ϵ, c_s, M) and slow-roll parameters constituting the quantity (Θ) of Equation (45) as a function of $\bar{\tau}_0$ can then translate into a scale-dependent amplification of the power spectrum. As we will see, this situation leads to a scale-dependent enhancement in the power spectrum realized through the 'constructive interference' of the time-dependent background parameters (ϵ, c_s, M).

An explicit realization

We now review a possible realization of this scenario, closely following the discussion presented in [64]. We focus on the generalized single-field framework discussed in Section 3.1, and we consider background dynamics that include simultaneous pronounced dips in the time-dependent profiles for the parameters ϵ, c_s, M^2. We then study the power spectrum by numerically solving numerically the Mukhanov–Sasaki equation for curvature perturbations (see Appendix C), and we compare the result with the analytical expressions discussed in Section 3.1 (see, e.g., (63)).

To analyze a representative scenario in this category, we parameterize the three time-dependent quantities ($X = \{\epsilon, c_s, M^2\}$) as [64]

$$\log_{10} X(N) = (n_{\rm a} - n_*) \tanh^2\left(\frac{N - N_*}{\sigma}\right) + n_*. \tag{64}$$

Each of these quantities tends to $10^{n_{\rm a}}$ asymptotically away from N_* in both directions, $|N - N_*| \gg \sigma$ and becomes equal to 10^{n_*} at N_* while staying around the neighborhood of this value for a number of e-folds determined by σ. Notice that we are interested in features of the inflationary dynamics that affect modes at scales much smaller than the CMB pivot scale. Hence, we can assume $N_* \gg 1$, where modes associated with CMB scales leave the horizon at $N \simeq 0$, while inflation ends at $N_{\rm end} = 60$. A representative set of parameter choices that leads to a localized decrease in the slow-roll parameters is presented in Table 3. The resulting background evolution and the behavior of z'/z are shown in the left and right panels of Figure 7, respectively. We observe from the right panel of the figure that $z'/z > 0$ is always satisfied during inflation, suggesting (as expected) that the decaying modes do not grow over time. Using the parameterization (64), we then numerically solve the Mukhanov–Sasaki equation, following Appendix C.

Table 3. Parameter choices that characterize the background evolution of the time-dependent parameters (ϵ, c_s, $\tilde{M} = M/M_{\rm pl}$).

X	$n_{\rm a}$	n_*	N_*	σ
ϵ	−2.0	−4.0	36.0	2.0
c_s	0.0	−2.0	35.5	2.5
$\tilde{M}^2$	0.0	−3.0	34.8	4.0

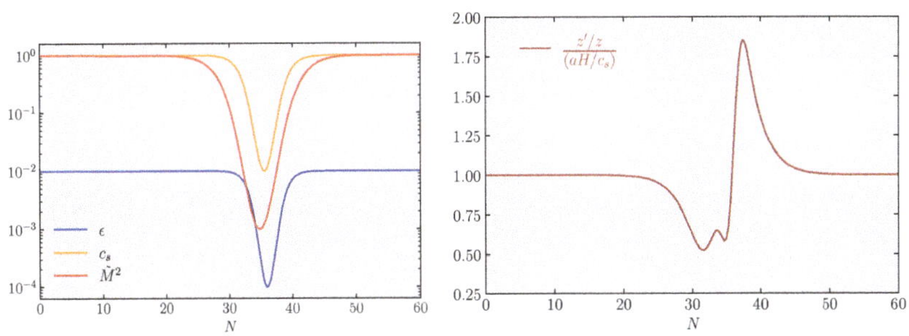

Figure 7. (**Left**) panel: evolution of $\epsilon, c_s, \tilde{M}^2 = M^2/M_{\rm pl}^2$ in e-folds characterized by the expression (64) with the parameter choices presented in Table 3. (**Right**) panel: time evolution of (46) in units of aH/c_s to illustrate the fact that the decaying modes do not grow for the background presented in the left panel.

Our results for $\mathcal{P}_{\mathcal{R}}(k, N_{\rm end})$ are presented in Figure 8. In the left panel, we notice that the expression (63) accurately describes the behavior of the power spectrum towards its peak, confirming our expectation that the constant growing mode is responsible for the enhancement. Notice the absence of a dip proceeding the growth (which characterizes the scenario shown in Figure 6). This is in agreement with our interpretation presented in the previous section; the dip is due to disruptive interference between 'growing' and 'decaying' modes, while in this context, the decaying mode is not active. On the other hand, as shown in the right panel of the figure, the power spectrum exhibits oscillations for scales following the peak. As we discussed in Note 24, this is due to multiple horizon crossing of modes within certain momentum scales, leading to excited states and oscillations in the spectrum.

We illustrate this phenomenon in Figure 9—for two neighboring modes labeled by red and purple dots in Figure 8—where we plot k versus $\sqrt{|z''/z|}$ as a function of e-folds. As shown in the right panel of Figure 9, although these modes are neighboring, they exhibit non-trivial behavior before their final horizon exit ($N \simeq 40$) so that their asymptotic values ($N \to N_{\rm end}$) differ considerably, giving rise to the sizable modulations in the late-time power spectrum (see Figure 8), as discussed in [64].

Besides [64], the ideas discussed in this subsection for enhancing the spectrum at small scales are further realized in [65]. Conceptually similar frameworks include a sound speed resonance scenario proposed in [66]—which can be explicitly realized within a DBI inflation model [67]—and a parametric resonance model discussed within the single-field canonical inflation framework [257].

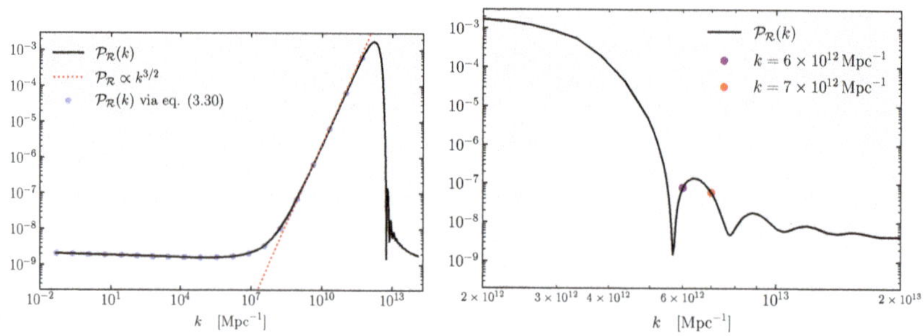

Figure 8. The full power spectrum of curvature perturbation for the background model presented in Figure 7 (**left**). The blue dots represent the accuracy of the Formula (63) in describing the rise of $\mathcal{P}_\mathcal{R}$ towards its peak. As shown by the red dotted line, the spectral index on the rise satisfies $n_s - 1 \lesssim 3/2$. The power spectrum for scales following the peak illustrates the oscillations in the amplitude at those scales (**right**).

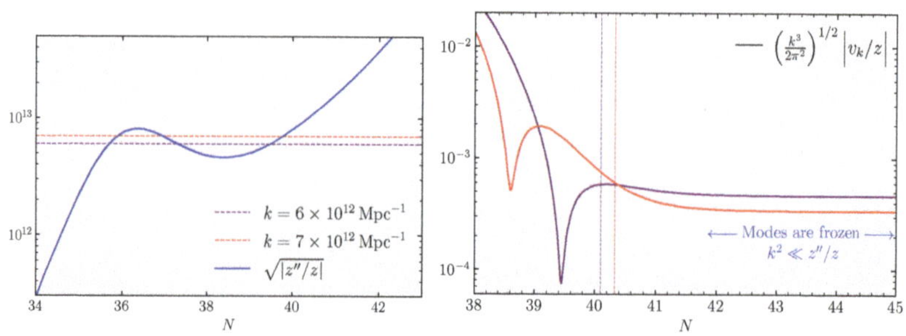

Figure 9. (**Left**): The occurrence of multiple horizon crossing for the neighboring modes labeled by red and purple dot in Figure 8. (**Right**): Evolution of the modes that exhibit multiple horizon crossing (same color coding as in the left panel). The vertical lines illustrate the final horizon crossing time for each mode.

3.4. Brief Summary

We find it remarkable that many distinct single-field models of inflation built with the aim of producing PBHs share common features in the properties of the resulting power spectrum because the enhancement of the curvature spectrum at small scales is the result of few mechanisms common to several scenarios. We identified the idea of resurrecting the decaying mode of curvature fluctuations at super Hubble scales by increasing the absolute value of the slow-roll parameter (η)—see Section 3.2—and the idea of having very rapid changes in the values of slow-roll parameters (keeping them relatively small)—see Section 3.3.

The resulting properties of the curvature spectrum might be essential for testing single-field models of inflation at small scales, thanks to their predictions for the PBH population properties, as well as for the spectrum of gravitational waves induced at the second order in perturbations [87,88]. In fact, the latter is a very interesting topic that is relevant and well-studied for single-field models; we refer the reader to [105] for a detailed review.

Nevertheless, the existing explicit scenarios of single-field inflation, which are able to generate an appreciable PBH abundance, typically suffer from severe fine-tunings on the choices of the parameters characterizing their Lagrangians, for example, in producing sufficiently flat plateau regions of their potential and in ensuring regular transitions between attractor and non-attractor eras during the inflation process (see e.g., the discussion in [50] and [258] for a scenario that can partially ameliorate the tuning involved). Moreover, as we learned in Section 2.2, the resulting PBH population can be very sensitive to the details of the spectrum profile and to the presence of primordial non-Gaussianities. The latter should, to a certain extent, be generated in many concrete single-field realizations (see Section 3.2) and render particularly delicate the task of making precise theoretical predictions. In fact, large non-Gaussianities might not only change the predictions for PBH production but can also impose restrictions on single-field model building due to large one-loop corrections [259–262] (see [263]).

It is then interesting to try to consider PBH inflationary scenarios in contexts in which more than one field acquires dynamics during inflation. The hope is to find qualitatively new ideas for the production of PBHs or, alternatively, more natural realizations of known mechanisms in order to amplify the primordial curvature spectrum at small scales. It is possible that multiple fields affect the predictions on the statistics of curvature fluctuations with respect to single-field models, with important implications for PBH formation. We review this topic in the following section.

4. Enhanced Primordial Power Spectrum in Multifield Models

Despite the remarkable agreement of single-field, slow-roll inflation with the CMB and the LSS data [4], the fundamental nature of inflation continues to elude us. In particular, it is very important to know whether additional fields took part in the dynamics of inflation, besides the scalar driving cosmic expansion. In fact, while current cosmological observations do not provide hints of isocurvature fluctuations associated with extra inflationary degrees of freedom, it might also be the case that additional fields play a role during epochs of inflation that are not well probed by current large-scale surveys. The formation of PBHs occurring at small scales can be sensitive to isocurvature fluctuations and represent a valuable probe of inflationary multifield dynamics. Multifield inflation can offer new possibilities for the production of PBHs, exploiting novel, distinctive ways of converting large gradients of background quantities into the spectrum of curvature fluctuations. Moreover, when used in tandem with the techniques discussed in the previous section, multifield models might alleviate some of the fine-tuning issues the occur in single-field scenarios (see the discussion in Section 3.4). For these reasons, in this section, we review a selected set of theoretical frameworks aimed at enhancing the curvature spectrum at small scales[25] by using tools that specifically involve more than one field during inflation, in particular the tachyonic behavior of field dynamics in some extra sectors during inflation.

First, in Section 4.1, we review PBH scenarios based on axions interacting with gauge fields during inflation, exploiting a mechanism based on particle production during inflation. This possibility is well-motivated by particle physics constructions, and it has been thoroughly explored in the literature; we make efforts to carefully review the state of the art, discussing opportunities and challenges for PBH formation in this context. Then, in Section 4.2, we review multifield inflationary scenarios in which the rise in the spectrum is a result of large isocurvature perturbations induced by a curved inflationary trajectory traversing a rich multifield moduli space.

Main References: In the discussion presented in Section 4.1, we benefit from the works presented in [79–82,85,86], while the material presented in Section 4.2 is based on [72,73].

4.1. Enhanced Scalar Perturbations from Axion-Gauge Field Dynamics

Axions [269,270] are pseudoscalar particles theoretically introduced in studies of particle physics theories beyond the Standard Model. First motivated by scenarios addressing the strong CP problem in terms of a Peccei–Quinn mechanism [271,272], they find natural realizations in string theory (see, e.g., [273]), as well as many useful applications in cosmology—see, e.g., [274] for a comprehensive review. The possibility of using the physics of axions to produce PBHs is very interesting, given their strong theoretical and experimental motivations that enable connections between particle physics, quantum gravity and cosmology. In this section, we review several representative ideas and their realizations in this context.

Axion fields, denoted here as χ, correspond to Goldstone bosons of Peccei–Quinn (PQ) global symmetry spontaneously broken at a scale dubbed f. Their Goldstone nature equips axions with a shift symmetry $\chi \rightarrow \chi$+constant, which is valid at all orders in perturbations. The shift symmetry makes them interesting inflation candidates, as their potential is naturally very flat [275]. The PQ symmetry suffers from a chiral anomaly, which breaks the shift symmetry and assigns axions a potential ($V(\chi)$) through non-perturbative contributions in field theory, or monodromy effects in string theory (see, e.g., [68] for a review). The chiral anomaly implies that axions interact with gauge vectors through higher-dimensional operators. This property is essential for our arguments. The typical structure of an axion Lagrangian considered for cosmological purposes reads

$$\mathcal{L} = -\frac{1}{2}\partial_\mu\chi\,\partial^\mu\chi - V(\chi) - \frac{1}{4}F_{\mu\nu}\,F^{\mu\nu} - \frac{g_{\rm cs}}{4f}\,\chi\,F_{\mu\nu}\,\tilde{F}^{\mu\nu}\,. \tag{65}$$

The axion χ is equipped by a kinetic term and a potential term ($V(\chi)$). The potential is often a polynomial function of the axion field, at least for small values of χ[26]. As anticipated, the Lagrangian (65) also includes a $U(1)$ gauge field (A_μ) with field strength ($F_{\mu\nu}$), coupling to the axion through a dimension-5 gauge-preserving operator

$$\mathcal{L}_{\rm int} = -\frac{g_{\rm cs}}{4f}\,\chi\,F_{\mu\nu}\,\tilde{F}^{\mu\nu}\,, \tag{66}$$

where f is the PQ symmetry-breaking scale, and $g_{\rm cs}$ is a dimensionless coupling constant. The dual field strength appearing in Equation (66) is $\tilde{F}^{\mu\nu} \equiv \eta^{\mu\nu\rho\sigma}F_{\rho\sigma}/(2\sqrt{-g})$, and the alternating symbol η is 1 for even permutations of its indices and -1 otherwise (starting with $\eta^{0123} = 1$). The dimension-5 operator (66) is especially important for our arguments. As we review below, this axion–vector coupling leads to a tachyonic instability in the gauge sector, which exponentially enhances one of the helicities of the gauge vector modes and produces a large amount of gauge quanta [280]. The energy adsorbed in the production of vector modes is carried away from the axion kinetic energy, which slows down its evolution; this is good news for axion inflationary models, since the axion can slowly roll, even along steep potentials [281]. Moreover, the inverse decay of enhanced gauge quanta into axion fluctuations (schematically, $A + A \rightarrow \chi$) can amplify curvature perturbations [282,283] and lead to a large curvature spectrum at small scales at a level able to produce PBHs [79]. We refer the reader to [284] for a review on axion inflation written one decade ago, which includes many of the topics mentioned above. We now focus on reviewing the consequences of these ideas for scenarios amplifying curvature fluctuations, in addition to covering opportunities and challenges put forward in the most recent literature on this subject.

Amplification of gauge field fluctuations

When considering a dynamical axion field with a time-dependent background profile ($\bar{\chi}(t)$), the dimension-5 axion-vector interaction of Equation (66) leads to a copious production of vector fluctuations. This property can be understood by considering the equation of motion (EoM) for the gauge field mode functions (See Appendix D) [281]

$$\partial_x^2 A_\pm + \left(1 \pm \frac{2\xi}{x}\right) A_\pm = 0, \qquad \xi \equiv -\frac{g_{cs}\,\dot{\bar{\chi}}}{2Hf} \tag{67}$$

where $A_\pm$ corresponds to the gauge vector polarizations, and we define a dimensionless time variable ($x \equiv -k\tau = k/(aH)$), as well as the effective dimensionless coupling (ξ) between the spectator axion and the gauge field. Without loss of generality, we work within the following conditions

$$\xi > 0 \quad \text{and} \quad \dot{\bar{\chi}} < 0\,. \tag{68}$$

Hence, the axion rolls along its potential from large positive to small values of $\bar{\chi} \geq 0$.

Notice from (67) that the dimension-5 operator of Equation (66) introduces a time-dependent mass term in the dispersion relation of the U(1) field, which changes sign depending on the helicity ($\lambda = \pm$). This property reflects the parity-violating nature of the dimension-5 interaction. When the gauge modes are deep inside the horizon (($x = k/(aH) \gg 1$)), the time-dependent correction term is negligible, and both vector polarizations obey a standard dispersion relation as in Minkowski space. However, as the modes stretch outside the horizon, the correction becomes dominant for $x = k/(aH) \lesssim 2\xi$, leading to an instability for one of the circular polarizations of the gauge fields. In our conventions (68), only the A_- state experiences a tachyonic instability, while A_+ is unaffected. The dynamics of the axion-like field controlled by the axion velocity ($\dot{\bar{\chi}} \neq 0$), therefore induces a significant production of helical vector fields.

The nature of gauge field production and its consequences as a source for scalar perturbations are sensitive to the scalar potential ($V(\chi)$), since this quantity determines the profile of the axion background velocity ($\dot{\bar{\chi}}$). The dynamics of the curvature perturbations, as generated through the axion-gauge field dynamics, depend on whether we identify the axion field as the inflation or whether it belongs to some extra spectator sector during inflation. In what comes next, we arrange our discussion so as to clearly distinguish between these possibilities. We consider the following scenarios:

1. **Section 4.1.1** As a first possibility, we identify the axion (χ) with ϕ that drives inflation: $\chi = \phi$. Then, the order parameter ($\dot{\bar{\phi}}$) controlling the gauge-field production increases over time, generating scalar perturbations through an inverse vector decay: $\delta A + \delta A \to \delta\phi$. We refer to this scenario as "Smooth Axion Inflation"; models that can be considered in this category are studied in [79–82]. Such scenarios suffer from dynamical instabilities associated with large back-reaction effects from the gauge fields on the axion evolution.
2. **Section 4.1.2** In certain axion-inflation models, the axion potential has special features located far away from the field range corresponding to scales affecting CMB physics. A sudden increase in the axion inflation velocity occurs at their location, with enhanced scalar perturbations amplified at small PBH scales through inverse vector decay. This possibility was first discussed in [81], while further developments were studied in [83,86]. We refer to scenarios producing PBHs by exploiting localized particle production as "Bumpy Axion Inflation", and we analyze how they address the instabilities mentioned above.
3. **Section 4.1.3** A final possibility corresponds to gauge field production in a hidden sector through the dynamics of an axion spectator field. This case also allows one to address the aforementioned dynamical instabilities, given that the back-reaction effects from the vector sector can be placed under control. Depending on the shape of the spectator axion potential, such a scenario can lead to localized peaks in the scalar curvature power spectrum [81,85]. We refer to this scenario as "Spectator axion-gauge field dynamics".

We build our discussion mainly in terms of representative, concrete examples, presenting the main results and relegating technical details to appendices.

4.1.1. Smooth Axion Inflation

In the first scenario, we identify the axion (χ) as the inflaton ϕ: $\chi \to \phi$ to study the behavior of scalar perturbations in a setup described by Lagrangian (65). We assume that the profile for the scalar potential ($V(\phi)$) is sufficiently flat to support inflation. In this case, the effective coupling (ξ), as defined in Equation (67), adiabatically increases during the inflationary process:

$$\xi \equiv g_{\rm cs}\sqrt{\frac{\epsilon}{2}}\frac{M_{\rm pl}}{f}, \tag{69}$$

where ϵ is the standard slow-roll parameter. The amplified gauge-field mode function can be analytically expressed as [281] (recall that the axion speed has a negative sign; see Equation (68)):

$$A_-(\tau,k) \simeq \frac{1}{\sqrt{2k}}\left(\frac{-k\tau}{2\xi}\right)^{1/4}\exp(\pi\xi - 2\sqrt{-2\xi k\tau}), \qquad \xi \equiv -\frac{g_{\rm cs}\dot{\phi}}{2Hf}. \tag{70}$$

The slowly changing time-dependent parameters (ξ and H) in Equation (70) are evaluated at the epoch of horizon crossing. The amplification of gauge field modes is maximized when the size of the mode is comparable to the horizon, $-k\tau \sim \mathcal{O}(1)$.

In fact, the analytic expression in (70) is valid within the interval $(8\xi)^{-1} \ll -k\tau \ll 2\xi$. For the values ($\xi \sim \mathcal{O}(1)$) we are be interested in, this range corresponds to a phase during which the gauge modes grow and then remain frozen to their maximal value before being diluted away by universe expansion. We proceed by discussing the time evolution of the relevant quantities in this setup. For concreteness, we focus on a representative example. For a more detailed account of the gauge field amplification by the slowly rolling scalar, we refer the reader to Section 2.1 of [285] or Appendix B of [286].

Background evolution in a concrete example

As the effective coupling (ξ) of Equation (70) increases during inflation, the enhanced vector modes eventually lead to a sizable back-reaction on the background evolution of the axionic inflation field. This fact can significantly affect the homogeneous dynamics of the system.

The back-reaction is mainly controlled by the vector-dependent friction term in the equation of motion for the homogeneous inflation field [281]. This phenomenon implies that the gauge field amplification occurs at the expense of the inflation velocity ($\dot{\phi}$). We consider a situation characterized by a transition between standard slow-roll dynamics in the early stages of inflation and a new attractor regime at late stages, when the gauge field enhancement dominates the Hubble friction [287] (see the cautionary remark towards the end of this section).

The aforementioned friction effect induced by the gauge-mode production can be analyzed considering modified Klein–Gordon and Friedmann equations:

$$\begin{aligned} \ddot{\phi} + 3H\dot{\phi} + V'(\phi) &= \frac{g_{\rm cs}}{f}\langle \vec{E}\cdot\vec{B}\rangle\,, \\ 3H^2M_{\rm pl}^2 &= \frac{1}{2}\dot{\phi}^2 + V(\phi) + \frac{1}{2}\left\langle \vec{E}^2 + \vec{B}^2\right\rangle, \end{aligned} \tag{71}$$

where we introduced 'electric' and 'magnetic' field contributions[27], as discussed in detail in Appendix D (see, in particular, Equation (A63)). Using the solutions (70) for gauge-field modes, the expectation values for the electric and magnetic fields can be computed analytically [281,286], finding (see Appendix D)

$$\langle \vec{E}\cdot\vec{B}\rangle \simeq 2.1\times 10^{-4}\,\frac{H^4}{\xi^4}\,e^{2\pi\xi}, \qquad \rho_A \equiv \frac{1}{2}\langle \vec{E}^2 + \vec{B}^2\rangle = 1.4\times 10^{-4}\,\frac{H^4}{\xi^3}\,e^{2\pi\xi}. \tag{72}$$

The quantity (ρ_A) corresponds the total energy density contained in the gauge field fluctuations. Using the evolution Equation (71), we can identify the conditions corresponding to a small back-reaction of the vector modes into the background evolution. They are $3H|\dot{\phi}| \gg g_{cs}\langle\vec{E}\cdot\vec{B}\rangle/f$ and $\rho_A \ll 3H^2M_{pl}^2$, which can be expressed as

$$\begin{aligned} \xi^{-3/2}e^{\pi\xi} &\ll 79\frac{\dot{\phi}}{H^2}, \quad &\longrightarrow \quad &\text{negligible back-reaction on } \phi \text{ equation}, \\ \xi^{-3/2}e^{\pi\xi} &\ll 146\frac{M_{pl}}{H}, \quad &\longrightarrow \quad &\text{negligible back-reaction on the Friedmann equation}. \end{aligned} \tag{73}$$

Assuming that inflation starts in a standard slow-roll regime, the relation $\dot{\phi} = \sqrt{2\epsilon}HM_{pl} \ll HM_{pl}$ implies that the first condition in (73) is more demanding than the second. When these conditions are not met, we enter into an inflationary phase characterized by a strong back-reaction of gauge modes, which goes beyond the standard slow-roll inflationary attractor.

To concretely illustrate the back-reaction of the gauge field production on the homogeneous background evolution of inflation, we focus on a modified version of the linear monodromy-type potential [276] that interpolates between $V \propto \phi$ and $V \propto \phi^2$ from large to small field values around the global minimum:

$$V(\phi) = \lambda M_{pl}^4\left[\sqrt{1+\frac{\phi^2}{M_{pl}^2}}-1\right], \tag{74}$$

where λ is a dimensionless parameter that fixes the overall amplitude of the potential. Assuming negligible back-reaction in the beginning of inflation, λ can determined by enforcing the standard normalization of the power spectrum in the slow-roll regime for a given choice of ϕ_{in} using

$$\mathcal{P}_{\mathcal{R}}(k_{cmb}) = \frac{H^2}{8\pi^2\epsilon M_{pl}^2} \simeq \frac{V(\phi_{in})}{24\pi^2\epsilon_V(\phi_{in})M_{pl}^4} \equiv 2.1\times10^{-9}. \tag{75}$$

We then numerically solve Equation (71) utilizing (72) by setting the coupling between the inflation and the gauge fields to its maximally allowed value[28] by Planck: $g_{cs}M_{pl}/f = 48$ (see e.g., [289]). Note that for a given initial value of the scalar field, this number can be translated to an initial value of the effective coupling, which we label as ξ_{cmb}:

$$\xi_{cmb} \simeq \frac{g_{cs}\,M_{pl}}{f}\frac{1}{2\phi_{in}}. \tag{76}$$

For $\phi_{in} = 10.5\,M_{pl}$ ($\xi_{cmb} = 2.5$), the results of the numerical computation are shown in Figure 10. In the left panel, we show the inflation profile as a function of e-folds until the end of inflation, where $\dot{H} = -H^2$. We can clearly observe that in the absence of coupling to the gauge fields ($g_{cs} = 0$) (black dashed line with $\xi = 0$), inflation lasts for about $\approx$48 e-folds. Turning on the interactions with gauge fields at the beginning of inflation (solid orange line with $\xi_{cmb} = 2.5$), the back-reaction induced by the vector fluctuations become noticeable around $\approx$36 e-folds, which, in turn, has the overall effect of extending the duration of inflation for about $\approx$12 e-folds compared to the standard slow-roll case ($\xi = 0$).

These findings are also supported by the right panel of Figure 10, where we compare the standard Hubble damping term ($3H\dot{\phi}$) with the friction term ($g_{cs}\langle\vec{E}\cdot\vec{B}\rangle/f$) induced by the gauge field production as a function of e-folds until the end of inflation. We observe that the standard Hubble friction dominates the dynamics at early times, but as the effective coupling (ξ) increases, the dynamics gradually evolve into a regime dominated by the back-reaction of the produced gauge quanta. For the parameter choices we adopt, the onset of this regime starts around $N_{br} \simeq 36$ e-folds, and at around $N \simeq \mathcal{O}(50)$ the system enters into a strong back-reaction regime wherein the dynamics become completely controlled by the gauge modes [281].

Therefore, in any viable model of axion inflation, back-reaction effects induced by gauge field production become important only towards the latest stages of inflation. For example, for a rough estimate of the orders of magnitude involved, we can adopt a linear potential ($V \propto \phi$). Assuming the dynamics are initially in the slow-roll regime, we have $\xi \simeq g_{\rm cs}\sqrt{2\epsilon_V}M_{\rm pl}/f \propto 1/\phi$ and $\phi_{\rm cmb}/\phi_{\rm br} = \xi_{\rm br}/\xi_{\rm cmb} \simeq 2.2$–$1.9$. For typical initial field values that can sustain enough inflationary evolution, we have $\phi_{\rm cmb} \sim \mathcal{O}(10)M_{\rm pl}$, which implies that $\phi_{\rm br} \sim \mathcal{O}(5)M_{\rm pl}$. Hence, back-reaction becomes dominant in the latest stages but before the end of inflation corresponding to $\phi_{\rm end} \sim \mathcal{O}(1)$.

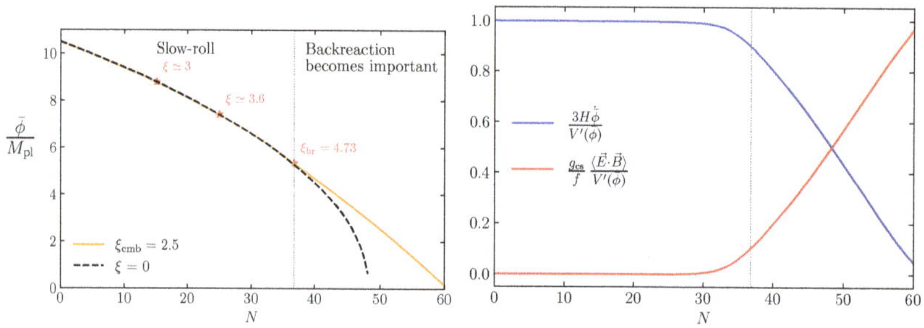

Figure 10. Field profile ($\bar{\phi}(N)$) with respect to e-folding number (N) in smooth-axion inflation for the initial choice ($\xi_{\rm cmb} = \xi(N=0) = 2.5$) (**left**). At early stages, the dynamics are in the standard slow-roll regime (light blue region), but as ξ increases, they enter into a stage dominated by the friction generated due to gauge field production (light red region), which has the effect of prolonging inflation with respect to the background evolution with $\xi = 0$. For the choice of $\xi_{\rm cmb} = 2.5$, the behavior of the Hubble damping term ($3H\dot{\phi}$) and the friction induced by the source term ($\propto \langle \vec{E}\cdot\vec{B}\rangle$) are shown in the (**right**) panel.

Scalar fluctuations sourced by gauge fields

As inflation progresses, particle production becomes increasingly efficient. Hence, the vector modes start to act as an additional source of inflation fluctuations ($\delta\phi$) due to the presence of trilinear coupling ($\delta\mathcal{L}_{\rm int} \propto \delta\phi F\tilde{F}$) (see Appendix D.2 and references therein). In particular, a tachyonic amplification of the vector fields leads—at the second order in perturbations—to an enhancement in the scalar sector. Since these contributions to the scalar fluctuations are statistically independent of their vacuum counterpart, the resulting curvature power spectrum acquires two separate contributions (see Appendix E) that can be parameterized as [79,281]

$$\begin{aligned}\mathcal{P}_{\mathcal{R}} &= \mathcal{P}_{\mathcal{R}}^{(\rm v)} + \mathcal{P}_{\mathcal{R}}^{(\rm s)},\\ &\simeq \frac{H^2}{8\pi^2\epsilon M_{\rm pl}^2} + \left(\frac{g_{\rm cs}}{f}\frac{\langle\vec{E}\cdot\vec{B}\rangle}{3H\beta\dot{\phi}}\right)^2\mathcal{F}^2.\end{aligned} \tag{77}$$

The first term in the second line denotes the standard vacuum contribution to the curvature power spectrum. The second accounts for the vector part, with the time-dependent quantities (β and $\mathcal{F}$) defined in Equations (A81) and (A99). In the weak back-reaction regime, the power spectrum of the curvature perturbation is dominated by the vacuum contribution in the first term of Equation (77). As inflation progresses, the effective coupling between the inflation and the vector fields increases; the source contribution to the power spectrum starts to kick in, and the power spectrum grows as $\mathcal{P}_{\mathcal{R}} \propto \langle\vec{E}\cdot\vec{B}\rangle^2 \sim e^{4\pi\xi}$ (the system is still in the weak back-reaction regime, where $\beta \simeq 1$). Then, the scalar fluctuations grow; they start influencing the gauge fields sources and lead to an increase in the damping factor ($\beta \sim \langle \vec{E} \cdot \vec{B} \rangle / 3H\dot{\phi}$). Therefore, the sourced contribution to the power spectrum eventually saturates towards late times during inflation. In fact, the time-dependent factor ($\mathcal{F}$) is introduced to capture the modified definition of the curvature perturbation in the strong back-reaction regime. As discussed in detail in [81], this factor is responsible for an order-one correction to the power spectrum amplitude towards the end of inflation.

We summarize these arguments with a plot (Figure 11) of the total power spectrum with respect to the e-folding number. We notice that as the effective coupling (ξ) between the scalar and the gauge fields increases along the inflationary trajectory, the power spectrum acquires a large contribution from the vector source. As desired, it grows towards smaller scales, eventually saturating towards a constant value. Since the source of the peak in the power spectrum originates from a trilinear coupling between the inflation and the gauge fields, the statistics of the curvature perturbation at those scales is non-Gaussian (see also [290]). The values of the curvature spectrum amplitude at the peak appearing in Figure 11 are sufficiently large to generate a sizable population of PBHs (recall the discussion in Section 2.3).

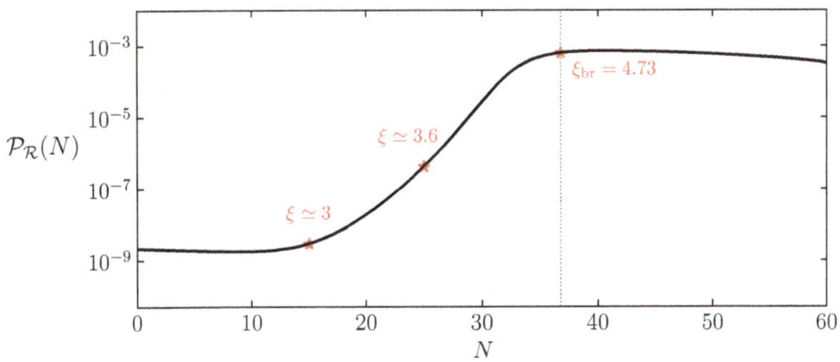

Figure 11. Total power spectrum (77) as a function of e-folds in smooth-axion inflation for the potential in (74) and $\xi(N=0) \equiv \xi_{\rm cmb} = 2.5$ as discussed in the main text.

Cautionary remarks: an instability in the inflation dynamics

We warn the reader that the findings reviewed in this section have been recently questioned. In particular, the authors of [290–294] studied the axion-gauge field dynamics for different choices of potentials and parameters, focusing on the strong back-reaction regime. By implementing sophisticated numerical techniques, these works go beyond the constant-velocity approximation assumed in our discussion above. Their findings indicate that after entering the strong back-reaction regime, the inflation velocity is characterized by oscillations of increasing amplitude, in sharp contrast to analytical studies approximating $\dot{\phi}$ (and ξ) as constant. Finally, the recent analytic study reported in [295] fully supports the numerical results. The physical source of this instability seems to be a delayed response of the vector source to changes in the axion velocity ($\dot{\phi}$) when the system enters a strong back-reaction regime. This phenomenon causes the aforementioned oscillations with increasing amplitude in the inflation velocity.

It is not clear at the moment whether these secular effects can be tamed by adding ingredients to the setup discussed in this section. A potential way out is to modify the inflationary dynamics in such a way as to push the strong vector back-reaction regime towards an epoch that does not affect the predictions of PBH formation in the scalar sector. This possibility can be achieved in scenarios in which the gauge field production is localized, leading to a pronounced peak in the scalar power spectrum before entering into a strong back-reaction regime (see Section 4.1.2). Such a way out does not avoid the instability but at least moves it towards epochs that do not affect our predictions for the production and properties of curvature perturbations. In Section 4.1.3, we instead discuss the possibility of avoiding a strong vector back-reaction regime altogether in scenarios in which the axion is a spectator field and does not drive inflation. Appropriate choices of the axion potential nevertheless lead to an enhancement of the scalar sector at a level sufficient for producing PBHs. All the opportunities we review next are possible in scenarios of inflation including axion–vector couplings, which have very rich ramifications and are well studied, given their particle physics motivations.

4.1.2. Bumpy Axion Inflation

In this section, motivated by the instability arguments discussed above, we consider scenarios with localized production of gauge fields able to enhance the curvature spectrum and produce PBHs before entering any strong back-reaction regime. As we will see, the mechanism is based on the ideas discussed in the previous Section 4.1.1 but also uses tools for the activation of the decaying mode that we introduced in Section 3.2.

For definiteness, we focus on a representative example, discussing its properties in the main text and referring to appendices for technical details. We assume that the shift symmetry of the axion is broken both by non-perturbative instanton corrections, as well as by a non-periodic monomial term in the potential[29] (see, e.g., [276,278,279,297–301]),

$$V(\phi) = \frac{1}{2}m^2\phi^2 + \Lambda^4\frac{\phi}{f}\sin\left(\frac{\phi}{f}\right), \tag{78}$$

where m, Λ, f are parameters of unit mass dimension. In fact, potentials such as those mentioned above characterized by modulations on top of a smooth profile are well-motivated and studied in theoretical constructions of axion inflation models in a variety of situations. For these reasons, we find it interesting to review their useful applications for our aim of producing PBHs.

We are interested in obtaining noticeable modulations of the functional form of the potential within the regime $\Lambda^4 \lesssim m^2 f^2$ (which we refer to as the "bumpy" regime in what follows). In particular, we exploit the aforementioned non-perturbative corrections to generate a roller-coaster profile to the otherwise smooth potential, where plateau-like regions are connected to each other by steep cliffs (see Figure 12). While plateau-like regions are suitable to sustain long enough inflation—and to generate nearly scale-invariant fluctuations at CMB scales—when reaching the cliff-like regions, the scalar velocity speeds up intermittently.

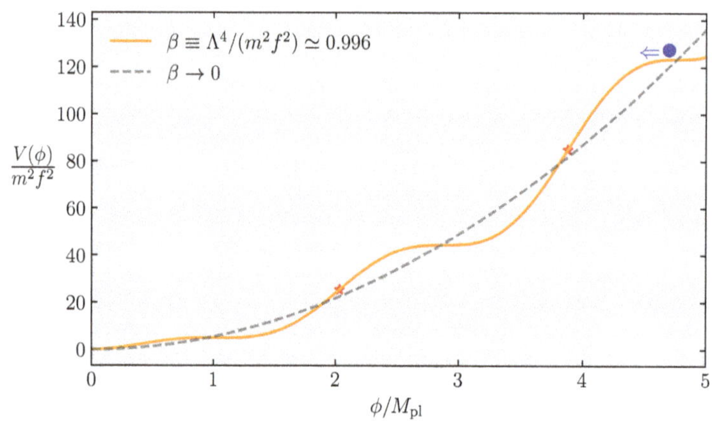

Figure 12. The shape of the potential (78) in the bumpy regime (orange curve) for $M_{\rm pl}/f = 3.3$. The red stars on the potential denotes the locations in field space where axion velocity and the slope of the potential are maximal.

During such brief stages, a localized production of gauge fields occurs. To illustrate this phenomenon explicitly, we rely on a numerical procedure along the same lines as that discussed with respect to single-field models of inflation. We numerically solve the background Equation (71) in the bumpy regime of the potential (78), neglecting possible back-reaction effects induced by gauge fields[30]. The resulting background evolution during inflation is represented in Figure 13 in terms of the first two slow-roll parameters $(\epsilon(N), \eta(N))$ for the parameter choices used in Figure 12.

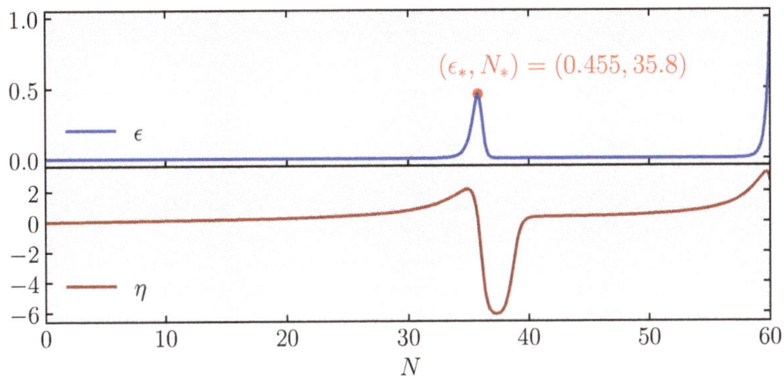

Figure 13. Evolution of the Hubble slow-roll parameters (ϵ and η) with respect to e-fold number for the bumpy axion inflation model described by potential (78). We numerically evaluate (71) using the parameter choices shown in Figure 12 with $\phi_{\rm in} \simeq 4.8$, ignoring back-reaction effects, as explained in the text.

Adopting the initial condition $\phi_{\rm in} \simeq 4.8 M_{\rm pl}$, the scalar field completes a total of 60 e-foldings of its evolution when traversing a single bump-like region, around which the slow-roll parameter (ϵ) reaches its maximal value of $\epsilon_* = 0.455$ at $N_* = 35.8$. Following the stage of maximal velocity for the scalar field, the background dynamics enter into a short non-attractor phase with $\eta \lesssim -6$, during which ϕ slows down before the dynamics re-enter into a final inflationary slow-roll stage ($\eta \ll 1$). Since the inflationary background proceeds through an epoch of slow-roll violation, in our calculations, we compute the vacuum contribution to the scalar power spectrum using the numerical methods discussed in Section 3.2 Appendix C. In fact, as anticipated above, in developing this example, we make use both of ideas based on the conversion into curvature fluctuations of gauge-field modes (as in the previous section) and on the activation of the would-be decaying mode of the inflation scalar sector (as discussed in Section 3.2).

Since while traveling the plateau-like regions, the scalar velocity is small, the only contributions to the scalar power spectrum arise from the cliff-like region of the potential (see Figure 12). Around the cliff, the scalar field velocity and the effective coupling (ξ) acquire the following profile [86]:

$$\xi(N) = \frac{g_{\rm cs}\,\delta}{1+\delta^2(N-N_*)^2}, \qquad \longrightarrow \qquad \xi_* \equiv \xi(N_*) = g_{\rm cs}\delta\,, \tag{79}$$

where ξ_* denotes the maximal value of the effective coupling. The dimensionless parameter ($\delta \simeq \sqrt{2/3}(M_{\rm pl}/f)(m/H)$) determines the number of e-folds during which the parameter ξ maintains its maximal value, i.e., $\Delta N \sim \delta^{-1}$. For $\delta \propto m$, the duration of such an epoch is inversely proportional to the restoring force provided by the potential; hence, it inversely depends on the mass parameter (m). A time-dependent profile for the effective coupling gives rise to a scale-dependent amplification of the gauge fields whose growing solutions correspond to the real part of the mode functions (see Equation (A61)), using Equation (A62) (see Appendix D.1 for more details).

As expected, the resulting sourced contribution to the power spectrum induced by the $\delta A + \delta A \to \delta\phi$ inherits the scale dependence of the gauge fields, and the curvature power spectrum can be parameterized as [86],

$$\mathcal{P}_{\mathcal{R}}(k) = \mathcal{P}_{\mathcal{R}}^{(\mathrm{v})}(k)\left[1 + \frac{H^2}{64\pi^2 M_{\mathrm{pl}}^2} f_{2,\mathcal{R}}\left(\xi_*, \frac{k}{k_*}, \delta\right)\right]. \tag{80}$$

The vacuum contribution ($\mathcal{P}_{\mathcal{R}}^{(\mathrm{v})}$) can be numerically calculated following some of the methods explained in Section 3. Due to the presence of a brief non-attractor phase, it is amplified by the activation of the would-be scalar decaying mode (see Section 3.2). The dimensionless function ($f_{2,\mathcal{R}}$) characterizes the contributions from the gauge fields and has a log-normal shape [86],

$$f_{2,\mathcal{R}}\left(\xi_*, \frac{k}{k_*}, \delta\right) \simeq f_{2,\mathcal{R}}^{c}[\xi_*, \delta] \exp\left[-\frac{1}{2\sigma_{2,\mathcal{R}}^2[\xi_*, \delta]} \ln^2\left(\frac{k}{k_* x_{2,\mathcal{R}}^{c}[\xi_*, \delta]}\right)\right]. \tag{81}$$

As suggested by the expression (81), the information about the location, width and height of the sourced signals in (80) depends on the motion of ϕ through the step-like features of its wiggly potential, particularly through ξ_* and δ dependence of the functions ($x_{2,\mathcal{R}}^{c}, \sigma_{2,\mathcal{R}}, f_{2,\mathcal{R}}^{c}$). We learn from (81) that the sourced signal is maximal at $k = k_* x_{2,\mathcal{R}}^{c}$, where it is elevated to $f_{2,\mathcal{R}}^{c}$, whereas $\sigma_{2,\mathcal{R}}$ controls the width of the signal. For the background evolution presented in Figure 13, the fitting functions describing the height, width and location of the sourced signal is calculated in [85] using the semianalytical techniques developed in [302].

Collecting all the results above, we present in Figure 14 the full power spectrum in Equation (80) in terms of the e-folding number by converting the scale dependence to e-fold dependence using the horizon crossing condition ($k = a(N)H(N)$). Due to non-trivial background evolution during which the slow-roll conditions are briefly violated, the vacuum contribution to the power spectrum (dotted lines) acquires a non-trivial shape, which shares the characteristic features of the single-field scenarios we discussed in Section 3.2. Notice in the presence of the dip due to the interference between growing and decaying modes of the curvature perturbation that immediately precedes a phase of steady growth of the spectrum.

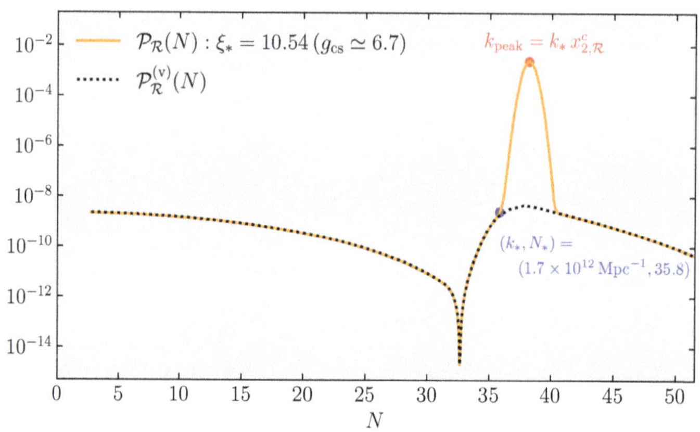

Figure 14. The total curvature power spectrum of Equation (80) as a function of N (orange curve) for the bumpy axion inflation model, the background evolution of which is shown in Figure 13. We make the following parameter choices: $\delta = 1.57$, $\xi_* = 10.54$ corresponding to $g_{\mathrm{cs}} \simeq 6.7$ (see, e.g., (79)).

However, in the present example, the duration of the non-attractor era is not sufficient to provide a prominent peak in the power spectrum solely given the presence of vacuum fluctuations of the curvature perturbation. At this point, the additional source provided by the gauge fields come to the rescue, giving rise to extra-steep growth, leading to the required power to generate a sizable peak at wave numbers $k_{\rm peak} = k_* x^c_{2,\mathcal{R}} \simeq 1.5 \times 10^{13}\,{\rm Mpc}^{-1}$ corresponding to peak PBH mass at the time of formation $M_{\rm PBH} \simeq 2.2 \times 10^{-13}\,M_\odot$ (see Equation (6)). Hence, in the presence of couplings to the Abelian gauge fields, the roller-coaster architecture of the potential provides an assisted amplification mechanism of the power spectrum. The curvature perturbation sourced by the vector fields and its vacuum counterpart help each another to generate a sufficiently pronounced peak to produce PBHs. Interestingly, the profile of the curvature spectrum as a function of momenta, in particular its growth rate and the details of the peak, are quite different from the corresponding predictions of single-field inflation, as investigated in Section 3. These differences make the two scenarios distinguishable.

In conclusion, the localized production scenario we presented in this section might be considered a possible way to address the instability issues associated with the background of the smooth-axion inflation discussed in Section 4.1.1. In fact, for the model we considered in this section, back-reaction effects become prominent for $\xi_* > 15.6$—see [86] for details—a much larger value than the phenomenological value ($\xi_* \simeq 10.5$) needed for the amplification of the power spectrum. Hence, we make predictions in a safe region of cosmological evolution.

Although this is promising in terms of building workable models of PBH production in axion inflation, the issue regarding the stability of the background is likely to reappear towards the end of bumpy axion inflation—in particular towards its final stages, when $\epsilon \to 1$, where $\xi \equiv g_{\rm cs}\sqrt{\epsilon/2}(M_{\rm pl}/f) \simeq \mathcal{O}(10)$. In the next section, we discuss an extension of the model presented here, in which a spectator axion sector generates a localized scalar enhancement through its coupling to an Abelian gauge sector. Since in this model, the dynamical evolution of the axion field stops long before inflation ends, we can avoid a strong back-reaction regime that leads to instabilities of the background configuration.

4.1.3. Spectator Axion-Gauge Field Dynamics

We continue to discuss possible methods to locally enhance the spectrum of scalar fluctuations in axion-gauge field models, with an eye to solving the dynamical instability problems mentioned above. We focus on scenarios in which the axion is a spectator field, i.e., does not directly drive inflation. We call σ the spectator axion field and couple it to an Abelian gauge sector. The Lagrangian is

$$\mathcal{L}_{\rm m} = \mathcal{L}_{\rm inf} - \frac{1}{2}\partial_\mu\sigma\,\partial^\mu\sigma - U(\sigma) - \frac{1}{4}F_{\mu\nu}\,F^{\mu\nu} - \frac{g_{\rm cs}}{4f}\,\sigma\,F_{\mu\nu}\,\tilde{F}^{\mu\nu}, \tag{82}$$

where $\mathcal{L}_{\rm inf} = -(\partial\phi)^2/2 - V(\phi)$ is the inflation Lagrangian, and $V(\phi)$ is its potential. We are interested in an inflationary setup whereby the spectator sector provides a subleading contribution to the total energy density during inflation. This implies that the energy densities of the scalar fields in the model (82) satisfy $\bar{\rho}_\sigma \ll \bar{\rho}_\phi$. Assuming small back-reaction from the gauge field fluctuations ($\rho_A \ll \bar{\rho}_\sigma$) (more on this later), the Friedmann equation simplifies to

$$3H^2M_{\rm pl}^2 \simeq \bar{\rho}_\phi + \bar{\rho}_\sigma \quad \longrightarrow \quad 3H^2M_{\rm pl}^2 \simeq V(\bar{\phi}), \tag{83}$$

so that the inflationary background is completely dictated by the inflation potential. We assume that $V(\phi)$ is flat enough to support sufficient inflation, but otherwise, we leave it unspecified, as the fine details of the inflation dynamics are not essential for what follows. Since we are working in a weak back-reaction system, we can avoid instabilities in the inflation sector such as that reviewed in Section 4.1.1.

From now on, we consider representative examples of axion potentials and their corresponding dynamics. The choice of our examples is motivated by their interesting ramifications for PBH production and for their motivations from high-energy physics. If the spectator axion (σ) is displaced from its global minimum, it can be dynamically active during inflation. Thanks to the perturbative shift symmetry of the axion sector, we expect the axion potential to be nearly flat; hence, the axion dynamics should occur in a slow-roll regime ($\ddot{\bar{\sigma}} \ll 3H\dot{\bar{\sigma}}$). In such a case, we can neglect back-reaction effects induced by gauge field fluctuations[31], and the background evolution of the spectator axion can be studied through the following equation,

$$\ddot{\bar{\sigma}} + 3H\dot{\bar{\sigma}} + U'(\bar{\sigma}) = 0, \qquad \longrightarrow \qquad 3H\dot{\bar{\sigma}} + U'(\bar{\sigma}) \simeq 0. \tag{84}$$

To realize localized gauge field production through the last term in (82), we consider two class of transiently rolling spectator axion models characterized by the following potentials [85,302]

$$U(\sigma) = \begin{cases} \Lambda^4\left[1 - \cos\left(\dfrac{\sigma}{f}\right)\right], & (\mathrm{M1}), \\ \mu^3\sigma + \Lambda^4\left[1 - \cos\left(\dfrac{\sigma}{f}\right)\right] \text{ and } \Lambda^4 \lesssim \mu^3 f & (\mathrm{M2}), \end{cases} \tag{85}$$

where μ and Λ are parameters of mass dimension one.

The first model (**M1**) features a standard (discrete) shift symmetric potential akin to natural inflation [275], where the size of the axion modulations is set by Λ. In this model, the axion dynamics span an interval bounded by the maximum ($\sigma = \pi f$) and the minimum ($\sigma = 0$) of the potential. Therefore, for large and small field values (early and late times, respectively), the axion rolls with small velocities. However, $\dot{\bar{\sigma}}$ obtains relatively large values at an intermediate time when σ passes through an inflection point ($\sigma_* = \sigma(t_*)$) with $U''(\sigma_*) = 0$, where the slope of the potential ($U'(\sigma)$) becomes maximal.

In the second model (**M2**), the axion field range is extended via a linear term [276,278] proportional to a soft symmetry breaking mass parameter (μ). We assume that the axion (σ) can probe the corresponding bumpy potential[32] in the $\Lambda^4 \lesssim \mu^3 f$ regime, similar to the axion inflation model we discussed in Section 4.1.3. In the plateau-like region(s) and towards the global minimum ($\sigma = 0$)[33], the spectator axion acquires very small velocities, but obtains a relatively large transient peak when the slope of the potential ($U'(\sigma)$) becomes maximal in the cliff-like region(s), in particular at the inflection point(s) denoted by $*$ in Figure 15.

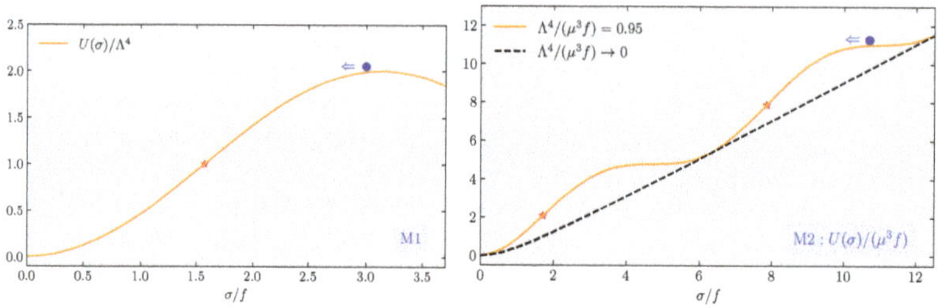

Figure 15. The shape of spectator axion potentials for M1 (**left**) and M2 (**right**). For both panels, the red stars indicate the location of the inflection point(s) at which the slope of the potential ($U'(\sigma)$) and, hence, the background velocity ($\dot{\bar{\sigma}}$) of σ become maximal (see e.g., Equation (84)).

The background evolution (σ) can be analytically derived in the slow-roll regime (84), starting from the expressions for the scalar potentials of Equation (85). For typical field ranges dictated by these potentials[34], the spectator axion velocity ($\dot{\sigma}$) and the effective coupling strength ($\xi = -g_{cs}\dot{\sigma}/(2Hf)$) obtain a peaked time-dependent profile [85,302]:

$$\xi(N) = \begin{cases} \dfrac{2\xi_*}{e^{\delta(N-N_*)} + e^{\delta(N_*-N)}}, & \delta \equiv \dfrac{\Lambda^4}{3H^2 f^2} \text{ and } \xi_* \equiv \dfrac{g_{cs}\,\delta}{2} \quad \text{(M1)} \\ \dfrac{\xi_*}{1+\delta^2(N-N_*)^2}, & \delta \equiv \dfrac{\mu^3}{3H^2 f} \text{ and } \xi_* \equiv g_{cs}\,\delta \quad \text{(M2)} \end{cases} \tag{86}$$

where N_* denotes the e-fold number when the axion passes through the inflection point (see Figure 15), and ξ_* is the maximal value of the effective coupling parameter at N_*. As in the bumpy axion inflation model, the width of the time-dependent peak in ξ depends on the dimensionless ratio (δ), which characterizes the mass of the spectator axion in its global minimum ($\delta \approx m_\sigma^2/H^2$). For a heavier axion (corresponding to larger δ), the restoring force towards the global minimum becomes very relevant; the axion σ traverses the inflection point faster, the result being a sharper peak in ξ. In other words, δ controls the acceleration ($\dot{\xi}/(\xi H) = \ddot{\sigma}/(\dot{\sigma}H) \sim \delta$) of σ as it rolls down on its potential. Notice that given our slow-roll approximation, we require $\delta \ll 3$.

The peaked structure of the ξ profile controls a critical scale ($k_* = a_* H_*$) characterizing the equation of motion (A56) corresponding to the scale of momenta leaving the horizon at an epoch when the axion velocity is maximal (i.e, when $\xi = \xi_*$). Since the mass of the U(1) field in (A56) is as tachyonic as possible around this scale, it gives rise to a scale-dependent growth of the gauge field fluctuations, where only modes with a size comparable to the horizon size at $N = N_*$, i.e $k \sim \mathcal{O}(1)a_* H_*$, are efficiently amplified. In Appendix D.1, we briefly review the parametric amplification of the gauge field modes corresponding to both models (**M1,2**) we are presenting in this section; for more details, we refer the reader to [85,285,302].

In the spectator axion models we are focusing on, the impact of the vector field production on the visible scalar fluctuations is encoded only indirectly by the presence of gravitational interactions [303]. In fact, although we consider a matter Lagrangian (82) where the spectator axion-gauge field sector is decoupled from the visible inflation sector, when integrating the non-dynamical lapse and shift metric fluctuations, we find a mass mixing between inflation ($\delta\phi$) and spectator axion ($\delta\sigma$) fluctuations. This phenomenon introduces the possibility that the gauge fields influence the curvature perturbation ($\mathcal{R} \simeq H\,\delta\phi/\dot{\phi}$) through a succession of inverse decays: $\delta A + \delta A \to \delta\sigma \to \delta\phi \propto \mathcal{R}$ (see Appendix E).

Therefore, the dynamics of this sourced contribution can be understood by first studying the influence of particle production on the spectator axion fluctuations ($\delta\sigma$) and then by computing their conversion to curvature perturbation ($\delta\phi \propto \mathcal{R}$). We sketch some of the steps in Appendix D.3; more detailed computations can be found in [85,302].

Taking into account the vacuum fluctuations within the inflation sector, the total power spectrum of curvature perturbation in the spectator axion-gauge field model can be expressed as [85,302],

$$\mathcal{P}_\mathcal{R}(k) = \mathcal{P}_\mathcal{R}^{(v)}(k) + \left[\epsilon_\phi \mathcal{P}_\mathcal{R}^{(v)}(k)\right]^2 \sum_i f_{2,\mathcal{R}}^{(i)}\left(\xi_*, \frac{k}{k_*}, \delta\right), \tag{87}$$

where $\mathcal{P}_\mathcal{R} = H^2/(8\pi^2\epsilon_\phi M_{\rm pl}^2)$ with $\epsilon_\phi \equiv \dot{\phi}^2/(2H^2 M_{\rm pl}^2)$ and the sum over i denotes contributions from each particle production location, e.g., for model **M1**, the sum is over only one of such regions, while for **M2**, there are two of them (due to our assumptions about the initial conditions of σ in the bumpy regime of the potential in (85)), as the spectator rolls along the potential towards its global minimum. Equation (87) teaches us that the part of the spectrum sourced by the gauge fields has an extra slow-roll suppression (ϵ_ϕ), in particular with respect to the direct coupling model we discussed in the previous section (see Equation (80)). However, the part that parameterizes the scale dependence of the sourced contribution, i.e., $f_{2,\mathcal{R}}$ in (87), exhibits the same log-normal scale dependence of the bumpy axion model in (81), the peak height, width and location $(f^c_{2,\mathcal{R}}, \sigma_{2,\mathcal{R}}, x^c_{2,\mathcal{R}})$ of which are calculated in [85,302] in terms of ξ_* for phenomenologically interesting values of δ that characterize the background evolution of the spectator axion.

To understand the full-scale dependence of the power spectrum (and, in particular, its vacuum contribution $(\mathcal{P}^{(\mathrm{v})}_{\mathcal{R}})$), we need to specify the scalar potential $(V(\phi))$ in the inflationary sector. Instead of specifying $V(\phi)$, we take a phenomenological approach to determine the important set of parameters summarizing the inflationary dynamics. For this purpose, first notice that assuming the effects introduced by the rolling of σ is negligible at CMB scales (early times during inflation), we have $n_s - 1 \simeq 2\eta_\phi - 6\epsilon_\phi$ and $r \simeq 16\epsilon_\phi$. Considering the latest Planck results, we adopt $r \lesssim 10^{-2}$ at CMB scales [4] to obtain $\epsilon_\phi \simeq 6.25 \times 10^{-4}$. However, the observed value of the spectral tilt gives $n_s - 1 \simeq -0.035$ [4]. Furthermore, neglecting higher-order slow-roll parameters, we assume that ϵ_ϕ remains constant throughout the inflation. These simplifying approximations enable us the describe the total power spectrum in the multifield scenarios we described above. Denoting the e-fold dependence of the Hubble parameter during slow-roll inflation as $H(N) = H_{\rm cmb}\, e^{-\epsilon_\phi N}$ (where the subscript cmb denotes the time at which the CMB pivot scale ($k_{\rm cmb} = 0.05\,\mathrm{Mpc}^{-1}$) exits the horizon), we can turn the the scale dependence of the vacuum scalar power spectrum into e-fold dependence as

$$\mathcal{P}^{(\mathrm{v})}_{\mathcal{R}}(k) \simeq \mathcal{A}_s\, \mathrm{e}^{-(1-\epsilon_\phi)(1-n_s)N}, \tag{88}$$

where $\mathcal{A}_s \equiv \mathcal{P}^{(\mathrm{v})}_{\mathcal{R}}(k_{\rm cmb}) = 2.1 \times 10^{-9}$, where $\epsilon_\phi \simeq 6.25 \times 10^{-4}$ and $1 - n_s \simeq 0.035$, as discussed above. Collecting all the information we presented above, we plot in Figure 16 the total scalar power spectrum (Equation (87)) for a representative set of parameter choices describing both of the spectator axion-gauge field models (**M1** and **M2**).

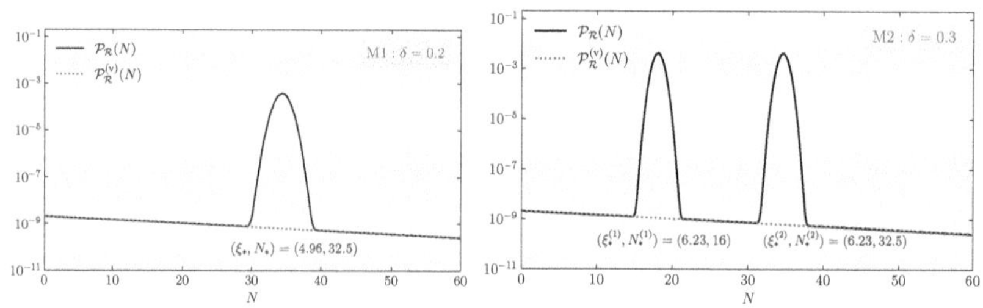

Figure 16. Total scalar power spectrum (black curves) as a function of e-folds during inflation for the spectator axion-gauge field models **M1** (**Left**) and **M2** (**Right**). In the left panel, we assume that $\dot{\sigma}$ is maximal at $N_* = 32.5$, where the effective coupling reaches $\xi(N_*) = 4.96$ corresponding to $g_{\rm cs} = 49.6$ using Equation (85). In the right panel, the spectator traverses two bumps before it settles to its minimum, and we set the inflection points at $N^{(1)}_* = 16$ and $N^{(1)}_* = 32.5$, where $\xi_* = 6.23$ ($g_{\rm cs} \simeq 20.7$).

We learn from Figure 16 that since the spectator dynamics do not significantly influence the inflationary background, the vacuum contribution (dotted lines) has a smooth red tilted power law form, which should be contrasted with the bumpy axion inflation model we discussed in the previous section. However, the transient roll of the spectator axion (σ) before settling to its global minimum triggers gauge field production as it probes the inflection point(s) on its potential (see Figure 15). This phenomenon generates an additional Gaussian bump in the scalar power spectrum. Notice that for larger δ values, the width of the corresponding bump decreases because $\dot{\sigma} \propto \xi$ is maximal for a shorter time interval, during which it can affect fewer gauge field modes. This explains why the sourced signals in the second model (**M2**) has a narrower width. As in the direct coupling cases with the gauge fields we considered earlier, the peak in the power spectrum originates from a cubic term, i.e., the last term in (82); therefore, the statistics of the curvature perturbation are expected to be non-Gaussian. This explains why the exemplifying scenarios we present in

Figure 16 with a peak power of $\mathcal{P}_{\mathcal{R}} \sim 10^{-3}$ are sufficient to generate large populations of PBHs at masses corresponding to the peak scales in Equation (7).

To close this section on axion-vector inflationary systems with the aim of producing PBHs, we can conclude that this possibility requires a fair amount of complex model building. Nevertheless, the physics of axion reached so far have a high degree of theoretical sophistication, thanks to the input and motivations from particle physics, quantum gravity and cosmology. The distinctive predictions we explored with respect to the properties of the curvature power spectrum, in particular its profile as a function of momenta and its shape around the peak, make these scenarios distinguishable from single-field inflation, with relevant ramifications for PBH phenomenology. It is certainly worthwhile to explore and apply existing efforts to explore their consequences with respect to the physics of PBH formation.

4.2. Strong Turns in the Multiscalar Field Space

An alternative approach to enhance scalar fluctuations at small scales is to exploit the dynamics of multifield inflation. Multifield inflation is a topic that has been thoroughly studied in the inflationary literature with a variety of motivations (see e.g., [304] for a review). In reviewing the applications to PBH production, we focus on a general class of non-linear sigma models that are well-motivated when embedding inflation high-energy physics. Although we go through representative concrete examples, the essence of the mechanism is again associated with a transient tachyonic instability in the scalar sector—along a direction orthogonal to the inflationary one—which is converted through suitable couplings into an amplification of curvature perturbations.

The generic two-derivative Lagrangian describing the dynamics of such a system of scalar fields (ϕ^I) minimally coupled to gravity is given by ($S \equiv \int d^4x\sqrt{-g}\,\mathcal{L}_m$),

$$\mathcal{L}_m = -\frac{1}{2}G_{IJ}(\phi)\partial_\mu\phi^I\partial^\mu\phi^J - V(\phi), \tag{89}$$

where the fields may interact through the potential ($V(\phi)$) and the field field space metric ($G_{IJ}(\phi)$). For an FLRW background characterized by the scale factor ($a(t)$) and Hubble parameter ($H(t)$), the equation of motion of the homogeneous fields is described by

$$\mathcal{D}_t\dot{\phi}^I + 3H\dot{\phi}^I + G^{IJ}V_{,J} = 0, \tag{90}$$

where the time field space covariant derivative of any field space vector (A^I) is defined as $\mathcal{D}_tA^I = \dot{A}^I + \Gamma^I{}_{JK}\dot{\phi}^J A^K$. Considering a two-field case for simplicity, the background trajectory can be split into adiabatic ($e^I_\sigma = \dot{\phi}^I/\dot{\sigma}$) and entropic ($e^I_s$) field bases that are orthogonal to each other. Here, $\sigma \equiv (G_{IJ}\dot{\phi}^I\dot{\phi}^J)^{1/2}$, which is related to the Hubble slow-roll parameter as $\epsilon \equiv -\dot{H}/H^2 = \dot{\sigma}^2/(2H^2M_{\rm pl}^2)$.

We find it worth mentioning that the vanilla multifield slow-roll trajectory corresponds to fields following standard gradient flow of the potential ($3H\dot{\phi}^I \simeq -V^{,I}$) corresponding to a field space trajectory that is approximately geodesic with $\mathcal{D}_t\dot{\phi}^I \simeq 0$ (see, e.g., (90)). However, a sufficiently long phase of inflation only requires a small acceleration along the unit vector tangent to the inflationary trajectory ($e_{\sigma\,I}\mathcal{D}_t\dot{\phi}^I \simeq 0$), and the acceleration pointing in the perpendicular direction—which parameterizes the "bending" of the inflationary trajectory—is generically not constrained and can therefore be large. The level of "bending" can be described by the dimensionless parameter ($\eta_\perp$) so that the orthogonal field bases evolve as

$$\mathcal{D}_t\, e^I_\sigma = H\eta_\perp e^I_s, \qquad \mathcal{D}_t\, e^I_s = -H\eta_\perp e^I_\sigma, \tag{91}$$

where $\eta_\perp = -e^I_s\, V_{,I}/(H\dot{\sigma})$ measures the acceleration of the trajectory perpendicular to its velocity [305,306]. In combination with the Hubble rate, a "turning rate" associated with the trajectory can be defined as $\Omega \equiv \eta_\perp H$. Therefore, an inflationary background with a strong turn ($\eta_\perp \gg 1$) satisfies $\Omega \gg H$. As we will show below, this is the regime of interest for the amplification of the scalar fluctuations in this multifield setup.

For this purpose, we consider the linear fluctuations around the background we described above, which is given by the following action [306–308]

$$S^{(2)} = \int \mathrm{d}^4x\, a^3 \left[M_{\rm pl}^2 \epsilon \left(\dot{\mathcal{R}}^2 - \frac{(\partial \mathcal{R})^2}{a^2} \right) + 2\dot{\sigma}\eta_\perp \dot{\mathcal{R}} Q_s + \frac{1}{2}\left(\dot{Q}_s^2 - \frac{(\partial Q_s)^2}{a^2} + m_s^2 Q_s^2 \right) \right], \quad (92)$$

where $\mathcal{R}$ is the comoving curvature perturbation[35], and Q_s is the instantaneous entropy perturbation, the mass of which is given by

$$m_s^2 = V_{;ss} - H^2\eta_\perp^2 + \epsilon H^2 M_{\rm pl}^2 R_{\rm fs} \quad (93)$$

were $V_{;ss} = e_s^I e_s^J V_{;IJ}$ is the projection of the covariant Hessian of the potential along the entropic direction, and $R_{\rm fs}$ is the field space curvature. As should be clear from the action (92), the dynamics of the fluctuations in this two-field model are completely dictated by three functions of time ($N = \ln(a)$), namely the Hubble rate ($H(N)$), the turning rate ($\eta_\perp(N)$) and the mass ($m_s^2(N)$) of the entropy mode. To demonstrate the amplification in the power spectrum via strong turns in curved multifield space, we follow the effective approach presented in [73] to consider a field space that undergoes a strong turn ($\eta_\perp \gg 1$) around an e-folding number (N_f) during inflation corresponding to a time well after CMB scales exited the horizon and assume a featureless Hubble rate ($H(N)$). We also assume that the time dependence of entropic mass is mainly controlled by the bending parameter, taking $m_s^2 = (b - \eta_\perp^2)H^2$, where b is chosen as a constant for simplicity.

The behavior of the bending parameter (and that of the entropic mass (m_s^2)) for typical strong turn cases is shown in Figure 17. An important quantity that determines the behavior of the fluctuations is the duration of the turn, which leads to either broad (left panel) or rapid turn (right panel) cases, as shown in the figure. Although the dynamics of the scalar perturbations are qualitatively different for broad and sudden (strong) turn cases, their behaviors share some common characteristics that can be attributed to the transient nature of the strong turn in field space. First of all, the modes that are still deep in the horizon at the end of the turn do not feel the presence of the feature; hence, their dynamics are of standard single-field type, where the power spectrum is given by

$$\mathcal{P}_{\mathcal{R}}^{(0)}(k) = \frac{H^2}{8\pi^2 \epsilon M_{\rm pl}^2}\bigg|_{k=aH}, \quad (94)$$

with a slight red-tilted scale dependence. For this purpose, we utilize the phenomenological model we discussed in the previous section to parameterize the scale dependence (time dependence) of the power spectrum as in Equation (88), with a constant $\epsilon \simeq 6.25 \times 10^{-4}$ and $1 - n_s \simeq 0.035$, along with $H_{\rm cmb} \simeq 10^{-5}$. On the other hand, large scales that leave the horizon well before the turn experience typical multifield dynamics, resulting in a transfer of power from entropic to curvature perturbations and enhancement in the power spectrum [304]. The corresponding amplification is sensitive to the value of the entropic mass (m_s) from the time of horizon exit until the turn. In the following, we assume a sizable m_s (i.e., b) before the turn so that the entropic fluctuations have decayed sufficiently before the onset of the turn so that the power spectrum is given by the standard result (94) for these scales. For scales that exit the horizon around the time of the turn, the entropic fluctuations exhibit transient (exponential) growth due to their tachyonic mass ($m_s^2 < 0$) around the turn, which is eventually transferred to the curvature perturbation through the kinetic coupling in (92) proportional to the bending ($\eta_\perp$) parameter. The exponential amplification of $\mathcal{P}_{\mathcal{R}}$ associated with these scales is a typical observational feature of strong turns, which we discuss briefly, focusing on broad and sharp turn cases separately.

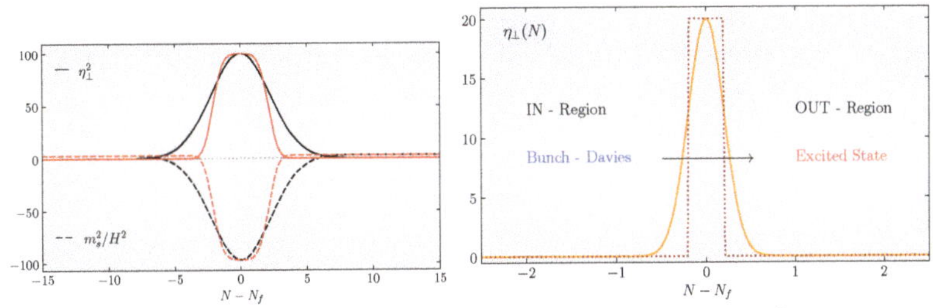

Figure 17. **(Left)**: schematic time evolution of the bending parameter ($\eta_\perp^2$) and entropic mass ($m_s^2 = b - \eta_\perp^2$) for a broad turn using a Gaussian ($\eta_\perp = \eta_\perp^{\rm max} e^{-y^2/(2\Delta^2)}$, $(\eta_\perp^{\rm max}, \Delta^2) = (10, 10)$) and smoothed top-hat profile ($\eta_\perp = \eta_\perp^{\rm max}({\rm erf}(y - \delta/2) - {\rm erf}(y + \delta/2))/2$, $(\eta_\perp^{\rm max}, \delta) = (10, 5)$) with $y = N - N_f$. **(Right)**: time dependence of the bending parameter ($\eta_\perp$) for representative sharp (rapid) turn examples with Gaussian and top-hat profiles.

Broad turns

For a strong turn in field space that lasts a sufficiently long time so that modes of interest (i.e., modes that are enhanced) have already exited the Hubble radius at the end of the turn, the dynamics of the curvature perturbation can be described by the single-field effective theory with an imaginary sound speed parameterizing the growth of the fluctuations [309,310]. From the perspective of two-field description, imaginary sound speed is just a manifestation of the transient instability induced by the strong non-geodesic motion in field space and the associated transient negative entropic mass (m_s^2) [311]. Therefore, the dynamics of mode functions are mainly characterized by two time scales: (i) $\tilde{N}$, corresponding to entropic mass crossing and the onset of instability; and (ii) $\tilde{N}$, denoting the sound horizon crossing of fluctuations after which the curvature perturbation ($\mathcal{R}$) becomes frozen:

$$\frac{k}{a(\tilde{N})} = \left|m_s(\tilde{N})\right|, \quad \& \quad \frac{k|c_s|}{a(\tilde{N})} = H(\tilde{N}), \tag{95}$$

where $|c_s|$ denotes the absolute value of the imaginary speed of sound describing the instability/growth in the fluctuations ($\mathcal{R}$) in the single EFT language. Since the background dynamics change mildly between the time of entropic mass crossing ($\tilde{N}$) and freeze-out ($\tilde{N}$) in the broad turn case, the exponential enhancement in the power spectrum can be parameterized as [73,311]

$$\mathcal{P}_{\mathcal{R}}(k) = \mathcal{P}_{\mathcal{R}}^{(0)}(k)\, \exp\left[\pi\eta_\perp\left(2 - \sqrt{3 + b/\eta_\perp^2}\right)\right]\Bigg|_{\tilde{N}_k}, \tag{96}$$

for all k modes that satisfies $k = a(\tilde{N})|m_s|$, while $m_s^2 < 0$. Notice that exponential enhancement in the power spectrum with respect to the base spectrum is proportional to the bending parameter ($\eta_\perp \gg 1$).

In Figure 18, we show the power spectrum profile as a function of e-folds (left panel) using the analytic Formula (96) for Gaussian bending profiles ($\eta_\perp = \eta_\perp^{\rm max} e^{(N-N_f)^2/(2\Delta^2)}$) with $(\eta_\perp^{\rm max}, \Delta^2) = (20, 10)$ (orange curve, Model 1) and $(\eta_\perp^{\rm max}, \Delta^2) = (20, 2)$ (black curve, Model 2) and $N_f = 30$. For both models, the range of scales and times affected by the turn where $m_s^2 < 0$ are highlighted by red and correspond to $2.4 \times 10^8 < k\,[{\rm Mpc}^{-1}] < 1.2 \times 10^{15}$ (Model 1) $1.7 \times 10^{10} < k\,[{\rm Mpc}^{-1}] < 1.6 \times 10^{13}$ (Model 2). As should be clear from the expression (96), the power spectrum reaches its maximal value for the scale for which entropic mass crossing overlaps with the peak of the bending parameter, i.e., when $N_f = \tilde{N}_{k_p}$ corresponding to

$$k_p \simeq k_f\, |(\eta_\perp^{\rm max})^2 - b|^{1/2}, \tag{97}$$

where $k_f = a(N_f)H(N_f) \simeq 5.24 \times 10^{11}\,\mathrm{Mpc}^{-1}$ is the scale that exits the horizon at N_f. To determine the slope of the power spectrum on the rise, one can compute the spectral index with respect to the base power spectrum to obtain

$$(n_s - 1) - (n_s - 1)_0 \simeq \pi(2 - \sqrt{3}) \frac{N_f - \tilde{N}}{\Delta^2 + N_f - \tilde{N}} \eta_\perp \tag{98}$$

which shows that stronger and/or less broad turns in field space lead to a steeper power spectrum. In particular, the expression (98) tells us that for broad and strong turn in field space, the slope of the power spectrum towards its peak can be much larger than the single-field models, where $n_s - 1 \lesssim 4$ [95,212,217,224].

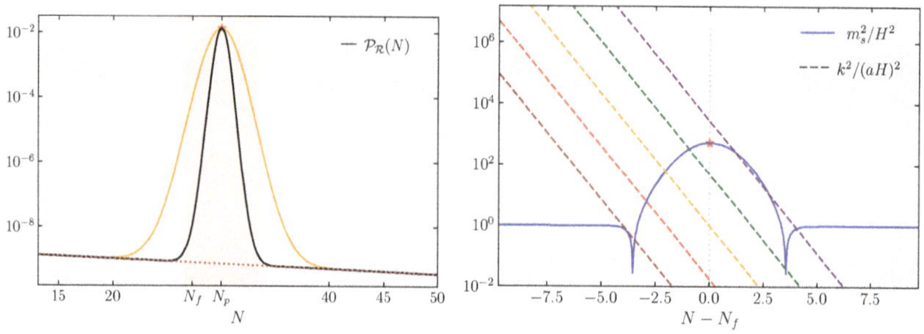

Figure 18. (**Left**): curvature power spectrum (normalized to $\mathcal{P}_\mathcal{R}(N = 0) = 2.1 \times 10^{-9}$) for broad strong turns and for two representative parameter choices using a Gaussian bending profile (see main text). (**Right**): migration of modes $k^2/(aH)^2$ for equally spaced $\ln(k)$ values around the turn to confirm the expectation that the power spectrum should fall more quickly after the peak (as compared to the left panel) because more modes experience entropic mass crossing for $N < N_f$ than for $N > N_f$ in the $m_s^2 < 0$ region highlighted in red.

We would like to note that for scales affected by the turn, the analytic profile presented in (96) is symmetric around the peak; however, numerical computations carried out in [73] show that the power spectrum typically exhibits a much quicker fall of behavior for scales following the peak compared to those preceding it. This can be understood by first recalling that the modes that are enhanced are those that go through entropic mass crossing (95) while the entropic mass is negative ($m_s^2 < 0$). The right panel of Figure 18 can then guide us for an intuitive understanding for the expected asymmetry in the power spectrum following its peak because a larger range of modes enjoys entropic mass crossing before the peak of $|m_s|$[36] than afterwards. We therefore conclude that the expression (96), along with the power spectrum shown in the left panel of Figure 18, only characterizes the behavior of the fluctuations qualitatively, and for more accurate results, numerical methods are required, as shown in [73].

Sudden turns

When the duration of the turn is shorter (a statement that we will make more precise below), the scales that are maximally enhanced are still those that experience entropic mass crossing during the turn; therefore, they are of order $k \sim k_f \eta_\perp^{\max}$ (see, e.g., Equation (97)). However, in contrast to the broad turn case, these modes get caught in the (rapid turn) feature while they are still deep inside the horizon, i.e., they satisfy $k > aH$ at the end of the turn. This situation effectively generates an initial excited state for these modes that can be studied analytically in the regime of sudden and strong (constant) turns ($\eta_\perp$) [72,312]. Along with a localized exponential amplification, the resulting power spectrum of curvature perturbation exhibits order-one oscillations in k, which is a common characteristic of sharp features [313,314], with a frequency set by the time of the turn. We will review these features below by focusing on the analytic model presented in [312][37]. For this purpose, we consider an inflationary trajectory with a top-hat[38]sudden turn profile (see, e.g., the right panel in Figure 17) characterized by three parameters: N_f, the central time of its location (measured with respect to the horizon exit time of the CMB pivot scale); its duration (δ) and typical (constant) value ($\eta_\perp^{\max}$) around N_f:

$$\eta_\perp(N) = \eta_\perp^{\max}\left[\theta\left(N - \left(N_f - \frac{\delta}{2}\right)\right) - \theta\left(N - \left(N_f + \frac{\delta}{2}\right)\right)\right], \tag{99}$$

along with an entropic mass profile during the turn that is given by

$$\frac{m_s^2}{H^2} = (\xi - 1)\eta_\perp^2(N) \tag{100}$$

where $\xi < 1$ is a constant parameter. Note from the parameterization of the bending parameter (99) that the sudden turn case we consider corresponds to $\delta > \ln(\eta_\perp^{\max})$.

In this setup, one can determine the power spectrum generated by sharp turns by studying the scattering problem of the two-field system using WKB methods [72]. In particular, this procedure includes matching operators (and their derivatives) describing the perturbations from the IN region, where the Bunch–Davies vacuum is assumed, passing through the turn and to the OUT region, resulting in an excited "initial" state as sketched in the right panel of Figure 17. The behavior of the entropic and curvature perturbations in both regions (IN and OUT) are standard, where the fields are dynamically decoupled from each other, while the turn sandwiched between the IN and OUT regions results in their exponential growth. Following this procedure, an analytic expression for the power spectrum was derived in [312] for scales satisfying $k/k_f < \sqrt{1-\xi}\eta_\perp$ as:

$$\frac{\mathcal{P}_\mathcal{R}(k)}{\mathcal{P}_\mathcal{R}^{(0)}} = \underbrace{\frac{e^{2\eta_\perp\delta}\,S}{2S^2(1+X+\sqrt{X(1+X)})}}_{\text{envelope}} \times \sin^2\left(e^{-\delta/2}\kappa\eta_\perp + \arctan(\kappa/S)\right) \tag{101}$$

where

$$S(k) = \sqrt{\sqrt{4\kappa^2 + \frac{(3+\xi)^2}{4}} - \left(\kappa^2 + \frac{(3+\xi)}{2}\right)}, \qquad \text{and} \quad X(k) = \frac{(3+\xi)^2}{16\kappa^2} \tag{102}$$

where $\kappa \equiv k/(k_f\eta_\perp)$ and $\eta_\perp$ denote the typical constant value of the bending parameter corresponding to $\eta_\perp^{\max}$ in the top-hat profile (99).

For the choice of $\xi = -3$—which corresponds to an effectively massless entropy mode on superhorizon scales (see Section 2 of [312] for a detailed discussion on this)—the profile of the curvature power spectrum[39] for two representative parameter choices characterizing the turn is shown in Figure 19. As described by the analytic expression (101), the power spectrum features characteristic oscillations ($\sin^2$ term) modulated by an envelope that is described as highlighted in the equation. The peak value of the envelope (see, e.g., the blue dashed lines in Figure 19) controls the maximal enhancement of the power spectrum with respect to the baseline power spectrum (94) (we assume it to be scale invariant for simplicity) which will occur roughly at the maximum of the function $S(k)$ defined in (102). Arguably, the oscillatory pattern of the power spectrum around the peak is much more interesting. In particular, the $\eta_\perp\delta \ll e^{-\delta/2}\eta_\perp$ $\sin^2$ term changes much faster than the envelope. On the other hand, in a sharp turn regime $e^{-\delta/2}\eta_\perp \gg 1$ so that the first term in the argument of the $\sin^2$ term also changes much faster than the second. These two observations imply that oscillations in the power spectrum are periodic to a considerable degree. The period of the maxima in the oscillations occurs roughly in $\Delta\kappa \approx \pi e^{\delta/2}/\eta_\perp$, corresponding to a linear frequency of

$$\Delta k \approx \pi e^{\delta/2}k_f \quad \longrightarrow \quad \omega \equiv \frac{2\pi}{\Delta k} \approx \frac{2e^{-\delta/2}}{k_f}. \tag{103}$$

Therefore, the scale (k_f) that exits the horizon at the center of the feature roughly set the frequency of oscillations in the power spectrum.

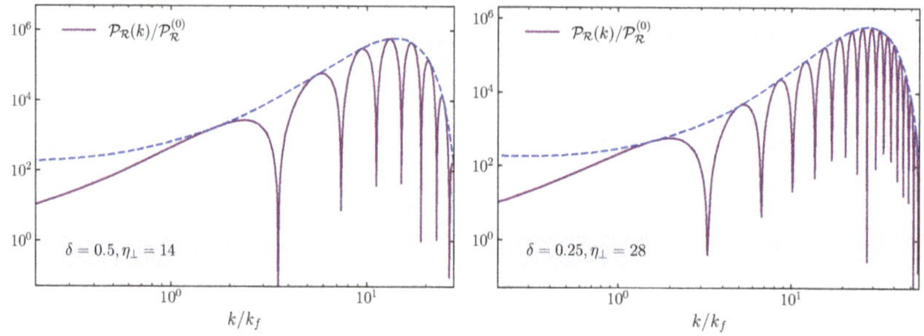

Figure 19. Shape of the curvature power spectrum using the analytic formula in Equation (101) for two representative sharp turn cases, where the turn in parameterized with a top-hat bending parameter of Equation (99). The dashed blue line represents the envelope characterized by the terms highlighted in Equation (101).

In summary, strong sudden turns in field space lead to a power spectrum that exhibits order-one oscillations, the amplitude of which is modulated by an exponentially enhanced envelope (101). The oscillations are fast, making their peaks almost periodic with a frequency given by (103). We note that these oscillations do not have an impact on the PBH mass spectrum ($\beta(M)$) (see, e.g., (15)) because the calculation of the variance ($\sigma^2(M)$) (of the density contrast) involves a smoothing procedure over scales comparable to the width of peaks in the power spectrum ($\mathcal{P}_\mathcal{R}$) [73]. Another issue of interest from the perspective of model building is the influence of non-Gaussianity. The periods of strong turns during inflation are known to generate non-Gaussianity of the flattened type [310,316,317]. The implications of this on the PBH distribution and model building (i.e., the required peak amplitude of $\mathcal{P}_\mathcal{R}$) is not known and subject to future research. We therefore would like to emphasize that in the strong turn examples we discussed in this section (see Figures 18 and 19), the peak amplitudes we provided are not chosen based on a particular bias on the basis of non-Gaussianity present in these models.

5. Outlook

Primordial black holes (PBHs), if they exist, can shed light on long-standing questions with respect to the nature of dark matter and on the mechanisms driving cosmic inflation. They can provide distinctive sources of gravitational waves that are potentially detectable with current or forthcoming gravitational wave experiments. For these reasons, the physics of PBHs offer promising opportunities for collaboration between cosmologists, astronomers and gravitational wave scientists.

In this review, we focused on theoretical aspects of PBH inflationary model building. We learned that generating PBHs from inflation is difficult but possible. We reviewed conceptual ideas for amplifying the primordial curvature spectrum at a level sufficient to trigger black hole formation. These mechanisms find realizations within single-field inflation through a conversion of pronounced gradients of homogeneous quantities into curvature perturbations or within multifield inflation, where curvature fluctuations are instead amplified through appropriate couplings with additional sectors, which are characterized by tachyonic instabilities. The required tuning of model parameters, or the degree of model sophistication required to realize these ideas can be demanding. However, it is certainly a worthwhile effort, given that experimental probes of PBHs are sensitive to the details of the curvature power spectrum, for example, through the properties of the resulting PBH population or through an induced stochastic gravitational wave background sourced by curvature fluctuations. In fact, different categories of models lead to distinct, potentially distinguishable predictions for the statistics of primordial fluctuations. Hence, a detailed analysis relating theoretical scenarios to cosmological and gravitational wave probes offers new precise tests of inflationary mechanisms complementary to traditional ones associated with the physics of the cosmic microwave background and of the large-scale structures of our universe.

Although much theoretical work has been done so far by the community, much more is needed to further investigate and clarify different aspects of the physics of inflation leading to PBH. At the level of model building, it will be important to clarify and address challenges associated with a severe tuning of model parameters or with the dynamical stability of PBH models requiring non-attractor, non-slow-roll phases of inflationary evolution. It will also be crucial to continue to characterize the rich and subtle properties of primordial fluctuations in PBH models of inflation with large enhancements of the primordial power spectrum at small scales. The to-do list includes a deeper analytic understanding of the properties of the curvature spectrum profile, as well as non-linearities and non-Gaussianities and their consequences for observable quantities. Such theoretical analysis will have relevant ramifications for designing appropriate cosmological probes of the physics of PBHs. Hopefully, a concerted effort of theory and experiments motivated by a deeper understanding of PBH physics will allow us to set new bounds or possibly make new discoveries with respect to the mechanisms driving inflation and the nature of dark matter.

Author Contributions: Conceptualization, O.Ö. and G.T.; formal analysis, O.Ö.; writing—original draft preparation, O.Ö.; writing—review and editing, O.Ö. and G.T.; visualization, O.Ö. All authors have read and agreed to the published version of the manuscript.

Funding: O.Ö. is supported by the "Juan de la Cierva" fellowship IJC2020-045803-I, the European Structural and Investment Funds and the Czech Ministry of Education, Youth and Sports (Project CoGraDS-CZ.02.1.01/0.0/0.0/15003/0000437) and by the Spanish Research Agency (Agencia Estatal de Investigación) through the Grant IFT Centro de Excelencia Severo Ochoa No CEX2020-001007-S funded by MCIN/AEI/10.13039/501100011033. G.T. is partially funded by the STFC grant ST/T000813/1.

Institutional Review Board Statement: Not applicable.

Informed Consent Statement: Not applicable.

Data Availability Statement: No data is reported.

Acknowledgments: We would like to thank Cristiano Germani and Shi Pi for discussions and comments. For the purpose of open access, the authors have applied a Creative Commons Attribution license to any author-accepted manuscript version arising from this work.

Conflicts of Interest: The authors declare no conflict of interest.

Appendix A. Background Cosmology: A Mini Review

Space-time metric. The central premise in modern cosmology is that as we look at the universe on large enough scales, it appears to be simpler and more uniform compared to the small scales. In other words, if we focus on sufficiently large scales, clumpy regions such as the distribution of galaxies appear to be isotropic and homogeneous. The large-scale spatial homogeneity and isotropy of the universe have been tested by a variety of observations such as large-scale structure (LSS) surveys [318,319]. However, perhaps the most important evidence supporting this claim comes from the almost uniform temperature of the CMB originating from different parts of the sky. For a first approximation, we can therefore assume the universe to be isotropic and homogeneous. The high degree of spatial symmetry uniquely determines the structure of space–time geometry, where physical distances are measured by the so-called Friedmann [320], Lemaître [321], Robertson [322] and Walker [323] (FLRW) line element[40],

$$ds^2 = -dt^2 + a^2(t)\left(\frac{dr^2}{1-Kr^2} + r^2\,d\Omega_2\right), \tag{A1}$$

where $d\Omega_2 = \left(d\theta^2 + \sin^2\theta\, d\phi^2\right)$ is the line element on the two-dimensional sphere (S_2), and $K = \{-1, 0, 1\}$ represent negative, zero and positive curvature of constant-time hypersurfaces respectively. Note that the symmetries of the universe allow us to describe the metric using just a single function of time ($a(t)$) and a constant parameter (K). The function $a(t)$ is called the scale factor, which parameterizes the size of the spatial slices at a given moment in time, and the Hubble rate ($H(t) \equiv \dot{a}(t)/a(t)$) describes the speed of expansion at a given moment in time. Here, an expanding universe ($H(t) > 0$) corresponds to a monotonically increasing scale factor ($a(t)$).

Dynamics of the universe. Dynamics of space–time are governed by the Einstein field equation,

$$R_{\mu\nu} - \frac{1}{2} g_{\mu\nu} R = \frac{1}{M_{\rm pl}^2} T_{\mu\nu}, \tag{A2}$$

where $R_{\mu\nu} \equiv \partial_\lambda \Gamma^\lambda_{\mu\nu} - \partial_\nu \Gamma^\lambda_{\mu\lambda} + \Gamma^\lambda_{\lambda\rho}\Gamma^\rho_{\mu\nu} - \Gamma^\rho_{\mu\lambda}\Gamma^\lambda_{\nu\rho}$ is the Ricci tensor build-out of Christoffel symbols, $R \equiv R^\mu_\mu = g^{\mu\nu} R_{\mu\nu}$ is the Ricci scalar and in terms of the metric and Christoffel symbols are given by

$$\Gamma^\sigma_{\mu\nu} \equiv \frac{1}{2} g^{\sigma\rho} \left(\partial_\mu g_{\rho\nu} + \partial_\nu g_{\mu\rho} - \partial_\rho g_{\mu\nu}\right). \tag{A3}$$

The Einstein equation in (A2) relates the geometry of space–time on the *l.h.s* to the matter content in the universe through the appearance of an energy momentum tensor ($T_{\mu\nu}$) on the *r.h.s.* As in the case of the space–time metric, homogeneity and isotropy restrict the possible choices of matter content and enforce the energy–momentum tensor to take the perfect fluid form,

$$\bar{T}_{\mu\nu} = (\bar{\rho} + \bar{P})\, U_\mu\, U_\nu + \bar{P}\, g_{\mu\nu}, \tag{A4}$$

where $\bar{\rho}$ and $\bar{P}$ are the background energy density and the pressure in the rest frame of the fluid, respectively, and $U_\mu = (-1, 0, 0, 0)$ is its time-like 4-velocity of the fluid relative to the observer. The evolution equation for the energy density deriving the expansion of the universe can be derived from the $\nu = 0$ component of covariant conservation law of the energy momentum tensor:

$$\nabla_\mu \bar{T}^\mu{}_0 = 0 \quad \Rightarrow \quad \dot{\bar{\rho}} + 3\frac{\dot{a}}{a}(\bar{\rho} + \bar{P}) = 0. \tag{A5}$$

On the other hand, using the FLRW metric (A1), Einstein field equations give us information about how space–time reacts to the matter content in the universe through the Friedmann,

$$H^2 = \frac{\bar{\rho}}{3M_{\rm pl}^2} - \frac{K}{a^2}, \tag{A6}$$

and acceleration equation,

$$\frac{\ddot{a}}{a} = -\frac{1}{6M_{\rm pl}^2}(\bar{\rho} + 3\bar{P}). \tag{A7}$$

The Equations (A5)–(A7) are the key in determining the evolution of the universe and its constituents at large scales, namely $a(t)$ and $\bar{\rho}(t), \bar{P}(t)$ (assuming knowledge of spatial curvature (K)). An important aspect of these equations is that they are not independent; for example, it is possible to combine the last two to obtain the first one. In practice, this implies that we need another ingredient to solve for three variables, i.e., $\bar{\rho}(t)$, $\bar{P}(t)$ and $a(t)$—or, equivalently, $H(t)$. A quantity that comes to the rescue in this context is the equation of state (EoS) parameter, which provides a linear relation between the pressure and energy density of the fluid(s) constituting the universe:

$$\bar{P}(t) = w\, \bar{\rho}(t), \tag{A8}$$

where $w =$ constant for each fluid contributing to the total pressure. Although this relation is not the most general form of $P(\rho)$ that is available to us, this parameterization is perfectly adequate in providing an accurate coarse-grained description of our universe through most of its history. In particular, it provides a simple analytic control for determining the dynamics of the different components of the universe and the evolution of the Hubble parameter (scale factor). In order to describe the history of the universe in a continuous manner, we typically consider multiple fluids that contribute to the energy density of the universe while satisfying the relation (A8). For example, labeling each EoS with w_i and assuming that a single component dominates the energy density for a given moment of time in the universe, one can easily integrate (A5) to obtain

$$\bar{\rho}_i(t) \propto a(t)^{-3(1+w_i)}. \tag{A9}$$

Cosmological inventory. We can classify different sources of cosmological evolution by their contribution to the pressure: (i) For relativistic gas of particles (radiation), pressure is about one-third of the energy density, with $w_r = 1/3$. Popular constituents of such a fluid are photons, neutrinos and gravitons. (ii) Non-relativistic pressureless dust (matter) ($w_m = 0$) such as dark matter (e.g., PBHs) and baryons. (iii) Cosmological constants ($w_\Lambda = -1$) such as vacuum energy. According to Equation (A9), the energy density of each of these fluids therefore evolves as $\bar{\rho}_r \propto a^{-4}$, $\bar{\rho}_m \propto a^{-3}$, $\bar{\rho}_\Lambda \propto a^0$. Therefore, in the future, the cosmological constant will dominate, while in the far past ($a \to 0$), the universe was radiation-dominated, and in between these two stages, there is a matter-dominated stage. Due to the inadequacy of explaining initial conditions of the observed CMB anisotropies, cosmologists tend to complete the picture we described above with a very early phase of accelerated expansion ($\ddot{a} > 0$) called inflation (see, e.g., [324] for a comprehensive review). During this phase, the EoS parameter also closely mimics the behavior of a cosmological constant ($w_{\rm inf} \simeq -1$). More precisely, the departure of the EoS from its value (-1) during inflation is characterized by a time-dependent slow-roll parameter ($\epsilon(t) = -\dot{H}/H^2 \ll 1$, where $w_{\rm inf} = -1 + 2\epsilon(t)/3$). In this framework, during inflation, $\epsilon \ll 1$, while inflation terminates when $\epsilon = 1$. After this stage, it is generically assumed that the constituent(s) (i.e., a scalar field or fields) that drive inflation decay to relativistic particles in a process called (p)reheating [121,122]. In this review, we assume that this process proceeds very efficiently so that soon after inflation ends, the universe is filled with relativistic particles and, consequently, evolves through the radiation-dominated (RD), matter-dominated (MD) and, finally, dark-energy-dominated (DED) phases. On the other hand, we refer to the full picture that arises by the inclusion of an early accelerated expansion as the *inflationary universe*, which is composed of the following consequent phases: inflation $\to$ RD $\to$ MD $\to$ DED.

Evolution of the comoving Hubble horizon. As we described in the main text, the evolution of the comoving horizon is crucial to understanding the causal evolution of fluctuation modes in the inflationary universe. To understand its time evolution in the picture we described above, we focus on its time derivative

$$\frac{\mathrm{d}}{\mathrm{d}t}(aH)^{-1} = -\frac{1}{aH^2}(H^2 + \dot{H}) = -\frac{\ddot{a}}{a^2H^2} = \frac{1}{a}\frac{(1+3w)}{2} \tag{A10}$$

where we used the acceleration Equation (A7), together with the Friedmann Equation (A6), focusing on flat FLRW slicing ($K = 0$)[41]. The expression mentioned above can be shaped into a form suitable for integration as $\mathrm{d}\ln((aH)^{-1}) = [(1+3w)/2]\,\mathrm{d}(\ln a)$, which immediately gives

$$(aH)^{-1} = H_0^{-1}\, a^{(1+3w)/2}, \tag{A11}$$

where we normalized the scale factor today as $a_0 = 1$. Notice that during inflation, $w_{\rm inf} \simeq -1$, and the comoving horizon decreases, while during the subsequent RD ($w_r = 1/3$) and MD ($w_m = 0$) phases, it grows with a scale factor.

Density parameters (Ω). To describe different energy components in the universe, cosmologists often parameterize radiation, matter and dark energy density relative to the critical energy density of spatially flat hypersurfaces using the following definitions (and dropping the over-bar notation to describe background quantities)

$$\rho_{c,0} \equiv 3H_0^2 M_{\rm pl}^2 \quad \rightarrow \quad \Omega_i \equiv \frac{\rho_{i,0}}{\rho_{c,0}} \tag{A12}$$

where subscript "0" denotes quantities evaluated today, and "i" indicates different kinds of fluids. Focusing on the post-inflationary era, we can then rewrite the Friedmann equation in terms of dimensionless density parameters as

$$3H(a)^2 M_{\rm pl}^2 = \rho_{r,0}\, a^{-4} + \rho_{m,0}\, a^{-3} + \rho_{\Lambda,0} \;\rightarrow\; \frac{H^2(a)}{H_0^2} = \Omega_r\, a^{-4} + \Omega_m\, a^{-3} + \Omega_\Lambda. \tag{A13}$$

According to the latest observations of the CMB anisotropies by the Planck collaboration, matter and dark energy density are determined to be [7],

$$\Omega_\Lambda = 0.6847 \pm 0.0073, \quad \Omega_m = 0.3153 \pm 0.0073. \tag{A14}$$

On the other hand, we can infer radiation density today by utilizing the transition time of the universe from RD to MD. This moment in our universe is commonly referred to as matter–radiation equality, which is defined by

$$\rho_r(a_{\rm eq}) = \rho_m(a_{\rm eq}) \quad \Rightarrow \quad a_{\rm eq} = \frac{\Omega_r}{\Omega_m}. \tag{A15}$$

Noting the relation between the red-shift parameter[42] ($z(t)$) and scale factor ($a(t) = (1+z(t))^{-1}$) and Planck's prediction ($z_{\rm eq} = 3402 \pm 26$) [7], together with Equation (A14), therefore gives

$$\Omega_r \simeq 9.26535 \times 10^{-5}, \tag{A16}$$

where we used the central values of the quantities ($\Omega_m, z_{\rm eq}$) determined by Planck.

Making a two-component fluid approximation around the time of matter radiation equality ($a \approx a_{\rm eq}$), the total energy density of the universe at $a = a_{\rm eq}$ can be related to matter density today via

$$\rho(a_{\rm eq}) \simeq \frac{\rho_{r,0}}{a_{\rm eq}^4} + \frac{\rho_{m,0}}{a_{\rm eq}^3} \simeq 2\,\frac{\rho_{m,0}}{a_{\rm eq}^3}. \tag{A17}$$

Cosmological parameters. Other cosmological parameters determined by the latest Planck data (2018) (TT,TE,EE + low E + lensing, % 68 CL) are as follows [7]:

$$\begin{aligned} \Omega_K &= 0.001 \pm 0.002, \qquad H_0 = 67.36 \pm 0.54\,[\mathrm{km\,s^{-1}Mpc^{-1}}], \\ \Omega_m h^2 &= 0.1430 \pm 0.0011, \qquad k_{\rm eq} = 0.010384 \pm 0.000081\,[\mathrm{Mpc^{-1}}]. \end{aligned} \tag{A18}$$

So far, have not mentioned thermodynamical properties such as temperature and entropy in the universe following inflation. In what follows, we will note some of the key formulas we will use in Section 2 without getting into the details of the derivation of these formulas. For an in-depth discussion on the contents we will introduce below, we refer the reader to Chapter 3 of the seminal books by Kolb and Turner [325] and Mukhanov [326] or to Baumann's recent book [128].

The total energy density of relativistic degrees of freedom that are in thermal equilibrium with the plasma can be related to the temperature of the plasma (T) (i.e., temperature of the photon gas) as

$$\rho_r = \frac{\pi^2}{30} g_*(T)\, T^4, \tag{A19}$$

where $g_*(T)$ is the effective number of relativistic degrees of freedom at the temperature (T) which is defined as

$$g_*(T) = \sum_{\text{bos.}} g_b + \frac{7}{8} \sum_{\text{fermi.}} g_f, \tag{A20}$$

where g_b and g_f denote the intrinsic degree of freedom (i.e., spin) for bosonic and fermionic species, respectively. The total entropy density of relativistic species is defined as

$$s(T) \equiv \frac{2\pi^2}{45} g_s(T)\, T^3, \tag{A21}$$

where g_s is the effective number of degrees of freedom in entropy. For species in thermal equilibrium (i.e., species that have the same temperature ($T_b = T_f = T$), $g_s(T) = g_*(T)$) because for each bosonic/fermionic degree of freedom, the entropy density is defined as $s \equiv (\rho + P)/T$ with a common denominator.

A very useful formula can be derived by using the conservation of total entropy in the universe ($S = sV \propto sa^3 \simeq constant.$), which can be utilized to relate the scale factor and the temperature of the plasma as

$$g_s(T)\, T^3\, a^3 \simeq constant. \quad \Longrightarrow \quad \frac{a(t_1)}{a(t_2)} = \frac{T_2}{T_1} \left(\frac{g_s(T_2)}{g_s(T_1)} \right)^{1/3}. \tag{A22}$$

Appendix B. A Simple Estimate for the Threshold of Collapse

In this short appendix, we present a simple analytic estimate of the characteristic value of the collapse threshold for PBH formation during the RDU based on the Jeans length argument in Newtonian gravity as first discussed in [13], closely following [30,127]. For this purpose, we take the background space–time after inflation to have the spatially flat ($K = 0$) FLRW form (A1) and therefore the evolution of the scale factor described by the Friedmann equation: $3H^2 M_{\rm pl}^2 = \bar{\rho}(t)$. Now consider a locally perturbed, spherically symmetric region in the universe that is initially outside the horizon and could eventually collapse to form a PBH upon horizon re-entry. The metric describing such a region can be written as

$$ds^2 = -dt^2 + a^2(t)\, e^{-2\mathcal{R}(\hat{r})} \left[d\hat{r}^2 + \hat{r}^2\, d\Omega_2 \right] \tag{A23}$$

where $a(t)$ is the scale factor, and $\mathcal{R} < 0$ is the non-linear generalization of the conserved comoving curvature perturbation defined on a super-Hubble scale [327]. At large distances ($\hat{r} \to \infty$), curvature perturbation is assumed to vanish ($\mathcal{R} \to 0$) so that the universe is described by the spatially flat FLRW metric. By making a coordinate redefinition ($r = \hat{r} e^{-\mathcal{R}(\hat{r})}$), the metric describing the spherical perturbed region (A23) can be transformed into the one describing a closed universe with positive spatial curvature (as in (A1)),

$$ds^2 = -dt^2 + a^2(t) \left(\frac{dr^2}{1 - K(r)\, r^2} + r^2\, d\Omega_2 \right), \tag{A24}$$

where the relation between perturbations of the two metrics is given by $K(r)\, r^2 = \hat{r}\mathcal{R}'(\hat{r})(2 - \hat{r}\mathcal{R}'(\hat{r}))$, showing that the local spatial curvature is given in terms of the first-order spatial (leading-order) derivatives of the curvature perturbation ($\mathcal{R}(\hat{r})$). Ignoring higher-order spatial derivatives of K (*i.e.*, $K' \sim \mathcal{R}''$) on sufficiently large scales, the evolution of the spherical region is given by the 00 component of the Einstein Equation (A2)

$$H^2 = \frac{\rho_{\rm tot}}{3M_{\rm pl}^2} - \frac{K(r)}{a^2} \tag{A25}$$

which is equivalent to the evolution of a closed universe (see e.g., (A6)) with a perturbation ($\delta\rho$) induced by spatial curvature ($K(r)$):

$$\frac{\delta\rho}{\rho} \equiv \frac{\rho_{\rm tot} - \rho}{\rho} = \frac{3M_{\rm pl}^2 K}{\bar{\rho} a^2} = \frac{K}{H^2 a^2}, \tag{A26}$$

where we make use of $\bar{\rho}(t) = 3H^2 M_{\rm pl}^2$. According to (A26), since $\bar{\rho} \propto a^{-4}$ during the RDU, perturbations are initially ($a \to 0$) small, i.e., $\delta\rho/\rho \ll 1$. As the universe evolves, a local spherical region with $K > 0$ will eventually stop expanding and collapse to form a PBH. This happens precisely when the right-hand side of (A25) becomes negative, i.e., at $t = t_{\rm c}$, where $3M_{\rm pl}^2 K = a^2 \rho_{\rm tot}$, corresponding to $\delta\rho/\rho = 1$ in (A26), assuming a homogeneous and isotropic universe. In the RDU, since perturbation modes that have a length scale smaller than the Jeans length ($k_J^{-1} \equiv c_s/aH$) cannot collapse, the smallest comoving scale that can undergo collapse at $t = t_{\rm c}$ is given by $k^2 = (aH)^2/c_s^2$, where c_s is the speed propagation of perturbation. Therefore, we have

$$\frac{\delta\rho}{\rho}(t_c) = \frac{K}{k^2}\frac{k^2}{a^2 H^2} = \frac{K}{c_s^2 k^2} = 1, \tag{A27}$$

which suggests the identification of $K = c_s^2 k^2$. Therefore, at the time of horizon re-entry ($k = a_{\rm f} H_{\rm f}$), the perturbations relevant for PBH formation should have a density contrast larger than

$$\left(\frac{\delta\rho}{\rho}(t_{\rm f})\right)_{\rm c} = \frac{K}{(a_{\rm f} H_{\rm f})^2} = c_s^2 \left(\frac{k}{a_{\rm f} H_{\rm f}}\right)^2 = c_s^2 = w. \tag{A28}$$

Appendix C. Solving the Mukhanov–Sasaki Equation: Numerical Procedure

In this appendix, we discuss important steps for the numerical evaluation of the Mukhanov–Sasaki (MS) equation. The aim is to obtain the power spectrum of curvature perturbation in the inflationary scenarios we discussed in Sections 3.2 and 3.3. Note that scenarios discussed in those sections can be captured by the generalized MS Equation (43) utilizing (44) and (59), as we describe below.

We start by scaling away the highly oscillatory contributions from the Bunch–Davies (BD) initial conditions (59). For this purpose, we define a dimensionless variable ($\bar{v}_k$) through

$$v_k(\bar{\tau}) = \frac{\bar{v}_k(\bar{\tau})}{\sqrt{2k}}\, e^{-ik\bar{\tau}}, \tag{A29}$$

to rewrite the MS Equation (43) as

$$\bar{v}_k''(\bar{\tau}) - 2ik\,\bar{v}_k'(\bar{\tau}) - \frac{z''}{z}\bar{v}_k(\bar{\tau}) = 0. \tag{A30}$$

For mode-by-mode evaluation, the e-folds (N) are more suitable variable for numerical integration, so we turn the derivatives with respect to $\bar{\tau}$ in (A30) into e-folds using $\mathrm{d}N = (aH/c_s)\mathrm{d}\bar{\tau}$ and obtain

$$\frac{\mathrm{d}^2 \bar{v}_k}{\mathrm{d}N^2} + \left(\left[1 - \epsilon - s\right] - 2i\,\frac{c_s k}{aH}\right)\frac{\mathrm{d}\,\bar{v}_k}{\mathrm{d}N} - \frac{c_s^2}{(aH)^2}\frac{z''}{z}\bar{v}_k = 0\,. \tag{A31}$$

We parameterize the scale factor and Hubble rate as

$$a(N) = a_{\rm end}\, e^{N-60}, \qquad H(N) = H_{\rm end}\, e^{-\int_{60}^{N} \epsilon(N')\mathrm{d}N'}, \tag{A32}$$

and note that

$$\frac{z''}{z} = \left(\frac{aH}{c_s}\right)^2 \left[(1-\epsilon-s)\left(1+\frac{\eta-s+\mu}{2}\right) + \left(1+\frac{\eta-s+\mu}{2}\right)^2 + \frac{1}{2}\left(\frac{\mathrm{d}\eta}{\mathrm{d}N} - \frac{\mathrm{d}s}{\mathrm{d}N} + \frac{\mathrm{d}\mu}{\mathrm{d}N}\right)\right]. \tag{A33}$$

In terms of the new variable, the Bunch–Davies initial conditions simplify considerably and can described as

$$\bar{v}_k(N)\big|_{\text{in}} = 1, \qquad \bar{v}'_k(N)\big|_{\text{in}} = 0. \tag{A34}$$

Provided that the background evolution (i.e., ϵ, η, μ, etc.) is known in terms of e-fold number, the MS equation can be solved numerically for each k mode in terms of the rescaled variable (A31) using the initial conditions (A34). In this respect, some care must be taken for the initialization time of individual modes because deep inside the horizon ($k^2 \gg z''/z$), the solution to (A31) is highly oscillatory, which would be costly for the numerical computation. Typically, it is enough to initialize the modes at some time such that they are sufficiently inside the horizon. For this purpose, we choose to evolve each mode by setting the initial time as

$$N_{\text{in}}^{(k)} = N_0^{(k)} - 4, \tag{A35}$$

where the "k" superscript indicates the intrinsic mode dependence for the choice of N_{in}, and $N_0^{(k)}$ denotes the horizon crossing time[43]. As argued in the main text (see the discussion around (62)), the latter can be obtained via

$$k^2 = \frac{z''}{2z}\bigg|_{N_0^{(k)}} \tag{A36}$$

using (A33) (or, equivalently, (44)), provided the background solution is known. Having obtained the individual mode evaluation from $N_{\text{in}}^{(k)}$ to $N_{\text{end}} = 60$, the power spectrum (55) can be described as

$$\mathcal{P}_{\mathcal{R}}(k, N_{\text{end}}) = \frac{H_{\text{end}}^2}{8\pi^2 \epsilon_{\text{end}} c_{s,\text{end}} M_{\text{pl}}^2} \left(\frac{c_{s,\text{end}}}{\tilde{M}_{\text{end}}}\right)^2 \left(\frac{k}{k_{\text{end}}}\right)^2 |\bar{v}_k(N_{\text{end}})|^2, \tag{A37}$$

where we defined $\tilde{M} \equiv M(N)/M_{\text{pl}}$, while $k_{\text{end}} = a_{\text{end}} H_{\text{end}}$ is the mode that exits the horizon at the end of inflation.

In order to set the overall normalization of the power spectrum, we need to determine H_{end} with respect to M_{pl}. We do so by requiring the normalization of the power spectrum indicated by Planck at the pivot scale ($k_{\text{cmb}} = 0.05\,\text{Mpc}^{-1}$) using $\mathcal{P}_{\mathcal{R}}(k_{\text{cmb}}, N_{\text{end}}) \simeq 2.1 \times 10^{-9}$. Notice that, from the general formulas we presented above, canonical single-field scenarios (see Section 3.2) can be recovered by making the following replacements: $c_s \to 1, s \to 0$, $M \to M_{\text{pl}}, \mu \to 0$ and $\mathrm{d}\tilde{\tau} \to \mathrm{d}\tau$ and $\mathrm{d}N = (aH)\,\mathrm{d}\tau$. A pedagogical Python notebook file that calculates the power spectrum using the prescription described above can be found through the github link.

The role of decaying modes in the canonical single-field scenarios. We can now verify whether the general expression (39) provides an accurate description for the enhancement of the power spectrum in the canonical single-field model discussed in Section 3.2. For this purpose, we assume that the leading solution to the curvature perturbation at superhorizon scales is given by the constant solution at horizon crossing ($\mathcal{R}_k(\tau) = \mathcal{R}_k^{(0)}$)

and plugging it into the last term in (39). We can then generate an iterative solution for the curvature perturbation as

$$\mathcal{R}_k(N_{\rm end}) = \mathcal{R}_k^{(0)}\left[1 + v_{\mathcal{R}}(k)\int_{N_0}^{N_{\rm end}} \frac{{\rm d}N'}{\tilde{z}^2\,(\tilde{a}\tilde{H})} - k^2 \int_{N_0}^{N_{\rm end}} \frac{{\rm d}N'}{\tilde{z}^2\,(\tilde{a}\tilde{H})}\int_{N_0}^{N'} \frac{{\rm d}N''\,\tilde{z}^2}{(\tilde{a}\tilde{H})}\right], \tag{A38}$$

where we switched to e-folds as a time variable, $v_{\mathcal{R}} \equiv \mathcal{R}'_k(N_0)/\mathcal{R}_k(N_0)$ is the fractional velocity of curvature perturbation at the horizon crossing epoch (N_0) and a tilde over quantities denotes a normalization with respect to their values at N_0, i.e., $\tilde{X} \equiv X(N)/\tilde{X}(N_0)$. Using the standard solution to the curvature perturbation during the initial slow-roll era, an analytic formula for the fractional velocity of curvature perturbation is obtained [95], which is valid for modes that leave the horizon during the initial slow-roll era:

$$v_{\mathcal{R}}(k) = -\frac{x_0^2}{1+x_0^2} - i\frac{x_0^3}{1+x_0^2}, \qquad \text{with} \qquad x_0 \equiv \frac{k}{a_0 H_0}. \tag{A39}$$

Similarly, for modes that cross the horizon during the initial slow-roll era, the amplitude of the curvature perturbation at horizon crossing results in

$$|\mathcal{R}_k^{(0)}|^2 = \frac{H_0^2}{8\pi^2\epsilon_0 M_{\rm pl}^2}\left(1 + \frac{k^2}{(a_0 H_0)^2}\right). \tag{A40}$$

Combining these results with a given background evolution (i.e., $\tilde{z}, \tilde{a}\tilde{H}$), the power spectrum of curvature perturbation for modes that leave the horizon can be described via the following definition

$$\mathcal{P}_{\mathcal{R}}(k, N_{\rm end}) = \frac{k^3}{2\pi^2}\left|\mathcal{R}_k(N_{\rm end})\right|^2, \tag{A41}$$

using (A38)–(A40). For a set of modes that exit the horizon during Phase 1 (the initial slow-roll era), the amplitude of the power spectrum obtained using the procedure we outlined is shown by blue dots in Figure 6.

Appendix D. Details of the Axion-Gauge Field Dynamics

In this appendix, we focus on the dynamics of the gauge fields according to the presence of a rolling scalar ($\chi(t)$) and the influence of these dynamics on the scalar perturbations for the models we discussed in Sections 4.1.1–4.1.3. The part of the action (65) relevant for this purpose is given by

$$S_{\rm GF} = \int {\rm d}^4x\sqrt{-g}\left[-\frac{1}{4}F_{\mu\nu}F^{\mu\nu} - \frac{g_{\rm cs}\,\chi}{8f}\frac{\eta^{\mu\nu\rho\sigma}}{\sqrt{-g}}F_{\mu\nu}F_{\rho\sigma}\right], \tag{A42}$$

where, using the definition of the field strength tensor ($F_{\mu\nu} = \partial_\mu A_\nu - \partial_\nu A_\mu$) and the totally antisymmetric nature of the symbol ($\eta^{\mu\nu\rho\sigma}$), we note the following identities

$$F_{\mu\nu}F^{\mu\nu} = 2\,(g^{\mu\rho}g^{\nu\sigma} - g^{\nu\rho}g^{\mu\sigma})\,\partial_\mu A_\nu\,\partial_\rho A_\sigma, \tag{A43}$$

$$\eta^{\mu\nu\rho\sigma}F_{\mu\nu}F_{\rho\sigma} = 4\,\eta^{\mu\nu\rho\sigma}\,\partial_\mu A_\nu\,\partial_\rho A_\sigma. \tag{A44}$$

Notice that apart from the gauge fields, these expressions involve the metric. Therefore, it is convenient to characterize the metric fluctuations in their most general form using ADM decomposition as

$${\rm d}s^2 = -N^2\,{\rm d}t^2 + \hat{g}_{ij}\Big({\rm d}x^i + N^i\,{\rm d}t\Big)\Big({\rm d}x^j + N^j\,{\rm d}t\Big), \tag{A45}$$

where $\hat{g}_{ij}$ is the three spatial metrics on constant-time hypersurfaces, N is the lapse function and N^i is the shift vector. In terms of these variables, the component of the metric with the upper indices can be expressed as

$$g^{00} = -\frac{1}{N^2}, \quad g^{0i} = g^{i0} = \frac{N^i}{N^2}, \quad g^{ij} = \hat{g}^{ij} - \frac{N^i N^j}{N^2}, \tag{A46}$$

where $\hat{g}^{ij}$ is the inverse of the spatial metric. Including the Einstein–Hilbert term ($\mathcal{L}_{\rm EH} = M_{\rm pl}^2 R/2$) with the action (A42) and the matter action associated with the scalar field(s) (χ), one can show that the second-order terms that include N, N^i do not contain more than one time derivative, implying that they can be identified as Lagrange multipliers to be solved in terms of the physical fluctuations of χ and A_μ [286,330].

To characterize the dynamics of gauge fields, we assume that vector fields start the linear order in perturbations (so that they do not exhibit a time-dependent background vev ($\langle \bar{A}_\mu(t) \rangle = 0$)) and adopt the Coulomb gauge[44] $\partial_i A_i = 0$ in the gauge sector along with the flat gauge choice in the scalar—gravitational sector

$$\begin{aligned} \chi(t,\vec{x}) &= \bar{\chi}(t) + \delta\chi(t,\vec{x}), \\ \hat{g}_{ij} &= a^2(t)\left[\,\delta_{ij} + h_{ij}\,\right], \\ \partial_i A_i &= 0, \end{aligned} \tag{A47}$$

where h_{ij} is the transverse, traceless tensor fluctuation of the metric, $\partial_i h_{ij} = h_{ii} = 0$. In this appendix, we will present the dynamics of the gauge fields by expanding the action (A42) up to the third order in fluctuations. Noting that the shift vector is order-one N^i in perturbations and expanding the lapse ($N = 1 + \delta N$), we compile all the information mentioned above to obtain the second- and third-order action involving gauge field fluctuations as

$$S_{\rm GF}^{(2)} = \int {\rm d}^4 x\, a^3 \left\{ \frac{1}{2a^2} \dot{A}_i \dot{A}_i - \frac{1}{2a^4} \partial_j A_i\, \partial_j A_i + \frac{g_{\rm cs}\, \dot{\bar{\chi}}}{a^3 f}\, \epsilon_{ijk}\, A_i\, \partial_j A_k \right\}, \tag{A48}$$

$$S_{\rm GF}^{(3,1)} = \int {\rm d}^4 x\, a^3 \left\{ g_{\rm cs} \frac{\delta\chi}{f} \left[\vec{E}.\vec{B} - \langle \vec{E}.\vec{B} \rangle \right] \right\} \tag{A49}$$

$$S_{\rm GF}^{(3,2)} = \int {\rm d}^4 x\, a^3 \left\{ -\frac{\delta N}{2} \left[\vec{E}^2 + \vec{B}^2 - \langle \vec{E}^2 + \vec{B}^2 \rangle \right] - \frac{N^i}{a^2} \dot{A}_j F_{ij} \right\} \tag{A50}$$

$$S_{\rm GF}^{(3,3)} = \int {\rm d}^4 x\, a^3 \left\{ -\frac{h_{ij}}{2} \left[E_i E_j + B_i B_j \right] \right\}, \tag{A51}$$

where we introduced electric and magnetic field notation using $E_i = -\dot{A}_i/a$, $B_i = \epsilon_{ijk} \partial_j A_k / a^2$. Note that we artificially introduced tadpole terms proportional to the expectation values involving $\vec{E}$ and $\vec{B}$ fields in the expressions (A49) and (A50). Later, we will subtract these terms in the action describing inflation (gravity plus inflation sector) in (A72) to consistently take into account the modifications that might arise in the background equations of the scalar field(s) ($\chi = \{\phi, \sigma\}$) and the Friedmann equation (see, e.g., Equation (71)).

The action labeled by $S_{\rm GF}^{(3,2)}$ parameterizes the cubic interactions (see, e.g., the first and the third term in (A50)) induced by the gravitational fluctuations. The influence of these terms on the observable scalar sector was studied in [283,332] with the conclusion that they can be neglected compared to the direct interaction term in (A49). This is because the vertices associated with the gravitationally induced cubic terms in (A50) is suppressed ($\mathcal{L}_{\rm GF}^{(3,2)} \propto (\sqrt{\epsilon}/M_{\rm pl})\, \delta\chi\, \mathcal{O}(A^2)$) with respect to that in (A49) unless $f \simeq M_{\rm pl}$ and $g_{\rm cs} \ll 1$[45] and therefore can be safely ignored compared to the latter. Finally, the cubic action in (A51) parameterizes the influence of the gauge fields on the tensor part of the metric. In this review, we will not study these effects. For the impact of gauge field production on the tensor fluctuations during inflation and its interesting parity violating effects, see [85,86,302,333–336] and the references therein.

To summarize, in the presence of rolling axion-like fields during inflation and the interaction ($\mathcal{L}_{\rm int} \propto \chi F\tilde{F}$), the dynamics of the gauge fields can be studied by focusing on the second-order action ($S^{(2)}_{\rm GF}$) (A48). On the other hand, the influence of the gauge fields on the scalar sector depends on whether we identify χ as the inflation (Sections 4.1.1 and 4.1.2) or as a spectator scalar rolling during inflation (Section 4.1.3). In the following, we will first introduce the basics of the gauge field production in the presence of a rolling scalar, discussing the different cases we cover in this review. The nature of the subsequent sourcing of (visible sector) scalar fluctuations by the vector fields depends on whether the scalar (χ) directly interacts with U(1) fields or not. We will cover each case following our discussion on the vector field production.

Appendix D.1. Gauge Field Production by Rolling Scalars

To understand the gauge field production by a rolling scalar (either an inflation or a spectator scalar), we focus on the second-order action (A48) and decompose the gauge field into its helicity modes ($\lambda = \pm$) in Fourier space using conformal time ($\mathrm{d}\tau = \mathrm{d}t/a$) as

$$\hat{A}_i(\tau,\vec{x}) = \int \frac{\mathrm{d}^3k}{(2\pi)^{3/2}}\, \mathrm{e}^{i\vec{k}\cdot\vec{x}} \sum_{\lambda=\pm} \epsilon_i^{(\lambda)}(\vec{k})\, \hat{A}_\lambda(\tau,\vec{k}), \tag{A52}$$

where the polarization vectors obey

$$\begin{aligned} &k_i\, \epsilon_i^{\pm}(\vec{k}) = 0, && \epsilon_{ijk}\, k_j\, \epsilon_k^{\pm}(\vec{k}) = \mp i\, |\vec{k}|\, \epsilon_i^{\pm}(\vec{k}), \\ &\epsilon_i^{\lambda}(\vec{k})\, \epsilon_i^{\lambda'}(\vec{k})^* = \delta^{\lambda\lambda'}, && \epsilon_i^{\lambda}(\vec{k})^* = \epsilon_i^{\lambda}(-\vec{k}) = \epsilon_i^{-\lambda}(\vec{k}) \end{aligned} \tag{A53}$$

and we defined

$$\hat{A}_\lambda(\tau,\vec{k}) = \Big[A_\lambda(\tau,k)\, a_\lambda(\vec{k}) + A^*_\lambda(\tau,k)\, a^\dagger_\lambda(-\vec{k})\Big], \tag{A54}$$

which satisfies $\hat{A}_\lambda(\tau,\vec{k})^\dagger = \hat{A}_\lambda(\tau,-\vec{k})$ so that $\hat{A}_i(\tau,\vec{x})$ is a hermitian operator. Finally, annihilation and creation operators satisfy the standard commutation relations

$$\Big[\hat{a}_\lambda(\vec{k}), \hat{a}^\dagger_{\lambda'}(\vec{k}')\Big] = \delta_{\lambda\lambda'}\delta(\vec{k}-\vec{k}'). \tag{A55}$$

Plugging the decomposition into (A48) and varying the action, the mode functions of the gauge field can be shown to satisfy

$$A''_\pm + k^2\left(1 \pm \frac{aH}{k} 2\xi\right) A_\pm = 0, \qquad \xi \equiv -\frac{g_{\rm cs}\,\dot{\chi}}{2Hf} \tag{A56}$$

It is clear from Equation (A56) that the dispersion relation of the gauge fields are modified in the presence of the last term in (A42). More importantly, the negative helicity modes (A_-) can experience a tachyonic instability for modes satisfying $k/(aH) < -\dot{\chi}/(Hf)$, while positive helicity modes stay in their vacuum. These facts reflect the parity-violating nature of the interaction ($\mathcal{L}_{\rm int} \propto \chi F\tilde{F}$). The behavior of the solutions according to Equation (A56) is sensitive to the velocity profile ($\dot{\chi}$) of the rolling scalar. In what follows, we discuss the different cases and the corresponding solutions within the context of Sections 4.1.1–4.1.3.

Production by a slowly rolling scalar. For a slowly rolling scalar field, we can treat $g_{\rm cs}\dot{\chi}/(Hf)$ as constant per Hubble time. In terms of the effective coupling (ξ), this adiabaticity condition can be parameterized as

$$\frac{\dot{\xi}}{\xi H} = \frac{\ddot{\chi}}{\dot{\chi}H} - \frac{\dot{H}}{H^2} \ll 1. \tag{A57}$$

In this case, the solution to the Equation (A56) that reduces to the Bunch–Davies vacuum solutions ($A_\pm = e^{-ik\tau}/\sqrt{2k}$) deep inside the horizon ($k \gg aH$) can be written in terms of the Coulomb functions

$$A_-(\tau,k) \simeq \frac{1}{\sqrt{2k}}[G_0(\xi,-k\tau) + iF_0(\xi,-k\tau)] \tag{A58}$$

where $\xi = -g_{cs}\dot{\chi}/(2Hf)$ and $-\tau = (aH)^{-1}$. Another simplification can be made, focusing on the $\xi \gg -k\tau$ regime (as we will see, particle production is only efficient for $\xi \sim \mathcal{O}(1)$ and takes place as $-k\tau \to 0$)

$$A_-(\tau,k) \simeq \sqrt{\frac{-k\tau}{2k}}\left[2e^{\pi\xi}\pi^{-1/2}K_1(\sqrt{-8\xi k\tau}) + ie^{-\pi\xi}\pi^{1/2}I_1(\sqrt{-8\xi k\tau})\right], \tag{A59}$$

where I_1 and K_1 are modified Bessel functions of the first and second kind, respectively. According to the solution presented above, one can realize that for the interesting case of $\xi \sim \mathcal{O}(1)$, field amplification occurs shortly after horizon crossing ($-k\tau \sim \mathcal{O}(1)$). Therefore, for a final simplification we can take the large argument limit ($(8\xi)^{-1} \ll -k\tau < 2\xi$) of the Bessel functions in (A59) to obtain

$$A_-(\tau,k) \simeq \frac{1}{\sqrt{2k}}\left(\frac{-k\tau}{2\xi}\right)^{1/4} e^{\pi\xi - 2\sqrt{-2\xi k\tau}}\left[1 + \frac{i}{2}e^{-2\pi\xi + 4\sqrt{-2\xi k\tau}}\right]. \tag{A60}$$

The real part of the solution (A60) is the growing solution, as $-k\tau \to 0$ and encodes the physical amplification of the negative helicity mode by the presence of a slowly rolling axion-like field. Within the approximations we undertake, the imaginary part of the solution (A60) represents the UV-divergent part as $-k\tau$ grows, which should precede the vacuum solution $A_- = e^{-ik\tau}/\sqrt{2k}$ deep inside the horizon. Therefore, ignoring the imaginary part typically amounts to throwing away the standard divergent (also present in flat space) peace of quantities such as $\langle \vec{E}.\vec{B}\rangle$ and $\langle \vec{E}^2 + \vec{B}^2\rangle$ (see below). For a detailed discussion of these issues, we refer the reader to [285] and the appendices of [85].

Production by a transiently rolling scalar. For the models we discuss in Section 4.1.2, the potential of the scalar (χ) has a feature around which the background velocity ($\dot{\chi}$) and the effective coupling (ξ) in the equation of motion of the gauge field modes (A56) have a transient peak with a maximal value (ξ_*) at $\tau = \tau_*$. A peaked time-dependent profile of $\xi = \xi(\tau)$, in turn, translates into a scale-dependent growth of the mode functions(A_-), where only modes that are in the vicinity of the scale ($k_* = a_*H_* = (-\tau_*)^{-1}$) that exits the horizon at $\tau = \tau_*$ are maximally amplified. For the time-dependent profiles we study in Sections 4.1.2 and 4.1.3, it is difficult to obtain a fully analytic solution describing the amplification of the gauge modes. However, an accurate description of the mode function at late times can be obtained by employing WKB approximation methods supplemented with numerical analysis [302], as we mention below. In particular, at late times ($\tau/\tau_* < 1$), the amplification of the mode functions can be parameterized in terms of a (real and positive) normalization factor as

$$A_-(\tau,k) \simeq \frac{1}{\sqrt{2k}}\left(\frac{-k\tau}{2\xi(\tau)}\right)^{1/4} N_A(\xi_*,x_*,\delta)\, e^{-2E(\tau)\sqrt{-2\xi_* k\tau}}\left\{1 + \frac{i\, e^{4E(\tau)\sqrt{-2\xi_* k\tau}}}{2N_A(\xi_*,x_*,\delta)^2}\right\} \tag{A61}$$

where we defined $x_* = -k\tau_* = k/k_*$ and $E(\tau)$ as a time-dependent function that asymptotes to zero at late times ($\tau/\tau_* \to 0$), the functional form of which depends on the model under consideration. For example, for the bumpy axion inflation discussed in Section 4.1.2 and its cousin spectator model (see Section 4.1.3), it is given by $E(\tau) = 1/(\delta|\ln(\tau/\tau_*)|)$. On the other hand, for the transiently rolling axion model with the standard cosine potential, one obtains $E(\tau) = \sqrt{2}(\tau/\tau_*)^{\delta/2}/(1+\delta)$. Notice that the solution (A61) reduces to (A60) of constant ξ if we make the following replacements: $E(\tau) \to 1$ and $N_A \to e^{\pi\xi}$. An important point that should be observed from the form of the solution (A61) is the dependence of the normalization (amplification) factor on the dimensionless parameter (δ), which can be derived in terms of the physical model parameters (see Sections 4.1.2 and 4.1.3). As we explain in the main text, this parameter is a measure of the scalar's mass around a global minimum ($\delta \approx m_\chi^2/H^2$) and therefore determines the rate at which the field rolls towards its minimum. In this sense, it determines the width of gauge field modes that take part in particle production; the faster the scalar traverses the region in its potential where the velocity is maximal (i.e., with larger acceleration($\dot{\xi}/(\xi H) \propto \ddot{\chi}/(\dot{\chi}H) \sim \delta$)), the less time each gauge field mode spends in the tachyonic region and the fewer of them will be excited by the rolling scalar, leading to a sharper distribution of excited modes. These arguments clarify the dependence of the normalization factor (N_A) in (A61) on the parameter (δ), along with ξ_* and $x_* = k/k_*$, which characterize the amplification and scale dependence of the particle production process, respectively.

At fixed ξ_* and δ, the scale dependence of the normalization factor can be obtained by numerically solving (A56) for a grid of $x_* = k/k_*$ values and matching these solutions to the WKB solution given at late times. In this way, one can confirm that N_A is given by a log-normal distribution for the models we consider in Sections 4.1.2 and 4.1.3:

$$N_A(\xi_*, x_*, \delta) \simeq N_A^c[\xi_*, \delta] \exp\left(-\frac{1}{2\sigma_A^2[\xi_*, \delta]} \ln^2\left(\frac{x_*}{q_A^c[\xi_*, \delta]}\right)\right), \tag{A62}$$

where the functions N_A^c, q_A^c and σ_A parameterizes the background dependence of gauge field production through their dependence on ξ_* and δ. In particular, accurate fitting formulas for these quantities can be obtained at fixed δ values in terms of ξ_*. For the parameter choices we adopt in this review, these formulas can be found in [85,86,285].

The "Electric" and "Magnetic" fields as sources. For future reference, we also note the electric and magnetic fields, which are related to the auxiliary potential (A_i) as

$$E_i(\tau, \vec{x}) = -\frac{1}{a^2}\partial_\tau A_i(\tau, \vec{x}), \qquad B_i(\tau, \vec{x}) = \frac{1}{a^2}(\vec{\nabla} \times \vec{A}(\tau, \vec{x}))_i = \frac{1}{a^2}\epsilon_{ijk}\,\partial_j A_k(\tau, \vec{x}). \tag{A63}$$

Utilizing the decomposition (A52), together with (A53) and (A54), we take into account only the growing part (i.e., the real part) of the solutions (corresponding to the physical amplification of vector fields caused by a rolling scalar) we derived in (A60) and (A61) to express the Fourier decomposition of the $\vec{E}$ and $\vec{B}$ fields as follows

$$\hat{E}_i(\tau, \vec{x}) = \int \frac{d^3k}{(2\pi)^{3/2}}\, e^{i\vec{k}\cdot\vec{x}}\, \hat{E}_i(\tau, \vec{k}), \qquad \hat{B}_i(\tau, \vec{x}) = \int \frac{d^3k}{(2\pi)^{3/2}}\, e^{i\vec{k}\cdot\vec{x}}\, \hat{B}_i(\tau, \vec{k}), \tag{A64}$$

where

$$\begin{aligned} \hat{E}_i(\tau, \vec{k}) &= -\sqrt{\frac{k}{2}}\frac{\epsilon_i^-(\vec{k})}{a(\tau)^2}\left(\frac{2\xi(\tau)}{-k\tau}\right)^{1/4} N_A(\xi_*, -k\tau_*, \delta)\exp\left[-E(\tau)\sqrt{-2\xi_* k\tau}\right]\hat{\mathcal{O}}_-(\vec{k}), \\ \hat{B}_i(\tau, \vec{k}) &= -\sqrt{\frac{k}{2}}\frac{\epsilon_i^-(\vec{k})}{a(\tau)^2}\left(\frac{-k\tau}{2\xi(\tau)}\right)^{1/4} N_A(\xi_*, -k\tau_*, \delta)\exp\left[-E(\tau)\sqrt{-2\xi_* k\tau}\right]\hat{\mathcal{O}}_-(\vec{k}), \end{aligned} \tag{A65}$$

where we defined the short-hand notation $\hat{\mathcal{O}}_\lambda(\vec{k}) = [a_\lambda(\vec{k}) + a_\lambda^\dagger(-\vec{k})]$. Note that the case of particle production through a slowly rolling scalar with $\xi \simeq$ cons. can be recovered from the formulas by making the following replacements: $\xi_* \to \xi$, $E(\tau) \to 1$ and $N_A \to e^{\pi\xi}$. Having studied the particle production in the gauge field sector, we now study the expectation values including electromagnetic fields before we discuss how these sources influence the fluctuations of the scalar sector.

Expectation values involving gauge fields. Noting again the decomposition of the vector fields (A52) (along with (A53) and (A54)) and the definition of their electromagnetic counterparts (A63), the energy density of the gauge fields (ρ_A and $\langle \vec{E}.\vec{B}\rangle$) can be expressed as

$$\rho_A \equiv \frac{1}{2}\langle \vec{E}^2 + \vec{B}^2\rangle = \int \mathrm{d}\ln k\, \frac{\mathrm{d}\rho_A}{\mathrm{d}\ln k},$$
$$\langle \vec{E}.\vec{B}\rangle = \int \mathrm{d}\ln k\, \frac{\mathrm{d}\langle \vec{E}.\vec{B}\rangle}{\mathrm{d}\ln k}. \tag{A66}$$

Taking into account only the amplified mode function (A_-) of the gauge field, the energy density (ρ_A and $\langle \vec{E}.\vec{B}\rangle$) per logarithmic wave number is given by

$$\frac{\mathrm{d}\rho_A}{\mathrm{d}\ln k} \simeq \frac{H^4}{8\pi^2}\, x^4 \left(\frac{2\xi}{x}+1\right) |\tilde{A}_-(x)|^2,$$
$$\frac{\mathrm{d}\langle \vec{E}.\vec{B}\rangle}{\mathrm{d}\ln k} \simeq -\frac{H^4}{8\pi^2}\, x^4\, \frac{\mathrm{d}}{\mathrm{d}x}|\tilde{A}_-(x)|^2, \tag{A67}$$

where we utilized [285],

$$A'_- = \sqrt{\frac{2k\xi}{-\tau}}\, A^*_- \quad \rightarrow \quad \frac{\mathrm{d}\tilde{A}_-}{\mathrm{d}x} = -\sqrt{\frac{2\xi}{x}}\, \tilde{A}^*_- \tag{A68}$$

defining the the following dimensionless variables: $x = -k\tau$ and $\sqrt{2k}\, A_-(\tau,k) \equiv \tilde{A}_-(x)$. Employing the solutions for A_- we derived in (A60) and (A61), one can use Formulas (A66) and (A67) to obtain ρ_A and $\langle \vec{E}.\vec{B}\rangle$. In particular, for the localized gauge field production models presented in Section 4.1.2 (see [86]) and Section 4.1.3 (see [85]), these formulas can be used to justify the negligible back-reaction of the gauge fields on the background evolution. We will not repeat these calculations here for the localized production case; however, below, we will derive the relevant formulas for the slowly rolling smooth axion inflation described in Section 4.1.1. For this purpose, we focus on the real part of the solution (A60) that corresponds to the physical amplification of the gauge field fluctuations. Plugging the solution into (A67) and (A66), we obtain

$$\rho_A = \frac{H^4\, e^{2\pi\xi}}{8\pi^2 (2\xi)^{1/2}} \int_0^{2\xi} \mathrm{d}x\, x^{7/2} \left(\frac{2\xi}{x}+1\right) e^{-4\sqrt{2\xi x}},$$
$$\langle \vec{E}.\vec{B}\rangle = \frac{H^4\, e^{2\pi\xi}}{4\pi^2} \int_0^{2\xi} \mathrm{d}x\, x^3 \left(1-\frac{1}{4\sqrt{2\xi x}}\right) e^{-4\sqrt{2\xi x}}, \tag{A69}$$

where we send the lower limits of the integrals to zero as the integrands converge in the $x \to 0$ limit. Similarly, since we are only focusing on the physical amplification of the gauge fields by throwing away the imaginary part of the solution (A60), the integrands vanish quickly deep in the UV ($x \to \infty$), so we can also send the upper limit of the integrals in (A69) to infinity ($2\xi \to \infty$). Finally, by making a change of variable to $4\sqrt{2\xi x} = y$, one can realize that the resulting integrals can be carried analytically; in fact, they are proportional to Gamma functions with integer arguments. In particular, we obtain

$$\rho_A = \frac{H^4}{\xi^3}\, e^{2\pi\xi}\, \frac{\Gamma(7)}{2^{19}\pi^2}\left[1+\frac{1}{2^6\xi^2}\frac{\Gamma(9)}{\Gamma(7)}\right],$$
$$\langle \vec{E}.\vec{B}\rangle = \frac{H^4}{\xi^4}\, e^{2\pi\xi}\, \frac{[\Gamma(8)-\Gamma(7)]}{2^{21}\pi^2}. \tag{A70}$$

Inserting the numerical values of the Gamma functions, these expressions give rise to Equation (72), which we provide in the main text.

Appendix D.2. Scalars Sourced by Vector Fields and the Direct Coupling Case: $\chi = \phi$

To understand the sourcing of scalar fluctuations by the gauge fields, we should consider the gravitational and inflation (that we refer to as $S_{\rm inf}$) action, in addition to $S_{\rm GF}$: $S_{\rm tot} = S_{\rm inf} + S_{\rm GF}$, where

$$S_{\rm inf} = \int {\rm d}^4x\sqrt{-g}\left\{\frac{M_{\rm pl}^2}{2}R - \frac{1}{2}\partial_\mu\phi\partial^\mu\phi - V(\phi)\right\}. \tag{A71}$$

Expanding the action in terms of the scalar fluctuations (δN, N^i and $\delta\phi$) around a background solution, one obtains the following linear and second-order actions,

$$S_{\rm inf}^{(1)} = \int {\rm d}^4x\, a^3\left\{\left[3H^2M_{\rm pl}^2 - \frac{1}{2}\dot{\bar\phi}^2 - V(\bar\phi) - \frac{1}{2}\langle\vec{E}^2 + \vec{B}^2\rangle\right]\delta N \right. \tag{A72}$$
$$\left. - \left[\ddot{\bar\phi} + 3H\dot{\bar\phi} + V'(\bar\phi) - \frac{g_{\rm cs}}{f}\langle\vec{E}.\vec{B}\rangle\right]\delta\phi\right\},$$

$$S_{\rm inf}^{(2)} = \frac{1}{2}\int {\rm d}^4x\, a^3\left\{\delta\dot\phi^2 - \frac{(\partial_i\delta\phi)^2}{a^2} - V''(\bar\phi)\,\delta\phi^2 - 3H^2M_{\rm pl}^2\delta N^2 - 2HM_{\rm pl}^2\partial_iN^i\delta N\right. \tag{A73}$$
$$\left. - 2V'(\bar\phi)\,\delta\phi\,\delta N - 2\dot{\bar\phi}\,\delta\dot\phi\,\delta N - 2\dot{\bar\phi}\,N^i\partial_i\delta\phi + \dot{\bar\phi}^2\delta N^2\right\}.$$

Notice that in $S_{\rm inf}^{(1)}$ (A72), we subtracted the tadpole terms ($\propto \langle\vec{E}.\vec{B}\rangle, \langle\vec{E}^2 + \vec{B}^2\rangle$) that we introduced earlier in Equations (A49) and (A50). By virtue of the background equations presented in Equation (71), these terms precisely cancel. The terms that contain gravitational fluctuations (δN and N^i) in the second-order action ($S_{\rm inf}^{(1)}$) (A73) induce additional mass contributions to the inflation fluctuations, which can be seen by varying this action with respect to δN and N^i and solving them in terms of $\delta\phi$ as

$$\delta N = -\sqrt{\frac{\epsilon}{2}}\frac{\delta\phi}{M_{\rm pl}}, \quad \partial_iN^i = \sqrt{\frac{\epsilon}{2}}\frac{1}{M_{\rm pl}}\left(\delta\dot\phi - \frac{\eta H}{2}\delta\phi\right), \tag{A74}$$

where we defined the Hubble slow-roll parameters as

$$\epsilon = -\frac{\dot H}{H^2} = \frac{\dot{\bar\phi}^2}{2H^2M_{\rm pl}^2}, \quad \text{and} \quad \eta = \frac{\dot\epsilon}{\epsilon H}. \tag{A75}$$

Similar to our discussion regarding the cubic terms induced by gravity within the gauge field sector (see e.g., (A50)), the influence of the lapse and shift on the second-order action of the inflation fluctuations can be typically ignored in the slow-roll regime (as well as for the non-attractor era associated with the model in Section 4.1.2, where $\epsilon \to 0$, $|\eta| \simeq \mathcal{O}(1)$), which is a situation that is generically referred as the decoupling limit of gravity within the literature. However, for completeness, we will keep them here. Plugging (A74) into the action (A73), we perform several integrations by parts, and by taking into account the source terms induced by the gauge fields ($\delta\chi \to \delta\phi$ in Equation (A49)), the equation of motion obeyed by the inflation fluctuations can be written as

$$\delta\ddot\phi + 3H\delta\dot\phi - \left(\vec\nabla^2 - m_{\rm eff}^2(t)\right)\delta\phi = \frac{g_{\rm cs}}{f}\left[\vec{E}.\vec{B} - \langle\vec{E}.\vec{B}\rangle\right] + \frac{g_{\rm cs}}{f}\frac{\partial\langle\vec{E}.\vec{B}\rangle}{\partial\dot{\bar\phi}}\delta\dot\phi \tag{A76}$$

where $\vec{\nabla}^2 = \partial_i\partial_i$ is the Laplacian in flat Euclidean space, and the effective time-dependent mass is defined as $m_{\rm eff}^2 = V''(\bar{\phi}) - (2\epsilon\eta + 6\epsilon - 2\epsilon^2)H^2$. Noting the second derivative of the potential in terms of the slow-roll parameters:

$$V''(\bar{\phi}) = H^2\left[-\frac{3\eta}{2} + \frac{5\epsilon\eta}{2} - \frac{1}{4}\eta^2 - \frac{\dot{\eta}}{2H} - 2\epsilon^2 + 6\epsilon\right], \tag{A77}$$

the effective time-dependent mass can be described fully in terms of the slow-roll parameters as

$$m_{\rm eff}^2 = H^2\left[\frac{9}{4} - \frac{1}{4}(\eta+3)^2 + \frac{\epsilon\eta}{2} - \frac{\dot{\eta}}{2H}\right]. \tag{A78}$$

In (A76), the last term is introduced to parameterize the additional sources that might arise in the strong back-reaction regime. In particular, as $\langle\vec{E}.\vec{B}\rangle$ grows during inflation (i.e., as the effective coupling ξ increases during inflation; see Equation (72)), this will first have the effect of sourcing the inflation perturbations through the first term on the right-hand side of (A76). As a result, the inflation perturbations start to grow, and eventually, the solution of the gauge field modes obtained by assuming a homogeneous inflation will no longer be valid and are expected to go from the solution (A60) to a more general solution ($A_-[\bar{\phi} + \delta\phi]$). The additional term on the right-hand side of (A76) is precisely introduced to capture the influence of this modified solution of the gauge fields on the inflation perturbations. Since gauge field production (and $\langle\vec{E}.\vec{B}\rangle$) is sensitive to the velocity of the homogeneous inflation mode, it is reasonable to expect the influence of this additional source to be proportional to $(\partial\langle\vec{E}.\vec{B}\rangle/\partial\dot{\bar{\phi}})\delta\dot{\phi}$, as shown in (A76) (see also [79,81,281] for a detailed discussion of this point). Recalling $\xi = -g_{\rm cs}\dot{\bar{\phi}}/(2Hf)$ and (72), we note

$$\frac{g_{\rm cs}}{f}\frac{\partial\langle\vec{E}.\vec{B}\rangle}{\partial\dot{\bar{\phi}}} \simeq \frac{g_{\rm cs}}{f}\frac{\langle\vec{E}.\vec{B}\rangle}{\dot{\bar{\phi}}}2\pi\xi, \tag{A79}$$

so the influence of the source term can be parameterized as an additional damping term in the equation of motion of the inflation fluctuations as

$$\delta\ddot{\phi} + 3H\beta\,\delta\dot{\phi} - \left(\vec{\nabla}^2 - m_{\rm eff}^2(t)\right)\delta\phi = \frac{g_{\rm cs}}{f}\left[\vec{E}.\vec{B} - \langle\vec{E}.\vec{B}\rangle\right], \tag{A80}$$

where

$$\beta = 1 - \frac{g_{\rm cs}}{f}\frac{\langle\vec{E}.\vec{B}\rangle}{3H\dot{\bar{\phi}}}2\pi\xi. \tag{A81}$$

Assuming that $\beta = 1$ amounts to neglecting the influence back-reaction effects of the inflation fluctuations on the gauge field solutions. This is, for example, the approach taken in [86] for the bumpy axion inflation we discussed in Section 4.1.2 with the reasoning that back-reaction effects are mild due to the localized nature of gauge field production (as is the resulting increase in inflation fluctuations) in the presence of a transiently increasing effective coupling (ξ) between scalar and gauge field sectors. However, in smooth axion inflation (Section 4.1.1), gauge field sources and the resulting scalar fluctuations continuously grow, which can eventually influence the dynamics of the fluctuations through an additional (positive) friction term in β (A81) as soon as the homogeneous dynamics of the inflation enter into the back-reaction regime with $g_{\rm cs}\langle\vec{E}.\vec{B}\rangle/(3H|\dot{\bar{\phi}}|f) \sim \mathcal{O}(1)$.

Armed with the equations of motion in real space, one can study vacuum and sourced solutions of the inflation perturbations in Fourier space by using the splitting ($\delta\phi = \delta\phi_{\rm v} + \delta\phi_{\rm s}$) and utilizing the standard Green function methods. We will not repeat these computations here; for the calculation of the scalar power spectrum relevant to PBH formation, interested readers can consult [79,81,281,283] in the context of smooth axion inflation and [86] in the context of bumpy axion inflation as discussed in Sections 4.1.1 and 4.1.2.

Appendix D.3. Scalars Sourced by Vector Fields and the Indirect Coupling Case: $\chi = \sigma$.

To capture the dynamics of scalar fluctuations in the models studied in Section 4.1.3, we extend the inflationary action with a spectator sector (σ) that interacts with the gauge fields as in (A42). The action that describes inflation is therefore given by

$$S_{\rm inf} = \int d^4x\sqrt{-g}\left\{\frac{M_{\rm pl}^2}{2}R - \frac{1}{2}\partial_\mu\phi\partial^\mu\phi - V(\phi) - \frac{1}{2}\partial_\mu\sigma\partial^\mu\sigma - U(\sigma)\right\}, \tag{A82}$$

where we assume that the spectator axion-like field does not contribute significantly to the background evolution during inflation, which is mainly driven by a sufficiently flat inflation potential (V) that we will leave unspecified.

Expanding the action in terms of the scalar fluctuations (δN, N^i and $\delta\phi, \delta\sigma$) around a background solution, we obtain the following linear and second-order actions,

$$S^{(1)}_{\rm inf} = \int d^4x\, a^3 \left\{\left[3H^2M_{\rm pl}^2 - \sum_a\left(\frac{1}{2}\dot{\bar{\phi}}_a^2 - V_a(\bar{\phi})\right) - \frac{1}{2}\langle \vec{E}^2 + \vec{B}^2\rangle\right]\delta N \right. \tag{A83}$$

$$\left. - \left[\ddot{\bar{\phi}} + 3H\dot{\bar{\phi}} + V'(\bar{\phi})\right]\delta\phi - \left[\ddot{\bar{\sigma}} + 3H\dot{\bar{\sigma}} + U'(\bar{\sigma}) - \frac{g_{\rm cs}}{f}\langle \vec{E}.\vec{B}\rangle\right]\delta\sigma\right\},$$

$$S^{(2)}_{\rm inf} = \frac{1}{2}\sum_a\int d^4x\, a^3\left\{\delta\dot{\phi}_a^2 - \frac{(\partial_i\delta\phi_a)^2}{a^2} - V_a''(\bar{\phi}_a)\,\delta\phi_a^2 - 3H^2M_{\rm pl}^2\delta N^2 - 2HM_{\rm pl}^2\partial_i N^i\delta N \right.$$

$$\left. - 2V_a'(\bar{\phi}_a)\,\delta\phi_a\,\delta N - 2\dot{\bar{\phi}}_a\,\delta\dot{\phi}_a\,\delta N - 2\dot{\bar{\phi}}_a\, N^i\partial_i\delta\phi_a + \dot{\bar{\phi}}_a^2\delta N^2\right\}, \tag{A84}$$

where the summation over a runs over fluctuations and background values of the following two fields: $\phi_a = \{\phi, \sigma\}$ and $V_a = \{V(\bar{\phi}), U(\bar{\sigma})\}$. Varying (A84) with respect to Lagrange multipliers δN and N^i, we obtain

$$2HM_{\rm pl}^2\,\delta N = \sum_a \dot{\bar{\phi}}_a\delta\phi_a, \tag{A85}$$

$$-2HM_{\rm pl}^2\,\partial_i N^i = \sum_a\left(\dot{\bar{\phi}}_a\delta\dot{\phi}_a + V_a'(\bar{\phi}_a)\delta\phi_a\right) + \left(6H^2M_{\rm pl}^2 - \sum_a\dot{\bar{\phi}}_a^2\right)\delta N. \tag{A86}$$

Plugging these solutions back into the actions, we obtain the following second-order action for scalar fluctuations [286,303],

$$S^{(2)}_{\rm inf} = \frac{1}{2}\int d^4x\, a^3\left\{\left[\delta\dot{\phi}_a^2 - \frac{(\partial_i\delta\phi_a)^2}{a^2}\right] - m_{ab}^2\delta\phi_a\delta\phi_b\right\}, \tag{A87}$$

$$m_{ab}^2 = V^{(\rm tot)}_{,ab} - \frac{1}{a^3}\frac{d}{dt}\left(\frac{a^3}{H}\frac{\dot{\bar{\phi}}_a\dot{\bar{\phi}}_b}{M_{\rm pl}^2}\right). \tag{A88}$$

where summation over repeated indices (a,b) is implied and $V^{(\rm tot)}_{,ab} \equiv \partial^2 V^{(\rm tot)}/(\partial\bar{\phi}_a\partial\bar{\phi}_b)$ with $V^{(\rm tot)} = \sum_a V_a(\bar{\phi}_a) = V(\bar{\phi}) + U(\bar{\sigma})$. We note that in deriving the expressions presented above, we used background equations of motion, ignoring the artificially introduced mean field expectation values involving gauge fields (noticing that these terms that appear in (A49), (A50) and (A83) sum up to zero):

$$\ddot{\bar{\phi}}_a + 3H\dot{\bar{\phi}}_a + V_a'(\bar{\phi}_a) = 0, \tag{A89}$$

$$-2\dot{H}M_{\rm pl}^2 = \sum_a\dot{\bar{\phi}}_a^2. \tag{A90}$$

Indeed, for the spectator models discussed in Section 4.1.3, due to the localized nature of the gauge field production, the back-reaction effect is small and can be ignored [81,85,285];

therefore, the background evolution can be studied in a consistent way by focusing on the equations in (A89), along with the Friedmann equation ($3H^2M_{\rm pl}^2 = \sum_a \dot{\bar{\phi}}_a^2 + V_a(\bar{\phi}_a)$). Taking into account the effects of gauge field sources ($\delta\chi \to \delta\sigma$ in Equation (A49)) on the spectator scalar fluctuations, the equations of motion for the scalar perturbations read as

$$\delta\ddot{\phi}_a + 3H\delta\dot{\phi}_a - \left(\vec{\nabla}^2 - V_a''\right)\delta\phi_a - \sum_b \frac{1}{a^3}\frac{\mathrm{d}}{\mathrm{d}t}\left(\frac{a^3}{H}\frac{\dot{\bar{\phi}}_a\dot{\bar{\phi}}_b}{M_{\rm pl}^2}\right)\delta\phi_b = J_a(\vec{E},\vec{B}), \tag{A91}$$

where $\delta\phi_a = (\delta\phi, \delta\sigma)^T$ and $J_a = (0, \frac{g_{\rm cs}}{f}\vec{E}.\vec{B})^T$. Notice that since we consider sum separable potentials ($V^{(\rm tot)} = V + U$), the first term in the mass matrix (A88) makes a diagonal contribution to the mass of each field in the equations of motion (A91). However, the presence of the second term in the mass matrix (A88), which is induced by the presence of gravitational fluctuations (δN and N^i) introduces mass mixing between scalar fluctuations ($\delta\phi_a - \delta\phi_b$ ($a \neq b$)) (i.e., through the last term in (A91)). In other words, although we consider a Lagrangian (A82) where the two scalar fields appear to be decoupled from each other, gravitational interaction will inevitably introduce a minimal communication channel between the physical fluctuations of the two scalar sectors. At leading order in the slow-roll expansion, the mass mixing ($a \neq b$) originates from the following terms in the Lagrangian

$$\mathcal{L}_{\rm mix} = \frac{1}{2a^3}\frac{\mathrm{d}}{\mathrm{d}t}\left(\frac{a^3}{H}\frac{\dot{\bar{\phi}}_a\dot{\bar{\phi}}_b}{M_{\rm pl}^2}\right)\delta\phi_a\delta\phi_b \quad\longrightarrow\quad \mathcal{L}_{\rm mix} \simeq 6\sqrt{\epsilon_\phi\epsilon_\sigma}\,H^2\delta\phi\delta\sigma, \tag{A92}$$

where $\epsilon_a = \dot{\bar{\phi}}_a^2/(2H^2M_{\rm pl}^2)$ is the slow-roll parameter of each field. Therefore, as long as both fields roll with a non-vanishing velocity ($\dot{\bar{\sigma}}, \dot{\bar{\phi}} \neq 0$) during inflation, their fluctuations can be converted to one another. In the presence of particle production in the gauge field sector, the mixing between the two scalar sectors is crucial in understanding the influence of the gauge field sources on the visible scalar sector fluctuations ($\delta\phi$). In order to see this explicitly, we focus on the leading order mixing in the slow-roll expansion to rewrite the system of equations in (A91) as

$$\begin{aligned} \delta\ddot{\phi} + 3H\delta\dot{\phi} - \left(\vec{\nabla}^2 - m_\phi^2\right)\delta\phi &\simeq 6\sqrt{\epsilon_\phi\epsilon_\sigma}\,H^2\delta\sigma, \\ \delta\ddot{\sigma} + 3H\delta\dot{\sigma} - \left(\vec{\nabla}^2 - m_\sigma^2\right)\delta\sigma &\simeq \frac{g_{\rm cs}}{f}\vec{E}.\vec{B}, \end{aligned} \tag{A93}$$

where $m_a^2 \simeq V_a'' - 6\epsilon_a H^2$ at leading order in the slow-roll expansion. We note that in the second line of (A93), we ignored mixing terms that can source spectator fluctuations ($\delta\sigma$) through $\delta\phi$, which is a subleading effect compared to the sourcing of $\delta\sigma$ by the enhanced gauge fields. This amounts to considering the main production channel of the observable scalar fluctuations schematically as $\delta A + \delta A \to \delta\sigma \to \delta\phi$. Using the equations of motion in real space (A93) and considering their sources in (A65), the observable power spectrum of scalar fluctuations can be carried out using Green's function methods in Fourier space following the detailed calculations presented in [85,302], along with the prescription we present in Appendix E regarding the curvature perturbation.

Appendix E. Curvature Perturbation

In this short appendix, we define the curvature perturbation ($\mathcal{R}$) on comoving slices and study its parametric dependence on the matter fluctuations for some of the scenarios we consider in the main text. We begin by noting the definition of comoving curvature perturbation in flat gauge, which reads as [324,337]:

$$\mathcal{R} = -\frac{H}{(\bar{\rho} + \bar{P})}\delta q_{\rm flat} \tag{A94}$$

where $\bar{\rho}$ and $\bar{P}$ are the total background energy density and pressure (see Appendix A), respectively, and $\delta q_{\rm flat}$ is the scalar momentum density in flat gauge. In terms of the perturbed energy momentum tensor, $\delta q_{\rm flat}$ is given by

$$\delta T^0_i = \partial_i \delta q_{\rm flat} \quad \longleftarrow \quad T_{\mu\nu} \equiv -\frac{2}{\sqrt{-g}} \frac{\delta S_m}{\delta g^{\mu\nu}}, \tag{A95}$$

where the definition of the unperturbed energy momentum tensor provided on the right-hand side of (A95) can be used for a given matter action describing the system. We note that in this definition, $\delta S_m / \delta g^{\mu\nu}$ represents the variation of the matter action with respect to the metric field. Anticipating that we will focus on different limits of a more complicated/general model, we start by parameterizing perturbed δT^0_i for the spectator axion model we discussed in Section 4.1.3 with the matter action given by the sum of (A82) and (A42). Noting that the last term in (A42) is topological and does not gravitate, we have

$$\delta T^0_i = g^{0\mu} \left(\partial_\mu \phi \partial_i \delta\phi + \partial_\mu \sigma \partial_i \delta\sigma + g^{\rho\sigma} F_{\mu\rho} F_{i\sigma} \right), \tag{A96}$$

where the last term characterizes the contribution from gauge fields, which is second-order in fluctuations. Following (A94)–(A96), it is clear that for a multisector inflationary model, curvature perturbation can, in principle, obtain contributions from all fields present in the matter Lagrangian. In what follows, we take (A96) as the main reference point to study different limits to derive an expression for the curvature perturbation ($\mathcal{R}$) relevant to some of the models we studied in the main text.

Canonical single-field inflation. This case corresponds to neglecting terms proportional to spectator and gauge field fluctuations in (A96). Noting that $\bar{\rho} + \bar{P} \rightarrow \dot{\phi}^2$ for single-field canonical inflation to linear order in inflation fluctuations, the curvature perturbation (A94) is given by

$$\mathcal{R} = \frac{H}{a\dot{\phi}} (a\delta\phi) \quad \leftrightarrow \quad \mathcal{R} = \frac{v}{z}, \tag{A97}$$

where we made the connection between the MS variable (see Section 3.1) and inflation fluctuations clear ($v = a\delta\phi$) using the definition of pump field in this case ($z = -a\dot{\phi}/H = a\sqrt{2\epsilon} M_{\rm pl}$). Upon canonical quantization of $\delta\phi$ (and hence $\mathcal{R}$), their corresponding Fourier space variables satisfy the same relation in (A97), and the dimensionless power spectrum at the end of inflation can be computed via (A41).

Smooth Axion Inflation and Bumpy Axion Inflation. For the models we studied in Sections 4.1.1 and 4.1.2, we instead focus on the curvature perturbation (ζ) on uniform-density gauge, which can be related to the density perturbation in flat gauge ($\delta\rho_{\rm flat}$) and comoving curvature perturbation ($\mathcal{R}$) on superhorizon scales as [337]

$$\zeta \simeq -\mathcal{R} = -\frac{H}{\dot{\bar{\rho}}} \delta\rho_{\rm flat}, \tag{A98}$$

where $\bar{\rho}$ is the total energy density of the axion gauge field system. Using background equations of the system, the time derivative of the total energy density is given by $\dot{\bar{\rho}} = -3H\dot{\bar{\phi}}^2 - 4H\rho_A$ [338] where ρ_A is the energy density of the gauge field as defined in (72). Noting this relation, to linear order[46] in the cosmological fluctuations, the comoving curvature perturbation ($\mathcal{R}$) can be obtained as [81]

$$\mathcal{R} = \frac{H}{\dot{\bar{\phi}}} \delta\phi \; \mathcal{F}, \qquad \mathcal{F} \equiv \frac{-\dot{\bar{\phi}}\, V'(\bar{\phi})}{H(3\dot{\bar{\phi}}^2 + 4\rho_A)}, \tag{A99}$$

where we used $\delta\rho_{\rm flat} \simeq V'(\bar{\phi})\delta\phi$, neglecting contributions to the total energy proportional to the kinetic energy of the inflation, as they should be small both in the slow-roll and the inflationary regime dominated by the friction provided by the gauge fields. In (A99), the factor $\mathcal{F}$ parameterizes the correction to the definition of the curvature perturbation in the strong back-reaction regime, which must be taken into account in the smooth axion inflation model we presented in the main text. On the other hand, the standard relation in (A97) applies to the regime of negligible back-reaction of the gauge field on the evolution of the inflation and expansion of the universe, i.e., when ρ_A is negligible and the relation $-3H\dot{\bar{\phi}} \simeq V'(\bar{\phi})$ is satisfied. This situation applies both at early times during the considered smooth axion model, as well as within the bumpy axion inflation model described in Section 4.1.2, where back-reaction effects were shown to be small around the peak of the scalar power spectrum. To summarize, in the strong back-reaction regime, the power spectrum of curvature perturbation can be computed using the standard relation (A97) times a factor of $\mathcal{F}$ correction.

Spectator axion-gauge field model. In the model presented in Section 4.1.3, in principle, we need to consider all the contributions to the curvature perturbation using Formulas (A94)–(A96). However, as explicitly checked in [85], the contribution to the curvature perturbation that is bilinear in the gauge fields can be safely ignored at late times during which we are interested in the correlators of $\mathcal{R}$. Further simplifications of the functional form of $\mathcal{R}$ arise due to the spectator nature of the axion sector (σ) and due to the assumption that it settles back to its global minimum long before the end of inflation, where $\dot{\bar{\sigma}} \to 0$. Noting $\bar{\rho} + \bar{P} = \dot{\bar{\phi}}^2 + \dot{\bar{\sigma}}^2 \simeq \dot{\bar{\phi}}^2$ and the fact that $\dot{\bar{\sigma}} \to 0$ long before inflation ends, the late-time curvature perturbation obtains the following form

$$\mathcal{R} = \frac{H}{a(\dot{\bar{\phi}}^2 + \dot{\bar{\sigma}}^2)}\Big(\dot{\bar{\phi}}\,(a\delta\phi) + \dot{\bar{\sigma}}\,(a\delta\sigma)\Big) \simeq \frac{H}{\dot{\bar{\phi}}}\delta\phi. \tag{A100}$$

Therefore, the standard expression $\mathcal{R} = (H/\dot{\bar{\phi}})\delta\phi$, which is valid in single-field inflation, still provides a very good approximation of the computation of late-time correlators of the curvature perturbation in this model.

Notes

1 See, e.g., the impact of extended mass functions on the PBH abundance constraints [27].

2 Ultralight PBHs with $M_{\rm pbh} \ll 10^{15}$g—although they cannot account for the DM density—may also have interesting observational effects if they come to dominate the energy budget of the universe before BBN (see, e.g., [36,37]).

3 Another inflationary background that exhibits similar features is called constant-roll inflation (see, e.g., [44,45]).

4 PBHs could also form in the post-inflationary universe through the collapse of cosmic strings [106–108] and domain walls [109–112], phase transitions [113,114], bubble collisions [115,116] or scalar-field fragmentation via instabilities [117,118]. We note that PBH formation in the post-inflationary universe can be triggered by bubble nucleation during inflation or instabilities generated in the final stage of inflation commonly referred as (p)reheating [119–122]. We will not dwell on these possibilities here; for a partial list of recent works in this line of research, see [123–125].

5 A detailed discussion on these topics can be found in Chapter 4 of Baumann's book [128].

6 Throughout this work, we assume an efficient reheating process at the end of inflation such that the universe becomes radiation-dominated shortly after inflation terminates. For a collage of interesting physics that might arise through the reheating stage and alternative post-inflationary histories, see the recent review [131].

7 A simple analytic argument that leads to this result is presented in Appendix B.

8 Other effects such as anisotropies [137] in the radiation fluid and a time-dependent equation of state parameter (w) [138] may also impact the estimate of the threshold for collapse.

9 Strictly speaking, $g_*(T) = g_s(T)$ is only satisfied when species are in thermal equilibrium at the same temperature. For a comprehensive overview of the thermal history of the universe after inflation, see Chapter 3 of [128].

10 We note that another common convention is to count e-folds with respect to the end of inflation, denoting the end point as $N_{\rm end} = 0$.

11 In contrast to the original approach by Press and Schechter [145], we do not take into account a symmetry factor of two on the right-hand side of (13), which accounts for all the mass in the universe, since it is not clear whether such a factor makes sense when considering non-symmetric PDFs of δ (e.g., non-Gaussian cases). Furthermore, the error introduced by omitting this factor is comparable to the other uncertainties in the computation of β, such as the fraction of the horizon mass that collapses to form a PBH (see, e.g., [146–148]).

12 Non-linearities that we neglect in the expression (16) can be important to understand intrinsic non-Gaussianity present in the PBH formation process (see, e.g., [150,151] and references therein.)

13 The variance (19) can be equivalently written as $\sigma^2 = \sigma^2(k)$ or $\sigma^2 = \sigma^2(M)$ using the relation between the peak scale of PBH formation and PBH mass (5).

14 Note that the power spectrum is the variance of curvature perturbation per logarithmic interval in k, i.e., $\langle \mathcal{R}^2 \rangle \equiv \int \mathrm{d}\ln k \, \mathcal{P}_{\mathcal{R}}(k)$. Therefore, the approximate signs ($\simeq$) in the expressions in (26) and (30) can be turned into an equality if we consider the β values defined in those expressions as the collapse fraction per $\mathrm{d}\ln k$ in the spectrum, namely $\beta = \beta(k)$.

15 One can also introduce a time-dependent mass in the action (35) which may arise through broken spatial translations as in solid [179] and supersolid [180,181] inflation. Another possibility is to include higher derivative terms in the quadratic action to modify the dispersion relation of curvature perturbation (see, e.g., [182]) as in ghost inflation [183]. We will not consider these possibilities here. For a discussion of PBH formation in solid and ghost inflation, see Sections 4 and 6 of [64,184].

16 In fact, if the time evolution of the pump field is known, up to second order in the gradient expansion, we can generate a solution for the curvature perturbation by replacing $\mathcal{R}_k(\bar{\tau}'')$ in the last integral of Equation (39) with the leading growing-mode solution of the homogeneous part of Equation (38), which we can identify as $\mathcal{R}_k^{(0)}$.

17 In particular, the standard decaying mode is given by the last term in (39), as it decays slowly, i.e., $\propto (-k\bar{\tau})^2$, compared to the second term.

18 Here, we assume that field rolls down on its potential from large to small values ($\dot{\phi} < 0$) with $V'(\phi)$ before it encounters the feature required for the enhancement. Since $V' < 0$ during the feature, there must be a point in the potential where the second derivative of the field vanishes ($V'' = 0$).

19 Note that for single-field inflation, we have $\epsilon \equiv \dot{\phi}^2/(2H^2M_{\rm pl}^2)$, and using η (40), we obtain $\eta = 2\ddot{\phi}/(\dot{\phi}H) + 2\epsilon$. Using the Klein–Gordon equation in the last expression yields the relation on the right-hand side of (49).

20 Note that evaluating the power spectrum at the end of inflation is necessary when modes evolve outside the horizon, as in the example background we are focusing on in this section.

21 In fact, the general procedure outlined in Appendix C can be generalized to accurately solve the Mukhanov–Sasaki equation using a broad class of single-field models of inflation. In the context of phenomenological models we discuss in this and the next section, jupyter notebook files that compute the power spectrum is are available at https://github.com/oozsoy/SingleFieldINF_Powerspec_PBH (accessed on 24 January 2023). We acknowledge the use of the following Python libraries: matplotlib [219], numpy [220], scipy [221], pandas [222] along with jupyter notebooks [223].

22 The distinct behavior of the cosmological correlators around the dip feature may also be probed by the 21 cm signal of the Hydrogen atom [230].

23 Notice that an exponential tail in the PDF of curvature fluctuations, similar to the tailarising in the context of a stochastic approach to inflation, has been recently shown [256] to be a property of all single-field models with potential up to quadratic. Such a feature is physically caused by the logarithmic relation between curvature fluctuation and the (Gaussian) inflation field fluctuations (see [256] for details).

24 Notice that if the background dynamics have localized features, phases of slow-roll violation might occur for some time interval, and this exact identification ceases to be valid. In particular, in this case, z''/z might lose its monotonic nature for some time interval, leading to multiple horizon crossings for a range of modes, resulting in oscillations of the spectrum [64]. For the time being, we set a discussion on this possibility aside and continue identifying the quantity z''/z as an effective horizon for the slow-roll violating scenarios we are interested in.

25 Some other notable models we do not cover in this context utilize (i) dissipative dynamics as in warm inflation-type scenarios discussed in [264–266], (ii) the conversion of resonantly produced spectator fluctuations to curvature perturbation during [75] or after inflation, e.g., in curvaton-type scenarios discussed in [267,268].

26 See e.g., [276–279] for string-theory-motivated constructions in the context of inflation.

27 Be aware that despite the terminology we adopt, the gauge fields (A_μ) do not need to correspond to Standard Model photons.

28 At large CMB scales described by modes leaving the horizon in the early stages of inflation, the value of the effective coupling (ξ) is restricted by existing information and constraints on the scalar power spectrum and non-Gaussianity [282,283,288]. Depending on priors and on the shape of the inflation potential, these constraints give $\xi_{\rm cmb} < 2.2$–2.5 [289].

29 In fact, the potential (78) is not unique for the purpose of generating a localized particle production scenario. In this context, any potential that exhibits (a) step-like feature(s) is sufficient (see e.g., [83,296]. However, the choice of the potential in (78) does provide some useful analytic control over the background evolution of the axion and the resulting amplification of the power spectrum, as discussed in [86].

30 As discussed in [86], this assumption is actually self-consistent due to the localized nature of the particle production.

31 For the phenomenological examples we will present in this section, this statement can be made precise, and back-reaction can be shown to be negligible [85,285].

32 In this work, by an appropriate choice of initial conditions and model parameters, we assume that σ traverses two step-like features on its potential before it settles to its global minimum.

33 The roll of σ towards the global minimum can be captured by modifying the monomial term as $\mu^3 sigma \to \mu^3 f[\sqrt{1+(\sigma/f)^2}-1]$ so that the axion potential (85) interpolates between $\mu^3\sigma$ and $(\mu^3/f)\sigma^2$ from large to small field ($\sigma/f \to 0$) values, respectively.

34 For **M2**, this translates into a single step-like region in the potential (see Figure 15).

35 In the comoving gauge with which we are operating, the field fluctuations are proportional to entropy fluctuation ($\delta\phi^I = Q_s e_s^I$), while the spatial part of the metric takes the form $g_{ij} = a^2 e^{-2\mathcal{R}}\delta_{ij}$.

36 Recall from our discussion above that entropic mass crossing at N_f corresponds to the peak of the power spectrum.

37 For a detailed analytic/numerical analysis of the behavior of the power spectrum from strong sharp turns, see [72,73].

38 More realistic models of the turn are expected to have a smooth time dependence of $\eta_\perp$ such as the Gaussian profile we considered earlier. As shown by the numerical evaluations in [312], the choice of top-hat profile does not introduce any qualitative difference but appears to be a convenient choice for analytic manipulations. For a detailed study of sudden turn trajectories in conjunction with PBH formation, see [315].

39 We also note that the presence of very light entropy mode (for the $\xi = -3$ case) leads to a non-negligible contribution to the power spectrum for scales that cross the horizon before the turn, i.e., for $k/k_f \to 0$. In particular, the analytic expression (101) does not perform well on these scales. For a more complete analytic formula including numerical analysis, see [312].

40 Note that for flat spatial hypersurfaces ($K = 0$), we can define new coordinates as $x = r\cos\phi\sin\theta$, $y = r\sin\phi\sin\theta$ and $z = r\cos\theta$ to turn the metric into a form that is commonly used in the literature ($\mathrm{d}s^2 = -\mathrm{d}t^2 + a^2(t)\delta_{ij}\mathrm{d}x^i\mathrm{d}x^j$).

41 Indeed, CMB data inform us that the spatial geometry of our universe is flat on large cosmological scales (see (A18)). On the other hand, since it dilutes slowly ($\propto a^{-2}$) compared to radiation and matter, we would expect it to dominate the energy density before DED, but this did not happen. This implies that the initial value of the curvature must either be tuned to be extremely small or that it should relax to small values through a dynamical mechanism. Inflation could be also a solution to this puzzle because during such an exponential expansion, any initially large curvature would be inflated away.

42 The red-shift parameter can be defined as the fractional shift in the physical wavelength (λ) of a photon emitted at a distant point and time (t) in the universe until today, i.e., $z(t) \equiv (\lambda_0 - \lambda(t))/\lambda(t) = a(t_0)/a(t) - 1 = 1/a(t) - 1$.

43 Unless modes undergo resonance and are excited deep inside the horizon, the choice of initial conditions in Equations (A34) and (A35) provide an accurate prescription for the initialization of the numerical evaluation. For the models we consider in Sections 3.2 and 3.3, this is indeed the case. For a model that leads to excited states inside the horizon, see [328,329].

44 We note that at linear order in fluctuations, the Coulomb gauge condition ($\partial_i A_i = 0$) is equivalent to setting the temporal component of the gauge field to zero, i.e., $A_0 = 0$. This can be seen by explicitly expanding the gauge field action (A42) to second order in A_0 and A_i, then solving for the non-dynamical A_0 mode (see also the discussion in Section 3 of [331]).

45 This regime is not interesting from a phenomenological point of view, as the gauge field production will be very weak in this case.

46 We note that $\mathcal{R}$ also takes up a contribution bilinear in gauge fields proportional to the absolute value of the Poynting vector ($\mathcal{R}_{(AA)} \propto a|\vec{E}\times\vec{B}|$). We expect this contribution to be negligible at the end of inflation (i.e., at the time for we are interested in the correlators of $\mathcal{R}$) because the gauge field production saturates on superhorizon scales, and the corresponding electromagnetic fields decay as $\vec{E}, \vec{B} \propto a^{-2}$. See, e.g., the discussion presented in [85,339] within similar contexts.

References

1. Guth, A.H. The Inflationary Universe: A Possible Solution to the Horizon and Flatness Problems. *Phys. Rev.* **1981**, *D23*, 347–356. [CrossRef]
2. Linde, A.D. A New Inflationary Universe Scenario: A Possible Solution of the Horizon, Flatness, Homogeneity, Isotropy and Primordial Monopole Problems. *Phys. Lett.* **1982**, *B108*, 389–393. [CrossRef]
3. Penzias, A.A.; Wilson, R.W. A Measurement of excess antenna temperature at 4080-Mc/s. *Astrophys. J.* **1965**, *142*, 419–421. [CrossRef]
4. Akrami, Y.; Arroja, F.; Ashdown, M.; Aumont, J.; Baccigalupi, C.; Ballardini, M.; Banday, A.J.; Barreiro, R.B.; Bartolo, N.; Basak, S.; et al. Planck 2018 results. X. Constraints on inflation. *Astron. Astrophys.* **2020**, *641*, A10. [CrossRef]
5. Akrami, Y.; Arroja, F.; Ashdown, M.; Aumont, J.; Baccigalupi, C.; Ballardini, M.; Banday, A.J.; Barreiro, R.B.; Bartolo, N.; Basak, S.; et al. Planck 2018 results. IX. Constraints on primordial non-Gaussianity. *Astron. Astrophys.* **2020**, *641*, A9. [CrossRef]
6. Bertone, G.; Hooper, D. History of dark matter. *Rev. Mod. Phys.* **2018**, *90*, 045002. [CrossRef]
7. Aghanim, N.; Arroja, F.; Ashdown, M.; Aumont, J.; Baccigalupi, C.; Ballardini, M.; Banday, A.J.; Barreiro, R.B.; Bartolo, N.; Basak, S.; et al. Planck 2018 results. VI. Cosmological parameters. *Astron. Astrophys.* **2020**, *641*, A6; Erratum in *Astron. Astrophys.* **2021**, *652*, C4. [CrossRef]
8. Bergström, L. Nonbaryonic dark matter: Observational evidence and detection methods. *Rep. Prog. Phys.* **2000**, *63*, 793. [CrossRef]
9. Bertone, G.; Hooper, D.; Silk, J. Particle dark matter: Evidence, candidates and constraints. *Phys. Rep.* **2005**, *405*, 279–390. [CrossRef]
10. Zel'dovich, Y.B.; Novikov, I.D. The Hypothesis of Cores Retarded during Expansion and the Hot Cosmological Model. *Sov. Astron. AJ (Engl. Transl.)* **1967**, *10*, 602.
11. Hawking, S. Gravitationally collapsed objects of very low mass. *Mon. Not. R. Astron. Soc.* **1971**, *152*, 75. [CrossRef]
12. Carr, B.J.; Hawking, S.W. Black holes in the early Universe. *Mon. Not. R. Astron. Soc.* **1974**, *168*, 399–415. [CrossRef]

13. Carr, B.J. The Primordial black hole mass spectrum. *Astrophys. J.* **1975**, *201*, 1–19. [CrossRef]
14. Chapline, G.F. Cosmological effects of primordial black holes. *Nature* **1975**, *253*, 251–252. [CrossRef]
15. Carr, B.J.; Rees, M.J. Can pregalactic objects generate galaxies? *Mon. Not. R. Astron. Soc.* **1984**, *206*, 801–818. [CrossRef]
16. Aubourg, E.; Bareyre, P.; Brehin, S.; Gros, M.; Lachieze-Rey, M.; Laurent, B.; Lesquoy, E.; Magneville, C.; Milsztajn, A.; Moscoso, L.; et al. Evidence for gravitational microlensing by dark objects in the galactic halo. *Nature* **1993**, *365*, 623–625. [CrossRef]
17. Tisserand, P.; Le Guillou, L.; Afonso, C.; Albert, J.N.; Andersen, J.; Ansari, R.; Aubourg, É.; Bareyre, P.; Beaulieu, J.P.; Charlot, X.; et al. Limits on the Macho Content of the Galactic Halo from the EROS-2 Survey of the Magellanic Clouds. *Astron. Astrophys.* **2007**, *469*, 387–404. [CrossRef]
18. Wyrzykowski, L.; Kozlowski, S.; Skowron, J.; Udalski, A.; Szymanski, M.K.; Kubiak, M.; Pietrzynski, G.; Soszynski, I.; Szewczyk, O.; Ulaczyk, K.; et al. The OGLE View of Microlensing towards the Magellanic Clouds. III. Ruling out sub-solar MACHOs with the OGLE-III LMC data. *Mon. Not. R. Astron. Soc.* **2011**, *413*, 493. [CrossRef]
19. Wyrzykowski, L.; Kozlowski, S.; Skowron, J.; Udalski, A.; Szymanski, M.K.; Kubiak, M.; Pietrzynski, G.; Soszynski, I.; Szewczyk, O.; Ulaczyk, K.; et al. The OGLE View of Microlensing towards the Magellanic Clouds. IV. OGLE-III SMC Data and Final Conclusions on MACHOs. *Mon. Not. R. Astron. Soc.* **2011**, *416*, 2949. [CrossRef]
20. Abbott, B.P.; Abbott, R.; Abbott, T.D.; Abernathy, M.R.; Acernese, F.; Ackley, K.; Adams, C.; Adams, T.; Addesso, P.; Adhikari, R.X.; et al. Observation of Gravitational Waves from a Binary Black Hole Merger. *Phys. Rev. Lett.* **2016**, *116*, 061102. [CrossRef]
21. Bird, S.; Cholis, I.; Muñoz, J.B.; Ali-Haïmoud, Y.; Kamionkowski, M.; Kovetz, E.D.; Raccanelli, A.; Riess, A.G. Did LIGO detect dark matter? *Phys. Rev. Lett.* **2016**, *116*, 201301. [CrossRef] [PubMed]
22. Clesse, S.; García-Bellido, J. The clustering of massive Primordial Black Holes as Dark Matter: measuring their mass distribution with Advanced LIGO. *Phys. Dark Univ.* **2017**, *15*, 142–147. [CrossRef]
23. Sasaki, M.; Suyama, T.; Tanaka, T.; Yokoyama, S. Primordial Black Hole Scenario for the Gravitational-Wave Event GW150914. *Phys. Rev. Lett.* **2016**, *117*, 061101; Erratum in *Phys. Rev. Lett.* **2018**, *121*, 059901. [CrossRef]
24. Niikura, H.; Takada, M.; Yasuda, N.; Lupton, R.H.; Sumi, T.; More, S.; Kurita, T.; Sugiyama, S.; More, A.; Oguri, M.; et al. Microlensing constraints on primordial black holes with Subaru/HSC Andromeda observations. *Nat. Astron.* **2019**, *3*, 524–534. [CrossRef]
25. Katz, A.; Kopp, J.; Sibiryakov, S.; Xue, W. Femtolensing by Dark Matter Revisited. *J. Cosmol. Astropart. Phys.* **2018**, *2018*, 005. [CrossRef]
26. Montero-Camacho, P.; Fang, X.; Vasquez, G.; Silva, M.; Hirata, C.M. Revisiting constraints on asteroid-mass primordial black holes as dark matter candidates. *J. Cosmol. Astropart. Phys.* **2019**, *2019*, 031. [CrossRef]
27. Carr, B.; Raidal, M.; Tenkanen, T.; Vaskonen, V.; Veermäe, H. Primordial black hole constraints for extended mass functions. *Phys. Rev. D* **2017**, *96*, 023514. [CrossRef]
28. Khlopov, M.Y. Primordial Black Holes. *Res. Astron. Astrophys.* **2010**, *10*, 495–528. [CrossRef]
29. García-Bellido, J. Massive Primordial Black Holes as Dark Matter and their detection with Gravitational Waves. *J. Phys. Conf. Ser.* **2017**, *840*, 012032. [CrossRef]
30. Sasaki, M.; Suyama, T.; Tanaka, T.; Yokoyama, S. Primordial black holes: Perspectives in gravitational wave astronomy. *Class. Quant. Grav.* **2018**, *35*, 063001. [CrossRef]
31. Carr, B.; Kuhnel, F. Primordial Black Holes as Dark Matter: Recent Developments. *Ann. Rev. Nucl. Part. Sci.* **2020**, *70*, 355–394. [CrossRef]
32. Carr, B.; Kohri, K.; Sendouda, Y.; Yokoyama, J. Constraints on primordial black holes. *Rep. Prog. Phys.* **2021**, *84*, 116902. [CrossRef]
33. Green, A.M.; Kavanagh, B.J. Primordial Black Holes as a dark matter candidate. *J. Phys. G* **2021**, *48*, 043001. [CrossRef]
34. Escrivà, A.; Kuhnel, F.; Tada, Y. Primordial Black Holes. *arXiv* **2022**, arXiv:2211.05767.
35. Page, D.N. Particle Emission Rates from a Black Hole: Massless Particles from an Uncharged, Nonrotating Hole. *Phys. Rev. D* **1976**, *13*, 198–206. [CrossRef]
36. Papanikolaou, T.; Vennin, V.; Langlois, D. Gravitational waves from a universe filled with primordial black holes. *J. Cosmol. Astropart. Phys.* **2021**, *2021*, 053. [CrossRef]
37. Papanikolaou, T. Gravitational waves induced from primordial black hole fluctuations: the effect of an extended mass function. *J. Cosmol. Astropart. Phys.* **2022**, *2022*, 089. [CrossRef]
38. Ivanov, P.; Naselsky, P.; Novikov, I. Inflation and primordial black holes as dark matter. *Phys. Rev.* **1994**, *D50*, 7173–7178. [CrossRef]
39. Starobinsky, A.A. Spectrum of adiabatic perturbations in the universe when there are singularities in the inflation potential. *JETP Lett.* **1992**, *55*, 489–494.
40. Dimopoulos, K. Ultra slow-roll inflation demystified. *Phys. Lett.* **2017**, *B775*, 262–265. [CrossRef]
41. Leach, S.M.; Liddle, A.R. Inflationary perturbations near horizon crossing. *Phys. Rev.* **2001**, *D63*, 043508. [CrossRef]
42. Leach, S.M.; Sasaki, M.; Wands, D.; Liddle, A.R. Enhancement of superhorizon scale inflationary curvature perturbations. *Phys. Rev.* **2001**, *D64*, 023512. [CrossRef]
43. Kinney, W.H. Horizon crossing and inflation with large eta. *Phys. Rev.* **2005**, *D72*, 023515. [CrossRef]
44. Martin, J.; Motohashi, H.; Suyama, T. Ultra Slow-Roll Inflation and the non-Gaussianity Consistency Relation. *Phys. Rev.* **2013**, *D87*, 023514. [CrossRef]

45. Motohashi, H.; Starobinsky, A.A.; Yokoyama, J. Inflation with a constant rate of roll. *J. Cosmol. Astropart. Phys.* **2015**, *2015*, 018. [CrossRef]
46. Garcia-Bellido, J.; Ruiz Morales, E. Primordial black holes from single field models of inflation. *Phys. Dark Univ.* **2017**, *18*, 47–54. [CrossRef]
47. Ezquiaga, J.M.; Garcia-Bellido, J.; Ruiz Morales, E. Primordial Black Hole production in Critical Higgs Inflation. *Phys. Lett. B* **2018**, *776*, 345–349. [CrossRef]
48. Germani, C.; Prokopec, T. On primordial black holes from an inflection point. *Phys. Dark Univ.* **2017**, *18*, 6–10. [CrossRef]
49. Ballesteros, G.; Taoso, M. Primordial black hole dark matter from single field inflation. *Phys. Rev.* **2018**, *D97*, 023501. [CrossRef]
50. Hertzberg, M.P.; Yamada, M. Primordial Black Holes from Polynomial Potentials in Single Field Inflation. *Phys. Rev. D* **2017**, *97*, 083509. [CrossRef]
51. Cicoli, M.; Diaz, V.A.; Pedro, F.G. Primordial Black Holes from String Inflation. *J. Cosmol. Astropart. Phys.* **2018**, *2018*, 034. [CrossRef]
52. Özsoy, O.; Parameswaran, S.; Tasinato, G.; Zavala, I. Mechanisms for Primordial Black Hole Production in String Theory. *J. Cosmol. Astropart. Phys.* **2018**, *2018*, 005. [CrossRef]
53. Mishra, S.S.; Sahni, V. Primordial Black Holes from a tiny bump/dip in the inflation potential. *J. Cosmol. Astropart. Phys.* **2020**, *2020*, 007. [CrossRef]
54. Dimopoulos, K.; Markkanen, T.; Racioppi, A.; Vaskonen, V. Primordial Black Holes from Thermal Inflation. *J. Cosmol. Astropart. Phys.* **2019**, *2019*, 046. [CrossRef]
55. Ballesteros, G.; Rey, J.; Taoso, M.; Urbano, A. Primordial black holes as dark matter and gravitational waves from single-field polynomial inflation. *J. Cosmol. Astropart. Phys.* **2020**, *2020*, 025. [CrossRef]
56. Inomata, K.; McDonough, E.; Hu, W. Amplification of primordial perturbations from the rise or fall of the inflation. *J. Cosmol. Astropart. Phys.* **2022**, *2022*, 031. [CrossRef]
57. Garcia-Bellido, J.; Linde, A.D.; Wands, D. Density perturbations and black hole formation in hybrid inflation. *Phys. Rev. D* **1996**, *54*, 6040–6058. [CrossRef]
58. Kawasaki, M.; Sugiyama, N.; Yanagida, T. Primordial black hole formation in a double inflation model in supergravity. *Phys. Rev. D* **1998**, *57*, 6050–6056. [CrossRef]
59. Yokoyama, J. Chaotic new inflation and formation of primordial black holes. *Phys. Rev. D* **1998**, *58*, 083510. [CrossRef]
60. Kawaguchi, T.; Kawasaki, M.; Takayama, T.; Yamaguchi, M.; Yokoyama, J. Formation of intermediate-mass black holes as primordial black holes in the inflationary cosmology with running spectral index. *Mon. Not. R. Astron. Soc.* **2008**, *388*, 1426–1432. [CrossRef]
61. Kohri, K.; Lyth, D.H.; Melchiorri, A. Black hole formation and slow-roll inflation. *J. Cosmol. Astropart. Phys.* **2008**, *2008*, 038. [CrossRef]
62. Frampton, P.H.; Kawasaki, M.; Takahashi, F.; Yanagida, T.T. Primordial Black Holes as All Dark Matter. *J. Cosmol. Astropart. Phys.* **2010**, *2010*, 023. [CrossRef]
63. Drees, M.; Erfani, E. Running Spectral Index and Formation of Primordial Black Hole in Single Field Inflation Models. *J. Cosmol. Astropart. Phys.* **2012**, *2012*, 035. [CrossRef]
64. Ballesteros, G.; Beltran Jimenez, J.; Pieroni, M. Black hole formation from a general quadratic action for inflationary primordial fluctuations. *J. Cosmol. Astropart. Phys.* **2019**, *2019*, 016. [CrossRef]
65. Kamenshchik, A.Y.; Tronconi, A.; Venturi, G. DBI inflation and warped black holes. *J. Cosmol. Astropart. Phys.* **2022**, *2022*, 051. [CrossRef]
66. Cai, Y.F.; Tong, X.; Wang, D.G.; Yan, S.F. Primordial Black Holes from Sound Speed Resonance during Inflation. *Phys. Rev. Lett.* **2018**, *121*, 081306. [CrossRef]
67. Chen, C.; Ma, X.H.; Cai, Y.F. Dirac-Born-Infeld realization of sound speed resonance mechanism for primordial black holes. *Phys. Rev. D* **2020**, *102*, 063526. [CrossRef]
68. Baumann, D.; McAllister, L. *Inflation and String Theory*; Cambridge Monographs on Mathematical Physics, Cambridge University Press: Cambridge, UK , 2015.
69. Randall, L.; Soljacic, M.; Guth, A.H. Supernatural inflation: Inflation from supersymmetry with no (very) small parameters. *Nucl. Phys. B* **1996**, *472*, 377–408. [CrossRef]
70. Clesse, S.; García-Bellido, J. Massive Primordial Black Holes from Hybrid Inflation as Dark Matter and the seeds of Galaxies. *Phys. Rev. D* **2015**, *92*, 023524. [CrossRef]
71. Brown, A.R. Hyperbolic Inflation. *Phys. Rev. Lett.* **2018**, *121*, 251601. [CrossRef] [PubMed]
72. Palma, G.A.; Sypsas, S.; Zenteno, C. Seeding primordial black holes in multifield inflation. *Phys. Rev. Lett.* **2020**, *125*, 121301. [CrossRef]
73. Fumagalli, J.; Renaux-Petel, S.; Ronayne, J.W.; Witkowski, L.T. Turning in the landscape: A new mechanism for generating Primordial Black Holes. *arXiv* **2020**, arXiv:2004.08369.
74. Braglia, M.; Hazra, D.K.; Finelli, F.; Smoot, G.F.; Sriramkumar, L.; Starobinsky, A.A. Generating PBHs and small-scale GWs in two-field models of inflation. *J. Cosmol. Astropart. Phys.* **2020**, *2020*, 001. [CrossRef]
75. Zhou, Z.; Jiang, J.; Cai, Y.F.; Sasaki, M.; Pi, S. Primordial black holes and gravitational waves from resonant amplification during inflation. *Phys. Rev. D* **2020**, *102*, 103527. [CrossRef]

76. Iacconi, L.; Assadullahi, H.; Fasiello, M.; Wands, D. Revisiting small-scale fluctuations in α-attractor models of inflation. *J. Cosmol. Astropart. Phys.* **2021**, *2022*, 007. [CrossRef]
77. Kallosh, R.; Linde, A. Dilaton-Axion Inflation with PBHs and GWs. *J. Cosmol. Astropart. Phys.* **2022**, *2022*, 037. [CrossRef]
78. Kawai, S.; Kim, J. Primordial black holes and gravitational waves from nonminimally coupled supergravity inflation. *Phys. Rev.* **2022**, *107*, 043523. [CrossRef]
79. Linde, A.; Mooij, S.; Pajer, E. Gauge field production in supergravity inflation: Local non-Gaussianity and primordial black holes. *Phys. Rev. D* **2013**, *87*, 103506. [CrossRef]
80. Bugaev, E.; Klimai, P. Axion inflation with gauge field production and primordial black holes. *Phys. Rev. D* **2014**, *90*, 103501. [CrossRef]
81. Garcia-Bellido, J.; Peloso, M.; Unal, C. Gravitational waves at interferometer scales and primordial black holes in axion inflation. *J. Cosmol. Astropart. Phys.* **2016**, *2016*, 031. [CrossRef]
82. Domcke, V.; Muia, F.; Pieroni, M.; Witkowski, L.T. PBH dark matter from axion inflation. *J. Cosmol. Astropart. Phys.* **2017**, *2017*, 048. [CrossRef]
83. Cheng, S.L.; Lee, W.; Ng, K.W. Primordial black holes and associated gravitational waves in axion monodromy inflation. *J. Cosmol. Astropart. Phys.* **2018**, *2018*, 001. [CrossRef]
84. Kawasaki, M.; Nakatsuka, H.; Obata, I. Generation of Primordial Black Holes and Gravitational Waves from Dilaton-Gauge Field Dynamics. *J. Cosmol. Astropart. Phys.* **2020**, *2020*, 007. [CrossRef]
85. Özsoy, O. Synthetic Gravitational Waves from a Rolling Axion Monodromy. *J. Cosmol. Astropart. Phys.* **2021**, *2021*, 040. [CrossRef]
86. Özsoy, O.; Lalak, Z. Primordial black holes as dark matter and gravitational waves from bumpy axion inflation. *J. Cosmol. Astropart. Phys.* **2021**, *2021*, 040. [CrossRef]
87. Ananda, K.N.; Clarkson, C.; Wands, D. The Cosmological gravitational wave background from primordial density perturbations. *Phys. Rev.* **2007**, *D75*, 123518. [CrossRef]
88. Baumann, D.; Steinhardt, P.J.; Takahashi, K.; Ichiki, K. Gravitational Wave Spectrum Induced by Primordial Scalar Perturbations. *Phys. Rev.* **2007**, *D76*, 084019. [CrossRef]
89. Kohri, K.; Terada, T. Semianalytic calculation of gravitational wave spectrum nonlinearly induced from primordial curvature perturbations. *Phys. Rev.* **2018**, *D97*, 123532. [CrossRef]
90. Nakama, T.; Silk, J.; Kamionkowski, M. Stochastic gravitational waves associated with the formation of primordial black holes. *Phys. Rev. D* **2017**, *95*, 043511. [CrossRef]
91. Cai, R.G.; Pi, S.; Sasaki, M. Gravitational Waves Induced by non-Gaussian Scalar Perturbations. *Phys. Rev. Lett.* **2019**, *122*, 201101. [CrossRef] [PubMed]
92. Unal, C. Imprints of Primordial Non-Gaussianity on Gravitational Wave Spectrum. *Phys. Rev. D* **2019**, *99*, 041301. [CrossRef]
93. Yuan, C.; Chen, Z.C.; Huang, Q.G. Log-dependent slope of scalar induced gravitational waves in the infrared regions. *Phys. Rev. D* **2020**, *101*, 043019. [CrossRef]
94. Cai, R.G.; Pi, S.; Sasaki, M. Universal infrared scaling of gravitational wave background spectra. *Phys. Rev. D* **2020**, *102*, 083528. [CrossRef]
95. Özsoy, O.; Tasinato, G. On the slope of the curvature power spectrum in non-attractor inflation. *J. Cosmol. Astropart. Phys.* **2020**, *2020*, 048. [CrossRef]
96. Pi, S.; Sasaki, M. Gravitational Waves Induced by Scalar Perturbations with a Lognormal Peak. *J. Cosmol. Astropart. Phys.* **2020**, *2020*, 037. [CrossRef]
97. Yuan, C.; Huang, Q.G. Gravitational waves induced by the local-type non-Gaussian curvature perturbations. *Phys. Lett. B* **2021**, *821*, 136606. [CrossRef]
98. Ragavendra, H.V. Accounting for scalar non-Gaussianity in secondary gravitational waves. *Phys. Rev. D* **2022**, *105*, 063533. [CrossRef]
99. Amaro-Seoane, P.; Audley, H.; Babak, S.; Baker, J.; Barausse, E.; Bender, P.; Berti, E.; Binetruy, P.; Born, M.; Bortoluzzi, D.; et al. Laser Interferometer Space Antenna *arXiv* **2017**, arXiv:1702.00786.
100. Barausse, E.; Berti, E.; Hertog, T.; Hughes, S.A.; Jetzer, P.; Pani, P.; Sotiriou, T.P.; Tamanini, N.; Witek, H.; Yagi, K.; et al. Prospects for Fundamental Physics with LISA. *Gen. Relat. Grav.* **2020**, *52*, 81. [CrossRef]
101. Lentati, L.; Taylor, S.R.; Mingarelli, C.M.; Sesana, A.; Sanidas, S.A.; Vecchio, A.; Caballero, R.N.; Lee, K.J.; Van Haasteren, R.; Babak, S.; et al. European Pulsar Timing Array Limits On An Isotropic Stochastic Gravitational-Wave Background. *Mon. Not. R. Astron. Soc.* **2015**, *453*, 2576–2598. [CrossRef]
102. Arzoumanian, Z.; Baker, P.T.; Blumer, H.; Bécsy, B.; Brazier, A.; Brook, P.R.; Burke-Spolaor, S.; Chatterjee, S.; Chen, S.; Cordes, J.M.; et al. The NANOGrav 12.5 yr Data Set: Search for an Isotropic Stochastic Gravitational-wave Background. *Astrophys. J. Lett.* **2020**, *905*, L34. [CrossRef]
103. Seto, N.; Kawamura, S.; Nakamura, T. Possibility of direct measurement of the acceleration of the universe using 0.1-Hz band laser interferometer gravitational wave antenna in space. *Phys. Rev. Lett.* **2001**, *87*, 221103. [CrossRef] [PubMed]
104. Kawamura, S.; Ando, M.; Seto, N.; Sato, S.; Musha, M.; Kawano, I.; Yokoyama, J.I.; Tanaka, T.; Ioka, K.; Akutsu, T.; et al. Current status of space gravitational wave antenna DECIGO and B-DECIGO. *Prog. Theor. Exp. Phys.* **2021**, *2021*, 05A105. [CrossRef]
105. Domènech, G. Scalar Induced Gravitational Waves Review. *Universe* **2021**, *7*, 398. [CrossRef]

106. Polnarev, A.; Zembowicz, R. Formation of Primordial Black Holes by Cosmic Strings. *Phys. Rev. D* **1991**, *43*, 1106–1109. [CrossRef] [PubMed]
107. Caldwell, R.R.; Casper, P. Formation of black holes from collapsed cosmic string loops. *Phys. Rev. D* **1996**, *53*, 3002–3010. [CrossRef]
108. Helfer, T.; Aurrekoetxea, J.C.; Lim, E.A. Cosmic String Loop Collapse in Full General Relativity. *Phys. Rev. D* **2019**, *99*, 104028. [CrossRef]
109. Rubin, S.G.; Khlopov, M.Y.; Sakharov, A.S. Primordial black holes from nonequilibrium second order phase transition. *Grav. Cosmol.* **2000**, *6*, 51–58.
110. Garriga, J.; Vilenkin, A.; Zhang, J. Black holes and the multiverse. *J. Cosmol. Astropart. Phys.* **2016**, *2016*, 064. [CrossRef]
111. Deng, H.; Garriga, J.; Vilenkin, A. Primordial black hole and wormhole formation by domain walls. *J. Cosmol. Astropart. Phys.* **2017**, *2017*, 050. [CrossRef]
112. Kusenko, A.; Sasaki, M.; Sugiyama, S.; Takada, M.; Takhistov, V.; Vitagliano, E. Exploring Primordial Black Holes from the Multiverse with Optical Telescopes. *Phys. Rev. Lett.* **2020**, *125*, 181304. [CrossRef] [PubMed]
113. Kodama, H.; Sasaki, M.; Sato, K. Abundance of Primordial Holes Produced by Cosmological First Order Phase Transition. *Prog. Theor. Phys.* **1982**, *68*, 1979. [CrossRef]
114. Jedamzik, K. Primordial black hole formation during the QCD epoch. *Phys. Rev. D* **1997**, *55*, 5871–5875. [CrossRef]
115. Moss, I.G. Singularity formation from colliding bubbles. *Phys. Rev. D* **1994**, *50*, 676–681. [CrossRef]
116. Kitajima, N.; Takahashi, F. Primordial Black Holes from QCD Axion Bubbles. *J. Cosmol. Astropart. Phys.* **2020**, *2020*, 060. [CrossRef]
117. Cotner, E.; Kusenko, A. Primordial black holes from supersymmetry in the early universe. *Phys. Rev. Lett.* **2017**, *119*, 031103. [CrossRef]
118. Cotner, E.; Kusenko, A.; Sasaki, M.; Takhistov, V. Analytic Description of Primordial Black Hole Formation from Scalar Field Fragmentation. *J. Cosmol. Astropart. Phys.* **2019**, *2019*, 077. [CrossRef]
119. Kofman, L.; Linde, A.D.; Starobinsky, A.A. Reheating after inflation. *Phys. Rev. Lett.* **1994**, *73*, 3195–3198. [CrossRef]
120. Shtanov, Y.; Traschen, J.H.; Brandenberger, R.H. Universe reheating after inflation. *Phys. Rev. D* **1995**, *51*, 5438–5455. [CrossRef]
121. Kofman, L.; Linde, A.D.; Starobinsky, A.A. Towards the theory of reheating after inflation. *Phys. Rev.* **1997**, *D56*, 3258–3295. [CrossRef]
122. Amin, M.A.; Hertzberg, M.P.; Kaiser, D.I.; Karouby, J. Nonperturbative Dynamics Of Reheating After Inflation: A Review. *Int. J. Mod. Phys. D* **2014**, *24*, 1530003. [CrossRef]
123. Ashoorioon, A.; Rostami, A.; Firouzjaee, J.T. Examining the end of inflation with primordial black holes mass distribution and gravitational waves. *Phys. Rev. D* **2021**, *103*, 123512. [CrossRef]
124. Martin, J.; Papanikolaou, T.; Vennin, V. Primordial black holes from the preheating instability in single-field inflation. *J. Cosmol. Astropart. Phys.* **2020**, *2020*, 024. [CrossRef]
125. Auclair, P.; Vennin, V. Primordial black holes from metric preheating: Mass fraction in the excursion-set approach. *J. Cosmol. Astropart. Phys.* **2021**, *2021*, 038. [CrossRef]
126. Byrnes, C.T.; Cole, P.S. Lecture notes on inflation and primordial black holes. *arXiv* **2021**, arXiv:2112.05716.
127. Franciolini, G. Primordial Black Holes: From Theory to Gravitational Wave Observations. Ph.D. Thesis, University of Geneva, Department of Theoretical Physics, Geneva, Switzerland, 2021. [CrossRef]
128. Baumann, D. *Cosmology*; Cambridge University Press: Cambridge, UK, 2022. [CrossRef]
129. Sasaki, M. Large Scale Quantum Fluctuations in the Inflationary Universe. *Prog. Theor. Phys.* **1986**, *76*, 1036. [CrossRef]
130. Mukhanov, V.F.; Feldman, H.A.; Brandenberger, R.H. Theory of cosmological perturbations. Part 1. Classical perturbations. Part 2. Quantum theory of perturbations. Part 3. Extensions. *Phys. Rep.* **1992**, *215*, 203–333. [CrossRef]
131. Allahverdi, R.; Amin, M.A.; Berlin, A.; Bernal, N.; Byrnes, C.T.; Delos, M.S.; Erickcek, A.L.; Escudero, M.; Figueroa, D.G.; Freese, K.; et al. The First Three Seconds: A Review of Possible Expansion Histories of the Early Universe. *arXiv* **2020**, arXiv:2006.16182. [CrossRef]
132. Harada, T.; Yoo, C.M.; Kohri, K. Threshold of primordial black hole formation. *Phys. Rev.* **2013**, *D88*, 084051. Erratum in *Phys. Rev.* **2014**, *D89*, 029903. [CrossRef]
133. Niemeyer, J.C.; Jedamzik, K. Dynamics of primordial black hole formation. *Phys. Rev. D* **1999**, *59*, 124013. [CrossRef]
134. Nakama, T.; Harada, T.; Polnarev, A.G.; Yokoyama, J. Identifying the most crucial parameters of the initial curvature profile for primordial black hole formation. *J. Cosmol. Astropart. Phys.* **2014**, *2014*, 037. [CrossRef]
135. Musco, I. Threshold for primordial black holes: Dependence on the shape of the cosmological perturbations. *Phys. Rev. D* **2019**, *100*, 123524. [CrossRef]
136. Musco, I.; De Luca, V.; Franciolini, G.; Riotto, A. Threshold for primordial black holes. II. A simple analytic prescription. *Phys. Rev. D* **2021**, *103*, 063538. [CrossRef]
137. Musco, I.; Papanikolaou, T. Primordial black hole formation for an anisotropic perfect fluid: Initial conditions and estimation of the threshold. *Phys. Rev. D* **2022**, *106*, 083017. [CrossRef]
138. Papanikolaou, T. Toward the primordial black hole formation threshold in a time-dependent equation-of-state background. *Phys. Rev. D* **2022**, *105*, 124055. [CrossRef]
139. Shibata, M.; Sasaki, M. Black hole formation in the Friedmann universe: Formulation and computation in numerical relativity. *Phys. Rev. D* **1999**, *60*, 084002. [CrossRef]

140. Germani, C.; Musco, I. Abundance of Primordial Black Holes Depends on the Shape of the Inflationary Power Spectrum. *Phys. Rev. Lett.* **2019**, *122*, 141302. [CrossRef] [PubMed]
141. Escrivà, A. PBH Formation from Spherically Symmetric Hydrodynamical Perturbations: A Review. *Universe* **2022**, *8*, 66. [CrossRef]
142. Escrivà, A.; Germani, C.; Sheth, R.K. Universal threshold for primordial black hole formation. *Phys. Rev. D* **2020**, *101*, 044022. [CrossRef]
143. Escrivà, A.; Germani, C.; Sheth, R.K. Analytical thresholds for black hole formation in general cosmological backgrounds. *J. Cosmol. Astropart. Phys.* **2021**, *2021*, 030. [CrossRef]
144. Akiyama, K.; Alberdi, A.; Alef, W.; Algaba, J.C.; Anantua, R.; Asada, K.; Azulay, R.; Bach, U.; Baczko, A.K.; Ball, D.; et al. First Sagittarius A* Event Horizon Telescope Results. I. The Shadow of the Supermassive Black Hole in the Center of the Milky Way. *Astrophys. J. Lett.* **2022**, *930*, L12. [CrossRef]
145. Press, W.H.; Schechter, P. Formation of galaxies and clusters of galaxies by selfsimilar gravitational condensation. *Astrophys. J.* **1974**, *187*, 425–438. [CrossRef]
146. Evans, C.R.; Coleman, J.S. Observation of critical phenomena and selfsimilarity in the gravitational collapse of radiation fluid. *Phys. Rev. Lett.* **1994**, *72*, 1782–1785. [CrossRef] [PubMed]
147. Niemeyer, J.C.; Jedamzik, K. Near-critical gravitational collapse and the initial mass function of primordial black holes. *Phys. Rev. Lett.* **1998**, *80*, 5481–5484. [CrossRef]
148. Koike, T.; Hara, T.; Adachi, S. Critical behavior in gravitational collapse of radiation fluid: A Renormalization group (linear perturbation) analysis. *Phys. Rev. Lett.* **1995**, *74*, 5170–5173. [CrossRef]
149. Germani, C.; Sheth, R.K. Nonlinear statistics of primordial black holes from Gaussian curvature perturbations. *Phys. Rev. D* **2020**, *101*, 063520. [CrossRef]
150. De Luca, V.; Franciolini, G.; Kehagias, A.; Peloso, M.; Riotto, A.; Ünal, C. The Ineludible non-Gaussianity of the Primordial Black Hole Abundance. *J. Cosmol. Astropart. Phys.* **2019**, *2019*, 048. [CrossRef]
151. Young, S.; Musco, I.; Byrnes, C.T. Primordial black hole formation and abundance: Contribution from the non-linear relation between the density and curvature perturbation. *J. Cosmol. Astropart. Phys.* **2019**, *2019*, 012. [CrossRef]
152. Young, S.; Byrnes, C.T.; Sasaki, M. Calculating the mass fraction of primordial black holes. *J. Cosmol. Astropart. Phys.* **2014**, *2014*, 045. [CrossRef]
153. Byrnes, C.T.; Copeland, E.J.; Green, A.M. Primordial black holes as a tool for constraining non-Gaussianity. *Phys. Rev.* **2012**, *D86*, 043512. [CrossRef]
154. Young, S.; Byrnes, C.T. Primordial black holes in non-Gaussian regimes. *J. Cosmol. Astropart. Phys.* **2013**, *2013*, 052. [CrossRef]
155. Passaglia, S.; Hu, W.; Motohashi, H. Primordial black holes and local non-Gaussianity in canonical inflation. *Phys. Rev. D* **2019**, *99*, 043536. [CrossRef]
156. Biagetti, M.; De Luca, V.; Franciolini, G.; Kehagias, A.; Riotto, A. The formation probability of primordial black holes. *Phys. Lett. B* **2021**, *820*, 136602. [CrossRef]
157. Atal, V.; Germani, C. The role of non-gaussianities in Primordial Black Hole formation. *Phys. Dark Univ.* **2019**, *24*, 100275. [CrossRef]
158. Taoso, M.; Urbano, A. Non-gaussianities for primordial black hole formation. *J. Cosmol. Astropart. Phys.* **2021**, *2021*, 016. [CrossRef]
159. Young, S. Peaks and primordial black holes: The effect of non-Gaussianity. *J. Cosmol. Astropart. Phys.* **2022**, *2022*, 037. [CrossRef]
160. Ferrante, G.; Franciolini, G.; Iovino, Junior., A.; Urbano, A. Primordial non-gaussianity up to all orders: Theoretical aspects and implications for primordial black hole models. *Phys. Rev. D* **2023**, *107*, 043520. [CrossRef]
161. Gow, A.D.; Assadullahi, H.; Jackson, J.H.P.; Koyama, K.; Vennin, V.; Wands, D. Non-perturbative non-Gaussianity and primordial black holes. *arXiv* **2022**, arXiv:2211.08348.
162. Lyth, D.H. The hybrid inflation waterfall and the primordial curvature perturbation. *J. Cosmol. Astropart. Phys.* **2012**, *2012*, 022. [CrossRef]
163. Musco, I.; Miller, J.C. Primordial black hole formation in the early universe: Critical behaviour and self-similarity. *Class. Quant. Grav.* **2013**, *30*, 145009. [CrossRef]
164. Byrnes, C.T.; Hindmarsh, M.; Young, S.; Hawkins, M.R.S. Primordial black holes with an accurate QCD equation of state. *J. Cosmol. Astropart. Phys.* **2018**, *2018*, 041. [CrossRef]
165. Carr, B.; Clesse, S.; García-Bellido, J.; Kühnel, F. Cosmic conundra explained by thermal history and primordial black holes. *Phys. Dark Univ.* **2021**, *31*, 100755. [CrossRef]
166. Ali-Haïmoud, Y. Correlation Function of High-Threshold Regions and Application to the Initial Small-Scale Clustering of Primordial Black Holes. *Phys. Rev. Lett.* **2018**, *121*, 081304. [CrossRef]
167. Young, S.; Byrnes, C.T. Initial clustering and the primordial black hole merger rate. *J. Cosmol. Astropart. Phys.* **2020**, *2020*, 004. [CrossRef]
168. De Luca, V.; Desjacques, V.; Franciolini, G.; Riotto, A. The clustering evolution of primordial black holes. *J. Cosmol. Astropart. Phys.* **2020**, *2020*, 028. [CrossRef]
169. De Luca, V.; Franciolini, G.; Riotto, A.; Veermäe, H. Ruling Out Initially Clustered Primordial Black Holes as Dark Matter. *Phys. Rev. Lett.* **2022**, *129*, 191302. [CrossRef]

170. Ali-Haïmoud, Y.; Kovetz, E.D.; Kamionkowski, M. Merger rate of primordial black-hole binaries. *Phys. Rev. D* **2017**, *96*, 123523. [CrossRef]
171. Ballesteros, G.; Serpico, P.D.; Taoso, M. On the merger rate of primordial black holes: Effects of nearest neighbours distribution and clustering. *J. Cosmol. Astropart. Phys.* **2018**, *2018*, 043. [CrossRef]
172. Vaskonen, V.; Veermäe, H. Lower bound on the primordial black hole merger rate. *Phys. Rev. D* **2020**, *101*, 043015. [CrossRef]
173. De Luca, V.; Franciolini, G.; Riotto, A. On the Primordial Black Hole Mass Function for Broad Spectra. *Phys. Lett. B* **2020**, *807*, 135550. [CrossRef]
174. Moradinezhad Dizgah, A.; Franciolini, G.; Riotto, A. Primordial Black Holes from Broad Spectra: Abundance and Clustering. *J. Cosmol. Astropart. Phys.* **2019**, *2019*, 001. [CrossRef]
175. De Luca, V.; Riotto, A. A note on the abundance of primordial black holes: Use and misuse of the metric curvature perturbation. *Phys. Lett. B* **2022**, *828*, 137035. [CrossRef]
176. Young, S. Constraining the Early Universe with Primordial Black Holes. Ph.D. Thesis, University of Sussex, Sussex, UK, 2016.
177. Garcia-Bellido, J.; Peloso, M.; Unal, C. Gravitational Wave signatures of inflationary models from Primordial Black Hole Dark Matter. *J. Cosmol. Astropart. Phys.* **2017**, *2017*, 013. [CrossRef]
178. Lyth, D.H.; Riotto, A. Particle physics models of inflation and the cosmological density perturbation. *Phys. Rep.* **1999**, *314*, 1–146. [CrossRef]
179. Endlich, S.; Nicolis, A.; Wang, J. Solid Inflation. *J. Cosmol. Astropart. Phys.* **2013**, *2013*, 011. [CrossRef]
180. Cannone, D.; Tasinato, G.; Wands, D. Generalised tensor fluctuations and inflation. *J. Cosmol. Astropart. Phys.* **2015**, *2015*, 029. [CrossRef]
181. Bartolo, N.; Cannone, D.; Ricciardone, A.; Tasinato, G. Distinctive signatures of space-time diffeomorphism breaking in EFT of inflation. *J. Cosmol. Astropart. Phys.* **2016**, *2016*, 044. [CrossRef]
182. Ashoorioon, A.; Rostami, A.; Firouzjaee, J.T. EFT compatible PBHs: Effective spawning of the seeds for primordial black holes during inflation. *J. High Energy Phys.* **2021**, *2021*, 087. [CrossRef]
183. Arkani-Hamed, N.; Creminelli, P.; Mukohyama, S.; Zaldarriaga, M. Ghost inflation. *J. Cosmol. Astropart. Phys.* **2004**, *2004*, 001. [CrossRef]
184. Ballesteros, G.; Céspedes, S.; Santoni, L. Large power spectrum and primordial black holes in the effective theory of inflation. *J. High Energy Phys.* **2022**, *2022*, 074. [CrossRef]
185. Drees, M.; Xu, Y. Overshooting, Critical Higgs Inflation and Second Order Gravitational Wave Signatures. *Eur. Phys. J. C* **2021**, *81*, 182. [CrossRef]
186. Cheong, D.Y.; Lee, S.M.; Park, S.C. Primordial black holes in Higgs-R^2 inflation as the whole of dark matter. *J. Cosmol. Astropart. Phys.* **2021**, *2021*, 032. [CrossRef]
187. Rasanen, S.; Tomberg, E. Planck scale black hole dark matter from Higgs inflation. *J. Cosmol. Astropart. Phys.* **2019**, *2019*, 038. [CrossRef]
188. Dalianis, I.; Kehagias, A.; Tringas, G. Primordial black holes from α-attractors. *J. Cosmol. Astropart. Phys.* **2019**, *2019*, 037. [CrossRef]
189. Dalianis, I.; Tringas, G. Primordial black hole remnants as dark matter produced in thermal, matter, and runaway-quintessence postinflationary scenarios. *Phys. Rev. D* **2019**, *100*, 083512. [CrossRef]
190. Cicoli, M.; Pedro, F.G.; Pedron, N. Secondary GWs and PBHs in string inflation: Formation and detectability. *J. Cosmol. Astropart. Phys.* **2022**, *2022*, 030. [CrossRef]
191. Armendariz-Picon, C.; Damour, T.; Mukhanov, V.F. k-inflation. *Phys. Lett. B* **1999**, *458*, 209–218. [CrossRef]
192. Garriga, J.; Mukhanov, V.F. Perturbations in k-inflation. *Phys. Lett.* **1999**, *B458*, 219–225. [CrossRef]
193. Solbi, M.; Karami, K. Primordial black holes and induced gravitational waves in k-inflation. *J. Cosmol. Astropart. Phys.* **2021**, *2021*, 056. [CrossRef]
194. Solbi, M.; Karami, K. Primordial black holes formation in the inflationary model with field-dependent kinetic term for quartic and natural potentials. *Eur. Phys. J. C* **2021**, *81*, 884. [CrossRef]
195. Teimoori, Z.; Rezazadeh, K.; Rasheed, M.A.; Karami, K. Mechanism of primordial black holes production and secondary gravitational waves in α-attractor Galileon inflationary scenario. *J. Cosmol. Astropart. Phys.* **2021**, *2021*, 018.
196. Ahmed, W.; Junaid, M.; Zubair, U. Primordial black holes and gravitational waves in hybrid inflation with chaotic potentials. *Nucl. Phys. B* **2022**, *984*, 115968. [CrossRef]
197. Kamenshchik, A.Y.; Tronconi, A.; Vardanyan, T.; Venturi, G. Non-Canonical Inflation and Primordial Black Holes Production. *Phys. Lett. B* **2019**, *791*, 201–205. [CrossRef]
198. Horndeski, G.W. Second-order scalar-tensor field equations in a four-dimensional space. *Int. J. Theor. Phys.* **1974**, *10*, 363–384. [CrossRef]
199. Kobayashi, T.; Yamaguchi, M.; Yokoyama, J. Generalized G-inflation: Inflation with the most general second-order field equations. *Prog. Theor. Phys.* **2011**, *126*, 511–529. [CrossRef]
200. Frolovsky, D.; Ketov, S.V.; Saburov, S. Formation of primordial black holes after Starobinsky inflation. *Mod. Phys. Lett. A* **2022**, *37*, 2250135. [CrossRef]
201. Fu, C.; Wu, P.; Yu, H. Primordial Black Holes from Inflation with Nonminimal Derivative Coupling. *Phys. Rev. D* **2019**, *100*, 063532. [CrossRef]

202. Heydari, S.; Karami, K. Primordial black holes in nonminimal derivative coupling inflation with quartic potential and reheating consideration. *Eur. Phys. J. C* **2022**, *82*, 83. [CrossRef]
203. Kawai, S.; Kim, J. Primordial black holes from Gauss-Bonnet-corrected single field inflation. *Phys. Rev. D* **2021**, *104*, 083545. [CrossRef]
204. Langlois, D.; Noui, K. Degenerate higher derivative theories beyond Horndeski: Evading the Ostrogradski instability. *J. Cosmol. Astropart. Phys.* **2016**, *2016*, 034. [CrossRef]
205. Crisostomi, M.; Koyama, K.; Tasinato, G. Extended Scalar-Tensor Theories of Gravity. *J. Cosmol. Astropart. Phys.* **2016**, *2016*, 044. [CrossRef]
206. Ben Achour, J.; Crisostomi, M.; Koyama, K.; Langlois, D.; Noui, K.; Tasinato, G. Degenerate higher order scalar-tensor theories beyond Horndeski up to cubic order. *J. High Energy Phys.* **2016**, *2016*, 100. [CrossRef]
207. Motohashi, H.; Hu, W. Primordial Black Holes and Slow-Roll Violation. *Phys. Rev.* **2017**, *D96*, 063503. [CrossRef]
208. Pi, S.; Wang, J. Primordial Black Hole Formation in Starobinsky's Linear Potential Model. *arXiv* **2022**, arXiv:2209.14183.
209. Inoue, S.; Yokoyama, J. Curvature perturbation at the local extremum of the inflation's potential. *Phys. Lett. B* **2002**, *524*, 15–20. [CrossRef]
210. Tzirakis, K.; Kinney, W.H. Inflation over the hill. *Phys. Rev.* **2007**, *D75*, 123510. [CrossRef]
211. Motohashi, H.; Mukohyama, S.; Oliosi, M. Constant Roll and Primordial Black Holes. *J. Cosmol. Astropart. Phys.* **2020**, *2020*, 002. [CrossRef]
212. Byrnes, C.T.; Cole, P.S.; Patil, S.P. Steepest growth of the power spectrum and primordial black holes. *J. Cosmol. Astropart. Phys.* **2019**, *2019*, 028. [CrossRef]
213. Ragavendra, H.V.; Saha, P.; Sriramkumar, L.; Silk, J. Primordial black holes and secondary gravitational waves from ultraslow roll and punctuated inflation. *Phys. Rev. D* **2021**, *103*, 083510. [CrossRef]
214. Ng, K.W.; Wu, Y.P. Constant-rate inflation: Primordial black holes from conformal weight transitions. *J. High Energy Phys.* **2021**, *2021*, 076. [CrossRef]
215. Karam, A.; Koivunen, N.; Tomberg, E.; Vaskonen, V.; Veermäe, H. Anatomy of single-field inflationary models for primordial black holes. *J. Cosmol. Astropart. Phys.* **2023**, *2023*, 013. [CrossRef]
216. Franciolini, G.; Urbano, A. Primordial black hole dark matter from inflation: The reverse engineering approach. *Phys. Rev. D* **2022**, *106*, 123519. [CrossRef]
217. Cole, P.S.; Gow, A.D.; Byrnes, C.T.; Patil, S.P. Steepest growth re-examined: Repercussions for primordial black hole formation. *arXiv* **2022**, arXiv:2204.07573.
218. Wands, D. Duality invariance of cosmological perturbation spectra. *Phys. Rev.* **1999**, *D60*, 023507. [CrossRef]
219. Hunter, J.D. Matplotlib: A 2D Graphics Environment. *Comput. Sci. Eng.* **2007**, *9*, 90–95. [CrossRef]
220. Harris, C.R.; Millman, K.J.; van der Walt, S.J.; Gommers, R.; Virtanen, P.; Cournapeau, D.; Wieser, E.; Taylor, J.; Berg, S.; Smith, N.J.; et al. Array programming with NumPy. *Nature* **2020**, *585*, 357–362. [CrossRef]
221. Virtanen, P.; Gommers, R.; Oliphant, T.E.; Haberland, M.; Reddy, T.; Cournapeau, D.; Burovski, E.; Peterson, P.; Weckesser, W.; Bright, J.; et al. SciPy 1.0: Fundamental algorithms for scientific computing in Python. *Nat. Methods* **2020**, *17*, 261–272. [CrossRef] [PubMed]
222. McKinney, W. Data Structures for Statistical Computing in Python. In Proceedings of the 9th Python in Science Conference, Austin, TX, USA, 28 June–3 July 2010; pp. 56–61. [CrossRef]
223. Kluyver, T.; Ragan-Kelley, B.; Pérez, F.; Granger, B.; Bussonnier, M.; Frederic, J.; Kelley, K.; Hamrick, J.; Grout, J.; Corlay, S.; et al. Jupyter Notebooks ? A publishing format for reproducible computational workflows. In *Positioning and Power in Academic Publishing: Players, Agents and Agendas*; Loizides, F., Scmidt, B., Eds.; IOS Press: Amsterdam, The Netherlands, 2016; pp. 87–90.
224. Carrilho, P.; Malik, K.A.; Mulryne, D.J. Dissecting the growth of the power spectrum for primordial black holes. *Phys. Rev. D* **2019**, *100*, 103529. [CrossRef]
225. Liu, J.; Guo, Z.K.; Cai, R.G. Analytical approximation of the scalar spectrum in the ultraslow-roll inflationary models. *Phys. Rev. D* **2020**, *101*, 083535. [CrossRef]
226. Tasinato, G. An analytic approach to non-slow-roll inflation. *Phys. Rev. D* **2021**, *103*, 023535. [CrossRef]
227. Özsoy, O.; Tasinato, G. CMB μT cross correlations as a probe of primordial black hole scenarios. *Phys. Rev. D* **2021**, *104*, 043526. [CrossRef]
228. Özsoy, O.; Tasinato, G. Consistency conditions and primordial black holes in single field inflation. *Phys. Rev. D* **2022**, *105*, 023524. [CrossRef]
229. Zegeye, D.; Inomata, K.; Hu, W. Spectral distortion anisotropy from inflation for primordial black holes. *Phys. Rev. D* **2022**, *105*, 103535. [CrossRef]
230. Balaji, S.; Ragavendra, H.V.; Sethi, S.K.; Silk, J.; Sriramkumar, L. Observing Nulling of Primordial Correlations via the 21-cm Signal. *Phys. Rev. Lett.* **2022**, *129*, 261301. [CrossRef]
231. Morse, M.J.P.; Kinney, W.H. Large-η constant-roll inflation is never an attractor. *Phys. Rev.* **2018**, *D97*, 123519. [CrossRef]
232. Suyama, T.; Tada, Y.; Yamaguchi, M. Revisiting non-Gaussianity in non-attractor inflation models in the light of the cosmological soft theorem. *Prog. Theor. Exp. Phys.* **2021**, *2021*, 073E02. [CrossRef]
233. Davies, M.W.; Carrilho, P.; Mulryne, D.J. Non-Gaussianity in inflationary scenarios for primordial black holes. *J. Cosmol. Astropart. Phys.* **2022**, *2022*, 019. [CrossRef]

234. Starobinsky, A.A. Stochastic de Sitter (inflationary) stage in the early universe. *Lect. Notes Phys.* **1986**, *246*, 107–126. [CrossRef]
235. Nambu, Y.; Sasaki, M. Stochastic Stage of an Inflationary Universe Model. *Phys. Lett. B* **1988**, *205*, 441–446. [CrossRef]
236. Kandrup, H.E. Stochastic inflation as a time dependent random walk. *Phys. Rev. D* **1989**, *39*, 2245. [CrossRef]
237. Nambu, Y. Stochastic Dynamics of an Inflationary Model and Initial Distribution of Universes. *Prog. Theor. Phys.* **1989**, *81*, 1037. [CrossRef]
238. Starobinsky, A.A.; Yokoyama, J. Equilibrium state of a selfinteracting scalar field in the De Sitter background. *Phys. Rev.* **1994**, *D50*, 6357–6368. [CrossRef]
239. Finelli, F.; Marozzi, G.; Starobinsky, A.A.; Vacca, G.P.; Venturi, G. Generation of fluctuations during inflation: Comparison of stochastic and field-theoretic approaches. *Phys. Rev. D* **2009**, *79*, 044007. [CrossRef]
240. Burgess, C.P.; Holman, R.; Tasinato, G.; Williams, M. EFT Beyond the Horizon: Stochastic Inflation and How Primordial Quantum Fluctuations Go Classical. *J. High Energy Phys.* **2015**, *2015*, 090. [CrossRef]
241. Vennin, V.; Starobinsky, A.A. Correlation Functions in Stochastic Inflation. *Eur. Phys. J. C* **2015**, *75*, 413. [CrossRef]
242. Burgess, C.P.; Holman, R.; Tasinato, G. Open EFTs, IR effects & late-time resummations: Systematic corrections in stochastic inflation. *J. High Energy Phys.* **2016**, *2016*, 153. [CrossRef]
243. Ando, K.; Vennin, V. Power spectrum in stochastic inflation. *J. Cosmol. Astropart. Phys.* **2021**, *2021*, 057. [CrossRef]
244. Biagetti, M.; Franciolini, G.; Kehagias, A.; Riotto, A. Primordial Black Holes from Inflation and Quantum Diffusion. *J. Cosmol. Astropart. Phys.* **2018**, *2018*, 032. [CrossRef]
245. Ezquiaga, J.M.; García-Bellido, J. Quantum diffusion beyond slow-roll: Implications for primordial black-hole production. *J. Cosmol. Astropart. Phys.* **2018**, *2018*, 018. [CrossRef]
246. Ballesteros, G.; Rey, J.; Taoso, M.; Urbano, A. Stochastic inflationary dynamics beyond slow-roll and consequences for primordial black hole formation. *J. Cosmol. Astropart. Phys.* **2020**, *2020*, 043. [CrossRef]
247. Cruces, D.; Germani, C.; Prokopec, T. Failure of the stochastic approach to inflation beyond slow-roll. *J. Cosmol. Astropart. Phys.* **2019**, *2019*, 048. [CrossRef]
248. Firouzjahi, H.; Nassiri-Rad, A.; Noorbala, M. Stochastic Ultra Slow Roll Inflation. *J. Cosmol. Astropart. Phys.* **2019**, *2019*, 040. [CrossRef]
249. Pattison, C.; Vennin, V.; Assadullahi, H.; Wands, D. Stochastic inflation beyond slow roll. *J. Cosmol. Astropart. Phys.* **2019**, *2019*, 031. [CrossRef]
250. Vennin, V. Stochastic Inflation and Primordial Black Holes. Habilitation Thesis , Universite Paris-Saclay, Bures-sur-Yvette, France, 2020.
251. Rigopoulos, G.; Wilkins, A. Inflation is always semi-classical: Diffusion domination overproduces Primordial Black Holes. *J. Cosmol. Astropart. Phys.* **2021**, *2021*, 027. [CrossRef]
252. Pattison, C.; Vennin, V.; Assadullahi, H.; Wands, D. Quantum diffusion during inflation and primordial black holes. *J. Cosmol. Astropart. Phys.* **2017**, *2017*, 046. [CrossRef]
253. Ezquiaga, J.M.; García-Bellido, J.; Vennin, V. The exponential tail of inflationary fluctuations: Consequences for primordial black holes. *J. Cosmol. Astropart. Phys.* **2020**, *2020*, 029. [CrossRef]
254. Figueroa, D.G.; Raatikainen, S.; Rasanen, S.; Tomberg, E. Non-Gaussian Tail of the Curvature Perturbation in Stochastic Ultraslow-Roll Inflation: Implications for Primordial Black Hole Production. *Phys. Rev. Lett.* **2021**, *127*, 101302. [CrossRef]
255. Figueroa, D.G.; Raatikainen, S.; Rasanen, S.; Tomberg, E. Implications of stochastic effects for primordial black hole production in ultra-slow-roll inflation. *J. Cosmol. Astropart. Phys.* **2022**, *2022*, 027. [CrossRef]
256. Pi, S.; Sasaki, M. Logarithmic Duality of the Curvature Perturbation. *arXiv* **2022**, arXiv:2211.13932.
257. Cai, R.G.; Guo, Z.K.; Liu, J.; Liu, L.; Yang, X.Y. Primordial black holes and gravitational waves from parametric amplification of curvature perturbations. *J. Cosmol. Astropart. Phys.* **2020**, *2020*, 013. [CrossRef]
258. Ballesteros, G.; Rey, J.; Rompineve, F. Detuning primordial black hole dark matter with early matter domination and axion monodromy. *J. Cosmol. Astropart. Phys.* **2020**, *2020*, 014. [CrossRef]
259. Kristiano, J.; Yokoyama, J. Why Must Primordial Non-Gaussianity Be Very Small? *Phys. Rev. Lett.* **2022**, *128*, 061301. [CrossRef]
260. Meng, D.S.; Yuan, C.; Huang, Q.G. One-loop correction to the enhanced curvature perturbation with local-type non-Gaussianity for the formation of primordial black holes. *Phys. Rev. D* **2022**, *106*, 063508. [CrossRef]
261. Kristiano, J.; Yokoyama, J. Ruling Out Primordial Black Hole Formation From Single-Field Inflation. *arXiv* **2022**, arXiv:2211.03395.
262. Inomata, K.; Braglia, M.; Chen, X. Questions on calculation of primordial power spectrum with large spikes: The resonance model case. *J. Cosmol. Astropart. Phys.* **2023**, *2023*, 011. [CrossRef]
263. Riotto, A. The Primordial Black Hole Formation from Single-Field Inflation is Not Ruled Out. *arXiv* **2023**, arXiv:2301.00599.
264. Arya, R. Formation of Primordial Black Holes from Warm Inflation. *J. Cosmol. Astropart. Phys.* **2020**, *2020*, 042. [CrossRef]
265. Bastero-Gil, M.; Díaz-Blanco, M.S. Gravity waves and primordial black holes in scalar warm little inflation. *J. Cosmol. Astropart. Phys.* **2021**, *2021*, 052. [CrossRef]
266. Ballesteros, G.; García, M.A.G.; Rodríguez, A.P.; Pierre, M.; Rey, J. Primordial black holes and gravitational waves from dissipation during inflation. *J. Cosmol. Astropart. Phys.* **2022**, *2022*, 006. [CrossRef]
267. Pi, S.; Sasaki, M. Primordial Black Hole Formation in Non-Minimal Curvaton Scenario. *arXiv* **2021**, arXiv:2112.12680.
268. Meng, D.S.; Yuan, C.; Huang, Q.G. Primordial black holes generated by the non-minimal spectator field. *arXiv* **2022**, arXiv:2212.03577.

269. Wilczek, F. Problem of Strong P and T Invariance in the Presence of Instantons. *Phys. Rev. Lett.* **1978**, *40*, 279–282. [CrossRef]
270. Weinberg, S. A New Light Boson? *Phys. Rev. Lett.* **1978**, *40*, 223–226. [CrossRef]
271. Peccei, R.D.; Quinn, H.R. CP Conservation in the Presence of Instantons. *Phys. Rev. Lett.* **1977**, *38*, 1440–1443. [CrossRef]
272. Peccei, R.D.; Quinn, H.R. Constraints Imposed by CP Conservation in the Presence of Instantons. *Phys. Rev. D* **1977**, *16*, 1791–1797. [CrossRef]
273. Svrcek, P.; Witten, E. Axions In String Theory. *J. High Energy Phys.* **2006**, *2006*, 051. [CrossRef]
274. Marsh, D.J.E. Axion Cosmology. *Phys. Rep.* **2016**, *643*, 1–79. [CrossRef]
275. Freese, K.; Frieman, J.A.; Olinto, A.V. Natural inflation with pseudo—Nambu-Goldstone bosons. *Phys. Rev. Lett.* **1990**, *65*, 3233–3236. [CrossRef] [PubMed]
276. McAllister, L.; Silverstein, E.; Westphal, A. Gravity Waves and Linear Inflation from Axion Monodromy. *Phys. Rev.* **2010**, *D82*, 046003. [CrossRef]
277. Silverstein, E.; Westphal, A. Monodromy in the CMB: Gravity Waves and String Inflation. *Phys. Rev. D* **2008**, *78*, 106003. [CrossRef]
278. McAllister, L.; Silverstein, E.; Westphal, A.; Wrase, T. The Powers of Monodromy. *J. High Energy Phys.* **2014**, *2014*, 123. [CrossRef]
279. Flauger, R.; McAllister, L.; Silverstein, E.; Westphal, A. Drifting Oscillations in Axion Monodromy. *J. Cosmol. Astropart. Phys.* **2017**, *2017*, 055. [CrossRef]
280. Anber, M.M.; Sorbo, L. N-flationary magnetic fields. *J. Cosmol. Astropart. Phys.* **2006**, *2006*, 018. [CrossRef]
281. Anber, M.M.; Sorbo, L. Naturally inflating on steep potentials through electromagnetic dissipation. *Phys. Rev.* **2010**, *D81*, 043534. [CrossRef]
282. Barnaby, N.; Peloso, M. Large Nongaussianity in Axion Inflation. *Phys. Rev. Lett.* **2011**, *106*, 181301. [CrossRef]
283. Barnaby, N.; Namba, R.; Peloso, M. Phenomenology of a Pseudo-Scalar Inflaton: Naturally Large Nongaussianity. *J. Cosmol. Astropart. Phys.* **2011**, *2011*, 009. [CrossRef]
284. Pajer, E.; Peloso, M. A review of Axion Inflation in the era of Planck. *Class. Quant. Grav.* **2013**, *30*, 214002. [CrossRef]
285. Peloso, M.; Sorbo, L.; Unal, C. Rolling axions during inflation: Perturbativity and signatures. *J. Cosmol. Astropart. Phys.* **2016**, *2016*, 001. [CrossRef]
286. Özsoy, O. On Synthetic Gravitational Waves from Multi-field Inflation. *J. Cosmol. Astropart. Phys.* **2018**, *2018*, 062. [CrossRef]
287. Barnaby, N.; Pajer, E.; Peloso, M. Gauge Field Production in Axion Inflation: Consequences for Monodromy, non-Gaussianity in the CMB, and Gravitational Waves at Interferometers. *Phys. Rev.* **2012**, *D85*, 023525. [CrossRef]
288. Meerburg, P.D.; Pajer, E. Observational Constraints on Gauge Field Production in Axion Inflation. *J. Cosmol. Astropart. Phys.* **2013**, *2013*, 017. [CrossRef]
289. Ade, P.A.R.; Aghanim, N.; Arnaud, M.; Arroja, F.; Ashdown, M.; Aumont, J.; Baccigalupi, C.; Ballardini, M.; Banday, A.J.; Barreiro, R.B.; et al. Planck 2015 results. XX. Constraints on inflation. *Astron. Astrophys.* **2016**, *594*, A20.
290. Caravano, A.; Komatsu, E.; Lozanov, K.D.; Weller, J. Lattice Simulations of Axion-U(1) Inflation. *arXiv* **2022**, arXiv:2204.12874.
291. Cheng, S.L.; Lee, W.; Ng, K.W. Numerical study of pseudoscalar inflation with an axion-gauge field coupling. *Phys. Rev. D* **2016**, *93*, 063510. [CrossRef]
292. Dall'Agata, G.; González-Martín, S.; Papageorgiou, A.; Peloso, M. Warm dark energy. *J. Cosmol. Astropart. Phys.* **2020**, *2020*, 032. [CrossRef]
293. Domcke, V.; Guidetti, V.; Welling, Y.; Westphal, A. Resonant backreaction in axion inflation. *J. Cosmol. Astropart. Phys.* **2020**, *2020*, 009. [CrossRef]
294. Gorbar, E.V.; Schmitz, K.; Sobol, O.O.; Vilchinskii, S.I. Gauge-field production during axion inflation in the gradient expansion formalism. *Phys. Rev. D* **2021**, *104*, 123504. [CrossRef]
295. Peloso, M.; Sorbo, L. Instability in axion inflation with strong backreaction from gauge modes. *J. Cosmol. Astropart. Phys.* **2023**, *2023*, 038. [CrossRef]
296. Cheng, S.L.; Lee, W.; Ng, K.W. Production of high stellar-mass primordial black holes in trapped inflation. *J. High Energy Phys.* **2017**, *2017*, 008. [CrossRef]
297. Kallosh, R.; Linde, A.; Vercnocke, B. Natural Inflation in Supergravity and Beyond. *Phys. Rev.* **2014**, *D90*, 041303. [CrossRef]
298. Kobayashi, T.; Oikawa, A.; Otsuka, H. New potentials for string axion inflation. *Phys. Rev.* **2016**, *D93*, 083508. [CrossRef]
299. Cabo Bizet, N.; Loaiza-Brito, O.; Zavala, I. Mirror quintic vacua: Hierarchies and inflation. *J. High Energy Phys.* **2016**, *2016*, 082. [CrossRef]
300. Parameswaran, S.; Tasinato, G.; Zavala, I. Subleading Effects and the Field Range in Axion Inflation. *J. Cosmol. Astropart. Phys.* **2016**, *2016*, 008. [CrossRef]
301. Bhattacharya, S.; Zavala, I. Sharp turns in axion monodromy: Primordial black holes and gravitational waves. *arXiv* **2022**, arXiv:2205.06065.
302. Namba, R.; Peloso, M.; Shiraishi, M.; Sorbo, L.; Unal, C. Scale-dependent gravitational waves from a rolling axion. *J. Cosmol. Astropart. Phys.* **2016**, *2016*, 041. [CrossRef]
303. Ferreira, R.Z.; Sloth, M.S. Universal Constraints on Axions from Inflation. *J. High Energy Phys.* **2014**, *2014*, 139. [CrossRef]
304. Bassett, B.A.; Tsujikawa, S.; Wands, D. Inflation dynamics and reheating. *Rev. Mod. Phys.* **2006**, *78*, 537–589. [CrossRef]
305. Groot Nibbelink, S.; van Tent, B.J.W. Density perturbations arising from multiple field slow roll inflation. *arXiv* **2000**, arXiv:hep-ph/0011325.
306. Groot Nibbelink, S.; van Tent, B.J.W. Scalar perturbations during multiple field slow-roll inflation. *Class. Quant. Grav.* **2002**, *19*, 613–640. [CrossRef]

307. Sasaki, M.; Stewart, E.D. A General analytic formula for the spectral index of the density perturbations produced during inflation. *Prog. Theor. Phys.* **1996**, *95*, 71–78. [CrossRef]
308. Langlois, D.; Renaux-Petel, S. Perturbations in generalized multi-field inflation. *J. Cosmol. Astropart. Phys.* **2008**, *2008*, 017. [CrossRef]
309. Garcia-Saenz, S.; Renaux-Petel, S.; Ronayne, J. Primordial fluctuations and non-Gaussianities in sidetracked inflation. *J. Cosmol. Astropart. Phys.* **2018**, *2018*, 057. [CrossRef]
310. Garcia-Saenz, S.; Renaux-Petel, S. Flattened non-Gaussianities from the effective field theory of inflation with imaginary speed of sound. *J. Cosmol. Astropart. Phys.* **2018**, *2018*, 005. [CrossRef]
311. Bjorkmo, T.; Ferreira, R.Z.; Marsh, M.C.D. Mild Non-Gaussianities under Perturbative Control from Rapid-Turn Inflation Models. *J. Cosmol. Astropart. Phys.* **2019**, *2019*, 036. [CrossRef]
312. Fumagalli, J.; Renaux-Petel, S.; Witkowski, L.T. Oscillations in the stochastic gravitational wave background from sharp features and particle production during inflation. *J. Cosmol. Astropart. Phys.* **2021**, *2021*, 030. [CrossRef]
313. Chluba, J.; Hamann, J.; Patil, S.P. Features and New Physical Scales in Primordial Observables: Theory and Observation. *Int. J. Mod. Phys. D* **2015**, *24*, 1530023. [CrossRef]
314. Slosar, A.; Chen, X.; Dvorkin, C.; Green, D.; Meerburg, P.D.; Silverstein, E.; Wallisch, B. Scratches from the Past: Inflationary Archaeology through Features in the Power Spectrum of Primordial Fluctuations. *Bull. Am. Astron. Soc.* **2019**, *51*, 98.
315. Anguelova, L. On Primordial Black Holes from Rapid Turns in Two-field Models. *J. Cosmol. Astropart. Phys.* **2021**, *2021*, 004. [CrossRef]
316. Fumagalli, J.; Garcia-Saenz, S.; Pinol, L.; Renaux-Petel, S.; Ronayne, J. Hyper-Non-Gaussianities in Inflation with Strongly Nongeodesic Motion. *Phys. Rev. Lett.* **2019**, *123*, 201302. [CrossRef]
317. Ferreira, R.Z. Non-Gaussianities in models of inflation with large and negative entropic masses. *J. Cosmol. Astropart. Phys.* **2020**, *2020*, 034. [CrossRef]
318. Wu, K.K.S.; Lahav, O.; Rees, M.J. The large-scale smoothness of the Universe. *Nature* **1999**, *397*, 225–230. [CrossRef]
319. Yadav, J.; Bharadwaj, S.; Pandey, B.; Seshadri, T.R. Testing homogeneity on large scales in the Sloan Digital Sky Survey Data Release One. *Mon. Not. R. Astron. Soc.* **2005**, *364*, 601–606. [CrossRef]
320. Friedmann, A. On the Possibility of a world with constant negative curvature of space. *Z. Phys.* **1924**, *21*, 326–332. [CrossRef]
321. Lemaitre, G. The expanding universe. *Ann. Soc. Sci. Brux. A* **1933**, *53*, 51–85. [CrossRef]
322. Robertson, H.P. Kinematics and World-Structure. *Astrophys. J.* **1935**, *82*, 284–301. [CrossRef]
323. Walker, A.G. On Milne's Theory of World-Structure. *Proc. Lond. Math. Soc.* **1937**, *42*, 90–127. [CrossRef]
324. Baumann, D. Inflation. In Proceedings of the Theoretical Advanced Study Institute in Elementary Particle Physics: Physics of the Large and the Small, Boulder, CO, USA, 1–26 June 2009; pp. 523–686. [CrossRef]
325. Kolb, E.W.; Turner, M.S. *The Early Universe*; CRC Press: Boca Raton, FL, USA, 1990; Volume 69. [CrossRef]
326. Mukhanov, V. *Physical Foundations of Cosmology*; Cambridge University Press: Oxford, UK, 2005.
327. Lyth, D.H.; Malik, K.A.; Sasaki, M. A General proof of the conservation of the curvature perturbation. *J. Cosmol. Astropart. Phys.* **2005**, *2005*, 004. [CrossRef]
328. Flauger, R.; McAllister, L.; Pajer, E.; Westphal, A.; Xu, G. Oscillations in the CMB from Axion Monodromy Inflation. *J. Cosmol. Astropart. Phys.* **2010**, *2010*, 009. [CrossRef]
329. Flauger, R.; Pajer, E. Resonant Non-Gaussianity. *J. Cosmol. Astropart. Phys.* **2011**, *2011*, 017. [CrossRef]
330. Maldacena, J.M. Non-Gaussian features of primordial fluctuations in single field inflationary models. *J. High Energy Phys.* **2003**, *2003*, 013. [CrossRef]
331. Adshead, P.; Giblin, J.T.; Scully, T.R.; Sfakianakis, E.I. Gauge-preheating and the end of axion inflation. *J. Cosmol. Astropart. Phys.* **2015**, *2015*, 034. [CrossRef]
332. Barnaby, N.; Moxon, J.; Namba, R.; Peloso, M.; Shiu, G.; Zhou, P. Gravity waves and non-Gaussian features from particle production in a sector gravitationally coupled to the inflaton. *Phys. Rev.* **2012**, *D86*, 103508. [CrossRef]
333. Sorbo, L. Parity violation in the Cosmic Microwave Background from a pseudoscalar inflaton. *J. Cosmol. Astropart. Phys.* **2011**, *2011*, 003. [CrossRef]
334. Cook, J.L.; Sorbo, L. Particle production during inflation and gravitational waves detectable by ground-based interferometers. *Phys. Rev. D* **2012**, *85*, 023534. [CrossRef]
335. Özsoy, O. Parity violating non-Gaussianity from axion-gauge field dynamics. *Phys. Rev. D* **2021**, *104*, 123523. [CrossRef]
336. Campeti, P.; Özsoy, O.; Obata, I.; Shiraishi, M. New constraints on axion-gauge field dynamics during inflation from Planck and BICEP/Keck data sets. *J. Cosmol. Astropart. Phys.* **2022**, *2022*, 039. [CrossRef]
337. Malik, K.A.; Wands, D. Cosmological perturbations. *Phys. Rep.* **2009**, *475*, 1–51. [CrossRef]
338. Notari, A.; Tywoniuk, K. Dissipative Axial Inflation. *J. Cosmol. Astropart. Phys.* **2016**, *2016*, 038. [CrossRef]
339. Caprini, C.; Guzzetti, M.C.; Sorbo, L. Inflationary magnetogenesis with added helicity: Constraints from non-gaussianities. *Class. Quantum Gravity* **2018**, *35*, 124003. [CrossRef]

universe

MDPI

Review

Review on Stochastic Approach to Inflation

Diego Cruces [1,2]

[1] Institut de Ciencies del Cosmos (ICCUB), Universitat de Barcelona, Martí i Franquès 1, 08028 Barcelona, Spain; dcruces@ub.edu; Tel.: +34-93-402-0817

[2] Departament de Física Quàntica i Astrofísica, Faculty of Physics, Universitat de Barcelona, Martí i Franquès 1, 08028 Barcelona, Spain

Abstract: We present a review on the state-of-the-art of the mathematical framework known as stochastic inflation, paying special attention to its derivation, and giving references for the readers interested in results coming from the application of the stochastic framework to different inflationary scenarios, especially to those of interest for primordial black hole formation. During the derivation of the stochastic formalism, we will emphasise two aspects in particular: the difference between the separate universe approach and the true long wavelength limit of scalar inhomogeneities and the generically non-Markovian nature of the noises that appear in the stochastic equations.

Keywords: stochastic inflation; primordial black holes; early universe; inflation

1. Introduction

Although cosmological inflation was first introduced as a possible solution to the hot Big Bang model problems [1,2], the study of vacuum fluctuations during this accelerated expansion of the universe also gives an explanation to the anisotropies observed in the Cosmic Microwave Background (CMB). The idea is that the different scales of fluctuations leave the observable universe during inflation, long after it, they reenter the observable universe at different times, being the scales reentering the cosmological horizon (horizon from now on) during the time of recombination the responsible for the CMB anisotropies. In fact, the almost gaussianity and scale invariance of the vacuum quantum fluctuations predicted by Slow Roll (SR) inflation has been remarkably well confirmed by the recent Plank satellite mission [3,4].

It is however true that, although inflation provides a causal mechanism to generate CMB anisotropies, the CMB is only accessible to a restricted range of scales, constraining the inflationary phase only during a limited time interval. In order to be accessible to smaller scales of inflation, that reenter the horizon before the recombination epoch, we must then seek for any other hint that help us to extend the time interval of inflation that we can constraint. Primordial Black Holes (PBHs) is one of those hints [5]. In addition to the fact that PBHs can probe the missing scales of inflation, they represent natural candidates not only for dark matter (DM) [6], but also as the seed of supermassive BHs at the center of massive galaxies [7] and even as the progenitors of some events that radiate the gravitational waves detected by the LIGO/VIRGO collaboration [8].

PBHs are expected to form if the amplitude of density perturbations from inflation is big enough such that when they reenter the horizon they will collapse into a Black Hole (BH). This is not the case for CMB scales, for which the SR approximation holds and where the amplitude of inhomogeneities is too small to form PBHs; however, for smaller scales, this amplitude is less constrained and it could grow until PBHs are possible to form, in this case, SR is not a good approximation and other inflationary regimes arise such as Ultra Slow Roll (USR) and Constant Roll (CR).

Since the over-densities responsible for the formation of PBHs must be large enough, they will be found at the tail of the probability distribution of density perturbations, which makes them exponentially rare [9]. At the same time, a large over-density at small scales can modify the large-scale dynamics of the universe in a non-perturbative way. Thus, in order to predict the abundance of PBHs, a precise statistical knowledge of inflationary perturbations is highly desirable.

The hope of the stochastic approach to inflation is that it incorporates quantum corrections to the inflationary dynamics in a non-perturbative way [10]. In this approach, wavelengths that are well outside the horizon are approximated in powers of spatial gradients rather than on amplitudes (as in linear theory). At the same time though, those modes are influenced by the quantum sector by receiving quantum-kicks from stochastic forces generated by the perturbative sub-horizon modes. The success of the stochastic formalism resides in the fact that it allows to reduce a quantum problem into a statistical one and it has been widely used in the literature [10–68].

Since this manuscript is mainly devoted to stochastic inflation, we will briefly review the long history behind it, which will also help us to understand why the paper is organized as it is.

Soon after Starobisnky first proposed the idea of treating the short-wavelength part of the field as a white noise which is included as a source term in the equation of motion for the classical (long-wavelength) field in [10], stochastic inflation immediately gained huge popularity [11–24], manly because of the possibility of solving the Fokker-Planck equation that govern the evolution of the probability distribution of the amplitude of the perturbations of the scalar field during inflation in an exact way for some specific form of the potential, or even for generic potentials with the help of the first time passage analysis [69] as firstly done in [24]. In the references above, the heuristic derivation of Starobinsky, which was only valid in single-field slow-roll inflationary models, was extended beyond SR and multifield inflation.

At the same time, Morikawa et al. put the stochastic formalism on a more firm ground [25,26], where the heuristic derivation of the equations of motion of stochastic inflation done by previous authors was refined. More concretely, they were able to derive the same stochastic equations integrating out the short-wavelength part of the scalar field in the path integral. Although this derivation lead to the same equations of motion as the heuristic argument followed by previous authors, it is very useful in order to understand what are the approximations that lead to the stochastic equations first presented by Starobinsky and its consequences.

The most important of these approximations is probably to compute the variance of the noises in a pure de-Sitter background. This approach has been vastly used in mort applications of the stochastic formalism until today. Examples of this can be found in [27,29–33,35,37–39,50,54,56], but more importantly, with the aid of this approach, the well-known δN formalism [70–73] was extended to an stochastic δN formalism [36,41,46,47,53,61,62,66] which allow us to compute the probability distribution of the curvature perturbation and hence to connect the results of stochastic inflation with PBH formation as done for example in [52,58,59,63,64,67].

However, computing the noises in a pure de-Sitter background has important consequences, in fact, it was checked in [35,37,40,45,47–49] that, under this approximation, stochastic formalism recovers the standard result of Quantum Field Theory (QFT) for *test* scalar fields on a fixed inflationary background, which give us a hint that thos approach does not correctly take into account the backreaction of the scalar field in the metric. As pointed out in [74], this method misses the coupling between the long- and short-wavelength sectors; more concretely, the background in which the noises should be computed is not exactly de-Sitter, but it is both slow-roll and stochastically corrected. This means that the results obtained with a stochastic formalism that computes the noises in a pure de-Sitter background should not be taken seriously if we are looking for deviations from de-Sitter shuch as in the spectral index. This important limitation of the stochastic formalism as presented by Starobinsky (which will be further studied in Section 6.1.1), led to some authors to present alternative ways of solving the equations of motion of stochastic inflation beyond the usual Fokker-Plank equation with de-Sitter noises, some examples are the recursive stategy presented in [42–44] or numerical codes as in [67].

Another important simplification that is usually performed in the stochastic framework is to split long- and short-wavelengths modes via a sharp window function (a step function). This choice leads to white noises which are very easy to handle both analytically and numerically. However, it was noted in [28] that this choice of window function leads to some problems in the asymptotic of the noise correlator at large spatial distances. This is why some works with smooth window functions that lead to coloured noises have also been proposed [34,68].

If we now go back to the historical controversies of stochastic inflation, we can find the issue of the time variable in which this formalism must be formulated. Already in one of the first works by Starobinsky [24], it was pointed out that the number of e-folds, (defined as $N = \int H dt$, where t is the cosmic time and H is the Hubble parameter) seem to be a very appropriate time variable. This choice was later justified by connecting stochastic inflation with results form QFT in curved space times [35,37,47]; however, as also pointed out by these references, the choice of N as time variable is only a consequence of writing the stochastic equations in terms only of the scalar field. In fact, one can check that the only way to write all the scalar degrees of freedom in terms solely of the scalar field is to use the uniform-N gauge in the separate universe approach [60], which justifies the use of N as time variable since, by definition, in this gauge it is unperturbed. It was not until recently [65] when the stochastic formalism was formulated in another gauge, making it clear that the choice of N as time variable is only a consequence of the gauge choice.

Finally, it was lately noticed that the stochastic formalism of inflation was leading to some inconsistencies at next-to-leading order in SR parameters even at linear level. For example, it was checked in [55] that the linearization of the stochastic equations do not exactly reproduce the well-known equation of motion of the scalar linear perturbations in the long-wavelength limit (the Mukhanov-Sasaki equation, see Section 4.1). Furthermore, in [57] it was demonstrated that, during an ultra-slow-roll (USR) inflationary regime, some noises that appear in the stochastic equations of motion were incompatible with the rest of the system. The origin of these inconsistencies can be traced back to the use of the separate universe approach in the construction of the stochastic formalism and they can be solved by including the momentum constraint in the separate universe approach [65].

This review will give a detailed derivation of the stochastic formalism to inflation rather than giving a list of results obtained within this formalism. During the derivation we will emphasise all controversial aspects that have been indicated in the above summary of the long history of the stochastic formalism, from the difference between the separate universe approach and the true long-wavelength limit of scalar inhomogeneities to the construction of the noises using an exact de-Sitter background.

The review is organized as follows: after presenting the basic inflationary concepts with the aid of the homogeneous picture of inflation in Section 2, we will start studying inflationary inhomogeneities, focusing on its long-wavelength behaviour. In order to do so we will present different approximations to the exact Arnowitt-Deser-Misner (ADM) equations of Section 3. These approximations are linear perturbation theory (in Section 4), gradient expansion (in Section 5) and the stochastic formalism (in Section 6), which combines the two approximation schemes presented before. Finally, in Section 7 we will give some conclusions.

2. Homogeneous Inflationary Scenarios and Slow-Roll Parameters

As we have already mentioned, PBHs represent natural candidates for dark matter (DM) (latest constraints on this idea can be found in [75]). However, one possibility to statistically generate enough PBHs for this to hold one needs, at least, a power spectrum of primordial curvature perturbations several order of magnitudes larger than the one observed in the cosmic microwave background (CMB). Note that this is not the only possibility; one could statistically generate enough PBHs with a large non-gaussianity in the probability distribution of fluctuations even without enhancing the power spectrum. This scenario is discussed in [76].

It is known that a period of Slow-Roll (SR), of which the predictions of the CMB are based upon, cannot lead to the appropriate power spectrum necessary to generate enough PBHs to match the DM abundance [77] (For the non-linear relation between the inflationary power spectrum and PBHS abundance, under the assumption of gaussianity, the interested reader can see [78]). Thus, one necessarily needs an inflationary epoch evolving beyond SR. A possibility is the introduction of an inflection point in the inflationary potential [79]. This leads the inflaton to undergo a so-called Ultra-Slow-Roll (USR) phase of inflation [80,81]. Taking into account that the statistics of PBHs from non-gaussian fluctuations has yet to be fully developed. The single field USR option with standard kinetic term seems then to be the best [76]. Such a system is described by the Einstein-Hilbert action with a minimally coupled scalar field i.e.,:

$$S = \frac{1}{2} \int \sqrt{-g} \Big[M_{PL}^2 R - (\nabla\phi)^2 - 2V(\phi) \Big] , \tag{1}$$

whose homogeneous solution is an universe described by a Friedman-Lemaitre-Robertson-Walker (FLRW) metric:

$$ds^2 = -dt^2 + a(t)^2 d\vec{x} \cdot d\vec{x} , \tag{2}$$

where $a(t)$ represents the scale factor.

The equation of motion of the scalar field in the universe described by (2) has the following equation of motion:

$$\ddot{\phi}^b + 3H^b \dot{\phi}^b + V_{\phi^b}\left(\phi^b\right) = 0 , \tag{3}$$

where $V_{\phi^b} \equiv \frac{\partial V(\phi^b)}{\partial \phi^b}$, $H^b \equiv \frac{\dot{a}}{a}$ is the Hubble parameter, and a dot denotes a derivative with respect to the cosmic time t. Finally, the super-script "b" stands for "background", the meaning of which will be clarified later on.

The Friedmann equation is

$$\left(H^b\right)^2 = \frac{1}{3M_{PL}^2} \left(\frac{\left(\dot{\phi}^b\right)^2}{2} + V\left(\phi^b\right) \right) . \tag{4}$$

The Slow-Roll (SR) parameters ϵ_i define the rate of change of the Hubble parameter:

$$\epsilon_1 \equiv -\frac{\dot{H}^b}{\left(H^b\right)^2} = \frac{\left(\dot{\phi}^b\right)^2}{2H^2 M_{PL}^2} \ll 1 ; \qquad \epsilon_{i+1} \equiv \frac{\dot{\epsilon}_i}{H\epsilon_i} \quad \text{with} \quad i \geq 1, \tag{5}$$

where, to write the final expressions, we have used the Friedmann equation and the equation of motion of the field.

We can now define different inflationary regimes depending on the values of the ϵ_is:

- Slow-Roll (SR): The field is slowly rolling down a potential with an almost constant velocity, which makes the acceleration negligible. In this case the equation of motion (3) is approximately

$$3H^b \dot{\phi}^b + V_\phi\left(\phi^b\right) \simeq 0 . \tag{6}$$

The SR parameters are much smaller than one ($\epsilon_i \ll 1$) and can be written in terms of the potential as

$$\epsilon_1^{SR} \simeq \frac{1}{2M_{PL}^2} \left(\frac{V_{\phi^b}}{V} \right)^2 ; \qquad \epsilon_2^{SR} \simeq \frac{2}{M_{PL}^2} \left(\frac{V_{\phi^b \phi^b}}{V} \right) - 4\epsilon_1^{SR} . \tag{7}$$

- Ultra-Slow-Roll (USR): The field is moving along an exactly flat potential $(V_\phi = 0)$, which makes the acceleration relevant. In this case the equation of motion (3) is

$$\ddot{\phi}^b + 3H^b\dot{\phi}^b = 0\,. \tag{8}$$

From (8) one can infer that the velocity of the field (and hence ϵ_1) exponentially decreases, which makes some $\epsilon_i \sim \mathcal{O}(1)$. More precisely:

$$\begin{aligned} \epsilon_i^{USR} &= -6 + 2\epsilon_1^{USR} && \text{when i even.} \\ \epsilon_i^{USR} &= 2\epsilon_1^{USR} && \text{when i>1 and odd.} \end{aligned} \tag{9}$$

As we will show later on, an exponential decrease of ϵ_1 makes the power spectrum of curvature perturbation increase.

- Both SR and USR are, at least approximately, sub-cases of Constant-Roll (CR). Here $\frac{\ddot{\phi}^b}{H^b\dot{\phi}^b} = \kappa$ where κ is a constant. SR is realized when $\kappa = 0$ while USR when $\kappa = -3$. We will not analyse further this generic case.

It is important to remark that, given a potential related to PBH formation, the SR and USR phases alternate. Thus, the equations of motion (6) and (8) will always be an approximation of the system.

3. Inhomogeneities during Inflation: The ADM Formalism

Although it is very useful in order to understand the inflationary dynamics, the homogeneous picture of inflation presented in Section 2 is not very realistic and some inhomogeneities must be introduced in order to explain the universe we currently observe. Thus, we must define a general metric which is not restricted to homogeneity and isotropy. An option which is very useful in the context of inflation is to work in the so-called Arnowitt-Deser-Misner (ADM) formalism [82]. This framework supposes that the four-dimensional space-time is foliated into a family of three-dimensional hypersurfaces Σ_t, labeled by its time coordinate t.

In order to work in the ADM formalism, it is convenient to write the metric as:

$$ds^2 = g_{\mu\nu}dx^\mu dx^\nu = -\alpha^2 dt^2 + \gamma_{ij}(dx^i + \beta^i dt)(dx^j + \beta^j dt)\,, \tag{10}$$

and the action (1) presented in the previous section becomes:

$$S = \frac{1}{2}\int \sqrt{\gamma}\Big[M_{PL}^2\Big(\alpha R^{(3)} + \alpha\Big(K_{ij}K^{ij} - K^2\Big)\Big) - 2\alpha V(\phi) + \alpha^{-1}\Big(\dot{\phi} - \beta^i\partial_i\phi - \alpha\gamma^{ij}\partial_i\phi\partial_j\phi\Big)\Big]\,, \tag{11}$$

where we have introduced many new terms:

- α is the lapse function, which measures the rate of flow of proper time with respect to coordinate time t as one moves normally to Σ_t.
- β^i is the shift vector, which measures how much the local spatial coordinate system shifts tangential to Σ_t, when moving from Σ_t to $\Sigma_{t+\delta t}$ along the normal direction to Σ_t.
- γ_{ij} represents the induced metric on the hypersurface Σ_t, that we we will decompose as $\gamma^{ij} = a^2 e^{2\zeta}\tilde{\gamma}^{ij}$ with $\det\tilde{\gamma}^{ij} = 1$ so that the scale factor a is explicitely present.
- $R_{ij}^{(3)}$ is the Ricci tensor of the spatial metric, and hence $R^{(3)} \equiv \gamma^{ij}R_{ij}^{(3)}$.
- Finally, K_{ij} is the extrinsic curvature ($K \equiv \gamma^{ij}K_{ij}$), which is defined as:

$$K_{ij} \equiv -\nabla_i n_j = -\frac{1}{2\alpha}\left(\partial_t\gamma_{ij} - D_i\beta_j - D_j\beta_i\right)\,, \tag{12}$$

where $n_i \equiv (-\alpha, 0, 0, 0)$ is the unit vector normal to the spatial hypersurfaces and ∇_μ and D_i are the covariant derivatives with respect to $g_{\mu\nu}$ and γ_{ij}, respectively.

It is also convenient to decompose the extrinsic curvature into its trace and traceless part as:

$$K_{ij} = \frac{\gamma_{ij}}{3}K + a^2 e^{2\zeta}\tilde{A}_{ij}\,, \tag{13}$$

where $\tilde{\gamma}^{ij}\tilde{A}_{ij} = 0$. Note that,in the homogeneous limit, i.e., when $\alpha = 1$, $\beta^i = 0$ and $\gamma^{ij} = a^2\delta_{ij}$ and hence the metric (10) reduces to (2), we can identify the extrinsic curvature with the background Hubble parameter, namely $K = -3\frac{\dot{a}}{a} = -3H^b$. We will then define a more general inhomogeneous Hubble parameter as $H \equiv -\frac{K}{3}$. This makes sense because K represents the expansion rate of the constant time hypersurfaces.

In the ADM formalism, the lapse function and the shift vector act as Lagrange multipliers for the action (11), and hence they generate two constraints: the Hamiltonian and the momentum constraints, which are, respectively:

$$R^{(3)} - \tilde{A}_{ij}\tilde{A}^{ij} + \frac{2}{3}K^2 = \frac{2}{M_{PL}^2}E\,, \tag{14}$$

$$D^j\tilde{A}_{ij} - \frac{2}{3}D_i K = \frac{1}{M_{PL}^2}J_i\,, \tag{15}$$

where $E \equiv T_{\mu\nu}n^\mu n^\nu$ and $J_i \equiv T_{\mu j}n^\mu \gamma_i^j$ and $T_{\mu\nu}$ is, in our case, the stress-energy tensor of the scalar field, i.e.,

$$T_{\mu\nu} = \nabla_\mu\phi\nabla_\nu\phi - \frac{1}{2}g_{\mu\nu}(\nabla^\alpha\phi\nabla_\alpha\phi + 2V(\phi))\,. \tag{16}$$

The two remaining variables, γ_{ij} and K_{ij} are the dynamical ones and their evolution equations are given by:

- For γ_{ij}:

$$(\partial_t - \beta^k\partial_k)\zeta + \frac{\dot{a}}{a} = -\frac{1}{3}(\alpha K - \partial_k\beta^k), \tag{17}$$

$$(\partial_t - \beta^k\partial_k)\tilde{\gamma}_{ij} = -2\alpha\tilde{A}_{ij} + \tilde{\gamma}_{ik}\partial_j\beta^k + \tilde{\gamma}_{jk}\partial_i\beta^k - \frac{2}{3}\tilde{\gamma}_{ij}\partial_k\beta^k. \tag{18}$$

- for K_{ij}:

$$(\partial_t - \beta^k\partial_k)K = \alpha\left(\tilde{A}_{ij}\tilde{A}^{ij} + \frac{1}{3}K^2\right) - D_k D^k\alpha + 4\pi G\alpha(E + S_k^k), \tag{19}$$

$$\begin{aligned}(\partial_t - \beta^k\partial_k)\tilde{A}_{ij} = {} & \frac{e^{-2\zeta}}{a^2}\left[\alpha\left(R_{ij}^{(3)} - \frac{\gamma_{ij}}{3}R^{(3)}\right) - \left(D_i D_j\alpha - \frac{\gamma_{ij}}{3}D_k D^k\alpha\right)\right] \\ & + \alpha(K\tilde{A}_{ij} - 2\tilde{A}_{ik}\tilde{A}_j^k) + \tilde{A}_{ik}\partial_j\beta^k + \tilde{A}_{jk}\partial_i\beta^k - \frac{2}{3}\tilde{A}_{ij}\partial_k\beta^k \\ & - \frac{8\pi G\alpha e^{-2\zeta}}{a^2}\left(S_{ij} - \frac{\gamma_{ij}}{3}S_k^k\right),\end{aligned} \tag{20}$$

where $S_{ij} = T_{ij}$ and $S_k^k = \gamma^{kl}S_{lk}$.

Finally, and although it can be recovered using the ADM equations just presented, it is also worthy to present here the Klein-Gordon equation for the evolution of the scalar field:

$$\frac{1}{\sqrt{-g}}\partial_\mu[\sqrt{-g}g^{\mu\nu}\partial_\nu\phi] - V_\phi = 0\,. \tag{21}$$

We will present more detail in the following section, but it is worth remarking here that the homogeneous equations of Section 2 are straightforwardly recovered setting $\alpha = 1$, $\beta^i = 0$ and $\gamma_{ij} = a^2\delta_{ij}$.

Since the ADM equations are equivalent to the Einstein equations but much easier to implement numerically, they are very useful to study inhomogeneous space-times in an exact way; however, the numerical methods capable of solving exactly these equations are computationally expensive [83]. Fortunately there exist some very useful approximations that can be done when studying the inflationary universe that, not only allow us to considerably simplify the numerical way of solving the ADM equations, but they even admit some analytical solutions as we will see in the following.

4. Linear Perturbation Theory

Since, as explained in the introduction, the deviations from an exactly homogeneous and isotropic FLRW universe that we observe are very small, it makes sense to solve the system of ADM Equations (14)–(21) by expanding them around a FLRW background. If this expansion is done up to first order, we call it linear perturbation theory. There are many reviews on this topic [84–88]; however, since we are taking a slightly different point of view from most of them, we will go through linear perturbation theory in some detail. With this in mind, the ADM metric (10) will be written as:

$$g_{\mu\nu} \simeq g^b_{\mu\nu} + \delta g_{\mu\nu}\,, \tag{22}$$

where $g^b_{\mu\nu}$ is the homogeneous and isotropic metric of Section 2 and $\delta g_{\mu\nu}$ represent the perturbation. This implies that the lapse function, the shift vector and the spatial metric are approximated as:

$$\begin{aligned} \alpha &\simeq 1 + A\,, \\ \beta_j &\simeq a B_j\,, \\ e^{2\zeta} &\simeq 1 + 2D\,, \\ \tilde{\gamma}_{ij} &\simeq \delta_{ij} - 2E_{ij} \end{aligned} \tag{23}$$

where E_{ij} is traceless. The reason why E_{ij} is traceless is because is the perturbation of $\tilde{\gamma}_{ij}$, which has unit determinant. Precisely, any matrix with unit determinant can be written as:

$$\tilde{\gamma}_{ij} = e^{-2M_{ij}}\,,$$

where M_{ij} is traceless. Note that the last two linearizations in (23) leads to

$$\gamma_{ij} \simeq a^2\left[(1+2D)\delta_{ij} - 2E_{ij}\right]. \tag{24}$$

Finally, the scalar field responsible for inflation must also be linearized, i.e.,

$$\phi \simeq \phi^b + \delta\phi\,. \tag{25}$$

It can be easily shown that, of the linear variables introduced above, A, D and $\delta\phi$ transform as scalars under rotations in the background space-time coordinates, B_i as a 3-vector and E_{ij} as a 3D-tensor. This does not mean that the only scalar components are A, D and $\delta\phi$. In fact, we know from Euclidean 3D vector calculus that a vector can be decomposed as:

$$B_i = B_i^S + B_i^V \qquad \text{with} \qquad \partial_i B_j^S - \partial_j B_i^S = 0 \qquad \text{and} \qquad \partial^i B_i^V = 0\,, \tag{26}$$

and hence

$$B_i^S = \partial_i B\,, \tag{27}$$

where B is some scalar field.

Similarly, for a tensor field we have

$$E_{ij} = E_{ij}^S + E_{ij}^V + E_{ij}^T\,, \tag{28}$$

where

$$
\begin{aligned}
E^S_{ij} &= \left(\partial_i\partial_j - \frac{1}{3}\delta_{ij}\nabla^2\right)E\,, \\
E^V_{ij} &= \frac{1}{2}(\partial_j E_i + \partial_i E_j) \qquad \text{with} \qquad \partial^i E_i = 0\,, \\
\partial^i E^T_{ij} &= 0\,, \qquad (29)\\
\delta^{ij} E^T_{ij} &= 0\,, \qquad (30)
\end{aligned}
$$

where E is again a scalar field.

The procedure explained above allow us to decompose the perturbations into a scalar, vector and tensor sector. These sectors have the characteristic of evolve independently one from each other at linear order in perturbation theory, which make them easier to handle. During this review we will be mostly focused on scalar perturbations of the metric since they are the ones that couple to the scalar field perturbation $\delta\phi$. This allows us to write the perturbed metric of (23) as:

$$
ds^2 = -(1+2A)dt^2 + 2a\partial_i B dx^i dt + a^2\left[(1+2D)\delta_{ij} - 2\left(\partial_i\partial_j - \frac{1}{3}\delta_{ij}\nabla^2\right)E\right]dx^i dx^j\,. \qquad (31)
$$

Similarly to what we have just done with the metric, we can split each one of the ADM equations presented in Section 3 into an homogeneous and a perturbed part as follows:

- Hamiltonian constraint (14).

$$
\begin{aligned}
0 =& R^{(3)} - \tilde{A}_{ij}\tilde{A}^{ij} + \frac{2}{3}K^2 - \frac{2}{M_{PL}^2}E \\
\simeq& \left[6\left(H^b\right)^2 - \frac{2}{M_{PL}^2}\left(\frac{\left(\dot{\phi}^b\right)^2}{2} + V(\phi^b)\right)\right] \\
&+\left[-12H^b\left(H^b A - \dot{D} + \frac{1}{3}\frac{\nabla^2}{a}B\right) - 4\frac{\nabla^2}{a^2}\left(D + \frac{1}{3}\nabla^2 E\right) - \frac{2}{M_{PL}^2}\left(\dot{\phi}^b\dot{\delta\phi} - \left(\dot{\phi}^b\right)^2 A + V_{\phi^b}\delta\phi\right)\right]. \qquad (32)
\end{aligned}
$$

Note that in (32) the term inside the first square brackets corresponds to the background Hamiltonian constraint (4) of Section 2, being the term inside the second square brackets the linearization of the Hamiltonian constraint. We will follow this same notation for the rest of the ADM equations.

- Momentum constraint (15).

$$
\begin{aligned}
0 &= D^j\tilde{A}_{ij} - \frac{2}{3}D_i K - \frac{1}{M_{PL}^2}J_i \\
&\simeq [0] \\
&+ \left[\partial_i\left(-2H^b A + 2\dot{D} + \frac{2}{3}\nabla^2\dot{E} + \frac{1}{M_{PL}^2}\dot{\phi}^b\delta\phi\right)\right]. \qquad (33)
\end{aligned}
$$

As it can be seen in (33), the momentum constraint do not have a contribution at the background level, which is logical due to the presence of spatial derivatives, which cannot appear in an exactly homogeneous and isotropic global background. However, it does have an homogeneous contribution in the perturbative sector due to the presence of a total spatial derivative, this is of crucial importance for the rest of the review.

- Evolution equation for ζ (17).

$$
\begin{aligned}
0 = (\partial_t - \beta^k\partial_k)\zeta + \frac{\dot{a}}{a} &= -\frac{1}{3}(\alpha K - \partial_k\beta^k) \\
&\simeq \left[-H^b + \frac{\dot{a}}{a}\right] \\
&+ \left[\dot{D} - H^b A - \delta H - \frac{1}{3}\frac{\nabla^2}{a}B\right],
\end{aligned} \tag{34}
$$

where we have used the identification $K \equiv -3H$ explained below Equation (13). From (34) we can easily identify the perturbation of the Hubble parameter, i.e.,

$$
H = H^b + \delta H = H^b + \left[-H^b A + \dot{D} - \frac{1}{3}\frac{\nabla^2}{a}B\right]. \tag{35}
$$

- Evolution equation for $\tilde{A}_{ij}$ (18).

$$
\begin{aligned}
0 = (\partial_t - \beta^k\partial_k)\tilde{\gamma}_{ij} &= -2\alpha\tilde{A}_{ij} + \tilde{\gamma}_{ik}\partial_j\beta^k + \tilde{\gamma}_{jk}\partial_i\beta^k - \frac{2}{3}\tilde{\gamma}_{ij}\partial_k\beta^k \\
&\simeq [0] \\
&+ \left[2\tilde{A}_{ij} - 2\left(\partial_i\partial_j - \frac{1}{3}\delta_{ij}\nabla^2\right)\left(\frac{B}{a} + \dot{E}\right)\right].
\end{aligned} \tag{36}
$$

- Evolution equation for K (19).

$$
\begin{aligned}
0 &= (\partial_t - \beta^k\partial_k)K = \alpha\left(\tilde{A}_{ij}\tilde{A}^{ij} + \frac{1}{3}K^2\right) - D_k D^k\alpha + 4\pi G\alpha(E + S^k_k) \\
&\simeq \left[-3\dot{H}^b - 3\left(H^b\right)^2 - \frac{1}{M_{PL}^2}\left(\left(\dot{\phi}^b\right)^2 - V(\phi^b)\right)\right] \\
&+ \Bigg[6\dot{H}^b A + 3H^b\dot{A} - 3\ddot{D} + \frac{\nabla^2}{a}\dot{B} + \frac{\nabla^2}{a^2}A + 6\left(H^b\right)^2 A - 6H^b\dot{D} + H^b\frac{\nabla^2}{a}B \\
&- \frac{1}{M_{PL}^2}\left(2\dot{\phi}^b\dot{\delta\phi} - 2\left(\dot{\phi}^b\right)^2 A - V_{\phi^b}\delta\phi\right)\Bigg].
\end{aligned} \tag{37}
$$

- Evolution equation for $\tilde{A}_{ij}$ (20).

$$
\begin{aligned}
0 &= (\partial_t - \beta^k\partial_k)\tilde{A}_{ij} - \frac{e^{-2\zeta}}{a^2}\left[\alpha\left(R^{(3)}_{ij} - \frac{\gamma_{ij}}{3}R^{(3)}\right) - \left(D_i D_j\alpha - \frac{\gamma_{ij}}{3}D_k D^k\alpha\right)\right] \\
&- \alpha(K\tilde{A}_{ij} - 2\tilde{A}_{ik}\tilde{A}^k_j) - \tilde{A}_{ik}\partial_j\beta^k - \tilde{A}_{jk}\partial_i\beta^k + \frac{2}{3}\tilde{A}_{ij}\partial_k\beta^k + \frac{8\pi G\alpha e^{-2\zeta}}{a^2}\left(S_{ij} - \frac{\gamma_{ij}}{3}S^k_k\right) \\
&\simeq [0] \\
&+ \left[\left(\partial_i\partial_j - \frac{1}{3}\delta_{ij}\nabla^2\right)\left(\frac{\dot{B}}{a} + 2\frac{H^b B}{a} + \ddot{E} + 3H^b\dot{E} + A + D + \frac{1}{3}\nabla^2 E\right)\right].
\end{aligned} \tag{38}
$$

- Evolution equation for the scalar field (21).

$$
\begin{aligned}
0 &= \frac{1}{\sqrt{-g}}\partial_\mu\left[\sqrt{-g}g^{\mu\nu}\partial_\nu\phi\right] - V_\phi \\
&\simeq \left[\ddot{\phi}^b + 3H^b\dot{\phi}^b + V_{\phi^b}(\phi^b)\right] \\
&+ \left[\ddot{\delta\phi} + 3H^b\dot{\delta\phi} + V_{\phi^b\phi^b}\delta\phi - \frac{\nabla^2}{a^2}\delta\phi + 2V_{\phi^b}A - \dot{\phi}^b\left(\dot{A} - 3\dot{D} + \frac{\nabla^2}{a}B\right)\right].
\end{aligned} \tag{39}
$$

Once we have seen how linear perturbation theory works and what are the linear equations that describe small inhomogeneities during an inflationary epoch, it is important to have a physical intuition about what a perturbation of the metric really means. By definition, a perturbation is the difference between the value of a quantity in the real and inhomogeneous space-time and its value on the idealized FLRW background. This seems trivial; however, in order to make such a comparison, it is necessary to compute these two values at the same space-time point. Since the quantities to compare live in different space-times, we require a pointwise correspondence between them, which is given by a coordinate system x^μ such that the point P^b in the background space-time and the point P in the perturbed space-time, which have the same coordinate values, correspond to each other.

The freedom in the choice among these coordinate systems is called the gauge choice. The way the different gauges are related in linear perturbation theory (for gauge transformations beyond linear perturbation theory see for example [89]) is via an infinitesimal gauge transformation of the coordinates:

$$\tilde{x}^\mu = x^\mu + \delta x^\mu \tag{40}$$

We can split the vector δx^μ into its time an space components $\delta x^\mu = (\lambda^0, \lambda^i)$, and following the same idea as when we decomposed the perturbations in the metric, λ^i can be written as $\lambda^i = \lambda^i_\perp + \partial^i \eta$, where $\lambda^i_\perp$ is a 3-dimensional divergenless vector and η is a scalar function. In terms of these functions, we can impose the gauge invariance of the perturbed metric (31) to deduce the gauge transformation of each one of the scalar variables in the metric:

$$\begin{aligned} D \to \tilde{D} &= D + aH^b\lambda^0 + \frac{1}{3}\nabla^2\eta\,, \\ A \to \tilde{A} &= A + aH^b\lambda^0 + a\dot{\lambda}^0\,, \\ E \to \tilde{E} &= E - \eta\,, \\ B \to \tilde{B} &= B + a\dot{\eta} - \lambda^0\,. \end{aligned} \tag{41}$$

Finally, the scalar field perturbation will transform as:

$$\delta\phi \to \tilde{\delta\phi} + a\dot{\phi}\lambda^0\,. \tag{42}$$

From (41) and (42) we can clearly see that the freedom on the choice of the gauge allow us to set two out of the five scalar perturbations to zero by choosing η and λ^0 accordingly. This reduces the scalar degrees of freedom to three (which further reduces to two when using the ADM equations for a single scalar field), which can be written in terms of gauge invariant, and hence physical, variables: the two Bardeen potentials [90]

$$\begin{aligned} \Psi &\equiv -D - \frac{1}{3}\nabla^2 E - aH^b(B + a\dot{E})\,, \\ \Phi &\equiv A + aH^b(B + a\dot{E}) + a\frac{d}{dt}(B + a\dot{E})\,, \end{aligned} \tag{43}$$

and the Mukhanov-Sasaki (MS) variable

$$Q \equiv \delta\phi - \frac{\dot{\phi}^b}{H^b}\left(D + \frac{1}{3}\nabla^2 E\right). \tag{44}$$

During this review we will pay special attention to the MS variable. In fact, it can be shown that, by rearranging the linearized ADM equations of (32)–(39), the MS variable follows a simple equation of motion which, written in terms of the SR parameters defined in Section 2, is:

$$\ddot{Q} + 3H^b\dot{Q} + \left[-\frac{\nabla^2}{a^2} + \left(H^b\right)^2\left(-\frac{3}{2}\epsilon_2 + \frac{1}{2}\epsilon_1\epsilon_2 - \frac{1}{4}\epsilon_2^2 - \frac{1}{2}\epsilon_2\epsilon_3\right)\right]Q = 0\,. \tag{45}$$

In order to solve (45), we must take into account that subhorizon scales during inflation are microscopic, and hence we must study the behavior of the inflationary scalar field using quantum field theory (QFT). The way to proceed is similar to what we do when quantizing the harmonic oscillator: $\hat{Q}(\mathbf{x},t)$, which is now a quantum operator, can be expressed in Fourier space as:

$$\hat{Q}(\mathbf{x},t) = \int \frac{d\mathbf{k}}{(2\pi)^{3/2}}\hat{\mathcal{Q}}_{\mathbf{k}}(t)\,, \tag{46}$$

where

$$\hat{\mathcal{Q}}_{\mathbf{k}}(t) = e^{-i\mathbf{k}\cdot\mathbf{x}}Q_{\mathbf{k}}(t)a_{\mathbf{k}} + e^{-i\mathbf{k}\cdot\mathbf{x}}Q^*_{\mathbf{k}}(t)a^\dagger_{\mathbf{k}}\,, \tag{47}$$

and $a_{\mathbf{k}}$ and $a^\dagger_{\mathbf{k}}$ are the usual creation and annihilation operators, that satisfy the usual commutation relation:

$$[a_{\mathbf{k}}, a^\dagger_{\mathbf{k}'}] = \delta^{(3)}(\mathbf{k}-\mathbf{k}') \tag{48}$$

With this construction, the variable $Q_{\mathbf{k}}$ of (47) is the solution of the MS Equation (45) in Fourier space, i.e.,:

$$\ddot{Q}_{\mathbf{k}} + 3H^b\dot{Q}_{\mathbf{k}} + \left[\frac{k^2}{a^2} + \left(H^b\right)^2\left(-\frac{3}{2}\epsilon_2 + \frac{1}{2}\epsilon_1\epsilon_2 - \frac{1}{4}\epsilon_2^2 - \frac{1}{2}\epsilon_2\epsilon_3\right)\right]Q_{\mathbf{k}} = 0 \tag{49}$$

If we impose the Bunch-Davies vacuum [91] to be recovered at early time, we can only write analitically a solution for (49) if a specific condition is satisfied. In order to see what is this condition we will define the variable z as $z = a\frac{\dot{\phi}^b}{H^b}$ and the conformal time τ as $ad\tau = dt$, with this, the analytical solution of (49) in terms of the conformal time only exists if

$$\nu^2 \equiv \frac{1}{4} + \tau^2\frac{1}{z}\frac{d^2z}{d\tau^2} \tag{50}$$

is a constant. It is easy to check that this is the case generically; in fact we can integrate by parts $\tau = \int\frac{dt}{a}$ to get:

$$\tau \simeq -\frac{1}{aH^b}(1+\mathcal{O}(\epsilon_1))\,.$$

See for example Appendix E of [65] for the derivation. The equation above, together with the fact that

$$\frac{1}{z}\frac{d^2z}{d\tau^2} = a^2\left(H^b\right)^2\left(2 - \epsilon_1 + \frac{3}{2}\epsilon_2 - \frac{1}{2}\epsilon_1\epsilon_2 + \frac{1}{4}\epsilon_2^2 + \frac{1}{2}\epsilon_2\epsilon_3\right),$$

makes ν defined in (50) to be a constant generically up to $\mathcal{O}(\epsilon_1)$. Note that, since in SR, ϵ_i is also a constant up to $\mathcal{O}(\epsilon_1)$, in this case we have that ν^{SR} is constant up to $\mathcal{O}\left(\left(\epsilon_i^{SR}\right)^2\right)$. The analytical solution in this case is:

$$Q_{\mathbf{k}} = e^{\frac{i}{2}\pi\left(\nu+\frac{1}{2}\right)}\frac{\sqrt{\pi}}{2a}\sqrt{-\tau}H_\nu^{(1)}(-k\tau)\,, \tag{51}$$

where $H_\nu^{(1)}$ is the Hankel function of first class.

In order to finish this reminder of linear perturbation theory, let us introduce a very useful quantity that characterizes the properties of the perturbations: the power spectrum. We will give a physical interpretation of the power spectrum later on and for the moment we will focus only on its mathematical definition. The power spectrum of a generic quantity $X(\mathbf{x},t)$ is defined as the Fourier transform of the two-point correlation function, i.e.:

$$\langle 0|X(\mathbf{x}_1,t)X(\mathbf{x}_2,t)|0\rangle = \int \frac{d\mathbf{k}}{(2\pi)^3}|X_{\mathbf{k}}(t)|^2 \equiv \int \frac{dk}{k}\mathcal{P}_X(k,t)\frac{\sin(kr)}{kr}\,, \tag{52}$$

where $r = |\mathbf{x}_1 - \mathbf{x}_2|$ and $k = |\mathbf{k}|$. From (52) it is clear that the power spectrum is defined as:

$$\mathcal{P}_X(k,t) \equiv \frac{k^3}{2\pi^2}|X_{\mathbf{k}}|^2\,. \tag{53}$$

Another interesting quantity derived from the power spectrum is the spectral index $n_s^X - 1$, which is defined as:

$$n_s^X - 1 \equiv \frac{d\log\mathcal{P}_X}{d\log k}\,. \tag{54}$$

4.1. Long Wavelength Limit of Linear Perturbation Theory

As explained in the introduction, many quantities of interest such as the power spectrum of the scalar tilt are usually computed in the long wavelength limit, or superhorizon scales. This makes sense because, as long as inflation is taking place, the exponential expansion of the universe stretch the perturbations of the quantum fields from microscopic to cosmological scales. In this way, the inhomogeneities that re-enter the horizon once inflation has finished and hence the ones of interest today, were in its long-wavelength limit when inflation ended. In this section we will take a close look to this limit and its physical consequences. We will study the behaviour of perturbations with characteristic wavelength L much larger that a local Hubble radius $H_l^{(-1)}$ in such a way that if we consider L to be infinitely large compared with $H_l^{(-1)}$, we can interpret the region inside $H_l^{(-1)}$ as a local universe without perturbations, i.e., homogeneous and isotropic. In other words, if $\frac{L}{H_l^{(-1)}} \to \infty$, or equivalently, if $\frac{k}{a_l H_l} \simeq k\tau_l \to 0$, then the region inside $H_l^{(-1)}$ represents a local FLRW universe. With this in mind, let us study the long wavelength limit of perturbations during inflation. In order to do so we will take two different point of view: the first one is very intuitive and we will call it the $k \to 0$ limit, the second one is slightly less intuitive but it is very useful when dealing with non-linear perturbations, this is the so-called linear separate universe approach of $k = 0$ case. Although they are very similar, the two point of view are not exactly equivalent as we will see in the following.

4.1.1. $K \to 0$ Limit

Since each spatial derivative introduces a factor k in Fourier space, it would be logical to think that in the $k \to 0$ limit of the ADM equations we must neglect any term that contains a spatial derivative (This will be done in the next subsection). However, by doing that, we would be neglecting terms that actually contribute in the $k \to 0$ limit.

In the following we will enumerate the two cases in which neglecting a term with a spatial derivative would be too naive:

1. Non-local terms.
   As an example, let us imagine a re-scaling of the spatial coordinates

   $$x^i \to \tilde{x}^i = (1+\delta)x^i\,. \tag{55}$$

   It is easy to realize that if we rewrite δ in terms of the transformations parameters of (41) we get:

   $$\nabla^2\eta = \delta\,. \tag{56}$$

   The fact that a transformation as the one described in (55) is perfectly allowed in a FLRW universe together with (56) and the transformations rules of the perturbed metric variables of (41) immediately tell us that variables like $\nabla^2 E$ cannot generically be neglected in the long wavelength limit even if they contain a Laplacian. In order to avoid this problem, we will only neglect terms that contain extra spatial derivatives, by extra we mean that we will neglect a term like $\partial_i X$ if and only if it is compared with X, but we will not neglect it if it appear alone.

2. Equations with overall spatial derivatives.
A clear example is the momentum constraint of (33). Because it contains a total spatial derivative, it gives non-trivial information even in the $k \to 0$ limit, namely:

$$H^b A - \dot{D} - \frac{1}{3}\nabla^2 \dot{E} - \frac{1}{2M_{PL}^2}\dot{\phi}^b \delta\phi = 0\,. \tag{57}$$

The same happens with the evolution equation for $\tilde{A}_{ij}$ (38).

Taking these two important point into account one can re-derive the equation of motion for the MS variable Q in the $k \to 0$ limit. The result is, as expected, the Equation (45) but without the Laplacian, i.e.,

$$\ddot{Q} + 3H^b\dot{Q} + \left(H^b\right)^2\left(-\frac{3}{2}\epsilon_2 + \frac{1}{2}\epsilon_1\epsilon_2 - \frac{1}{4}\epsilon_2^2 - \frac{1}{2}\epsilon_2\epsilon_3\right)Q = 0\,. \tag{58}$$

If we want to know the solution for Equation (58) we can simply take the $k\tau \to 0$ limit of the solution (51) obtained before (we will do that later on); however, this would be restricted to ν (defined in (50)) being a constant. In this section we will solve independently the long wavelength equation of motion for Q without any assumption, i.e., valid at all orders in slow-roll parameters. The price to pay is that, since the initial conditions are given at sub-horizon scales and (58) is only valid at super-horizon scales, we will not specify any initial condition here.

It can be checked that (58) can be written as a total derivative as follows:

$$\frac{2M_{PL}^2 H^b}{a^3\dot{\phi}^b}\frac{d}{dt}\left[\frac{a^3\dot{\phi}^b}{2M_{PL}^2}\left(\frac{\dot{Q}}{H^b} - \frac{\epsilon_2}{2}Q\right)\right] = 0\,. \tag{59}$$

For convenience, we will define the comoving curvature perturbation as:

$$\mathcal{R}_c \equiv -\frac{H^b}{\dot{\phi}^b}Q\,. \tag{60}$$

It terms of $\mathcal{R}_c$, (59) takes a very simple form:

$$a^3\epsilon_1\dot{\mathcal{R}}_c = C_1\,, \tag{61}$$

where C_1 is a constant. Therefore, the solution of the MS equation in the long wavelength limit (58) is

$$Q = \frac{\dot{\phi}^b}{H^b}\mathcal{R}_c = C_2\frac{\dot{\phi}^b}{H^b} + C_1\frac{\dot{\phi}^b}{H^b}\int\frac{dt}{a^3\epsilon_1}\,, \tag{62}$$

where C_2 is also a constant. Solution (62) is the well-known exact solution in the $k \to 0$ limit for the single component scalar field case [85,92]. The term proportional to C_1 is usually known as decaying mode, name inherited from its SR behaviour where $\dot{\phi}^b$ is roughly constant, since $a \sim e^{-H^b t}$ during inflation, we can clearly see that the term proportional to C_1 decays as $e^{-3H^b t}$ during SR. However, this is not the case beyond SR, for example in USR we have $\dot{\phi}^b \sim e^{-3H^b t}$ (see (8)) and hence $\epsilon_1 \sim e^{-6H^b t}$, which makes the term proportional to C_1 be approximately constant whereas the one proportional to C_2 decays, contrary to what happens in SR.

4.1.2. $K = 0$ or Linear Separate Universe Approach

The second point of view to solve linear perturbation theory in the long wavelength limit is the linear separate universe approach [93–95], in this case we will forget for a moment about the perturbed ADM equations and we will take advantage of the fact that, in the long wavelength limit, each local patch represents an homogeneous and isotropic universe such that its line element is:

$$ds_l^2 = -dt_l^2 + a_l(t_l)^2\delta_{ij}dx_l^i dx_l^j\,, \tag{63}$$

where, as before, the subscript l stands for local. As a consequence, the constraints and equations of motion will be the same as the ones presented in Section 2, i.e.,

$$\begin{aligned} 3M_{PL}^2 H_l^2 &= \frac{1}{2}\left(\frac{d}{dt_l}\phi_l\right)^2 + V(\phi_l)\,, \\ \frac{d}{dt_l}H_l + H_l^2 &= -\frac{1}{3M_{PL}^2}\left[\left(\frac{d}{dt_l}\phi_l\right)^2 - V(\phi_l)\right], \\ \frac{d^2\phi^l}{dt_l^2} &+ 3H_l\frac{d}{dt_l}\phi_l + V_{\phi_l}(\phi_l) = 0\,. \end{aligned} \tag{64}$$

The first thing to realize here is that the momentum constraint is missing. This is because in the separate universe approach we are not only assuming that each patch is homogeneous and isotropic (we were also making this assumption in the $k \to 0$ limit where the momentum constraint plays a role), but we are also assuming that each patch evolve independently from each other. Since the momentum constraint gives information about the interaction between the different FLRW patches due to the presence of a spatial derivative, it makes sense that it does not appear in the linear separate universe approach. Before proceeding, let us clarify here that when we talk about the information about the interaction between different patches encoded in the momentum constraint, which makes these patches evolve in a not completely independent way, we do not claim that what happens in one patch will affect the others, but rather that all the patches as an ensemble must satisfy some conditions (given by the momentum constraint) that are absent if we look to a single patch but that are important when comparing them.

Within this approximation, we can only see the effect of perturbations if we compare different patches between them or with the global background. Knowing that the differences must be perturbative, we can follow our results from linear perturbation theory and write:

$$\begin{aligned} H_l &\simeq H^b + \delta H = H^b - H^b A + \dot{D} - \frac{1}{3}\frac{\nabla^2}{a}B\,, \\ dt_l &\simeq (1+A)dt\,, \\ \phi_l &\simeq \phi^b + \delta\phi\,. \end{aligned} \tag{65}$$

The term $\frac{1}{3}\frac{\nabla^2}{a}B$ can be set to zero without loss of generality, this is because since each patch is independent from each other, we can always choose an orthogonal threading for all of them, in which β_i (and hence B) is zero. Another way of reasoning is that, as checked in the $k \to 0$ limit of perturbation theory, a term with a Laplacian can only be important in the long wavelength limit if it contains non-local information, but at the same time, non-local terms would give some information about the interaction between local patches, which is in contradiction with the absence of the momentum constraint (and hence with the separate universe assumption) as explained above.

If we also take into account that the scalar perturbation related with the traceless part of the spatial metric (i.e., E) does not appear in (65) so neither in (64), we can set

$$\nabla^2 B^{sep} = \nabla^2 E^{sep} = 0\,. \tag{66}$$

in the separate universe approach (as indicated with the superscript "sep"). Note that, although (66) coincides with the Newtonian (or longitudinal) gauge, this is not a gauge choice, but a consequence of the separate universe approach.

With this in mind, we can write the equations in (64) in terms of the global background and perturbations over it as follows:

$$
\begin{aligned}
-6\left(H^b\right)^2 A^{sep} + 6H^b \dot{D}^{sep} &= \frac{1}{M_{PL}^2}\left(\dot{\phi}^b \dot{\delta\phi}^{sep} - \left(\dot{\phi}^b\right)^2 A^{sep} + V_{\phi^b}\delta\phi^{sep}\right), \\
6\dot{H}^b A^{sep} + 3H^b \dot{A}^{sep} - 3\ddot{D}^{sep} + 6\left(H^b\right)^2 A^{sep} - 6H^b \dot{D}^{sep} &= \frac{1}{M_{PL}^2}\left(2\dot{\phi}^b \dot{\delta\phi}^{sep} - \left(\dot{\phi}^b\right)^2 A^{sep} - V_{\phi^b}\delta\phi^{sep}\right), \\
\ddot{\delta\phi}^{sep} + 3H^b \dot{\delta\phi}^{sep} + V_{\phi^b\phi^b}\delta\phi^{sep} &= -2V_{\phi^b}A^{sep} + \dot{\phi}^b\left[\dot{A}^{sep} - 3\dot{D}^{sep}\right].
\end{aligned} \tag{67}
$$

As one can see from (67), the linear separate universe approach does not use the whole system of ADM equations presented in Section 4, it only uses the long-wavelength version of: (a) the Hamiltonian constraint (32), (b) the evolution equation for the trace of the extrinsic curvature (37) and (c) the equation of motion of the scalar field (39). All of them setting both $\nabla^2 B^{sep}$ and $\nabla^2 E^{sep}$ to zero accordingly with the separate universe assumption.

In order to see what are the consequences of this reduction in the number of equations and variables when comparing with the $k \to 0$ limit of Section 4.1.1, we can follow a similar procedure to what we did when defining the MS variable and write down a "separate-universe" gauge invariant variable [96,97], meaning that it is gauge invariant under time reparametrizations.

$$
Q^{sep} \equiv \delta\phi^{sep} - \frac{\dot{\phi}^b}{H^b} D^{sep}. \tag{68}
$$

Similarly to what we did with the MS Equation (45), we can rearrange the equations of the linear separate universe approach (67) and write a single equation of motion for Q^{sep}, the result, in terms of the SR parameters, is:

$$
\ddot{Q}^{sep} + 3H^b\left(1 + \frac{\epsilon_1\epsilon_2}{3(3-\epsilon_1)}\right)\dot{Q}^{sep} + \left(H^b\right)^2\left(-\frac{3}{2}\epsilon_2 + \frac{1}{2}\epsilon_1\epsilon_2 - \frac{1}{4}\epsilon_2^2 - \frac{1}{2}\epsilon_2\epsilon_3 - \frac{\epsilon_1\epsilon_2^2}{2(3-\epsilon_1)}\right)Q^{sep} = 0. \tag{69}
$$

Comparing (69) with (58) we can now clearly identify two extra terms that appear when assuming that each local patch evolves independently from each other, i.e., when using the separate universe approach. These terms are $\mathcal{O}(\epsilon_1\epsilon_2)$ [55], being strongly dependent on the inflationary regime, for example, they are $\mathcal{O}(\epsilon_1^{USR})$ in USR (where $\epsilon_2 \simeq -6$) and $\mathcal{O}\left(\left(\epsilon_1^{SR}\right)^2\right)$ in SR.

In order to better quantify this difference we will solve (69) in a similar way as done with (58) and compare the results. We can again write (69) as a total derivative as follows

$$
\frac{2V}{3a^3 H^b \dot{\phi}^b}\frac{d}{dt}\left[\frac{3a^3\left(H^b\right)^2\dot{\phi}^b}{2V}\left(\frac{\dot{Q}^{sep}}{H^b} - \frac{\epsilon_2}{2}Q^{sep}\right)\right] = 0. \tag{70}
$$

Defining a "separate universe" comoving curvature perturbation as

$$
\mathcal{R}_c^{sep} \equiv -\frac{H^b}{\dot{\phi}^b}Q^{sep}, \tag{71}
$$

we can write (70) as

$$
\frac{3a^3\left(\dot{\phi}^b\right)^2}{2V} = C_1', \tag{72}
$$

where C_1' is a constant. Hence, the solution for Q^{sep} is:

$$Q^{sep} = C_2' \frac{\dot{\phi}^b}{H^b} + C_1' \frac{\dot{\phi}^b}{H^b} \int \left(\frac{1}{a^3 \epsilon_1} - \frac{1}{3a^3} \right) . \tag{73}$$

We can see that the solutions for Q^{sep} of (73) and for Q of (62) differ by an extra term in the decaying mode, which is obviously due to the difference of $\mathcal{O}(\epsilon_1 \epsilon_2)$ in the equation of motion. The importance of this extra term depends on the inflationary regime, being more important beyond SR, where the term proportional to C_1' does not decay (see discussion below (62)).

Since the main difference between the linear separate universe approach and the $k \to 0$ limit is the inclusion of the non-trivial information of the momentum constraint at large scales, which is not taken into account in the former, one could think that the two solutions will coincide if we impose the solution (73) to satisfy the momentum constraint; however, this is not the case. In fact, in the following we are going to check that the mode proportional to C_1', which contains the new extra term, does not represent the $k \to 0$ limit of some solution to the perturbations equations with $k \neq 0$. This should not be a surprise since the correct solution for the MS variable in the long wavelength limit is given by (62). In order to check that, let us rearrange equations (65) to write

$$H^b A^{sep} - \dot{D}^{sep} - \frac{\dot{\phi}^b}{2M_{PL}^2} \delta\phi^{sep} = -\frac{H^b \dot{\phi}^b}{2V} \left(\frac{\dot{Q}^{sep}}{H^b} - \frac{\epsilon_2}{2} Q^{sep} \right) . \tag{74}$$

If we want solution (73) to be the $k \to 0$ solution to the perturbation equations with $k \neq 0$ we must be sure that our solution satisfies the momentum constraint (although it is lost in the separate universe approach.) because, as we know, this constraint gives non-trivial information even at large scales. What would be the momentum constraint in the linear separate universe approach is

$$H^b A^{sep} - \dot{D}^{sep} - \frac{\dot{\phi}^b}{2M_{PL}^2} \delta\phi^{sep} = 0 , \tag{75}$$

which, comparing with (74), immediately implies:

$$\frac{\dot{Q}^{sep}}{H^b} - \frac{\epsilon_2}{2} Q^{sep} = 0 . \tag{76}$$

From condition (76) and the equation of motion (70) we can see that, if the momentum constraint of (75) is satisfied at initial time, it will be always satisfied so one could think that the problem of ignoring the momentum constraint in the linear separate universe approach is simply solved by an appropriate choice of initial conditions; however, the only mode that satisfies (76) is the mode proportional C_2' in the solution (73), which means that the mode proportional to C_1' therein does not represent a $k \to 0$ solution of the perturbation equations with $k \neq 0$ and it must be set to zero. However, setting $C_1' = 0$ means that we are losing the term proportional to C_1 in the correct solution of (62), which can be important beyond SR. Thus, imposing that the momentum constraint must be satisfied in the separate universe approach does not give the correct solution for the long-wavelength limit of the MS variable and hence we have to use the whole set of ADM equations in the long wavelength limit to describe the correct dynamics (at all orders in SR parameters) of the MS variable at super-horizon scales. Let us indicate here that although there exist some ways to recover the correct $k \to 0$ limit of the MS variable using only the linear separate universe approach (see [96,97]), we will not study them here because they are restricted to linear perturbation theory and, as we will see later on, we are planning to use the long wavelength limit of inhomogeneities beyond perturbation theory.

For clarity purposes, let us summarize the findings of this section until now:

- Although both the $k \to 0$ and the linear separate universe approach consider local homogeneous and isotropic universes, the $k \to 0$ limit allow some interaction between them whereas the linear separate universe approach assume that they evolve independently. Mathematically speaking, the linear separate universe approach ignore possible non-local terms and equations with overall spatial derivatives, both present in the $k \to 0$ limit.
- As a consequence, the equation of motion for Q^{sep} differ by terms of $\mathcal{O}(\epsilon_1\epsilon_2)$ from the equation of motion for Q in the long wavelength limit. This difference induces an extra decaying term, namely:

$$Q^{sep} = Q(k \to 0) + C_1' \frac{\dot{\phi}^b}{H^b} \int \frac{1}{3a^3} dt \tag{77}$$

- The difference (77) always decays so one could think that it can be safely ignored; however, its importance strongly depends on the inflationary regime. For example, in USR we have $Q^{sep} - Q(k \to 0) \sim \mathcal{O}(\epsilon_1)$ so if we want to be precise enough when studying the long-wavelength limit of perturbation theory, we should use the $k \to 0$ limit rather than the linear separate universe approach.
- Finally, imposing the momentum constraint to be satisfied in the separate universe approach not only does not solve the difference between Q^{sep} and $Q(k \to 0)$, but it makes it worse.

4.2. Linear Perturbation Theory and Pbhs

In order to finish this section we will study the behaviour of the power spectrum of the inflationary regimes that can lead to the formation of PBHs in the long wavelength limit (The reader interested in PBHs as probes for the physics of the very early universe in a more detailed way can read the following reviews: [98,99]). We will do that in a very qualitative way and at leading order in SR parameters, this is why we will simply use the $k\tau \to 0$ limit of solution (51), where the initial conditions are properly specified. If we expand the Hankel function of first order we have:

$$\lim_{k\tau \to 0} Q_{\mathbf{k}} = \lim_{k\tau \to 0} e^{\frac{i}{2}(\nu+\frac{1}{2})} \frac{\sqrt{\pi}}{2a} \sqrt{-\tau} H_\nu^{(1)}(-k\tau) \simeq -ie^{\frac{i}{2}(\nu+\frac{1}{2})} \frac{2^{\nu-1}}{a\sqrt{\pi}} \sqrt{-\tau} (-k\tau)^{-\nu} \Gamma[\nu], \tag{78}$$

where $\Gamma[\nu]$ is the Euler gamma.

For the formation of PBH we are interested in the power spectrum of the comoving curvature perturbation, i.e.,

$$\mathcal{P}_{\mathcal{R}_c} = \frac{k^3}{2\pi^2} \left(\frac{H^b}{\dot{\phi}^b} \right)^2 |Q_{\mathbf{k}}|^2 \simeq \frac{\left(H^b\right)^4}{\pi^3 \left(\dot{\phi}^b\right)^2} \left(\frac{k}{2} \right)^{3-2\nu} \left(\frac{1}{aH^b} \right)^{3-2\nu} \Gamma[\nu]^2, \tag{79}$$

where we have used $\tau \simeq -\frac{1}{aH^b}$.

For USR and SR we have $\nu \simeq \frac{3}{2}$ and hence the power spectrum is roughly scale invariant:

$$\mathcal{P}_{\mathcal{R}_c}^{SR} \simeq \mathcal{P}_{\mathcal{R}_c}^{USR} \simeq \frac{\left(H^b\right)^4}{4\pi^2 \left(\dot{\phi}^b\right)^2} = \frac{\left(H^b\right)^2}{8\pi^2 M_{PL}^2 \epsilon_1}. \tag{80}$$

However, and although the k dependence of the comoving curvature power spectrum for USR and SR is roughly the same, and hence $n_s^{\mathcal{R}_c} - 1 \sim \mathcal{O}(\epsilon_1)$ for both, this is not true for the time dependence. In fact we have

$$\mathcal{P}_{\mathcal{R}_c}^{SR} \sim \text{constant}, \qquad \mathcal{P}_{\mathcal{R}_c}^{USR} \sim e^{6H^b t}. \tag{81}$$

An exponential growth of the power spectrum at super-horizon scales as it happens in USR has important consequences for the formation of PBH. In fact, by definition of the power spectrum, $\mathcal{P}_{\mathcal{R}_c}$ give us the variance of the probability distribution that follow the amplitude of the perturbations $\left(\mathcal{R}_{\mathbf{k}_1}\right)_c$ with characteristic wavenumber $\mathbf{k}_1$, therefore, a growth in the power spectrum can be interpreted as a growth in the variance or, equivalently, a spreading of the probability distribution. This means that high values for the amplitude of the perturbations placed at the tail of the probability distribution are much more probables if the power spectrum grows. Finally, and since the high values for the amplitude of the curvature perturbations are the ones that collapse to form a PBH when they re-enter the horizon, we can conclude that inflationary regimes beyond SR favor the creation of PBH.

The argument given above is very hand-waving but it is enough to remark the importance of inflationary regimes beyond SR when talking about PBH. However, if we want to make actual predictions about the mass and abundances of PBH we need to study inflationary dynamics in a much more precise way, in fact, the tail of the probability distribution for the amplitude of perturbations, which is where the perturbations responsible for the formation of PBH are located, is very sensitive to any small change in the power spectrum (or in the higher order correlators as the bispectrum). This is the reason why a description of the perturbations at all orders in SR is highly desirable, making the linear separate universe approach not the best approximation to study the tail of the probability distribution of the amplitude of perturbations generated during a inflationary regime beyond SR, i.e., to study the formation of PBH. The reason is that, generically, $Q^{sep} - Q(k \to 0) \sim \mathcal{O}(\epsilon_1)$ as explained above.

Finally, another problem arises when dealing with inhomogeneities large enough to form a PBH, namely, non-linear or even non-perturbative effects can play an important role, reason why we need to go beyond linear perturbation theory. The rest of the review is devoted to the study of inhomogeneities generated during inflation beyond linear perturbation theory.

5. Gradient Expansion

The gradient expansion is a non-perturbative, in terms of the amplitude of the inhomogeneities, expansion of the ADM equations valid when the characteristic wavelength of the inhomogeneities L is much larger that the Hubble horizon H_I^{-1} [100–109]. Since this is the same approximation that we do when studying the long wavelength behaviour of the linear perturbations, we will follow the same idea and we will take advantage of the fact that, when $L \gg H_I^{-1}$, the universe is locally homogeneous and isotropic to define an expansion parameter $\sigma \equiv \frac{H_I^{-1}}{L}$ such that at leading order in σ, each local patch of the universe of size (σH_I^{-1}) (we will call this scale the coarse-grained scale) is approximately described by a FLRW universe. Higher order terms in the σ will instead capture local inhomogeneities.

Contrary to the linear theory approach to cosmological perturbations, the gradient expansion is valid for any amplitude of local over-densities as long as the patch is taken small enough for the gradients to be negligible. Note that this assumption on which the gradient expansion is based on implies that a patch can be found such that any spatial gradient would only introduce an order σ. In other words, for any generic function X, $\partial_i X \sim X \times \mathcal{O}(\sigma)$. This is because a function which is approximately homogeneous in local coordinates can be written as $X(t, \sigma x^i)$ with $\sigma \ll 1$. Thus, we have

$$\partial_i X(t,\sigma x^i) = \sigma \frac{\partial}{\partial(\sigma x^i)} X(t,\sigma x^i) = \sigma \frac{\partial}{\partial(\sigma x^i)} X(t,\sigma x^i)\Big|_{\sigma x^i=0} + \mathcal{O}(\sigma^2)\,,$$

and since $\frac{\partial}{\partial(\sigma x^i)} X(t,\sigma x^i)\big|_{\sigma x^i=0}$ can be of the same order as $X(t,\sigma x^i)$ we can generically write:

$$\partial_i X \sim X \times \mathcal{O}(\sigma)\,.$$

When we were studying linear perturbation theory, the way of relating different patches was to define a global background over which each patch represented a perturbation. In this case we do not have a physical background metric because we are no longer perturbing anything; however we will still define a fictitious global background metric with coordinates t and x^i, i.e.,

$$ds_b^2 = -dt^2 + a(t)^2\delta_{ij}dx^i dx^j \,. \tag{82}$$

We are then interested in writing each local FLRW patch, which at leading order in σ and in terms of local coordinates is simply (63), in terms of the coordinates t and x^i. At leading order in gradient expansion and considering only scalar perturbations (note that, if we want to study inflation in a fully non-perturbative way, we should also take into account vector and tensor perturbations. This is because, although they are independent at linear order in perturbation theory, this is no longer true at higher orders. The reason why we do not include vector and tensor perturbations here is because, although at this level it would be straightforward, it is not possible when applying gradient expansion to stochastic inflation as we will see) we have:

$$ds_l^2 = -\,{}_{(0)}\alpha^2 dt^2 + {}_{(0)}\gamma_{ij}\left(dx^i + {}_{(0)}\beta^i dt\right)\left(dx^j + {}_{(0)}\beta^j\right), \tag{83}$$

with the conditions:

1. ${}_{(0)}\alpha = {}_{(0)}\alpha(t)\,,$
2. ${}_{(0)}\beta^i = b(t)x^i\,,$
3. ${}_{(0)}\gamma_{ij} = \gamma(t)\delta_{ij} = a(t)^2 e^{2\,{}_{(0)}\zeta(t)}\delta_{ij}\,,$

where we are using the subscript $_{(0)}$ to remind the reader that we are at leading order in gradient expansion It is important to remark here that the leading order in gradient expansion of each quantity can be different, for example, the leading order for α and ζ is $\mathcal{O}(\sigma^0)$ whereas the leading order for β^i is $\mathcal{O}(\sigma^{-1})$. Having a quantity whose leading order is gradient expansion is $\mathcal{O}(\sigma^{-1})$ could seem problematic; however this is only telling us that β^i is generically a non-local quantity, in fact, its linearization give us the non-local variable B (see (23)) studied in Section 4.1.1. Furthermore, we will see that β^i always appear together with a spatial derivative in the equations of motion.

One could be worried about the fact that a homogeneous and isotropic metric contains terms outside the diagonal; however, following [110] we know that a space-time is homogeneous and isotropic if:

1. All constant time hypersurfaces Σ_t are constant curvature spaces. In our case the hypersurfaces Σ_t are simply Euclidean and this condition is trivially satisfied.
2. The extrinsic curvature of the hypersurfaces is homogeneous and isotropic. Using the definition of extrinsic curvature of (12) together with the conditions for ${}_{(0)}\alpha$, ${}_{(0)}\beta^i$ and ${}_{(0)}\gamma_{ij}$ specified below (83), we can see that the extrinsic curvature only depends on time and hence this condition is also satisfied.

Different functions for ${}_{(0)}\alpha$, ${}_{(0)}\beta^i$ and ${}_{(0)}\zeta$ will give different FLRW patches as long as they satisfy the conditions given below (83). We can then relate the different locally homogeneous and isotropic patches by knowing the different non-perturbative functions for ${}_{(0)}\alpha$, ${}_{(0)}\beta^i$ and ${}_{(0)}\zeta$ that lead to each one of them. Of course, in the same way as in perturbation theory, the value of ${}_{(0)}\alpha$, ${}_{(0)}\beta^i$ and ${}_{(0)}\zeta$ will depend on the gauge choice and on the solution for the ADM equations.

We can generalize the idea of writing a FLRW metric in terms of x^i and t to writing quasi-FLRW metrics using the same coordinates. Following the ADM formalism we can write a metric valid at all orders in gradient expansion. With the identification $\gamma_{ij} = a(t)^2 e^{2\zeta}\tilde{\gamma}_{ij}$ we have:

$$ds^2 = g_{\mu\nu}dx^\mu dx^\nu = -\alpha^2 dt^2 + a(t)^2 e^{2\zeta}\tilde{\gamma}_{ij}(dx^i + \beta^i dt)(dx^j + \beta^j dt)\,, \tag{84}$$

where the leading order in gradient expansion for each variable is:

$$\begin{aligned} {}_{(0)}\alpha \sim \mathcal{O}(\sigma^0)\,, \qquad & {}_{(0)}\zeta \sim \mathcal{O}(\sigma^0)\,, \qquad {}_{(0)}\beta^i \sim \mathcal{O}(\sigma^{-1})\,, \\ \tilde{\gamma}_{ij} - \delta_{ij} \sim \mathcal{O}(\sigma)\,, \qquad & {}_{(0)}\phi \sim \mathcal{O}(\sigma^0)\,. \end{aligned} \tag{85}$$

The last term has been added to take into account the expansion of the scalar field, which is generically non-zero at the background level. It is important to realize here that the condition $\tilde{\gamma}_{ij} - \delta_{ij} \sim \mathcal{O}(\sigma)$ implies a further condition on the scalar part of $\tilde{\gamma}_{ij}$, in fact, using the expansion of the exponential of a matrix we can write

$$\tilde{\gamma}_{ij} = e^{-2M_{ij}} \simeq \delta_{ij} - 2M_{ij} + \mathcal{O}(\sigma^2)\,. \tag{86}$$

Now, M_{ij} must be traceless by definition (see see the paragraph under (23)). Focusing only in the scalar part of M_{ij} we can then write

$$\tilde{\gamma}_{ij} - \delta_{ij} \simeq -2\left(\partial_i\partial_j - \frac{1}{3}\delta_{ij}\nabla^2\right)C + \mathcal{O}(\sigma^2)\,, \tag{87}$$

where C is a scalar function. This immediataly implies that $\left(\partial_i\partial_j - \frac{1}{3}\delta_{ij}\nabla^2\right)C \sim \mathcal{O}(\sigma)$. Note that, as we will see later on, this condition is not in contradiction with $\nabla^2 C \sim \mathcal{O}(\sigma^0)$.

5.1. $\mathcal{O}(\sigma^0)$ in Gradient Expansion

Having specified the leading order gradient expansion for each one of the non-perturbative variables, we can now expand the ADM equations in terms of σ. As an example we will expand the Hamiltonian constraint at $\mathcal{O}(\sigma^0)$ using the spatially flat gauge, i.e.,

$$(\gamma_{\rm f})_{ij} = a^2\delta_{ij}\,, \tag{88}$$

where the subscript f stands for "flat". Since in this gauge we also have $R^{(3)}_{\rm f} = 0$ we can write the Hamiltonian constraint from (14) as:

$$-(A_{\rm f})_{ij}(A_{\rm f})^{ij} + \frac{2}{3}K_{\rm f}^2 - \frac{2}{M_{PL}^2}(T_{\rm f})_{\mu\nu}(n_{\rm f})^\mu(n_{\rm f})^\nu = 0\,, \tag{89}$$

where $(n_{\rm f})^\mu = g^{\mu\nu}(n_{\rm f})_\nu = \left(\frac{1}{\alpha_{\rm f}}, -\frac{(\beta_{\rm f})^i}{\alpha_{\rm f}}\right)$.

Equation (89) can be written in terms of the metric variables $\alpha_{\rm f}$ and $(\beta_{\rm f})^i$ of (84) and the scalar field $\phi_{\rm f}$. Using the results for $(A_{\rm f})_{ij}$ and $K_{\rm f}$ given by (18) and (17) respectively we have:

$$\begin{aligned} &-\frac{1}{4\alpha_{\rm f}^2}\left[\delta^{ik}\partial^j(\beta_{\rm f})_k + \delta^{jk}\partial^i(\beta_{\rm f})_k - \frac{2}{3}\delta^{ij}\partial^k(\beta_{\rm f})_k\right]\left[\delta_{ik}\partial_j(\beta_{\rm f})^k + \delta_{jk}\partial_i(\beta_{\rm f})^k - \frac{2}{3}\delta_{ij}\partial_k(\beta_{\rm f})^k\right] \\ &+\frac{2}{3}\left(-3\frac{H^b}{\alpha_{\rm f}} + \frac{1}{\alpha_{\rm f}}\partial_k(\beta_{\rm f})^k\right)^2 \\ &-\frac{2}{M_{PL}^2}\left[\frac{\dot{\phi}_{\rm f}^2}{2\alpha_{\rm f}^2} - \frac{\dot{\phi}_{\rm f}(\beta_{\rm f})^i\partial_i\phi_{\rm f}}{\alpha_{\rm f}^2} + \frac{(\beta_{\rm f})^i(\beta_{\rm f})^j\partial_i\phi_{\rm f}\partial_j\phi_{\rm f}}{2\alpha_{\rm f}^2} + \frac{\partial^i\phi_{\rm f}\partial_i\phi_{\rm f}}{2a^2} + V(\phi_{\rm f})\right] = 0\,, \end{aligned} \tag{90}$$

which is valid at all orders in gradient expansion. Keeping only the $\mathcal{O}(\sigma^0)$ terms we get:

$$\frac{2}{3}\left(-3\frac{H^b}{{}_{(0)}\alpha_{\rm f}}+\frac{1}{\alpha_{\rm f}}\partial_k\left({}_{(0)}\beta_{\rm f}\right)^k\right)^2$$
$$-\frac{2}{M_{PL}^2}\left[\frac{\left({}_{(0)}\dot{\phi}_{\rm f}\right)^2}{2\left({}_{(0)}\alpha_{\rm f}\right)^2}-\frac{{}_{(0)}\dot{\phi}_{\rm f}\left({}_{(0)}\beta_{\rm f}\right)^i{}_{(0)}(\partial_i\phi_{\rm f})}{\left({}_{(0)}\alpha_{\rm f}\right)^2}+\frac{\left({}_{(0)}\beta_{\rm f}\right)^i\left({}_{(0)}\beta_{\rm f}\right)^j{}_{(0)}(\partial_i\phi_{\rm f})\,{}_{(0)}(\partial_j\phi_{\rm f})}{2\left({}_{(0)}\alpha_{\rm f}\right)^2}+V(\phi_{\rm f})\right]=0\,, \tag{91}$$

where by ${}_{(0)}(\partial_i\phi_{\rm f})$ we mean $\sigma\frac{\partial}{\partial(\sigma x^i)}\phi(t,\sigma x^i)\big|_{\sigma x^i=0}$ as explained before.

Note that the first line of (90) is $\mathcal{O}(\sigma)$ because of the condition of ${}_{(0)}\beta^i$ below Equation (83). Following this procedure, it is straightforward to write all the ADM equations at order $\mathcal{O}(\sigma^0)$; however, we know from linear perturbation theory that there are equations with global spatial derivatives that play a role in the $k\to 0$ limit, which translated to the gradient expansion way of thinking means that there are equations that contain an overall factor σ and hence in order to extract its contribution at $\mathcal{O}(\sigma^0)$ they must be written up to $\mathcal{O}(\sigma)$. We know that the momentum constraint is one of these equations so let us analyze it. In spatially flat gauge, the momentum constraint (15) is:

$$\partial^j\left(\tilde{A}_{\rm f}\right)_{ij}-\frac{2}{3}\partial_i K_{\rm f}=\frac{1}{M_{PL}^2}J_i \tag{92}$$

If we expand (92) up to $\mathcal{O}(\sigma)$ we get:

$$-\frac{2}{3}\,{}_{(0)}\partial_i\left(-3\frac{H^b}{\alpha_{\rm f}}\right)=\frac{1}{M_{PL}^2}\left(-\frac{1}{{}_{(0)}\alpha_{\rm f}}\,{}_{(0)}\dot{\phi}_{\rm f}\,{}_{(0)}(\partial_i\phi_{\rm f})+\frac{\left({}_{(0)}\beta_{\rm f}\right)^k}{{}_{(0)}\alpha_{\rm f}}\,{}_{(0)}(\partial_k\phi_{\rm f})\,{}_{(0)}(\partial_i\phi_{\rm f})\right). \tag{93}$$

An important aspect regarding the gradient expansion is worthy to remark at this point, note that, in the derivation of (93) we have used $\partial^j\left(\tilde{A}_{\rm f}\right)_{ij}\sim\mathcal{O}(\sigma)$, which seems in contradiction with $\left(\tilde{A}_{\rm f}\right)_{ij}\sim\mathcal{O}(\sigma)$ (as we saw when deriving the Hamiltonian constraint at $\mathcal{O}(\sigma^0)$) and the fact that each spatial gradient introduces an order in σ. However, in this case we have an exception due to the traceless nature of $\left(\tilde{A}_{\rm f}\right)_{ij}$. Let us see why: from the condition 2 below (83) we have $(\beta_{\rm f})^i\simeq b(t,\sigma x^l)x^i$ and from (18) in spatially flat gauge:

$$\partial^j\left(\tilde{A}_{\rm f}\right)_{ij}=\delta_{ik}\partial_j\partial^j(\beta_{\rm f})^k+\partial_i\partial_k(\beta_{\rm f})^k-\frac{2}{3}\partial_i\partial_k(\beta_{\rm f})^k=\frac{4}{3}\partial_i\partial_k(\beta_{\rm f})^k=4\partial_i b(t,\sigma x^l)+\mathcal{O}(\sigma^2)\,,$$

which is clearly $\mathcal{O}(\sigma)$.

The linear version of (93) obviously corresponds to the linear momentum constraint in spatially flat gauge, i.e.,

$$\partial_i\left(H^b A_{\rm f}\right)=\partial_i\left(\frac{\dot{\phi}^b}{2M_{PL}^2}\delta\phi_{\rm f}\right). \tag{94}$$

As already stressed during Section 4.1.1, although (94) is an equation that appears at $\mathcal{O}(\sigma)$, it contains information at $\mathcal{O}(\sigma^0)$, namely

$$H^b A_{\rm f}=\frac{\dot{\phi}^b}{2M_{PL}^2}\delta\phi_{\rm f}\,. \tag{95}$$

Unfortunately, things are not that easy when dealing with the fully non-perturbative momentum constraint at $\mathcal{O}(\sigma)$, in fact, if we look at (93), we will easily realize that the total spatial derivative is no longer present. How the $\mathcal{O}(\sigma^0)$ information is encoded in (93) in a fully non-perturbative way is beyond the scope of this review, nevertheless, it is very important to be aware that we must take this information into account if we want to correctly describe the non-perturbative and long wavelength dynamics of the inhomogeneities, otherwise we would be making a mistake equivalent as when using the linear separate universe approach of Section 4.1.2 instead of the $k \to 0$ limit of Section 4.1.1. Nevertheless, due to its usefulness, in the following section we will present the non-perturbative version of the linear separate universe approach.

5.2. Non-Perturbative Separate Universe Approach

The $\mathcal{O}(\sigma^0)$ expression for the Hamiltonian constraint (91) seems very difficult to solve, and we have just argued the difficulties that arise when trying to extract the $\mathcal{O}(\sigma^0)$ information from the $\mathcal{O}(\sigma)$ momentum constraint. This is why a further approximation in addition to the $\mathcal{O}(\sigma^0)$ gradient expansion is usually performed in the literature [20,111,112]. This new approximation is the separate universe approach, which assumes that each FLRW patch of the universe evolve independently one from each other. We have already studied the linear version of the separate universe approach in Section 4.1.2, where we have seen that this assumption implies that neither non-local terms nor the momentum constraint will be present, which is equivalent to state that $\beta^i \sim \mathcal{O}(\sigma^3)$ as done in [109]. Within this approximation, the Hamiltonian constraint in spatially flat gauge of (91) takes a much simpler form:

$$\left(\frac{H^b}{{}_{(0)}\alpha_{\rm f}^{sep}}\right)^2 = \frac{1}{3M_{PL}^2}\left[\frac{\left({}_{(0)}\dot{\phi}_{\rm f}^{sep}\right)^2}{2\left({}_{(0)}\alpha_{\rm f}^{sep}\right)^2} + V\left({}_{(0)}\phi_{\rm f}^{sep}\right)\right], \tag{96}$$

which reminds us of the background Hamiltonian constraint (4). In fact, this coincidence extends also to the equation of motion of the scalar field as one could expect since we are assuming every FLRW to evolve independently from the others. A famous application of the separate universe approach is the δN formalism in order to compute the uniform-density curvature perturbation in a non perturbative way [70–73]. As a reminder, the linear uniform-density curvature perturbation is defined as

$$\mathcal{R}_{ud} \equiv -\frac{H^b}{\dot{\rho}^b}\delta\rho_{\rm f},$$

where ρ is the energy density of the scalar field, which in inflation is $\rho = 3M_{PL}^2\left(H^b\right)^2$ and $\delta\rho_{\rm f}$ is the perturbation of the energy density in spatially flat gauge.

In this section we have presented the non-perturbative generalization of the long wavelength linear perturbation theory presented in Section 4, being the $\mathcal{O}(\sigma^0)$ in gradient expansion of Section 5.1 the generalization of the $k \to 0$ limit for scalar perturbations of Section 4.1.1 and being the separate universe approach the non-linear generalization of the $k = 0$ (or linear separate universe approach) case of Section 4.1.2. For this reason, it is obvious that the $\mathcal{O}(\sigma^0)$ gradient expansion and the separate universe approach will give different "decaying" terms, being the ones of the $\mathcal{O}(\sigma^0)$ gradient expansion the correct ones, as it happened in linear theory.

Using the results from linear theory again, we can conclude that the error made when using the separate universe approach instead of the $\mathcal{O}(\sigma^0)$ gradient expansion will be generically of $\mathcal{O}(\epsilon_1)$, strongly depending on the inflationary regime. This conclusion is deduced of course assuming that higher orders in perturbation theory will not spoil the results at leading order, assumption that can be problematic if the inhomogeneity under study is non-perturbative.

We will finish this section by reminding the reader that setting initial conditions for long wavelength perturbations is problematic because we cannot use the Bunch-Davies vacuum, which is only well defined when $\frac{k}{aH} \to \infty$. This is why, although the gradient expansion is a very useful way to study inhomogeneities in a non-perturbative way, we need some other tool to set the initial conditions.

6. Stochastic Approach to Inflation

The stochastic approach to inflation combines the two approximations schemes presented until now to study the evolution of inhomogeneities in a non-perturbative way. The idea is to split the variables of interest (let us say X) into two parts: an infrared (IR) part that contains all the inhomogeneities with characteristic wavelength larger that some coarse-grained scale $(\sigma H)^{-1}$ (σ is the same parameter as the one used in gradient expansion) and a ultraviolet (UV) part, which encompasses inhomogeneities with characteristic scale smaller than $(\sigma H)^{-1}$ (or characteristic wavenumber k bigger than $\sigma a H$).

Since the UV part starts evolving well inside the Hubble horizon, we will assume that it is perturbatively small. Thus, one can use linear perturbation theory to describe it, where initial conditions are well defined. The IR part instead can be large; however, since the IR part only contains long wavelengths, the gradient expansion can be used there. As we will see, whenever an UV mode exits the coarse-grained scale, it will act as a kick for the IR part, solving the initial condition problem of gradient expansion. Note that, although we will only study stochastic inflation in the context of Einstein's gravity, the only requirements for the construction of a stochastic formalism are the possibility of a separation between IR and UV modes and a well-behaved perturbative expansion. There is then no reason a priori to think that the stochastic framework can not be applied to modifications of Einstein's gravity such as massive gravity [113,114].

The stochastic formalism is then a mathematical framework that, in principle, allows us to study the inhomogeneities generated during inflation in a non-perturbative way, reason why it is widely used when studying PBHs formation (see for example [115,116]). To see how stochastic inflation works and why it is called "stochastic" we will derive the formalism step by step. From Section 5 it should be clear now that we will have different stochastic equations depending on if we are using the separate universe approach or the $\mathcal{O}(\sigma^0)$ gradient expansion. The stochastic formalism that uses the separate universe approach is the most widely used in the literature (see for example [10,14–19,23,27,35,37,38, 51–53,63,66,115,116]) so we will start with its derivation. After that, we will take advantage of the stochastic equations just derived to present a stochastic formalism based on the $\mathcal{O}(\sigma^0)$ gradient expansion [65].

Before starting with the derivation and, since it is not as trivial as the spatially flat gauge, let us present the gauge we will be working with: the uniform-N gauge. We define the number of e-folds N as the integrated expansion rate of Σ_t hypersurfaces, i.e.,

$$N \equiv -\frac{1}{3}\int K dt_l \,, \tag{97}$$

where $K \equiv -3H_l$ (being H_l the local Hubble parameter) is the extrinsic curvature defined in (12), which coincides with the expansion rate of Σ_t hypersurfaces as shown in Section 3 and t_l is the local time of (63). In terms of the variables and coordinates of the ADM metric (84) N can be written as:

$$N \equiv \int \left(H^b + \dot{\zeta} - \frac{1}{3} D_i \beta^i \right) dt \,. \tag{98}$$

The uniform-N gauge is defined such that N coincides with the number of e-folds defined in the global background of (82), i.e., $N \equiv \int H^b dt$. From (98), this immediately implies $\zeta_{\delta N} = 0$ and $(\beta_{\delta N})^i = 0$ (where the subscript δN specifies the gauge), or $D_{\delta N} = 0$ and $B_{\delta N} = 0$ in its linear limit (see Equation (23)). The reason behind this gauge choice will be clarified along the following two subsections.

6.1. Stochastic Formalism Based on the Separate Universe Approach

As we have already indicated, this is the stochastic formalism most widely used in the literature due to its simplicity; however, as we will see, it has some problems. First of all, it is very important to realize that in the separate universe approach, both the uniform-N gauge and the spatially flat gauge are equivalent, which leads to many authors to use the uniform-N gauge for the IR part and the spatially flat gauge for the UV part [60]. Let us see why:

As we know, in the separate universe approach, due to the absence of non-local terms, both $(\tilde{\gamma}^{sep})_{ij} = \delta_{ij}$ and $\partial_i(\beta^{sep})^i = 0$ are automatically satisfied (see (66)). This means that, under this assumption, we have to add the condition $\partial_i\left(\beta_{\mathrm{f}}^{sep}\right)^i = 0$ to the spatially flat gauge where $(\tilde{\gamma}_{\mathrm{f}}) = \delta_{ij}$ and $\zeta_{\mathrm{f}} = 0$. In the same way, we have to add the condition $\left(\tilde{\gamma}_{\delta N}^{sep}\right)_{ij} = \delta_{ij}$ to the uniform-N gauge, where $\partial_i(\beta_{\delta N})^i = 0$ and $\zeta_{\delta N} = 0$. The main consequence of this is that, under the separate universe condition, we can express all the scalar fluctuations only in terms of the field inhomogeneities not only in the spatially flat gauge, but also in the uniform-N gauge. This can be clearly seen when looking at the linear "separate-universe" gauge invariant MS variable of (68). For non-linear generalizations of this variables see [117,118].

Before continuing, let us remind the reader that the equivalence between these two gauges in only valid under the separate universe approach, which, as seen in Section 4.1.2, it generically fails at $\mathcal{O}(\epsilon_1)$.

During this section, and although it is not compulsory, we will work using the number of e-folds $N \equiv \int H^b dt$ as time variable (Note that we can use any time variable we want because we will use the coordinates of a fictitious global background, i.e., the coordinates of (84). If we would instead use local coordinates (as in (63)), we would be interested in using an unperturbed time variable, being N the natural choice in the uniform-N gauge). Another important aspect is that, in order not to overload notation, we will suppress the subscript δN indicating the gauge we are using, such that in the following, unless otherwise stated, a variable without a subscript that indicates the gauge is a variable in the uniform-N gauge.

As indicated before, the stochastic formalism presented in this subsection is based on the separate universe approach for the IR part and on linear perturbation theory for the UV part. To illustrate this we will consider in detail the equation of motion for the trace the extrinsic curvature (19) in uniform-N gauge, which can be written using the variables of the ADM metric (84) as:

$$-3\frac{H^b}{\alpha}\frac{\partial}{\partial N}\left(\frac{H^b}{\alpha}\right) - \left(\frac{H^b}{2\alpha}\right)^2\frac{\partial\tilde{\gamma}_{ij}}{\partial N}\frac{\partial\tilde{\gamma}^{ij}}{\partial N} - 3\left(\frac{H^b}{\alpha}\right)^2 + D_k D^k\alpha - \frac{1}{M_{PL}^2}\left(\left(\frac{H^b}{\alpha}\right)^2\left(\frac{\partial\phi}{\partial N}\right)^2 - V(\phi)\right) = 0. \quad (99)$$

The first thing to do is to split the variables of interest into their IR and UV part. In this case we only have two variables to split:

$$\begin{aligned} \alpha &= \alpha^{IR} + \alpha^{UV}, \\ \phi &= \phi^{IR} + \phi^{UV}. \end{aligned} \quad (100)$$

In (100) we are not considering $\frac{\partial\tilde{\gamma}_{ij}}{\partial N}$ as a variable of interest not only because $\tilde{\gamma}_{ij} = \delta_{ij}$ in the separate universe approach, but also because $\frac{\partial\tilde{\gamma}_{ij}}{\partial N}\frac{\partial\tilde{\gamma}^{ij}}{\partial N} \sim \mathcal{O}(\sigma^2)$ in gradient expansion and quadratic in perturbation theory so it does not play any role even if we were using $\mathcal{O}(\sigma^0)$ gradient expansion.

Due to the perturbative nature of the UV variables, we will expand (99) keeping only linear terms in UV and isolate them in the right hand side of the equation getting

$$-3\frac{H^b}{\alpha^{IR}}\frac{\partial}{\partial N}\left(\frac{H^b}{\alpha^{IR}}\right)-3\left(\frac{H^b}{\alpha^{IR}}\right)^2+D^kD_k\alpha^{IR}-\frac{1}{M_{PL}^2}\left(\left(\frac{H^b}{\alpha^{IR}}\right)^2\left(\frac{\partial\phi^{IR}}{\partial N}\right)^2-V\left(\phi^{IR}\right)\right)$$
$$=-3\frac{\left(H^b\right)^2}{(\alpha^{IR})^3}\frac{\partial\alpha^{UV}}{\partial N}+\left(\frac{9\left(H^b\right)^2}{(\alpha^{IR})^4}\frac{\partial\alpha^{IR}}{\partial N}-\frac{6H^b}{(\alpha^{IR})^3}\left(\frac{\partial H^b}{\partial N}\right)\right)\alpha^{UV}-\frac{6\left(H^b\right)^2}{(\alpha^{IR})^3}\alpha^{UV}-\frac{\nabla^2}{a^2}\alpha^{UV}$$
$$+\frac{1}{M_{PL}^2}\left[2\left(\frac{H^b}{\alpha^{IR}}\right)^2\frac{\partial\phi^{IR}}{\partial N}\frac{\partial\phi^{UV}}{\partial N}-2\frac{\left(H^b\right)^2}{(\alpha^{IR})^3}\left(\frac{\partial\phi^{IR}}{\partial N}\right)^2\alpha^{UV}-V_\phi\left(\phi^{IR}\right)\phi^{UV}\right]. \tag{101}$$

Note that this is the same as we did in linear perturbation theory of Section 4 but using the metric (63) (or equivalently (84)) as background metric.

Now, since the IR variables are well outside the Hubble horizon, we will use the separate universe approach for them. Since $\alpha^{IR} \sim \mathcal{O}(\sigma^0)$ and hence $D_kD^k\alpha^{IR} \sim \mathcal{O}(\sigma^2)$ we have:

$$-3\frac{H^b}{{}_{(0)}\alpha^{IR}}\frac{\partial}{\partial N}\left(\frac{H^b}{{}_{(0)}\alpha^{IR}}\right)-3\left(\frac{H^b}{{}_{(0)}\alpha^{IR}}\right)^2-\frac{1}{M_{PL}^2}\left(\left(\frac{H^b}{{}_{(0)}\alpha^{IR}}\right)^2\left(\frac{\partial\,{}_{(0)}\phi^{IR}}{\partial N}\right)^2-V\left({}_{(0)}\phi^{IR}\right)\right)$$
$$=-3\frac{\left(H^b\right)^2}{\left({}_{(0)}\alpha^{IR}\right)^3}\frac{\partial\alpha^{UV}}{\partial N}+\left(\frac{9\left(H^b\right)^2}{\left({}_{(0)}\alpha^{IR}\right)^4}\frac{\partial\,{}_{(0)}\alpha^{IR}}{\partial N}-\frac{6H^b}{\left({}_{(0)}\alpha^{IR}\right)^3}\left(\frac{\partial H^b}{\partial N}\right)\right)\alpha^{UV}-\frac{6\left(H^b\right)^2}{\left({}_{(0)}\alpha^{IR}\right)^3}\alpha^{UV}$$
$$-\frac{\nabla^2}{a^2}\alpha^{UV}+\frac{1}{M_{PL}^2}\left[2\left(\frac{H^b}{{}_{(0)}\alpha^{IR}}\right)^2\frac{\partial\,{}_{(0)}\phi^{IR}}{\partial N}\frac{\partial\phi^{UV}}{\partial N}-2\frac{\left(H^b\right)^2}{\left({}_{(0)}\alpha^{IR}\right)^3}\left(\frac{\partial\,{}_{(0)}\phi^{IR}}{\partial N}\right)^2\alpha^{UV}-V_\phi\left({}_{(0)}\phi^{IR}\right)\phi^{UV}\right], \tag{102}$$

where we have inserted an extra subindex $_{(0)}$ to indicate that we are at leading order in gradient expansion.

Using Fourier analysis we can now define more rigorously the IR and UV modes. If we choose the Heaviside theta as a window function (The choice of the Heaviside theta as window function in the stochastic formalism is the most common one. However it can lead to some problems as indicated in [28]) we have the following decomposition for a generic function X.

$$X^{IR}(t,\mathbf{x}) \equiv \int\frac{d\mathbf{k}}{(2\pi)^{3/2}}\Theta(\sigma a_l(N)H_l(N)-k)\mathcal{X}^{IR}_{\mathbf{k}}(t,\mathbf{x}),$$
$$X^{UV}(t,\mathbf{x}) \equiv \int\frac{d\mathbf{k}}{(2\pi)^{3/2}}\Theta(k-\sigma a_l(N)H_l(N))\mathcal{X}^{UV}_{\mathbf{k}}(t,\mathbf{x}), \tag{103}$$

where, similarly as in linear perturbation theory (see (47)),$\mathcal{X}^{UV}_{\mathbf{k}}(t,\mathbf{x})$ is define as the following hermitian operator:

$$\mathcal{X}^{UV}_{\mathbf{k}}(t,\mathbf{x}) = e^{-i\mathbf{k}\cdot\mathbf{x}}X_{\mathbf{k}}(N)a_{\mathbf{k}} + e^{i\mathbf{k}\cdot\mathbf{x}X^*_{\mathbf{k}}a^\dagger_{\mathbf{k}}}, \tag{104}$$

where $X_{\mathbf{k}}(N)$ is the solution of the evolution equation for the perturbation X over the local background defined by (83) and $a_{\mathbf{k}}$ and $a^\dagger_{\mathbf{k}}$ are the usual creation and annihilation operators which follow the commutation relation given in (48).

Note that, in the spirit of gradient expansion, the splitting is done in the local cosmological coarse-grained scale $(\sigma H_l)^{-1}$, which generically differs form the one of the background, for example in uniform-N gauge we have $H_l = \frac{H^b}{_{(0)}\alpha^{IR}}$. Inserting the definition of X^{UV} of (103) into (102) we get:

$$
\begin{aligned}
&-3\frac{H^b}{_{(0)}\alpha^{IR}}\frac{\partial}{\partial N}\left(\frac{H^b}{_{(0)}\alpha^{IR}}\right)-3\left(\frac{H^b}{_{(0)}\alpha^{IR}}\right)^2-\frac{1}{M_{PL}^2}\left(\left(\frac{H^b}{_{(0)}\alpha^{IR}}\right)^2\left(\frac{\partial\,_{(0)}\phi^{IR}}{\partial N}\right)^2-V\left(_{(0)}\phi^{IR}\right)\right)\\
&3\frac{\left(H^b\right)^2}{\left(_{(0)}\alpha^{IR}\right)^3}\frac{\partial}{\partial N}\left(\sigma a\frac{H^b}{_{(0)}\alpha^{IR}}\right)\int\frac{d\mathbf{k}}{(2\pi)^{3/2}}\delta\left(k-\sigma a\frac{H^b}{_{(0)}\alpha^{IR}}\right)\boldsymbol{\alpha}_{\mathbf{k}}^{UV}\\
&-\frac{2}{M_{PL}^2}\left(\frac{H^b}{_{(0)}\alpha^{IR}}\right)^2\frac{\partial\,_{(0)}\phi^{IR}}{\partial N}\frac{\partial}{\partial N}\left(\sigma a\frac{H^b}{_{(0)}\alpha^{IR}}\right)\int\frac{d\mathbf{k}}{(2\pi)^{3/2}}\delta\left(k-\sigma a\frac{H^b}{_{(0)}\alpha^{IR}}\right)\varphi_{\mathbf{k}}^{UV}\\
&+\int\frac{d\mathbf{k}}{(2\pi)^{3/2}}\Theta\left(k-\sigma a\frac{H^b}{_{(0)}\alpha^{IR}}\right)\Bigg\{-3\frac{\left(H^b\right)^2}{\left(_{(0)}\alpha^{IR}\right)^3}\frac{\partial\boldsymbol{\alpha}_{\mathbf{k}}^{UV}}{\partial N}\\
&+\left(\frac{9\left(H^b\right)^2}{\left(_{(0)}\alpha^{IR}\right)^4}\frac{\partial\,_{(0)}\alpha^{IR}}{\partial N}-\frac{6H^b}{\left(_{(0)}\alpha^{IR}\right)^3}\left(\frac{\partial H^b}{\partial N}\right)\right)\boldsymbol{\alpha}_{\mathbf{k}}^{UV}-\frac{6\left(H^b\right)^2}{\left(_{(0)}\alpha^{IR}\right)^3}\boldsymbol{\alpha}_{\mathbf{k}}^{UV}+\frac{k^2}{a^2}\boldsymbol{\alpha}_{\mathbf{k}}^{UV}\\
&\frac{1}{M_{PL}^2}\left[2\left(\frac{H^b}{_{(0)}\alpha^{IR}}\right)^2\frac{\partial\,_{(0)}\phi^{IR}}{\partial N}\frac{\partial\varphi_{\mathbf{k}}^{UV}}{\partial N}-\frac{2\left(H^b\right)^2}{\left(_{(0)}\alpha^{IR}\right)^3}\left(\frac{\partial\,_{(0)}\phi^{IR}}{\partial N}\right)^2\boldsymbol{\alpha}_{\mathbf{k}}^{UV}-V_\phi\left(_{(0)}\phi^{IR}\right)\varphi_{\mathbf{k}}^{UV}\right]\Bigg\},
\end{aligned}
\tag{105}
$$

where $\boldsymbol{\alpha}_{\mathbf{k}}^{UV}$ and $\varphi_{\mathbf{k}}^{UV}$ are operators defined as in (104).

The right-hand side of (105) has two different terms:

1. The second integral (terms multiplying the Heaviside theta) is the evolution equation for the extrinsic curvature linearized over a local FLRW patch defined by $_{(0)}\alpha^{IR}$ and $_{(0)}\phi^{IR}$. Once the Bunch-Davies vacuum is chosen for that patch, this term will be automatically satisfied so it can be consistently set to zero. Note that the solution of this part equalized to zero is precisely what give us the functions $X_{\mathbf{k}}$ in (104).
2. The first two integrals, proportional to a Dirac delta, can be seen as boundary conditions and hence they will act as the initial conditions missing when using only gradient expansion.

We then get:

$$
\begin{aligned}
&-3\frac{H^b}{_{(0)}\alpha^{IR}}\frac{\partial}{\partial N}\left(\frac{H^b}{_{(0)}\alpha^{IR}}\right)-3\left(\frac{H^b}{_{(0)}\alpha^{IR}}\right)^2-\frac{1}{M_{PL}^2}\left(\left(\frac{H^b}{_{(0)}\alpha^{IR}}\right)^2\left(\frac{\partial\,_{(0)}\phi^{IR}}{\partial N}\right)^2-V\left(_{(0)}\phi^{IR}\right)\right)\\
&=-3\frac{\left(H^b\right)^2}{\left(_{(0)}\alpha^{IR}\right)^3}\xi_3+\frac{2}{M_{PL}^2}\left(\frac{H^b}{_{(0)}\alpha^{IR}}\right)^2\frac{\partial\,_{(0)}\phi^{IR}}{\partial N}\xi_1,
\end{aligned}
\tag{106}
$$

where we have defined ξ_1 and ξ_3 as:

$$
\begin{aligned}
\xi_1 &\equiv -\frac{\partial}{\partial N}\left(\sigma a\frac{H^b}{_{(0)}\alpha^{IR}}\right)\int\frac{d\mathbf{k}}{(2\pi)^{3/2}}\delta\left(k-\sigma a\frac{H^b}{_{(0)}\alpha^{IR}}\right)\varphi_{\mathbf{k}}^{UV},\\
\xi_3 &\equiv -\frac{\partial}{\partial N}\left(\sigma a\frac{H^b}{_{(0)}\alpha^{IR}}\right)\int\frac{d\mathbf{k}}{(2\pi)^{3/2}}\delta\left(k-\sigma a\frac{H^b}{_{(0)}\alpha^{IR}}\right)\boldsymbol{\alpha}_{\mathbf{k}}^{UV}.
\end{aligned}
\tag{107}
$$

For the moment we will not characterize the quantities ξ_1 and ξ_3. If we now follow the procedure for the evolution equation of the trace of the extrinsic curvature just explained with the Hamiltonian constraint we get:

$$6\left(\frac{H^b}{{}_{(0)}\alpha^{IR}}\right)^2 - \frac{2}{M_{PL}^2}\left[\left(\frac{H^b}{{}_{(0)}\alpha^{IR}}\right)^2\left(\frac{\partial\,{}_{(0)}\phi^{IR}}{\partial N}\right)^2 + V\left({}_{(0)}\phi^{IR}\right)\right] = \frac{2}{M_{PL}^2}\left(\frac{H^b}{{}_{(0)}\alpha^{IR}}\right)^2\frac{\partial\,{}_{(0)}\phi^{IR}}{\partial N}\xi_1, \tag{108}$$

which can be solved for $\left(\frac{H^b}{{}_{(0)}\alpha^{IR}}\right)^2$, i.e.,

$$\left(\frac{H^b}{{}_{(0)}\alpha^{IR}}\right)^2 = \frac{V\left({}_{(0)}\phi^{IR}\right)}{3M_{PL}^2 - \frac{1}{2}\left(\frac{\partial\,{}_{(0)}\phi^{IR}}{\partial N}\right)^2 - \frac{\partial\,{}_{(0)}\phi^{IR}}{\partial N}\xi_1}. \tag{109}$$

Finally, the stochastic equation of motion for the field is obtained in the same way:

$$\begin{aligned}&\frac{\partial^2\,{}_{(0)}\phi^{IR}}{\partial N^2} + \left(3 + \frac{\frac{\partial}{\partial N}\left(\frac{H^b}{{}_{(0)}\alpha^{IR}}\right)}{\frac{H^b}{{}_{(0)}\alpha^{IR}}}\right)\frac{\partial\,{}_{(0)}\phi^{IR}}{\partial N} + \frac{V_\phi\left({}_{(0)}\phi^{IR}\right)}{\left(\frac{H^b}{{}_{(0)}\alpha^{IR}}\right)^2}\\ &= -\frac{\partial\xi_1}{\partial N} - \xi_2 - \left(3 + \frac{\frac{\partial}{\partial N}\left(\frac{H^b}{{}_{(0)}\alpha^{IR}}\right)}{\frac{H^b}{{}_{(0)}\alpha^{IR}}}\right)\xi_1 + \frac{\partial\,{}_{(0)}\phi^{IR}}{\partial N}\frac{\xi_3}{{}_{(0)}\alpha^{IR}},\end{aligned} \tag{110}$$

where ξ_2 is defined similarly to ξ_1 and ξ_3:

$$\xi_2 \equiv -\frac{\partial}{\partial N}\left(\sigma a\frac{H^b}{{}_{(0)}\alpha^{IR}}\right)\int\frac{d\mathbf{k}}{(2\pi)^{3/2}}\delta\left(k - \sigma a\frac{H^b}{{}_{(0)}\alpha^{IR}}\right)\frac{\partial\varphi_{\mathbf{k}}^{UV}}{\partial N}. \tag{111}$$

As anticipated before, the usage and uniform-N gauge and the separate universe approach ensures that all the scalar inhomogeneities are encoded in the scalar field. This becomes clearer once we realize that we can write the system (106)–(110) in terms only of the scalar field. Inserting (106) and (109) into (110) and neglecting ξ_i^2 terms because they are quadratic in perturbation theory we get:

$$\begin{aligned}&\frac{\partial^2\,{}_{(0)}\phi^{IR}}{\partial N^2} + \left(3 - \frac{1}{2M_{PL}^2}\left(\frac{\partial\,{}_{(0)}\phi^{IR}}{\partial N}\right)^2\right)\frac{\partial\,{}_{(0)}\phi^{IR}}{\partial N} + \left(3M_{PL}^2 - \frac{1}{2}\left(\frac{\partial\,{}_{(0)}\phi^{IR}}{\partial N}\right)^2\right)\frac{V_\phi\left({}_{(0)}\phi^{IR}\right)}{V\left({}_{(0)}\phi^{IR}\right)}\\ &-\frac{\partial\xi_1}{\partial N} - \xi_2 - \left[3 - \frac{1}{2M_{PL}^2}\left(\frac{\partial\,{}_{(0)}\phi^{IR}}{\partial N}\right)^2 - \frac{1}{M_{PL}^2}\left(\frac{\partial\,{}_{(0)}\phi^{IR}}{\partial N}\right)^2 - \frac{V_\phi\left({}_{(0)}\phi^{IR}\right)}{V\left({}_{(0)}\phi^{IR}\right)}\left(\frac{\partial\,{}_{(0)}\phi^{IR}}{\partial N}\right)\right]\xi_1,\end{aligned} \tag{112}$$

which can be conveniently written if we use an auxiliary variable ${}_{(0)}\pi^{IR}$:

$$\begin{aligned}&{}_{(0)}\pi^{IR} = \frac{\partial\,{}_{(0)}\phi^{IR}}{\partial N} + \xi_1,\\ &\frac{\partial\,{}_{(0)}\pi^{IR}}{\partial N} + \left(3 - \frac{\left({}_{(0)}\pi^{IR}\right)^2}{2M_{PL}^2}\right){}_{(0)}\pi^{IR} + \left(3M_{PL}^2 - \frac{\left({}_{(0)}\pi^{IR}\right)^2}{2}\right)\frac{V_\phi\left({}_{(0)}\phi^{IR}\right)}{V\left({}_{(0)}\phi^{IR}\right)} = -\xi_2.\end{aligned} \tag{113}$$

6.1.1. Characterization of the Noises

The interpretation of ξ_1 and ξ_2 (note that ξ_3 no longer appears in the final equation of motion) as classical noises is not trivial because they are, strictly speaking, quantum operators. In order to see how they are effectively classical, we can compute the two-point correlation function of ξ_1 for example at equal space point, the result is:

$$\langle 0|\xi_1(N_1)\xi_1(N_2)|0\rangle = \frac{\partial}{\partial N}\left(\sigma a \frac{H^b}{{}_{(0)}\alpha^{IR}}\right)\left(\sigma a \frac{H^b}{{}_{(0)}\alpha^{IR}}\right)^2 \left|\delta\phi(N_1)_{k=\left(\sigma a \frac{H^b}{{}_{(0)}\alpha^{IR}}\right)}\right|^2 \delta(N_1 - N_2), \tag{114}$$

where we have used (104) together with the commutation relation (48). From (114) we see that $\delta\phi_{\mathbf{k}}$, which is the solution for the field perturbation over the local patch of size $\left(\sigma a \frac{H^b}{{}_{(0)}\alpha^{IR}}\right)^{-1}$, is evaluated at the coarse-grained scale, i.e., well outside the Hubble horizon. It can then be shown that at those scales, any perturbation that started from a coherent vacuum state has evolved into a highly squeezed state [119,120], which means that we can consider $\left|\delta\phi(N)_{k=\left(\sigma a \frac{H^b}{{}_{(0)}\alpha^{IR}}\right)}\right|^2$ as the power spectrum of a classical random variable, whose time evolution is consistent with probabilities conserved along classical trajectories.

Once this is clarified, we are now in position to describe ξ_1 as a classical white noise (Its "white" nature is due to the presence of $\delta(N_1 - N_2)$ in the two-point correlator. Note that this is a consequence of the the choice of the Heaviside theta function as Window function, any other choice would lead to coloured noises, which are much more difficult to deal with, both analytically and numerically) with variance given in (114). Furthermore, since the field fluctuations are Gaussian to a good level of approximation, the variance computed in (114) is enough to fully characterize ξ_1. Finally, in order to characterize the system (113) we also need:

$$\langle 0|\xi_1(N_1)\xi_2(N_2)|0\rangle = \langle 0|\xi_2(N_1)\xi_1(N_2)|0\rangle^* =$$

$$\frac{\partial}{\partial N}\left(\sigma a \frac{H^b}{{}_{(0)}\alpha^{IR}}\right)\left(\sigma a \frac{H^b}{{}_{(0)}\alpha^{IR}}\right)^2 \left(\delta\phi(N_1)_{k=\left(\sigma a \frac{H^b}{{}_{(0)}\alpha^{IR}}\right)} \frac{\partial\delta\phi^*(N_1)_{k=\left(\sigma a \frac{H^b}{{}_{(0)}\alpha^{IR}}\right)}}{\partial N}\right)\delta(N_1 - N_2),$$

$$\langle 0|\xi_1(N_1)\xi_1(N_2)|0\rangle = \frac{\partial}{\partial N}\left(\sigma a \frac{H^b}{{}_{(0)}\alpha^{IR}}\right)\left(\sigma a \frac{H^b}{{}_{(0)}\alpha^{IR}}\right)^2 \left|\frac{\partial\delta\phi(N_1)_{k=\left(\sigma a \frac{H^b}{{}_{(0)}\alpha^{IR}}\right)}}{\partial N}\right|^2 \delta(N_1 - N_2). \tag{115}$$

The characterization of ξ_1 and ξ_2 as white noises give us now a intuitive picture of the physics behind the stochastic formalism. As explained before, different functions for ${}_{(0)}\alpha^{IR}$ and ${}_{(0)}\phi^{IR}$ (in uniform-N gauge) describe the evolution of different FLRW patches, the way of getting these different functions is now clear if we see ξ_1 and ξ_2 as random variables. For example, the evolution of a specific patch, let us call it ${}^1FLRW$, will be given by ${}^1_{(0)}\alpha^{IR}$ and ${}^1_{(0)}\phi^{IR}$, whose specific form will be determined by the random values that the noises ${}^1\xi_1$ and ${}^1\xi_2$ will pick at each time step. Now, if we want to describe a second patch ${}^2FLRW$, we just have to solve again the stochastic equation with different random values for the noises ${}^2\xi_1$ and ${}^2\xi_2$, always satisfying the statistics described by (114) and (115). Like this, we are then able to describe the evolution of an ensemble of FLRW patches by solving many times the same stochastic equation with different random values for the noises. The correlators between these patches are then simply described by statistical moments of the IR variables.

The stochastic system is already fully characterized by (113)–(115); however, this system is very difficult to solve because, in order to compute the variance of the noises ξ_1 and ξ_2, we need to solve the perturbation equations and compute $\delta\phi_{\mathbf{k}}$ over a stochastic background every time step, which makes the systen non-Markovian. Although there are numerical algorithms capable of doing so [67], it would be very convenient if we were able to write an analytical solution for $\delta\phi_{\mathbf{k}}$ in terms of the IR variables, that would make the system Markovian and easily solvable. Let us try to solve the equation for $\delta\phi_{\mathbf{k}}$ over the stochastic local background.

We know that, as a consequence of the separate universe approach used here, the linearized equation for the field perturbation in the uniform-N gauge will be the same as the equation for the "separate-universe" gauge invariant variable Q^{sep} defined in (68), in other words, $Q^{sep} = \delta\phi_{\mathrm{f}} = \delta\phi_{\delta N}$ under the separate universe assumption. Now, since the linearized equation for Q^{sep} only gives the correct solution for the true gauge-invariant MS variable Q defined in (44) if we set $\epsilon_1 = 0$ as checked in Section 4.1.2, the equation we will try to solve here is the linearized equation for the gauge invariant quantity Q over a local background in which all the SR parameters have been set to zero (note that, in order for this approximation to work we need $-\frac{3}{2}\epsilon_2 - \frac{1}{4}\epsilon_2^2 \sim \mathcal{O}(\epsilon_1)$, which is true for SR and USR. This is because we are setting the combination of SR parameters multiplying Q in (45) to zero), in other words, we will try to solve the equation for Q as if the local background were an exact de-Sitter, which is:

$$\frac{\partial^2 \delta\phi_{\mathbf{k}}}{\partial N^2} + 3\frac{\partial \delta\phi_{\mathbf{k}}}{\partial N} + \left({}_{(0)}\alpha^{IR}\right)^2 \frac{k^2}{a^2\left(H^b\right)^2}\delta\phi_{\mathbf{k}} = 0\,, \tag{116}$$

or, using the Hamiltonian constraint of (109):

$$\frac{\partial^2 \delta\phi_{\mathbf{k}}}{\partial N^2} + 3\frac{\partial \delta\phi_{\mathbf{k}}}{\partial N} + \frac{k^2}{a^2}\left[\frac{3M_{PL}^2 - \frac{\left({}_{(0)}\pi^{IR}\right)^2}{2}}{V\left({}_{(0)}\phi^{IR}\right)}\right]\delta\phi_{\mathbf{k}} = 0\,. \tag{117}$$

Now, both the functions ${}_{(0)}\pi^{IR}$ and $V\left({}_{(0)}\phi^{IR}\right)$ are stochastic so, in order to know them we must already know the value for the variable ξ_1. This makes Equation (117) not analitically solvable, more concretely, the stochastic system of (113) is non-Markovian, meaning that the value of the noises at each time N will depend on the whole evolution of the stochastic patch up to N.

The only option remaining to solve the stochastic system of (113) seems then to be numerically; however, a further and very important approximation is usually done, which consists on assuming that the IR (and stochastic) quantities do not differ much from their global background (and deterministic) counterpart, namely $Y^{IR}X^{UV} \simeq Y^b X^{UV} + \mathcal{O}\left(\left(X^{UV}\right)^2\right)$. Here X^{UV} and Y^{IR} are any UV and IR functions, we then define Y^b as the equivalent background function of Y^{IR}. Under this approximation, the last term of (116) is:

$$k^2\left(\frac{{}_{(0)}\alpha^{IR}}{aH^b}\right)^2\delta\phi_{\mathbf{k}} \simeq \left(\frac{k}{aH^b}\right)^2\delta\phi_{\mathbf{k}} + \mathcal{O}\left(\delta\phi_{\mathbf{k}}^2\right), \tag{118}$$

where we have substituted ${}_{(0)}\alpha^{IR}$ by its background value, i.e., 1. Under this approximation we can write a analytical solution for (116), that, once evaluated at coarse-grained scale, will correspond to the long-wavelength limit of solution (51) with $\nu = \frac{3}{2}$, i.e.,

$$\left|\delta\phi(N)_{k=\left(\sigma a H^b\right)}\right|^2 = \frac{\left(H^b\right)^2}{2\left(\sigma a H^b\right)^3}\,. \tag{119}$$

and therefore the correlators (114) and (115) can be written as follows:

$$\langle \xi_1(N_1)\xi_1(N_2)\rangle = \left(\frac{H^b}{2\pi}\right)^2 \delta(N_1 - N_2)\,, \tag{120}$$

$$\langle \xi_1(N_1)\xi_2(N_2)\rangle = \langle \xi_2(N_1)\xi_2(N_2)\rangle = 0\,, \tag{121}$$

Under this approximation, the stochastic system of (113) simplifies considerably and becomes a Markovian process with additive noises, meaning that ξ_1 depends only on time, which is true as long as we are using the global background to characterize it. The system is:

$$\begin{aligned} {}_{(0)}\pi^{IR} &= \frac{\partial\, {}_{(0)}\phi^{IR}}{\partial N} + \frac{H^b}{2\pi}\xi(N)\,, \\ \frac{\partial\, {}_{(0)}\pi^{IR}}{\partial N} &+ 3\, {}_{(0)}\pi^{IR} + M_{PL}^2 \frac{V_\phi\left({}_{(0)}\phi^{IR}\right)}{V\left({}_{(0)}\phi^{IR}\right)} = 0\,, \end{aligned} \tag{122}$$

where $\langle \xi(N_1)\xi(N_2)\rangle = \delta(N_1 - N_2)$ and we have dropped $\mathcal{O}(\epsilon_1)$ terms in order to be consistent with the computation of the noises and the reasoning above (116). Note that in SR the acceleration of the field is also of higher order in SR parameters so in this case the stochastic equation would simplify even more:

$$\frac{\partial\, {}_{(0)}\phi^{IR}}{\partial N} + M_{PL}^2 \frac{V_\phi\left({}_{(0)}\phi^{IR}\right)}{V\left({}_{(0)}\phi^{IR}\right)} = -\frac{H^b}{2\pi}\xi(N)\,. \tag{123}$$

Although the approximation used in order to arrive to (122) seems very useful, it has important consequences, in fact, it is equivalent to state that any $Y^{IR} - Y^b \sim \mathcal{O}(X^{UV})$. Thus, we immediately see that if this approximation holds, the system in (122) can only reproduce the results of linear theory at leading order in SR parameters, and hence it does not give any non-perturbative (or even non-linear) information. In fact, (122) is slightly inconsistent. The point is that, by the same approximation adopted on the right hand side, the left hand side should also be linearized (in fact the linearization of the left-hand side of (122) is sometimes performed when recursive methods are used in order to solve the stochastic system as it can be seen in [42–44]). This inconsistency can however give some information when comparing the results from the stochastic formalism with the ones of linear perturbation theory, in fact, since we know that the correlations functions calculated with the stochastic system of (122) will coincide, up to second order in perturbation theory, to the ones calculated in linear perturbation theory with QFT methods, any inconsistency between the two approaches will signal the break-down of perturbation theory.

Since the stochastic Equation (116) and its deterministic counterpart with ${}_{(0)}\alpha^{IR} = 1$ have the same structure, one could be tempted to write the solution of (116) in a similar way as the solution (119) but substituting H^b by $\frac{H^b}{{}_{(0}\alpha^{IR}}$, i.e:

$$\left|\delta\phi(N)_{k=\left(\sigma a \frac{H^b}{{}_{(0}\alpha^{IR}}\right)}\right|^2 = \left(\frac{H^b}{{}_{(0}\alpha^{IR}}\right)^2 \frac{1}{2\left(\sigma a \frac{H^b}{{}_{(0}\alpha^{IR}}\right)^3}\,, \tag{124}$$

and hence the stochastic system of (113) would approximately be:

$$
\begin{aligned}
{}_{(0)}\pi^{IR} &= \frac{\partial\,{}_{(0)}\phi^{IR}}{\partial N} + \frac{H^b}{2\,{}_{(0)}\alpha^{IR}\pi}\xi(N)\,,\\
\frac{\partial\,{}_{(0)}\pi^{IR}}{\partial N} &+ 3\,{}_{(0)}\pi^{IR} + 3M_{PL}^2\frac{V_\phi\left({}_{(0)}\phi^{IR}\right)}{V\left({}_{(0)}\phi^{IR}\right)} = 0\,.
\end{aligned}
\tag{125}
$$

The system of (125) represents a Markovian process with non-additive noises (the term proportional to ξ does no longer depends only on time), which is, provided that (124) holds, able to describe the inhomogeneities generated during inflation in a non-perturbative way. Unfortunately, we cannot trivially generalize the solution (119) for a deterministic equation to a solution (124) for a stochastic equation, this is due to the differences between stochastic and deterministic integrals (see for example [121]). This is why, although it is the most common approach found in the literature, we will not trust (125) to describe non-linear inflationary effects in this review.

Having discarded this option, we are left with three different ways of solving the stochastic system of (113):

1. We can use the system (113) where the noises are computed numerically over the stochastic local background, for example, using the algorithm described in [67].
    - **Pros:** It describe non-linear inflationary dynamics.
    - **Cons:** It is very difficult to solve due to the non-Markovianity of the process. Furthermore, it is only valid up to leading order in SR parameters due to the use of the separate universe approach.
2. We can use the system of (122)
    - **Pros:** It is a Markovian process with additive noises for which even analytical solutions can be obtained (see Section VI of [65]).
    - **Cons:** It only describes linear perturbations whenever they are approximately described by solution (119), which is not the case during some interesting regimes for PBH formation, such as a SR-USR transition where $\nu \neq \frac{3}{2}$. It is then even less precise than the linear separate universe approach.
3. Finally, we can solve the MS equation over a global background at all orders in SR parameters, i.e., we can solve (45) and characterize the noises as in (114) with this solution. This can be done because, as we have already indicated, under the separate universe assumption $Q = \delta\phi_{\delta N}$. Once the noises are characterized in this way, we can use the stochastic system (113) and solve the dynamics.
    - **Pros:** It is a Markovian process with additive noises able to describe the linear dynamics of the inflationary perturbations even when they are not approximately described by (119).
    - **Cons:** It is not capable of describing any non-linear effects and it is inconsistent generically at leading order in ϵ_1 due to the use of the separate universe approach. This inconsistency can be clearly seen in the term proportional to ξ_1 in (112), which in the background can be written as $\left(3 - \epsilon_1 + \frac{\epsilon_1\epsilon_2}{3-\epsilon_1}\right)$ and not $(3 - \epsilon_1)$ as it should be if it came from the MS equation. In fact $\left(3 - \epsilon_1 + \frac{\epsilon_1\epsilon_2}{3-\epsilon_1}\right)$ is precisely the term that appears in the equation for Q^{sep} (69) derived in Section 4.1.2, making it clear that this inconsistency is a consequence of the separate universe approach.

Due to the problems that the stochastic formalism presented in this section has (namely its validity only up to leading order in ϵ_1 due to the separate universe assumption and its difficulty to describe non-linear dynamics), it does not seem very promising if we want to describe the non-perturbative dynamics of the inohomogeneities generated during inflationary regimes of interest for PBH formation with enough precision. However, in the next subsection we will present a stochastic formalism (firstly presented in [65]) which is at least able to reproduce linear perturbation theory at all orders in SR parameters, and with the potential to correctly describe the non-perturbative dynamics of the inhomogeneities during any inflationary regime.

However, before doing so, and in order to finish this subsection, it is important to know that the stochastic system of (113) can be straightforwardly derived by splitting into a IR and an UV only the scalar field in the separate universe equations of (64). In this case we would have used the metric (63) for the local patch instead of the metric (83) as we have done here. We have not chosen this option in this review because of three main reasons:

1. It does not make it clear that we have used gradient expansion, and hence the problem of not using the momentum constraint that we solve in the next subsection is difficult to remark.
2. Since it only works if the time variable is unperturbed, it could lead us to think that the number of e-folds N is the only allowed time variable for a stochastic formalism that describes all the scalar inhomogeneities in terms of the inflaton field. On the contrary, the derivation used in this review is valid for any time variable and makes it clear that the description of inhomogeneities in terms solely of the inflaton field is only a gauge choice.
3. It does not explicitly obtains the linear equations used for the characterization of $\delta\phi_{\mathbf{k}}$, more concretely, for example this derivation would miss the $\frac{k^2}{a^2}\alpha^{UV}_{\mathbf{k}}$ term that multiplies the Heaviside theta in (105).

6.2. Stochastic Formalism Based on $\mathcal{O}(\sigma^0)$ Gradient Expansion

As many times claimed during the review, the separate universe approach generically fails to give the correct long-wavelength evolution of the inhomogeneities at $\mathcal{O}(\epsilon_1)$. The $\mathcal{O}(\sigma^0)$ gradient expanision solves this problem by including both non-local terms and the momentum constraint, this is why in this section we will construct a stochastic formalism based on $\mathcal{O}(\sigma^0)$ gradient expansion.

First of all, it is important to remark that some of the affirmations we did about the uniform-N gauge at the beginning of Section 6.1 are no longer correct, more concretely, we cannot longer study the scalar inhomogeneities in terms solely of the inflaton field. This is clear form linear perturbation theory where the MS variable can be written as

$$Q = \delta\phi_{\mathrm{f}} = \delta\phi_{\delta N} - \frac{\partial\phi^b}{\partial N}\frac{1}{3}\nabla^2 E_{\delta N}\,, \tag{126}$$

so if we insist on using the uniform-N gauge we must also take into account the contribution from E when studying scalar perturbations.

We could also use spatially flat gauge in this case and forget about E; however, in this case we should take into account all the terms proportional to $\left({}_{(0)}\beta_{\mathrm{f}}\right)^i$ that appear for example in (90), this is why we will keep using the uniform-N gauge, where $\beta^i = 0$.

One can easily check that the stochastic equations for the evolution of the extrinsic curvature (106), for the Hamiltonian constraint (109) and for the evolution of the field (110) do not change when including E and hence the stochastic system is still the one given by (113). The only difference will then be given by the inclusion of the momentum constraint (15), which, in uniform-N gauge can be written as:

$$D^j\left(-\frac{H^b}{2\alpha}\frac{\partial\tilde{\gamma}_{ij}}{\partial N}\right) - \frac{2}{3}D_i K = -\frac{1}{M_{PL}^2\alpha}\frac{\partial\phi}{\partial N}\partial_i\phi\,, \tag{127}$$

where we have used the evolution Equation for $\tilde{\gamma}_{ij}$ (18) to eliminate $\tilde{A}_{ij}$. Splitting (127) into IR and UV and using the decomposition of $\tilde{\gamma}_{ij}$ explained around (87) to keep only $\mathcal{O}(\sigma)$ terms in the IR part (remember that the $\mathcal{O}(\sigma^0)$ information from the momentum constraint can only be extracted if we write the momentum constraint up to $\mathcal{O}(\sigma)$), we can write the stochastic equation for the momentum constraint:

$$ {}_{(0)}\partial_i\left(\frac{\partial}{\partial N}\left(\frac{1}{3}\nabla^2 C^{IR}\right)\right) - \frac{{}_{(0)}\partial_i\alpha^{IR}}{{}_{(0)}\alpha^{IR}} + \frac{\partial\,{}_{(0)}\phi^{IR}}{\partial N}\frac{{}_{(0)}\partial_i\phi}{2M_{PL}^2} = -\partial_i\xi_4\,, \tag{128} $$

where ξ_4 is defined similarly to ξ_1, ξ_2 and ξ_3, i.e.,

$$ \xi_2 \equiv -\frac{\partial}{\partial N}\left(\sigma a\frac{H^b}{{}_{(0)}\alpha^{IR}}\right)\int\frac{d\mathbf{k}}{(2\pi)^{3/2}}\delta\left(k-\sigma a\frac{H^b}{{}_{(0)}\alpha^{IR}}\right)\left(-\frac{k^2}{3}C^{UV}_{\mathbf{k}}\right). \tag{129} $$

With the addition of the stochastic Equation (128) to the system of (113) obtained before, we have a stochastic formalism able to describe the non-linear evolution of scalar inhomogeneities at all orders in SR parameters.

However, it is not all good news: firstly, since the construction of the gradient expansion in Section 5, we have been neglecting possible interactions scalar-tensor or scalar-vector, reason why the stochastic formalism constructed here will not take these interactions into account either. Secondly, we do not exactly know how to extract the $\mathcal{O}(\sigma^0)$ information form (128) in a fully non linear way. Finally, we do not know which is the combination of ${}_{(0)}\phi^{IR}$ and $\nabla^2 C^{IR}$ that give us the correct and non-perturbative and gauge invariant quantity that describe scalar inhomogeneities, i.e., we do not have a non-linear generalization of the MS variable. The first issue is beyond the scope of this review. With respect to the third one, it is true that a non-linear gauge invariant variable at leading order in gradient expansion has been defined in [117,118] as:

$$ \partial_i Q^{NL} \equiv \partial_i\phi + \frac{1}{\alpha}\frac{\partial\phi}{\partial N}\partial_i\zeta $$

However, the variable above does not include the term proportional to $\nabla^2 E$ in its linearization. Reason why it can be only interpreted as a non-linear generalization of Q^{sep}. This is the reason we will not use the non-linear variable defined above and we will solve the second and third issues, at least approximately, by imposing that both the momentum constraint and the non-linear generalization of the MS variable match their linear counterpart when the global background is subtracted. In this way, the $\mathcal{O}(\sigma^0)$ information of (128) can be straightforwardly extracted and the whole system of stochastic equations based on the $\mathcal{O}(\sigma^0)$ gradient expansion and hence valid at all orders in SR parameters is:

$$ \begin{aligned} &{}_{(0)}\pi^{IR} = \frac{\partial\,{}_{(0)}\phi^{IR}}{\partial N} + \xi_1\,, \\ &\frac{\partial\,{}_{(0)}\pi^{IR}}{\partial N} + \left(3 - \frac{\left({}_{(0)}\pi^{IR}\right)^2}{2M_{PL}^2}\right){}_{(0)}\pi^{IR} + \left(3M_{PL}^2 - \frac{\left({}_{(0)}\pi^{IR}\right)^2}{2}\right)\frac{V_\phi\left({}_{(0)}\phi^{IR}\right)}{V\left({}_{(0)}\phi^{IR}\right)} = -\xi_2\,, \\ &\frac{\partial}{\partial N}\left(\frac{1}{3}\,{}_{(0)}\nabla^2 C^{IR}\right) - \left(H^b\sqrt{\frac{3M_{PL}^2 - \frac{\left({}_{(0)}\pi^{IR}\right)^2}{2}}{V\left({}_{(0)}\phi^{IR}\right)}} - 1\right) + \frac{1}{2_{PL}^2}\frac{\partial\,{}_{(0)}\phi^{IR}}{\partial N}\left({}_{(0)}\phi^{IR} - \phi^b\right) = -\xi_4\,, \end{aligned} \tag{130} $$

where we have used the Hamiltonian constraint to eliminate ${}_{(0)}\alpha^{IR}$ in the last equation. Note that ξ_1,ξ_2 and ξ_4 are constructed in the uniform-N gauge, which is no longer equivalent to the spatially flat gauge.

To see the gauge transformation between spatially flat and uniform-N gauges in linear theory one can see [60], where it is claimed that the differences between those two gauges is always of higher order in gradient expansion; however, this conclusion is reached by considering the value of ϵ_1 at horizon crossing (ϵ_1^* there) to be constant with k, which is generically not a good approximation beyond SR. In fact in [67] it is shown numerically that the difference between $\delta\phi_{\rm f}$ and $\delta\phi_{\delta N}$ can be $\mathcal{O}(\epsilon_1)$ in regimes of interest for PBH formation, in agreement with the differences between the separate universe approach and $\mathcal{O}(\sigma^0)$ gradient expansion remarked in this paper.

Finally, as suggested in [122–124], we can define the non-linear counterpart of the MS variable of (44) at leading order in gradient expansion as:

$$Q^{IR} = {}_{(0)}\phi^{IR} - \phi^b - \frac{\partial\, {}_{(0)}\phi^{IR}}{\partial N}\frac{1}{3}\,{}_{(0)}\nabla^2 C^{IR}\,, \tag{131}$$

where we remind the reader that ${}_{(0)}\phi^{IR}$ and ${}_{(0)}\nabla^2 C^{IR}$ are both in the uniform N gauge.

The stochastic formalism based on the $\mathcal{O}(\sigma^0)$ gradient expansion presented in this section was introduced for the first time, although in a different gauge, in [65], where the authors show numerically that, when computing the noises over a global background, i.e., when using the approximation described around (118), the two point correlator computed using the stochastic formalism perfectly matches with the one computed in linear perturbation theory at all orders in SR parameters, as expected. This already represents an important improvement with respect to any stochastic formalism based on the separate universe approach as the one explained in Section 6.1.

6.3. Stochastic Formalism Versus Linear Perturbation Theory

Once the stochastic formalism has been properly introduced, the simplest way to know if there exists any important non-perturbative effect during inflation is to compare the results obtained within the stochastic framework with the results coming from linear perturbation theory explained in Section 4. In order to do so, it is crucial to realize that the results coming from the stochastic formalism are in real space whereas the result for the linear MS variable in terms of Hankel function of (51) is in Fourier space. Due to the difficulty when trying to express stochastic correlators in Fourier space, we will compare the stochastic two-point correlator with the linear two-point correlator, both in real space.

As explained below (115), the correlators between different FLRW patches are described by statistical moments of IR variables, so the two point correlator of the scalar inhomogeneities is:

$$\langle\, {}_{(0)}\phi^{IR}\, {}_{(0)}\phi^{IR}\rangle = Var\Big(\, {}_{(0)}\phi^{IR}\Big)\,, \tag{132}$$

if we are using the separate universe approach or

$$\langle Q^{IR} Q^{IR}\rangle = Var\Big(Q^{IR}\Big)\,, \tag{133}$$

if we are using $\mathcal{O}(\sigma^0)$ in gradient expansion.

On the other hand, the two-point correlator in linear perturbation theory is defined as (see (52)):

$$\langle Q(N)Q(N)\rangle = \int_{\sigma a(N=0)H^b(N=0)}^{\sigma a(N)H^b(N)} \frac{dk}{k}\mathcal{P}_Q(k,N)\,, \tag{134}$$

where we are now using the power spectrum evaluated at the same spatial point

$$\mathcal{P}_Q(k,N) \equiv \frac{k^3}{2\pi^2}|Q_{\mathbf{k}}(N)|^2\,, \tag{135}$$

where $Q_{\mathbf{k}}$ in uniform-N gauge is given by (126). The limits of (134) correspond to the selection of modes inside the coarse-grained scale defined by $k = \sigma a(N)H^b(N)$ and they are necessary to correctly compare the two point correlators obtained in both formalisms, in fact, in the stochastic formalism the IR part of the field recieves stochastic kicks from $N = 0$ onwards. Thus the first k-mode from which the IR field recieves a kick is the one with $k = \sigma a(N = 0)H^b(N = 0)$.

Whenever $\mathcal{P}_Q(k,N)$ does not depend on N, one can do a very useful approximation, which consist on evaluating the power spectrum at coarse-grained scale crossing, i.e., at $k=\sigma a H^b$, and assume that this value does not change with time. This would allow us to write (134) as:

$$\langle Q(N)Q(N)\rangle = \int_0^N \mathcal{P}\left(k=\sigma a(N')H^b(N')\right)dN' , \tag{136}$$

In this case one could write the power spectrum as the derivative with respect to the number of e-folds N of the correlator in real space.

$$\mathcal{P}_Q(k) = \frac{d}{dN}\langle Q(N)Q(N)\rangle . \tag{137}$$

However, and although it has been sometimes wrongly used for a time-dependent power spectrum [33], this technique cannot be used if the power spectrum evolves with time, which makes it only valid at zeroth order in $\mathcal{O}(\epsilon_1)$.

Finally, and due to the difficulty in the definition of non-linear gauge invariant variables (remember for example that the "trivial" non-linear generalization of the MS variable of (131) is not guaranteed to be the correct one), some authors have used the so-called stochastic-δN formalism to compute the non-linear curvature perturbation, which is very useful when dealing with PBHs, by studying the statistics in the number of e-folds [36,41,46,50,51]. However, and similarly to what happens with the usual δN formalism, the stochastic-δN formalism uses the separate universe approach and hence it is generically only valid at leading order in SR parameters.

7. Conclusions

The stochastic approach to inflation seems the most promising way to study, in a non-perturbative way, the probability distribution of the scalar inhomogeneities generated during inflation and responsible for the creation of PBHs. However, as remarked during the review, there is still a lot of work to do in that direction. Indeed, although very useful, the most common and widely used stochastic formalism has some difficult problems to solve, which can be summed up in three points:

1. As we have seen during the review, the stochastic formalism uses a gradient expansion for the IR part and a perturbative expansion for the UV part in such a way that the IR part, due to the large wavelength of the characteristic inhomogeneities that form this sector, can be described as a local FLRW universe. This description, that we called $\mathcal{O}(\sigma^0)$ gradient expansion, relates the different local FLRW patches via non-local terms and the momentum constraint, and describes at all orders in SR parameters the correct dynamics of long wavelength scalar inhomogeneities. However, it presents some problems such as the extraction of the $\mathcal{O}(\sigma^0)$ information from the momentum constraint.
   This is the reason why it is usually assumed that the different local FLRW universes evolve independently from each other, which is an assumption known as the separate universe approach. Under this approximation, the problem with the momentum constraint disappears (the momentum constraint itself disappears); nevertheless, we checked in Section 4.1.2 that this assumption fails to describe the correct long-wavelength dynamics of scalar perturbations generically at $\mathcal{O}(\epsilon_1)$ already in its linear limit. For this reason, the stochastic formalism commonly used presented in Section 6.1, which is based on the separate universe approach will fail to describe the non-perturbative dynamics of the scalar inhomogeneities at $\mathcal{O}(\epsilon_1)$.
   A stochastic formalism that does not uses the separate universe approach can also be constructed, as we did in Section 6.2. However, this option is not without its difficulties, more concretely, it has problems when extracting the long wavelength information from the momentum constraint and when describing the scalar inhomogeneities with a gauge invariant generalization of the MS variable.

2. On the other hand, the UV part of the inhomogeneities is assumed to behave perturbatively, having as a background the local FLRW background defined by the IR part. Since the UV part acts as a stochastic noise for the IR part and the IR part is necessary to characterize the UV noises, the stochastic approach to inflation is generically a non-Markovian process, meaning that the value of the noises depend on the whole history of the local patch over which they are computed.
   This does not represent a huge problem when solving the stochastic equations numerically; however, in order to have analytical results we need to do some other approximation that makes the process Markovian. This approximation consists of computing the stochastic noises over the global background instead of over the local one. Unfortunately, doing so is equivalent to assume that all the IR inhomogeneities are linear in perturbation theory.
3. Finally, and although we have not paid too much attention to this problem, any stochastic approach that aims to describe the non-perturbative behaviour of scalar inhomogeneities at the long wavelength limit should also take into account scalar-vector and scalar-tensor interactions, which no longer decouple beyond linear perturbation theory.

Funding: This research received no external funding.

Acknowledgments: I would like to thank Cristiano Germani for our weekly discussions and his constant support and encouragement. I would also like to thank Misao Sasaki, Vincent Vennin and Danilo Artigas for the many correspondences and help in understanding the gradient expansion method and the stochastic formalism. I am supported by the Spanish MECD fellowship PRE2018-086135. The group I belong to is also supported by the Unidad de Excelencia Maria de Maeztu Grants No. MDM-2014-0369 and CEX2019-000918-M, and the Spanish national grants FPA2016-76005-C2-2-P, PID2019-105614GB-C22 and PID2019-106515GB-I00.

Conflicts of Interest: The authors declare no conflict of interest.

References

1. Starobinsky, A.A. A New Type of Isotropic Cosmological Models without Singularity. *Phys. Lett. B* **1980**, *91*, 99–102. [CrossRef]
2. Guth, A.H. The Inflationary Universe: A Possible Solution to the Horizon and Flatness Problems. *Phys. Rev. D* **1981**, *23*, 347–356. [CrossRef]
3. Aghanim, N.; Akrami, Y.; Ashdown, M.; Aumont, J.; Baccigalupi, C.; Ballardini, M.; Bay, A.J.; Barreiro, R.B.; Bartolo, N.; Basak, S.; et al. [Planck], Planck 2018 results. VI. Cosmological parameters. *Astron. Astrophys.* **2020**, *641*, A6. Erratum in *Astron. Astrophys.* **2021**, *652*, C4. [CrossRef]
4. Akrami, Y.; Arroja, F.; Ashdown, M.; Aumont, J.; Baccigalupi, C.; Ballardini, M.; Banday, A.J.; Barreiro, R.B.; Bartolo, N.; Basak, N.; et al. Planck 2018 results. X. Constraints on inflation. *Astron. Astrophys.* **2020**, *641*, A10. [CrossRef]
5. Carr, B.J.; Hawking, S.W. Black holes in the early Universe. *Mon. Not. Roy. Astron. Soc.* **1974**, *168*, 399–415. [CrossRef]
6. Blais, D.; Kiefer, C.; Polarski, D. Can primordial black holes be a significant part of dark matter? *Phys. Lett. B* **2002**, *535*, 11–16. [CrossRef]
7. Bean, R.; Magueijo, J. Could supermassive black holes be quintessential primordial black holes? *Phys. Rev. D* **2002**, *66*, 63505. [CrossRef]
8. Sasaki, M.; Suyama, T.; Tanaka, T.; Yokoyama, S. Primordial Black Hole Scenario for the Gravitational-Wave Event GW150914. *Phys. Rev. Lett.* **2016**, *117*, 61101. Erratum in *Phys. Rev. Lett.* **2018**, *121*, 59901. [CrossRef]
9. Germani, C.; Musco, I. Abundance of Primordial Black Holes Depends on the Shape of the Inflationary Power Spectrum. *Phys. Rev. Lett.* **2019**, *122*, 141302. [CrossRef]
10. Starobinsky, A.A.; Stochastic de Sitter (inflationary) stage in the early universe. *Lect. Notes Phys.* **1986**, *246*, 107–126. [CrossRef]
11. Bardeen, J.M.; Bublik, G.J. Quantum fluctuactions and inflation. *Class. Quantum Grav.* **1987**, *4*, 573. [CrossRef]
12. Rey, S.J. Dynamics of Inflationary Phase Transition. *Nucl. Phys. B* **1987**, *284*, 706–728. [CrossRef]
13. Goncharov, A.S.; Linde, A.D.; Mukhanov, V.F. The Global Structure of the Inflationary Universe. *Int. J. Mod. Phys. A* **1987**, *2*, 561–591. [CrossRef]
14. Nambu, Y.; Sasaki, M. Stochastic Stage of an Inflationary Universe Model. *Phys. Lett. B* **1988**, *205*, 441–446. [CrossRef]
15. Nambu, Y.; Sasaki, M. Stochastic Approach to Chaotic Inflation and the Distribution of Universes. *Phys. Lett. B* **1989**, *219*, 240–246. [CrossRef]
16. Kandrup, H.E. Stochastic inflation as a time dependent random walk. *Phys. Rev. D* **1989**, *39*, 2245. [CrossRef]
17. Nakao, K.i.; Nambu, Y.; Sasaki, M. Stochastic Dynamics of New Inflation. *Prog. Theor. Phys.* **1988**, *88*, 1041. [CrossRef]

18. Nambu, Y. Stochastic Dynamics of an Inflationary Model and Initial Distribution of Universes. *Prog. Theor. Phys.* **1989**, *81*, 1037 [CrossRef]
19. Mollerach, S.; Matarrese, S.; Ortolan, A.; Lucchin, F. Stochastic inflation in a simple two field model. *Phys. Rev. D* **1991**, *44*, 1670–1679. [CrossRef]
20. Salopek, D.S.; Bond, J.R.; Nonlinear evolution of long wavelength metric fluctuations in inflationary models. *Phys. Rev. D* **1990**, *42*, 3936–3962. [CrossRef]
21. Salopek, D.S.; Bond, J.R. Stochastic inflation and nonlinear gravity. *Phys. Rev. D* **1991**, *43*, 1005–1031. [CrossRef] [PubMed]
22. Habib, S. Stochastic inflation: The Quantum phase space approach. *Phys. Rev. D* **1992**, *46*, 2408–2427. [CrossRef] [PubMed]
23. Linde, A.D.; Linde, D.A.; Mezhlumian, A. From the Big Bang theory to the theory of a stationary universe. *Phys. Rev. D* **1994**, *49*, 1783–1826. [CrossRef] [PubMed]
24. Starobinsky, A.A.; Yokoyama, J. Equilibrium state of a selfinteracting scalar field in the De Sitter background. *Phys. Rev. D* **1994**, *50*, 6357–6368. [CrossRef]
25. Morikawa, M. Dissipation and fluctuation of quantum fields in expanding universes. *Phys. Rev. D* **1990**, *42*, 1027–1034. [CrossRef]
26. Hosoya, A.; Morikawa, M.; Nakayama, K. Stochastic dynamics of scalar field in the inflationary universe. *Int. J. Mod. Phys. A* **1989**, *4*, 2613–2625. [CrossRef]
27. Casini, H.; Montemayor, R.; Sisterna, P. Stochastic approach to inflation. 2. Classicality, coarse graining and noises. *Phys. Rev. D* **1999**, *59*, 63512. [CrossRef]
28. Winitzki, S.; Vilenkin, A. Effective noise in stochastic description of inflation. *Phys. Rev. D* **2000**, *61*, 84008. [CrossRef]
29. Afshordi, N.; Brandenberger, R.H. Super Hubble nonlinear perturbations during inflation. *Phys. Rev. D* **2001**, *63*, 123505. [CrossRef]
30. Geshnizjani, G.; Afshordi, N. Coarse-grained back reaction in single scalar field driven inflation. *J. Cosmol. Astropart. Phys.* **2005**, *1*, 11. [CrossRef]
31. Tsamis, N.C.; Woodard, R.P. Stochastic quantum gravitational inflation. *Nucl. Phys. B* **2005**, *724*, 295–328. [CrossRef]
32. Martin, J.; Musso, M. Solving stochastic inflation for arbitrary potentials. *Phys. Rev. D* **2006**, *73*, 43516. [CrossRef]
33. Kunze, K.E. Perturbations in stochastic inflation. *J. Cosmol. Astropart. Phys.* **2006**, *7*, 14. [CrossRef]
34. Breuer, H.P.; Kunze, K.E. Stochastic inflation with coloured noise. *AIP Conf. Proc.* **2006**, *841*, 314. [CrossRef]
35. Finelli, F.; Marozzi, G.; Starobinsky, A.A.; Vacca, G.P.; Venturi, G. Generation of fluctuations during inflation: Comparison of stochastic and field-theoretic approaches. *Phys. Rev. D* **2009**, *79*, 44007. [CrossRef]
36. Enqvist, K.; Nurmi, S.; Podolsky, D.; Rigopoulos, G.I. On the divergences of inflationary superhorizon perturbations. *J. Cosmol. Astropart. Phys.* **2008**, *4*, 25. [CrossRef]
37. Finelli, F.; Marozzi, G.; Starobinsky, A.A.; Vacca, G.P.; Venturi, G. Stochastic growth of quantum fluctuations during slow-roll inflation. *Phys. Rev. D* **2010**, *82*, 64020. [CrossRef]
38. Clesse, S. Hybrid inflation along waterfall trajectories. *Phys. Rev. D* **2011**, *83*, 63518. [CrossRef]
39. Martin, J.; Vennin, V. Stochastic Effects in Hybrid Inflation. *Phys. Rev. D* **2012**, *85*, 43525. [CrossRef]
40. Garbrecht, B.; Rigopoulos, G.; Zhu, Y. Infrared correlations in de Sitter space: Field theoretic versus stochastic approach. *Phys. Rev. D* **2014**, *89*, 63506. [CrossRef]
41. Fujita, T.; Kawasaki, M.; Tada, Y.; Takesako, T. A new algorithm for calculating the curvature perturbations in stochastic inflation. *J. Cosmol. Astropart. Phys.* **2013**, *12*, 36. [CrossRef]
42. Levasseur, L.P. Lagrangian formulation of stochastic inflation: Langevin equations, one-loop corrections and a proposed recursive approach. *Phys. Rev. D* **2013**, *88*, 83537. [CrossRef]
43. Levasseur, L.P.; Vennin, V.; Brandenberger, R. Recursive Stochastic Effects in Valley Hybrid Inflation. *Phys. Rev. D* **2013**, *88*, 83538. [CrossRef]
44. Levasseur, L.P.; McDonough, E. Backreaction and Stochastic Effects in Single Field Inflation. *Phys. Rev. D* **2015**, *91*, 63513. [CrossRef]
45. Garbrecht, B.; Gautier, F.; Rigopoulos, G.; Zhu, Y. Feynman Diagrams for Stochastic Inflation and Quantum Field Theory in de Sitter Space. *Phys. Rev. D* **2015**, *91*, 63520. [CrossRef]
46. Fujita, T.; Kawasaki, M.; Tada, Y. Non-perturbative approach for curvature perturbations in stochastic δN formalism. *J. Cosmol. Astropart. Phys.* **2014**, *10*, 30. [CrossRef]
47. Vennin, V.; Starobinsky, A.A. Correlation Functions in Stochastic Inflation. *Eur. Phys. J. C* **2015**, *75*, 413. [CrossRef]
48. Onemli, V.K. Vacuum Fluctuations of a Scalar Field during Inflation: Quantum versus Stochastic Analysis. *Phys. Rev. D* **2015**, *91*, 103537. [CrossRef]
49. Burgess, C.P.; Holman, R.; Tasinato, G. Open EFTs, IR effects \& late-time resummations: systematic corrections in stochastic inflation. *J. High Energy Phys.* **2016**, *1*, 153. [CrossRef]
50. Vennin, V.; Assadullahi, H.; Firouzjahi, H.; Noorbala, M.; Wands, D. Critical Number of Fields in Stochastic Inflation. *Phys. Rev. Lett.* **2017**, *118*, 31301. [CrossRef]
51. Assadullahi, H.; Firouzjahi, H.; Noorbala, M.; Vennin, V.; Wands, D. Multiple Fields in Stochastic Inflation. *J. Cosmol. Astropart. Phys.* **2016**, *16*, 43. [CrossRef]
52. Pattison, C.; Vennin, V.; Assadullahi, H.; Wands, D. Quantum diffusion during inflation and primordial black holes. *J. Cosmol. Astropart. Phys.* **2017**, *10*, 46. [CrossRef]

53. Firouzjahi, H.; Nassiri-Rad, A.; Noorbala, M. Stochastic Ultra Slow Roll Inflation. *J. Cosmol. Astropart. Phys.* **2019**, *1*, 40. [CrossRef]
54. Pattison, C.; Vennin, V.; Assadullahi, H.; Wands, D. The attractive behaviour of ultra-slow-roll inflation. *J. Cosmol. Astropart. Phys.* **2018**, *8*, 48. [CrossRef]
55. Cruces, D.; Germani, C.; Prokopec, T. Failure of the stochastic approach to inflation beyond slow-roll. *J. Cosmol. Astropart. Phys.* **2019**, *3*, 48. [CrossRef]
56. Kitajima, N.; Tada, Y.; Takahashi, F. Stochastic inflation with an extremely large number of *e*-folds. *Phys. Lett. B* **2020**, *800*, 135097. [CrossRef]
57. Prokopec, T.; Rigopoulos, G. ΔN and the stochastic conveyor belt of ultra slow-roll inflation. *Phys. Rev. D* **2021**, *104*, 83505. [CrossRef]
58. Kuhnel, F.; Freese, K. On Stochastic Effects and Primordial Black-Hole Formation. *Eur. Phys. J. C* **2019**, *79*, 954. [CrossRef]
59. Ezquiaga, J.M.; García-Bellido, J.; Vennin, V. The exponential tail of inflationary fluctuations: consequences for primordial black holes. *J. Cosmol. Astropart. Phys.* **2020**, 3, 29. [CrossRef]
60. Pattison, C.; Vennin, V.; Assadullahi, H.; Wands, D. Stochastic inflation beyond slow roll. *J. Cosmol. Astropart. Phys.* **2019**, *7*, 31. [CrossRef]
61. Firouzjahi, H.; Nassiri-Rad, A.; Noorbala, M. Stochastic nonattractor inflation. *Phys. Rev. D* **2020**, *102*, 123504. [CrossRef]
62. Ando, K.; Vennin, V. Power spectrum in stochastic inflation. *J. Cosmol. Astropart. Phys.* **2021**, *4*, 57. [CrossRef]
63. Ballesteros, G.; Rey, J.; Taoso, M.; Urbano, A. Stochastic inflationary dynamics beyond slow-roll and consequences for primordial black hole formation. *J. Cosmol. Astropart. Phys.* **2020**, *8*, 43. [CrossRef]
64. Vennin, V. Stochastic inflation and primordial black holes. *arXiv* **2009**, arXiv:2009.08715 .
65. Cruces, D.; Germani, C. Stochastic inflation at all order in slow-roll parameters: Foundations. *Phys. Rev. D* **2022**, *105*, 23533. [CrossRef]
66. Pattison, C.; Vennin, V.; Wands, D.; Assadullahi, H. Ultra-slow-roll inflation with quantum diffusion. *J. Cosmol. Astropart. Phys.* **2021**, *2021*, doi: 10.1088/1475-7516/2021/04/080 . [CrossRef]
67. Figueroa, D.G.; Raatikainen, S.; Rasanen, S.; Tomberg, E. Implications of stochastic effects for primordial black hole production in ultra-slow-roll inflation. *J. Cosmol. Astropart. Phys.* **2021**, *2021*, 80. [CrossRef]
68. Mahbub, R.; De, A. Smooth coarse-graining and colored noise dynamics in stochastic inflation. *arXiv* **2022**, arXiv:2204.03859 .
69. Gihman, I.; Skorohod, A. *Stochastic Differential Equations*; Springer: Berlin/Heidelberg, Germany; New York, NY, USA, 1972.
70. Starobinsky, A.A. Dynamics of Phase Transition in the New Inflationary Universe Scenario and Generation of Perturbations. *Phys. Lett. B* **1982**, *117*, 175–178. [CrossRef]
71. Sasaki, M.; Stewart, E.D. A General analytic formula for the spectral index of the density perturbations produced during inflation. *Prog. Theor. Phys.* **1996**, *95*, 71–78. [CrossRef]
72. Lyth, D.H.; Rodriguez, Y. The Inflationary prediction for primordial non-Gaussianity. *Phys. Rev. Lett.* **2005**, *95*, 121302. [CrossRef] [PubMed]
73. Sugiyama, N.S.; Komatsu, E.; Futamase, T. δN formalism. *Phys. Rev. D* **2013**, *87*, 23530. [CrossRef]
74. Boyanovsky, D.; de Vega, H.J.; Holman, R.; Lee, D.S.; Singh, A. Dissipation via particle production in scalar field theories. *Phys. Rev. D* **1995**, *51*, 4419–4444. [CrossRef]
75. Carr, B.; Kohri, K.; Sendouda, Y.; Yokoyama, J. Constraints on primordial black holes. *Rept. Prog. Phys.* **2021**, *84*, 116902. [CrossRef] [PubMed]
76. Atal, V.; Germani, C. The role of non-gaussianities in Primordial Black Hole formation. *Phys. Dark Univ.* **2019**, *24*, 100275. [CrossRef]
77. Motohashi, H.; Hu, W. Primordial Black Holes and Slow-Roll Violation. *Phys. Rev. D* **2017**, *96*, 63503. [CrossRef]
78. Germani, C.; Sheth, R.K. Nonlinear statistics of primordial black holes from Gaussian curvature perturbations. *Phys. Rev. D* **2020**, *101*, 63520. [CrossRef]
79. Germani, C.; Prokopec, T. On primordial black holes from an inflection point. *Phys. Dark Univ.* **2017**, *18*, 6–10. [CrossRef]
80. Kinney, W.H. Horizon crossing and inflation with large eta. *Phys. Rev. D* **2005**, *72*, 23515. [CrossRef]
81. Martin, J.; Motohashi, H.; Suyama, T. Ultra Slow-Roll Inflation and the non-Gaussianity Consistency Relation. *Phys. Rev. D* **2013**, *87*, 23514. [CrossRef]
82. Arnowitt, R.L.; Deser, S.; Misner, C.W. The Dynamics of general relativity. *Gen. Rel. Grav.* **2008**, *40*, 1997. [CrossRef]
83. Palenzuela, C. Introduction to Numerical Relativity. *Front. Astron. Space Sci.* **2020**, *7*, 58. [CrossRef]
84. Kodama, H.; Sasaki, M. Cosmological Perturbation Theory. *Prog. Theor. Phys. Suppl.* **1984**, *78*, 1–166. [CrossRef]
85. Mukhanov, V.F.; Feldman, H.A.; Brandenberger, R.H. Theory of cosmological perturbations. Part 1. Classical perturbations. Part 2. Quantum theory of perturbations. Part 3. Extensions. *Phys. Rept.* **1992**, *215*, 203–333. [CrossRef]
86. Malik, K.A.; Wands, D. Cosmological perturbations. *Phys. Rept.* **2009**, *475*, 1–51. [CrossRef]
87. Durrer, R. Cosmological perturbation theory. *Lect. Notes Phys.* **2004**, *653*, 31–70. [CrossRef]
88. Riotto, A. Inflation and the theory of cosmological perturbations. *ICTP Lect. Notes Ser.* **2003**, *14*, 317–413.
89. Bruni, M.; Matarrese, S.; Mollerach, S.; Sonego, S. Perturbations of space-time: Gauge transformations and gauge invariance at second order and beyond. *Class. Quant. Grav.* **1997**, *14*, 2585–2606. [CrossRef]
90. Bardeen, J.M. Gauge Invariant Cosmological Perturbations. *Phys. Rev. D* **1980**, *22*, 1882–1905. [CrossRef]

91. Bunch, T.S.; Davies, P.C.W. Quantum Field Theory in de Sitter Space: Renormalization by Point Splitting. *Proc. Roy. Soc. Lond. A* **1978**, *360*, 117. [CrossRef]
92. Polarski, D.; Starobinsky, A.A. Spectra of perturbations produced by double inflation with an intermediate matter dominated stage. *Nucl. Phys. B* **1992**, *385*, 623–650. [CrossRef]
93. Wands, D.; Malik, K.A.; Lyth, D.H.; Liddle, A.R. A New approach to the evolution of cosmological perturbations on large scales. *Phys. Rev. D* **2000**, *62*, 43527. [CrossRef]
94. Lyth, D.H.; Wands, D. Conserved cosmological perturbations. *Phys. Rev. D* **2003**, *68*, 103515. [CrossRef]
95. Artigas, D.; Grain, J.; Vennin, V. Hamiltonian formalism for cosmological perturbations: the separate-universe approach. *J. Cosmol. Astropart. Phys.* **2022**, *2*, 1. [CrossRef]
96. Kodama, H.; Hamazaki, T. Evolution of cosmological perturbations in the long wavelength limit. *Phys. Rev. D* **1998**, *57*, 7177–7185. [CrossRef]
97. Sasaki, M.; Tanaka, T. Superhorizon scale dynamics of multiscalar inflation. *Prog. Theor. Phys.* **1998**, *99*, 763–782. [CrossRef]
98. Khlopov, M.Y. Primordial Black Holes. *Res. Astron. Astrophys.* **2010**, *10*, 495–528. [CrossRef]
99. Dolgov, A.D. Massive and supermassive black holes in the contemporary and early Universe and problems in cosmology and astrophysics. *Usp. Fiz. Nauk.* **2018**, *188*, 121–142. [CrossRef]
100. Lyth, D.H.; Malik, K.A.; Sasaki, M. A General proof of the conservation of the curvature perturbation. *J. Cosmol. Astropart. Phys.* **2005**, *5*, 4. [CrossRef]
101. Deruelle, N.; Langlois, D. Long wavelength iteration of Einstein's equations near a space-time singularity. *Phys. Rev. D* **1995**, *52)*, 2007–2019. [CrossRef]
102. Iguchi, O.; Ishihara, H.; Soda, J. Inhomogeneity of spatial curvature for inflation. *Phys. Rev. D* **1997**, *55*, 3337–3345. [CrossRef]
103. Khalatnikov, I.M.; Kamenshchik, A.Y.; Martellini, M.; Starobinsky, A.A. Quasiisotropic solution of the Einstein equations near a cosmological singularity for a two fluid cosmological model. *J. Cosmol. Astropart. Phys.* **2003**, *3*, 1. [CrossRef]
104. Parry, J.; Salopek, D.S.; Stewart, J.M. Solving the Hamilton-Jacobi equation for general relativity. *Phys. Rev. D* **1994**, *49*, 2872–2881. [CrossRef] [PubMed]
105. Soda, J.; Ishihara, H.; Iguchi, O. Hamilton-Jacobi equation for Brans-Dicke theory and its long wavelength solution. *Prog. Theor. Phys.* **1995**, *94*, 781–794. [CrossRef]
106. Nambu, Y.; Taruya, A. Application of gradient expansion to inflationary universe. *Class. Quant. Grav.* **1996**, *13*, 705–714. [CrossRef]
107. Leach, S.M.; Sasaki, M.; Wands, D.; Liddle, A.R. Enhancement of superhorizon scale inflationary curvature perturbations. *Phys. Rev. D* **2001**, *64*, 23512. [CrossRef]
108. Tanaka, Y.; Sasaki, M. Gradient expansion approach to nonlinear superhorizon perturbations. *Prog. Theor. Phys.* **2007**, *117*, 633–654. [CrossRef]
109. Tanaka, Y.; Sasaki, M. Gradient expansion approach to nonlinear superhorizon perturbations. II. A Single scalar field. *Prog. Theor. Phys.* **2007**, *118*, 455–473. [CrossRef]
110. Katanaev, M.O. Killing vector fields and a homogeneous isotropic universe. *Phys. Usp.* **2016**, *59*, 689–700. [CrossRef]
111. Rigopoulos, G.I.; Shellard, E.P.S. The separate universe approach and the evolution of nonlinear superhorizon cosmological perturbations. *Phys. Rev. D* **2003**, *68*, 123518. [CrossRef]
112. Tanaka, T.; Urakawa, Y. Anisotropic separate universe and Weinberg's adiabatic mode. *J. Cosmol. Astropart. Phys.* **2021**, *7*, 51. [CrossRef]
113. de Rham, C.; Gabadadze, G.; Tolley, A.J. Resummation of Massive Gravity. *Phys. Rev. Lett.* **2011**, *106*, 231101 [CrossRef] [PubMed]
114. Kodama, H.; Arraut, I. Stability of the Schwarzschild–de Sitter black hole in the dRGT massive gravity theory. *Prog. Theor. Exp. Phys.* **2014**, *2014*, 23E02. [CrossRef]
115. Clesse, S.; García-Bellido, J. Massive Primordial Black Holes from Hybrid Inflation as Dark Matter and the seeds of Galaxies. *Phys. Rev. D* **2015**, *92*, 23524. [CrossRef]
116. Tada, Y.; Vennin, V. Statistics of coarse-grained cosmological fields in stochastic inflation. *J. Cosmol. Astropart. Phys.* **2022**, *2*, 21. [CrossRef]
117. Langlois, D.; Vernizzi, F. Evolution of non-linear cosmological perturbations. *Phys. Rev. Lett.* **2005**, *95*, 91303. [CrossRef]
118. Rigopoulos, G.I.; Shellard, E.P.S. Non-linear inflationary perturbations. *J. Cosmol. Astropart. Phys.* **2005**, *10*, 6. [CrossRef]
119. Kiefer, C.; Polarski, D. Why do cosmological perturbations look classical to us? *Adv. Sci. Lett.* **2009**, *2*, 164–173. [CrossRef]
120. Grishchuk, L.P.; Sidorov, Y.V. Squeezed quantum states of relic gravitons and primordial density fluctuations. *Phys. Rev. D* **1990**, *42*, 3413–3421. [CrossRef]
121. Kloeden, P.E.; Platen, E. *Numerical Solution of Stochastic Differential Equations*; Springer: Berlin/Heidelberg, Germany, 1992. [CrossRef]
122. Takamizu, Y.i.; Kobayashi, T. Nonlinear superhorizon curvature perturbation in generic single-field inflation. *Prog. Theor. Exp. Phys.* **2013**, *2013*, 063E03. [CrossRef]
123. Takamizu, Y.I.; Mukohyama, S.; Sasaki, M.; Tanaka, Y. Non-Gaussianity of superhorizon curvature perturbations beyond δ N formalism. *J. Cosmol. Astropart. Phys.* **2010**, *6*, 19. [CrossRef]
124. Wang, J. Construction of the conserved ζ via the effective action for perfect fluids. *Annals Phys.* **2015**, *362*, 223–238. [CrossRef]

universe

MDPI

Review

PBH Formation from Spherically Symmetric Hydrodynamical Perturbations: A Review

Albert Escrivà [1,2,3]

1 Service de Physique Théorique, Université Libre de Bruxelles (ULB), Boulevard du Triomphe, CP225, 1050 Brussels, Belgium; albert.escriva@ulb.be
2 Institut d'Estudis Espacials de Catalunya (IEEC), Edifici Nexus I, C/ Gran Capità, 2–4, Desp. 201, 08034 Barcelona, Spain
3 Institut de Ciències del Cosmos, Universitat de Barcelona, Martí i Franquès 1, 08028 Barcelona, Spain

Abstract: Primordial black holes, which could have been formed in the very early Universe due to the collapse of large curvature fluctuations, are currently one of the most attractive and fascinating research areas in cosmology for their possible theoretical and observational implications. This review article presents the current results and developments on the conditions for primordial black hole formation from the collapse of curvature fluctuations in spherical symmetry on a Friedman–Lemaître–Robertson–Walker background and its numerical simulation. We review the appropriate formalism for the conditions of primordial black hole formation, and we detail a numerical implementation. We then focus on different results regarding the threshold and the black hole mass using different sets of curvature fluctuations. Finally, we present the current state of analytical estimations for the primordial black hole formation threshold, contrasted with numerical simulations.

Keywords: primordial black holes; early Universe cosmology; numerical general relativity

1. Introduction

One of the great mysteries in science is the composition of dark matter, which accounts for 27% of the present Universe [1]. Although there are different theories and candidates that try to explain it [2], still, the answer remains elusive. One of the most promising possibilities is Primordial Black Holes (PBHs), i.e., Black Holes (BH) generated at earlier than star formation times and therefore not of stellar origin. For the current observational status and constraints of PBHs in the form of dark matter, we refer the reader to [3–8].

PBHs were first considered in [9–11]. They could have formed in the very early Universe due to the gravitational collapse of cosmological perturbations. Under this scenario, PBHs could have been generated as a consequence of high non-linear peaks in the primordial distribution of density perturbations, leaving open the possibility that PBHs could constitute a significant fraction of Dark Matter (DM) [12–38].

PBHs with a size smaller than $M_{\rm PBH} < 10^{-20}\ {\rm M}_{\odot}$ would have already evaporated due to Hawking radiation [39]. Therefore, those with higher masses can account for dark matter. In addition, they may be responsible for seeding supermassive black holes at the centre of galaxies [40,41], generating large-scale structures through Poisson statistics [42], or changing the thermal history of the Universe [43].

Although PBHs were theorised a long time ago, it has not been until now that their popularity has increased exponentially. The current big interest in PBHs (and their golden age) is due to the first detection of gravitational waves from a black hole merger by the LIGO/Virgo collaborations [44]. Soon after this remarkable discovery, it was suggested that the constituents of the black hole merger could have had a primordial origin [19,45]. Later on, several analyses of the data signals suggested that the population of black holes detected could have a primordial origin [27,46–50] (see also the references therein). On the other hand, recent results from the NANOGrav experiment [51] have also been connected with PBHs [26,52–59].

If formed by the collapse of inflationary perturbations, the abundance of PBHs is exponentially sensitive to the threshold of the gravitational collapse δ_c [60] (where δ_c is the minimum amplitude of the gravitational potential peak related to the perturbation undergoing gravitational collapse and leading to a BH). In order to obtain the necessary precision on δ_c, a numerical analysis of PBH formation is an obvious way to go.

Numerical simulations of PBH formation started some time ago, the first one being performed in [61], where the non-linear behaviour of the gravitational collapse was studied. Later on, in [62,63], δ_c was computed and a scaling behaviour for the PBH mass was found whenever the amplitude of the perturbation was close to δ_c [62–65]. The value of the scaling exponent found was consistent with the one given in the literature and computed from a perturbative treatment [66,67] or from numerical simulations [68] in asymptotically flat spacetime. While the constant of proportionality appearing in the scaling law depends on the specific shape of the perturbation considered, the scaling exponent is a universal quantity only dependent on the type of fluid. Earlier simulations of PBHs were mainly based on the application of a Lagrangian hydrodynamic code based on finite differences and developed from an earlier work [69], solving Misner–Sharp equations [70], which describe the motion of a relativistic fluid in a curved spacetime. An inconvenience of this procedure is the appearance of a coordinate singularity after forming the apparent horizon, which breaks down the simulation. To resolve that issue, Hernandez–Misner equations [71] (which are the Misner–Sharp equations in null coordinates) are used to evade the formation of an apparent horizon and follow the subsequent evolution to determine the value of the black hole mass. The method is based on [72]. As was shown in [73], the formation of PBHs from the collapse of initial fluctuations at super-horizon scales does not form shocks (a discontinuous solution) in comparison with the earlier result of [64], where shock waves were found, but using initial conditions on sub-horizon scales.

On the other hand, even if a numerical code allows computing the threshold, the realisation of a numerical simulation can be quite expensive to be actually useful for statistical applications such as the calculations of PBH abundances. Therefore, an analytical expression would drastically simplify the problem. There are some analytical estimations of δ_c (e.g., [74,75]) that were based on analytical models under certain assumptions. However, numerical simulations have shown that, even for a fixed equation of state, δ_c is not universal [63,64,73,76–80]. The main argument is that δ_c is dependent on the specific detail of the initial curvature fluctuation [79], i.e., on the scale dependence or "shape" of the perturbation, something that these analytical estimations do not take into account.

Some works have tried to characterise the threshold for black hole formation in terms of the initial density perturbation profile. In [81], the amplitude of the curvature fluctuation and its second radial derivate at the centre were used to characterise the initial configuration profiles, showing that the probability of PBH formation is sensitive to them. Later on, in [82], numerical simulations were used to argue the formation of a PBH from an over-density peak in an FLRW Universe assuming the perturbation to be initially at super-cosmological scales, and it only depends on two master parameters: the integral of the initial curvature perturbation and the edge of the over-density distribution. The paper [79] recently refined the arguments of [82] by showing that these parameters may be more conveniently given in terms of the amplitude of the "gravitational potential" at its maximum ($r = r_m$), as already noticed in [77], and that it mainly depends on the functional form (shape) of the gravitational potential. More precisely, in [79], the threshold δ_c was identified with the peak of the "compaction function" [77] at super-horizon scales. The compaction function closely resembles the Schwarzschild gravitational potential and is defined as twice the local excess-mass over the co-moving areal radius.

Following the aim to have a more accurate analytical estimation for the threshold of PBH formation, recently, in [83], it was argued that the threshold for primordial black hole formation should be quite insensitive to the physics beyond r_m: the threshold is the amplitude for which a "black hole" with zero mass would be formed, taking an infinite time. This is an asymptotically and ideal limit. Therefore, all the over-density beyond r_m will be diffused away while that just in the proximity of r_m will block the collapse.

With those arguments, it was shown that during a radiation-dominated epoch (equation of state $p = w\rho$ with $w = 1/3$), with a very good approximation, there exists a universal threshold value for the volume-averaged compaction function, which is actually shape-independent. Taking into account that the volume average is dominated by scales near the maximum of the compaction function, in [83], it was shown that it is enough to parametrise the profile dependence of δ_c by the curvature shape around the compaction function at its maximum. This result was used to make an analytical approximation for δ_c taking into account the shape dependence on the curvature fluctuation, which matches that found in simulations to within a few percent [83].

From the results in [83], there was the question of whether or not this universality on the averaged critical compaction function is generic and not only an accident of radiation. Thus, in [84], the universality question beyond the case of a radiation fluid was addressed. In [84], the arguments of [83] were generalised to provide a new analytical estimation for the threshold in terms of the equation of state and the shape profile. This analytical formula was tested in front of numerical simulations and was found with an accuracy better than a 6%.

The other half-part of the problem is the determination of the final PBH mass. When a perturbation collapses, it will firstly form an apparent horizon with an associated mass $M_{\mathrm{PBH},i}$. After that, due to the surrounding fluid of the FLRW background, there is an accretion process up to a final mass $M_{\mathrm{PBH},f}$ when the PBH is formed. As we have indicated, for fluctuations whose amplitude is near the threshold (the critical regime), a scaling law was found for the PBH mass in terms of the amplitude of the perturbation [62–65]. Although the importance of this result was pointed out in the context of critical collapse [85], currently, it is substantially relevant because, as indicated in the current literature [60], the PBHs with a higher probability of formation are those that are formed in the critical regime. The probability distribution function is exponentially suppressed beyond the threshold, and therefore, the scaling law can be directly used for estimations of PBH abundances.

On the other hand, some studies have also considered the maximum size of PBHs at the formation time and the accretion effect from the surrounding background in the final mass of the PBH $M_{\mathrm{PBH},f}$. It was already noticed in [74,86] that the accretion effect for small PBHs should be small. Specifically, in [86], some analytical expressions were derived for the upper bound of a PBH formed by the collapse of hydrodynamical perturbations. In [87], the $M_{\mathrm{PBH},f}$ was computed numerically for different curvature profiles and the accretion effect was quantified, which could be substantially large for PBHs formed beyond the critical regime $O(10\%)$.

In conclusion, considering the recent developments in the field and future perspectives, we present this review paper to give the current state-of-the-art regarding the analytical, numerical, and theoretical results about PBH formation from the collapse of hydrodynamical perturbations on an FLRW background under spherical symmetry, currently the most common and popular scenario in the literature [39]. We hope that this review paper allows the scientific community to be introduced to the topic of PBH formation and understand the main insights.

We want to tell the reader that there are fascinating and valuable review papers on the PBH topic with a different perspective than the one considered here. For an interesting historical review of the dark matter candidates, we refer the reader to [2]. Regarding reviews that focus on gravitational waves applied to PBHs and the different constraints to account for dark matter, we have [3,14,15,39,88–91]. Another interesting review with potential applications to PBHs is [92], which is focused on scenarios with deviations from radiation domination in the early Universe.

This review paper is organised as follows: In Section 2, we present the basics of PBH formation. In Section 3, we give some basic details about the statistical estimation of

PBH abundances. In Section 4, we give the necessary ingredients to set up the formalism of PBH formation, in particular: the differential equations that describe the collapse of the perturbations, the initial conditions, the definition of the compaction function, and the threshold criteria. In Section 5, we discuss and comment on the different numerical techniques employed in the literature, and we focus on a specific method that is publicly available. In Section 6, we give numerical results regarding the threshold and PBH mass in terms of the specific profiles using numerical simulations. In Section 7, we discuss in detail the different developments for the analytical estimation of the threshold. Finally, in Section 8, we enumerate other scenarios of PBH formation not explicitly considered in this review.

2. Basics of PBH Formation

As we pointed out in the previous Section 1, PBHs could have been formed in the very early Universe during radiation domination due to the gravitational collapse of large curvature perturbations generated during inflation [9,10].

Under this scenario, the collapse or dispersion of those hydrodynamic perturbations depends on the perturbation's strength: If its greater than a given threshold, the perturbation will collapse and form a BH after the perturbation re-enters the cosmological horizon. Otherwise, if it is lower, it will disperse because of pressure gradients preventing the collapse (a schematic picture can be found in Figure 1). Both things could happen, i.e., the perturbation could undergo gravitational collapse and subsequently bounce. In this case, the fluid is dispersed and collapsed continuously, making rarefaction waves (we will see this in more detail in Section 6). This behaviour is particularly evident when the initial strength of the perturbation is very close to its threshold value. Therefore, pressure gradients play a crucial role in the collapse of the perturbation. If pressure gradients are strong enough, they will prevent gravitational collapse, which implies having a large threshold compared to the situation when the pressure gradients are small.

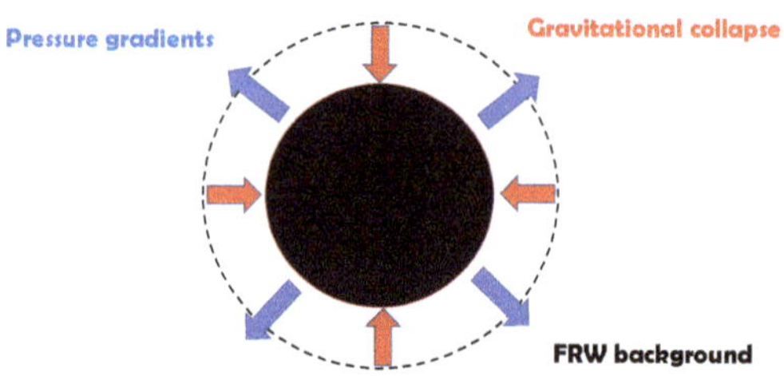

Figure 1. Schematic picture of the collapse/dispersion of a perturbation on an FLRW background.

A big effort has been made during the past few decades to find the correct PBH formation criteria. The first estimation of a threshold was given by [9], using a Jeans length argument and Newtonian gravity. The criterion for the formation of a BH used was that the size of an over-density at the maximum expansion should be larger than the Jeans length, but also smaller than the particle horizon. This translates to the requirement that the peak value of the density contrast at scales smaller than the cosmological horizon must be at least w (where w is the equation of state of the perfect fluid that fills the FLRW Universe) in order to collapse.

A simple, intuitive picture of PBH formation and its threshold connected with the Jeans length argument can be found in [15], and we summarise it here. Consider a cosmological perturbation at super-horizon scales. The spacetime metric can be written as a closed FLRW Universe (we will see this in more detail in Section 4) with a radial dependence on $K(r)$,

$$ds^2 = -dt^2 + a^2(t)\left[\frac{dr^2}{1-K(r)r^2} + r^2 d\Omega^2\right]. \tag{1}$$

where $a(t)$ is the scale factor. If we ignore the radial derivative of $K(r)$, the time–time component of the Einstein equations is given by,

$$H^2 + \frac{K(r)}{a^2} = \frac{8\pi}{3}\rho, \tag{2}$$

which basically corresponds to the FLRW equation, but with a small non-homogeneity term given by $K(r)$. The previous equation allows defining a density contrast as,

$$\frac{\delta\rho}{\rho_b} = \frac{\rho - \rho_b}{\rho_b} = \frac{3K}{8\pi\rho_b a^2} = \frac{K}{(aH)^2}, \tag{3}$$

where ρ_b is the energy density of the background. A region will collapse instead of expanding when $K > 0$ if we ignore the radial dependence on $K(r)$. This precisely happens when $3K/a^2 = 8\pi\rho$, which corresponds roughly to when the Hubble scale becomes equal to the length scale of the curved region. In this case, the assumption made with Equation (1) is no longer valid, but still can be assumed to obtain some threshold estimation.

When $\delta\rho(t_c)/\rho_b = 1$ is the time when a homogeneous and isotropic Universe would stop expanding, we therefore assume that this epoch is the time of black hole formation $t = t_c$, in particular,

$$\frac{\delta\rho}{\rho_b}(t_c) = \frac{K}{k^2 c_s^2} = 1. \tag{4}$$

The criterion is that perturbations cannot collapse if they have smaller scales than the Jeans length, which immediately implies that $k^2 = a^2H^2/c_s^2$, where $c_s = \sqrt{w}$ is the sound speed of the fluid (perfect fluid in our case) and k is the wavelength mode associated with the perturbation. Therefore, the condition that should satisfy the density contrast at the time when the perturbation re-enters the horizon ($H^2(t_m)a^2(t_m) = k^2$) is given by:

$$\left(\frac{\delta\rho}{\rho_b}(t_m)\right)_c = \frac{K}{H^2(t_m)a^2(t_m)} = c_s^2\frac{k^2}{H^2(t_m)a^2(t_m)} = c_s^2 = w, \tag{5}$$

which is basically the threshold value estimation obtained in [9].

The second analytical estimation came years later in [75]. The authors considered a compensated "three-zone" model, i.e., a central over-dense region followed by an under-dense layer (which compensates the over-density), and finally, the FLRW background, to estimate the threshold for BH formation. Following an argument about the sound waves at the maximum expansion of the perturbation, it was found that the threshold of the over-density (at $r = 0$) when it renters the horizon must be $(3(1+w)/(5+3w)) \cdot \sin^2(\pi\sqrt{w}/(1+3w))$.

Later on, however, it became clear that the threshold depends on the shape of the curvature perturbation [63,65,82]. Moreover, the threshold also depends on the specific equation of state of the fluid since pressure gradients play a role in its determination: the larger is w, the stronger are the pressure gradients, and therefore, the higher is the threshold.

In [77], a new criterion for PBH formation was introduced. Importantly, it was found that the peak of the compaction function (the average mass excess on a given volume) was a good criterion for PBH formation, and it currently has become the standard definition of the threshold, which we call δ_c. Prior to this criterion, some other measures for the threshold "definition" were not empty of some issues. See [79] for a historical perspective about that.

In [79], simulations were performed for a radiation fluid with a set of curvatures profiles leading to a range of threshold values given by $0.41 \lesssim \delta_c \leq 2/3$ using the criterion of the compaction function's peak as the threshold definition. Later on, in [83], the minimum threshold value was refined, obtaining 0.4 rather than 0.41 in the case of a radiation fluid. This difference was due to the choice of unphysical profiles in some extreme limit in [79].

Although δ_c is a profile-dependent quantity, in [83,84], it was shown that there exists an (approximately) universal value given by the averaged critical compaction function, which allows building a sufficiently accurate analytical expression for δ_c, which only depends on the type of fluid and the curvature around the peak of the compaction function.

On the other hand, if a perturbation is sufficiently strong, it will collapse and form an apparent horizon. Later on, it will follow a process of accretion of energy density from the surrounded FLRW background until reaching a stationary state, when the final mass of the PBH is achieved. As was shown in [62,63], the final PBH mass follows a critical scaling law when the amplitude of the perturbation is close to its threshold value (the critical regime),

$$M_{BH} = \mathcal{K} M_H (\delta - \delta_c)^{\gamma}, \tag{6}$$

with values of γ consistent with the previous numerical computation [85,93] $\gamma \approx 0.356$. The scaling exponent is universal and only depends on the type of fluid, not on the initial condition. M_H is the mass of the cosmological horizon at the time of horizon crossing (when the length scale of the perturbations equals the cosmological horizon). The constant $\mathcal{K}$ depends on each initial condition used. As in the case of the threshold, the values of $\mathcal{K}$ are different depending on what the initial conditions are, but with values $O(1)$ [94].

The simulations in [63] were performed for $\delta - \delta_c \gtrsim 10^{-5}$, which was not sufficient to test the critical regime up to very small values, i.e., up to machine precision $\delta - \delta_c \approx 10^{-15}$. In [73], the scaling law up to machine precision was finally verified, and this was tested with an explicit example of a Gaussian perturbation.

3. The Abundance of PBHs

This section gives a brief review of the two approaches commonly used in the literature to perform PBH statistics (formed in a radiation-dominated Universe) and estimate their abundances. In this review, we do not focus on the statistical estimation of PBH abundances, for which there are extensive works in this direction, with different approaches and methods. If the reader is interested, we suggest checking these references and the ones therein [60,95–101,101–114]. However, to show this topic at the basic level, we basically describe the Press–Schechter [115] and the peak theory [116] procedures. This section aims to clearly point out the relation between the threshold and PBH mass on the estimation of PBH abundances, especially why the threshold should be known with enough accuracy to be useful for statistics.

In both cases, we consider the amplitude of the peak of density contrast in the comoving slicing at super-horizon scales (usually denoted in the literature as $\Delta \equiv \delta\rho(r=0,t)/\rho_b(t)$ [60,117]), as a statistically distributed variable. The power spectrum associated with the density contrast is defined as,

$$P_{\Delta}(k,t)\delta_D(k,k') = \frac{1}{(2\pi)^3} < \Delta(k,t)\Delta(k',t) >, \tag{7}$$

where δ_D is the Dirac delta and the moments j of P_Δ are given by,

$$\sigma_j^2(t) = \int \frac{k^2 dk}{2\pi^2} P_{\Delta}(k,t) k^{2j}. \tag{8}$$

In the Press–Schechter formalism, we make two assumptions: (i) the density contrast field Δ is a Gaussian variable; (ii) perturbations with $\Delta > \Delta_c$ will collapse and form a PBH. Therefore, basically, we integrate the probability distribution $P(\Delta)$ over the range $\Delta_c \leq \Delta < \Delta_{\rm max}$, where $\Delta_{\rm max}$ is the maximum allowed value. In practice, we integrate up to $\Delta_{\rm max} \to \infty$ since the probability distribution is a rapidly decreasing function above Δ_c, and therefore does not change the result and allows simplifying the computation. It is important to notice that $\Delta_c \neq \delta_c$, since we are comparing the critical density contrast with the peak of the critical compaction function (as we will see later, it is related to the averaged density contrast).

Finally, consider a Gaussian distribution probability distribution,

$$P(\Delta) = \frac{1}{\sqrt{2\pi}\sigma_0} e^{-\frac{\Delta^2}{2\sigma_0^2}}; \tag{9}$$

the abundance of PBHs can be computed as,

$$\begin{aligned} \beta = \frac{\rho_{\text{PBH}}}{\rho_{\text{tot}}} &= 2\int_{\Delta_c}^{\infty} \frac{M_{\text{PBH}}}{M_H} P(\Delta) d\Delta = 2\int_{\Delta_c}^{\infty} \mathcal{K}(\Delta-\Delta_c)^{\gamma} P(\Delta) d\Delta \\ &= \frac{1}{\sqrt{\pi}} 2^{-\frac{1+\gamma}{2}} \mathcal{K}\Delta_c \sigma_0^{-1+\gamma} \Gamma(1+\gamma) \Gamma_U\left(1+\frac{\gamma}{2}, \frac{3}{2}, \frac{\Delta_c^2}{2\sigma_0^2}\right) e^{-\frac{\Delta_c^2}{2\sigma_0^2}}, \end{aligned} \tag{10}$$

where $\Gamma_U(a,b,c)$ is the confluent hypergeometric function, [1] and the factor 2 at the beginning of the integral put by hand is introduced to avoid the under-counting that happens in Press–Schechter theory. It has been assumed that the scaling law is accurate always even when $\Delta \gg \Delta_c$, something not true, as we will show in Section 6. However, since the probability distribution is exponentially smaller for large Δ, Equation (10) is accurate. In the case of considering a monochromatic mass spectrum, i.e., all the PBHs formed to have the same mass (in particular, the horizon mass M_H), the abundance is given by [2],

$$\beta = \bar{\mathcal{K}}\,\text{erf}\left(\frac{\Delta_c}{\sqrt{2}\sigma_0}\right). \tag{11}$$

This can be obtained by simply taking the limit $\gamma \to 0$ and $\mathcal{K} = \bar{\mathcal{K}}$ in Equation (10).

The procedure in the peak theory approach is a bit different. It makes statistics on counting the numbers of over-threshold peaks on the over-density; another approach used the curvature fluctuation ζ peaks instead [102,118].

We summarise the procedure performed in [60] to estimate the abundances through peak theory.

First of all, we define the variable $\nu = \Delta/\sigma_0$, for which in the rare peak assumption, we have that $\nu \gg 1$. The number of rare peaks is given by [116],

$$\mathcal{N}(\nu) = \frac{1}{a(t_f)^3}\frac{1}{4\pi^2}\left(\frac{\sigma_1}{3\sqrt{\sigma_0}}\right)^3 \nu^3 e^{-\nu^2/2}\theta(\nu-\nu_c), \tag{12}$$

where t_f is the time when the PBHs are formed. It is clear that we consider only that peaks higher than the threshold value ν_c will contribute to the formation of PBHs. Therefore, the abundance of PBHs measured at formation with respect to the background energy-density can be computed as,

$$\beta = \int_{\nu_c}^{\infty} \frac{\rho_{\text{PBH}}(\nu)}{\rho_b(t_f)} d\nu = \int_{\nu_c}^{\infty} \frac{M_{\text{PBH}}(\nu)\mathcal{N}(\nu)}{\rho_b(t_f)} d\nu. \tag{13}$$

The scaling law mass, in terms of ν, is given by,

$$M_{\text{PBH}} = \mathcal{K} M_H \left(\frac{\tilde{\sigma}_0}{a_m^2 H_m^2}\right)^{\gamma} (\nu-\nu_c)^{\gamma}, \tag{14}$$

where $\tilde{\sigma}_0 = \sigma_0 a^2 H^2$ and t_m is the time of horizon crossing. Taking into account that numerical simulations have shown $a_f \approx a_m \approx 3$ [79], a_f is weakly dependent on ν, as pointed out in [60], and using the saddle point approximation [119] $\nu_s \approx \nu_c + \gamma/\nu_c$, finally, we obtain,

$$\beta \approx \sqrt{\frac{2}{\pi}} \mathcal{K} \left(\frac{k_*}{a_m H_m}\right)^3 \left(\frac{\tilde{\sigma}_0}{a_m^2 H_m^2}\right)^{\gamma} \nu_c^{1-\gamma} \gamma^{\gamma+1/2} e^{-\nu_c^2/2}, \tag{15}$$

where $k_* = \sigma_1/(\sqrt{3}\sigma_0)$. From Equations (10), (11) and (15), we clearly see the exponential dependence on the threshold for PBH formation and the linear dependence on the constant $\mathcal{K}$ associated with the scaling law. This is the reason why an accurate determination of the threshold is essential, since the abundances depend exponentially on it. These two methods are however only an approximation of the true statistics, which consider the fact that each statistical realisation of the profiles has a different threshold. This was developed in [95], so we will not discuss it here as it is beyond the scope of this review paper. However, as can be seen in [95], one finds again the generic behaviour that β is exponentially sensitive to the threshold. We also mention that, in principle, a window function has to be used to estimate PBH abundances correctly. It is necessary to relate the power spectrum in Fourier space to the probability distribution function in real space by coarse-graining the cosmological perturbations. The necessity of considering a window function is remarkably important in the case of having a broad power spectrum (where several wavelength scales k are involved). Despite that, the choice of the window function is not unique, and the result depends on the choice used. We refer the reader to [105,111,120] to check the details.

4. Cosmological Setup for PBH Formation

This section reviews the needed ingredients to simulate the formation of PBHs in an FLRW Universe under spherical symmetry and filled by a perfect fluid. We will see what differential equations describe the gravitational collapse, the consistent initial conditions for PBH formation we need to use, and the boundary conditions to supply. Later on, we will study the formation of the apparent horizon and characterise it in terms of the expansion of the congruences. Finally, we will describe the process of the accretion of the PBH mass from the FLRW background and the estimation of the final mass.

4.1. Misner–Sharp Equations

The Misner–Sharp equations [70] describe the motion of a relativistic fluid with spherical symmetry. This corresponds to the Einstein field equations written in the comoving gauge (the gauge we use in this review paper). To obtain them, first of all, we need to consider a perfect fluid with the energy-momentum tensor,

$$T^{\mu\nu} = (p+\rho)u^{\mu}u^{\nu} + pg^{\mu\nu}, \tag{16}$$

and with the following spacetime metric in spherical symmetry:

$$ds^2 = -A(r,t)^2dt^2 + B(r,t)^2dr^2 + R(r,t)^2d\Omega^2, \tag{17}$$

where $R(r,t)$ is the areal radius, $A(r,t)$ is the lapse function, and $d\Omega^2 = d\theta^2 + \sin^2(\theta)d\phi^2$ is the line element of a two-sphere. The components of the four-velocity u^{μ} are given by $u^t = 1/A$ and $u^i = 0$ for $i = r, \theta, \phi$, since we are considering comoving coordinates (comoving gauge). Throughout the review paper, we use units $G_N = c = 1$.

Solving the Einstein field equations for this problem, the following quantities appear:

$$\begin{aligned} \frac{1}{A(r,t)}\frac{\partial R(r,t)}{\partial t} &\equiv D_t R \equiv U(r,t), \\ \frac{1}{B(r,t)}\frac{\partial R(r,t)}{\partial r} &\equiv D_r R \equiv \Gamma(r,t), \end{aligned} \tag{18}$$

where D_t and D_r are the proper time and distance derivatives, respectively. U is the radial component of the four-velocity associated with an Eulerian frame (so not comoving), which measures the radial velocity of the fluid with respect to the origin of the coordinates. The Misner–Sharp mass $M(r,t)$ is introduced as:

$$M(R) \equiv \int_0^R 4\pi\tilde{R}^2\rho\, d\tilde{R}, \tag{19}$$

which is related with Γ, U and R though the constraint:

$$\Gamma = \sqrt{1 + U^2 - \frac{2M}{R}}, \tag{20}$$

where Γ is called the generalised Lorentz factor, which includes the gravitational potential energy and kinetic energy per unit mass. This also can be seen in the Newtonian limit,

$$\Gamma \sim 1 + \frac{1}{2}\frac{v^2}{c^2} - \frac{GM}{c^2 R}. \tag{21}$$

Finally, the differential equations governing the evolution of a spherically symmetric collapse of a perfect fluid in general relativity are:

$$D_t U = -\left[\frac{\Gamma}{(\rho + p)} D_r p + \frac{M}{R^2} + 4\pi R p\right], \tag{22}$$

$$D_t R = U, \tag{23}$$

$$D_t \rho = -\frac{(\rho + p)}{\Gamma R^2} D_r (U R^2), \tag{24}$$

$$D_t M = -4\pi R^2 U p, \tag{25}$$

$$D_r M = 4\pi \Gamma \rho R^2, \tag{26}$$

$$D_r A = \frac{-A}{\rho + p} D_r p. \tag{27}$$

We refer the reader to [70] to see the details of the derivation. The boundary conditions in the Misner–Sharp equation are $R(r = 0, t) = 0$, leading to $U(r = 0, t) = 0$ and $M(r = 0, t) = 0$. Then, by spherical symmetry, we have $D_r p(r = 0, t) = 0$, and also, we consider a perfect fluid $p = w\rho$.

At $r \to \infty$, we should recover the solution of the FLRW background. However, in a numerical simulation, we have to handle the discretisation of time and space. Therefore, the condition $D_r p(r = r_f, t) = 0$ is used (where r_f is the outer point of the grid) to match with the FLRW solution and avoid reflections from density waves. On the other hand, Equation (26) is called the Hamiltonian constraint, which is commonly used to check the correct solution of the numerical equations. For the case $p = w\rho$, the lapse equation in Equation (27) can be solved considering $A(r_f, t) = 1$ to match with the FLRW spacetime,

$$A(r, t) = \left(\frac{\rho_b(t)}{\rho(r, t)}\right)^{\frac{w}{w+1}}, \tag{28}$$

where $\rho_b(t) = \rho_0 (t_0/t)^2$ is the energy density of the FLRW background and $\rho_0 = 3H_0^2/8\pi$.

Actually, the Misner–Sharp equations can be written in a more advantageous way to perform numerical simulations using the definitions of Equation (18),

$$\dot{U} = -A\left[\frac{w}{1+w}\frac{\Gamma^2}{\rho}\frac{\rho'}{R'} + \frac{M}{R^2} + 4\pi R w \rho\right], \tag{29}$$

$$\dot{R} = AU, \tag{30}$$

$$\dot{\rho} = -A\rho(1 + w)\left(2\frac{U}{R} + \frac{U'}{R'}\right), \tag{31}$$

$$\dot{M} = -4\pi A w \rho U R^2, \tag{32}$$

$$M' = 4\pi \rho R^2 R', \tag{33}$$

where the dot $\dot{U}$ represents the time derivate $\partial U/\partial t$ and ρ' the radial derivative $\partial \rho/\partial r$.

4.2. The Gradient Expansion Approximation, Initial Conditions, and Compaction Function

The gradient expansion method [121] (also called long-wavelength approximation) was used in [122] to set up consistent initial conditions for PBH formation on an FLRW background at super-horizon scales. First of all, let us consider a cosmological perturbation at super-horizon scales, i.e., with a length scale R_m much larger than the Hubble horizon. We can define a parameter ϵ to relate the two scales: the Hubble horizon R_H and the length scale of the perturbation R_m,

$$\epsilon = \frac{R_H(t)}{R_m(t)}, \tag{34}$$

where $R_H = 1/H$. It is clear that at super-horizon scales, we will have $\epsilon \ll 1$. The gradient expansion method allows expanding these inhomogeneities in the spatial gradient in terms of ϵ. In the limit $\epsilon \to 0$, the spacetime locally corresponds to the FLRW metric, when the perturbation is smoothed out at sufficiently large scales R_m.

Consider a general spacetime metric in the $3+1$ Arnowitt–Deser–Misner (ADM) formalism [123,124],

$$ds^2 = -\alpha^2 dt^2 + \gamma_{ij}(dx^i + \beta^i dt)(dx^j + \beta^j dt); \tag{35}$$

in general, the spatial part of the metric can be decomposed into:

$$\gamma_{ij} = a^2(t) e^{2\zeta(t,x^i)} \tilde{\gamma}_{ij} \tag{36}$$

where γ_{ij}, β_i, and α are the spatial metric, shift-vector, and lapse function, respectively. ζ is usually called the curvature fluctuation. Under the gradient expansion method, it was shown in [77,125–127] that $\beta = O(\epsilon)$, $\alpha = 1 + O(\epsilon^2)$, and $\tilde{\gamma}_{ij} = \delta_{ij} + O(\epsilon^2)$. Therefore, in spherical symmetry and in the limit $\epsilon \to 0$, the metric of Equation (35) can be written as,

$$ds^2 = -dt^2 + a^2(t) e^{2\zeta(\tilde{r})}(d\tilde{r}^2 + \tilde{r}^2 d\Omega^2). \tag{37}$$

Notice the time independence of $\zeta(\tilde{r})$ since at super-horizon scales, it is shown that $\dot{\zeta} = O(\epsilon^2)$. It is important to point out that although, we consider the comoving gauge, other gauges are possible. In particular, in [77], the constant mean curvature gauge was used instead. However, as was shown in [76,126], the differences between the two gauges in ζ are of $O(\epsilon^2)$.

Finally, the metric Equation (37) can be also written in other coordinates as an FLRW metric with a non-constant curvature $K(r)$,

$$ds^2 = -dt^2 + a^2(t)\left[\frac{dr^2}{1-K(r)r^2} + r^2 d\Omega^2\right]. \tag{38}$$

The change of coordinates $r, \tilde{r}$ between the two metrics and the conversion between $K(r), \zeta(\tilde{r})$ are given by the following relations, which were obtained in several works previously (not necessarily in the context of PBH formation for all of them) [79,81,109,128]:

$$\begin{aligned}
r &= \tilde{r} e^{\zeta(\tilde{r})}, \\
\frac{dr}{d\tilde{r}} &= e^{\zeta(\tilde{r})}\left[1 + \tilde{r}\zeta'(\tilde{r})\right], \\
K(r)r^2 &= -\tilde{r}\zeta'(\tilde{r})\left[2 + \tilde{r}\zeta'(\tilde{r})\right], \\
\zeta(\tilde{r}) &= \int_\infty^r \left(1 - \frac{1}{\sqrt{1-K(\hat{r})\hat{r}^2}}\right)\frac{d\hat{r}}{\hat{r}}, \\
\tilde{r} &= r\,\exp\left[\int_\infty^r \left(\frac{1}{\sqrt{1-K(\hat{r})\hat{r}^2}} - 1\right)\frac{d\hat{r}}{\hat{r}}\right]
\end{aligned} \tag{39}$$

Therefore, the cosmological perturbation will be characterised by the curvature perturbation $K(r), \zeta(\tilde{r})$. The Misner–Sharp equations can be solved at leading order in $\epsilon \ll 1$ using the long wavelength approximation as was performed in [122],

$$\begin{aligned}
A(r,t) &= 1+\epsilon^2(t)\tilde{A}, \\
R(r,t) &= a(t)r(1+\epsilon^2(t)\tilde{R}), \\
U(r,t) &= H(t)R(r,t)(1+\epsilon^2(t)\tilde{U}), \\
\rho(r,t) &= \rho_b(t)(1+\epsilon^2(t)\tilde{\rho}), \\
M(r,t) &= \frac{4\pi}{3}\rho_b(t)R(r,t)^3(1+\epsilon^2(t)\tilde{M}),
\end{aligned} \tag{40}$$

where for $\epsilon \to 0$, we recover the (FLRW) solution. The perturbations of the tilde variables at first order in gradient expansion were computed in [122], which we summarise here:

$$\begin{aligned}
\tilde{\rho} &= \frac{3(1+w)}{5+3w}\left[K(r)+\frac{r}{3}K'(r)\right]r_m^2, \\
\tilde{U} &= -\frac{1}{5+3w}K(r)r_m^2, \\
\tilde{A} &= -\frac{w}{1+w}\tilde{\rho}, \\
\tilde{M} &= -3(1+w)\tilde{U}, \\
\tilde{R} &= -\frac{w}{(1+3w)(1+w)}\tilde{\rho}+\frac{1}{1+3w}\tilde{U}.
\end{aligned} \tag{41}$$

The perturbations in the $\tilde{r}$ coordinate with $\zeta(\tilde{r})$ are [79],

$$\begin{aligned}
\tilde{\rho} &= -\frac{2(1+w)}{5+3w}\frac{\exp[2\zeta(\tilde{r}_m)]}{\exp[2\zeta(\tilde{r})]}\left[\zeta''(\tilde{r})+\zeta'(\tilde{r})\left(\frac{2}{\tilde{r}}+\frac{\zeta'(\tilde{r})}{2}\right)\right]\tilde{r}_m^2, \\
\tilde{U} &= \frac{1}{5+3w}\frac{\exp[2\zeta(\tilde{r}_m)]}{\exp[2\zeta(\tilde{r})]}\zeta'(\tilde{r})\left[\frac{2}{\tilde{r}}+\zeta'(\tilde{r})\right]\tilde{r}_m^2.
\end{aligned} \tag{42}$$

From Equations (41) and (42), an expression for the density contrast at super-horizon scales in terms of the curvature fluctuations can be obtained [79], given by:

$$\begin{aligned}
\frac{\delta\rho}{\rho_b}(r) &= \frac{3(1+w)}{5+3w}\left(\frac{1}{aH}\right)^2\left[K(r)+\frac{r}{3}K'(r)\right], \\
\frac{\delta\rho}{\rho_b}(\tilde{r}) &= -\frac{3(1+w)}{5+3w}\left(\frac{1}{aH}\right)^2\exp[-2\zeta(\tilde{r})]\left[\zeta''(\tilde{r})+\zeta'(\tilde{r})\left(\frac{2}{\tilde{r}}+\frac{\zeta'(\tilde{r})}{2}\right)\right].
\end{aligned} \tag{43}$$

The areal radius of the length scale R_m is given by $R_m(t) = a(t)r_m$ or also $R_m(t) = a(t)\tilde{r}_m e^{\zeta(\tilde{r}_m)}$ with the $\tilde{r}$ coordinate taking into account Equation (39). As was shown in [129], where the gradient expansion method beyond first order was applied using an iterative scheme (see [129] for more terms in the ϵ expansion in comparison with Equation (40)), the long-wavelength approximation at first order is accurate enough if a sufficiently small ϵ parameter is taken.

When we take $\epsilon = 0$, we recover the background solution equations: $H(t) = H_0 t_0/t$, $a(t) = a_0(t/t_0)^\alpha$, and $R_H(t) = R_H(t_0)(t/t_0)$ where $a_0 = a(t_0)$, $H_0 = H(t_0) = \alpha/t_0$, and $R_H(t_0) = 1/H_0$. Moreover, we define $\alpha = 2/3(1+w)$. We also consider as the initial condition that $r_m = \Lambda R_H(t_0)$, where Λ is some number (usually taken as $O(10)$ in simulations). A time scale given when $\epsilon(t_m) = 1$ is commonly defined in the literature, which precisely corresponds to the time when the Hubble horizon equals the length scale of the perturbation (the time of horizon crossing) $t_m = t_0(a_0 r_m/R_H(t_0))^{1/(1-\alpha)}$. Notice that t_0 is the initial time, and it is assumed to be a time where the perturbation is at super-horizon scales.

On the other hand, the amplitude of a cosmological perturbation can be measured as the averaged density contrast within a spherical region:

$$\bar{\delta}(R) = \frac{1}{V}\int_0^R 4\pi \frac{\delta\rho}{\rho_b}\hat{R}^2\, d\hat{R}, \tag{44}$$

where $\delta\rho = \rho - \rho_b$ and $V = 4\pi R^3/3$.

Related to that, in [77], the compaction function $\mathcal{C}(r)$ was defined, which gives a measure of the mass excess on a given volume. In particular,

$$\mathcal{C}(r,t) = \frac{2[M(r,t) - M_b(r,t)]}{R(r,t)}, \tag{45}$$

where $M_b(r,t)$ is the mass of the FLRW background in the volume V. The compaction function can also be written in terms of $\bar{\delta}$, as was shown in [76],

$$\mathcal{C}(R) = \bar{\delta}(RH)^2. \tag{46}$$

At leading order in ϵ,

$$\bar{\delta}(r,t) = \left(\frac{1}{aHr_m}\right)^2 \delta(r), \tag{47}$$

where:

$$\delta(r) = f(w)K(r)r_m^2, \tag{48}$$

$$f(w) = 3(1+w)/(5+3w). \tag{49}$$

From the above definitions, we finally define the compaction function at super-horizon scales at leading order of gradient expansion as:

$$\mathcal{C}(r,t) \simeq \mathcal{C}(r) = f(w)K(r)r^2 = \frac{r^2}{r_m^2}\delta(r), \tag{50}$$

which yields $\mathcal{C}(r_m) = \delta(r_m) = \delta_m$, i.e., $\delta_m = f(w)K(r_m)r_m^2$. The previous equation can be obtained introducing the initial condition of Equation (40) into Equation (45) and making an expansion in ϵ. Notice in Equation (50) the time-independence of the compaction function at super-horizon scales at leading order in gradient expansion, since the curvature fluctuations remain frozen at super-horizon scales, as we have mentioned before. The compaction function at super-horizon scales is an essential magnitude, since it allows defining the threshold. In particular, we now define the location of the maximum of $\mathcal{C}(r)$ as r_m, and its value $\mathcal{C}_{\max} = \mathcal{C}(r_{\rm m})$ is used as a criterion for PBH formation, as was proposed in [77], while r_m ($\tilde{r}_m$ in the case of using the $\tilde{r}$ coordinate) is considered the length scale of the perturbation. The threshold for PBH formation will correspond to the peak value of the critical compaction function, i.e., $\mathcal{C}_c(r_m) = \delta_c$. Therefore, we define the "amplitude" of the perturbation as $\delta_m = \mathcal{C}(r_m)$. Perturbations with $\delta_m > \delta_c$ will collapse and form a PBH. In the opposite case, perturbations with $\delta_m < \delta_c$ will disperse on the FLRW background with no black hole formation. Notice that from Equation (38), we have a bound on the amplitude value δ_m at super-horizon scales given by $\delta_{\rm m,max} = f(w)$, since $K_{\max}(r_m)r_m^2 = 1$.

Because of the previous definitions, the value of r_m is given by the solution of $\mathcal{C}'(r) = 0$:

$$K(r_m) + \frac{r_m}{2}K'(r_m) = 0. \tag{51}$$

The definition of the compaction function at super-horizon scales using the $\zeta(\tilde{r})$ perturbation instead of $K(r)$ leads to a slightly different definition,

$$\mathcal{C}(\tilde{r}) = f(w)\left[1 - \left(1 + \tilde{r}\zeta'(\tilde{r})\right)^2\right], \tag{52}$$

which yields the following condition for the peak $\tilde{r}_m$:

$$\zeta'(\tilde{r}_m) + \tilde{r}_m \zeta''(\tilde{r}_m) = 0, \tag{53}$$

and therefore, $\delta_m = -f(w)[2 + \tilde{r}_m \zeta'(\tilde{r}_m)]\tilde{r}_m \zeta'(\tilde{r}_m)$. Interestingly, in [79], $\bar{\delta}(r_m, t) = 3\delta\rho(r_m, t)/\rho_b$ was found at super-horizon scales, which relates the averaged mass excess with the local density contrast at r_m.

In this review paper, we focus on type I PBHs, which are those that fulfil that the areal radius R is a monotonic function, i.e., $R' > 0$. Instead, type II PBHs satisfy $R' < 0$ [130]. Type II PBHs are still unexplored in the literature from a numerical point of view or even with a general analytical treatment.

4.3. Horizon Formation

If an initial perturbation at super-horizon scales has an amplitude δ_m bigger than its threshold δ_c, the perturbation will continue growing and, at some point, a trapped surface will be formed. This indicates the onset of gravitational collapse.

To identify when trapped surfaces are formed, we have to consider the expansion $\Theta^{\pm}$ of null geodesics' congruences $k^{\pm}$, orthogonal to a spherical surface Σ. The expansion $\Theta^{\pm}$ is defined as $\Theta^{\pm} \equiv h^{\mu\nu}\nabla_\mu k^{\pm}_\nu$, where $h^{\mu\nu}$ is the spacetime metric induced on Σ. There are two congruences: we call them inwards k^{+}_μ and outwards k^{-}_μ, whose components are $k^{\pm}_\mu = (A, \pm B, 0, 0)$ with $k^{+} \cdot k^{-} = -2$.

In the case of flat spacetime, $\Theta^{-} < 0$ and $\Theta^{+} > 0$, and these surfaces Σ are called normal surfaces. On the other hand, if $\Theta^{\pm} < 0$, the surface is called trapped, while if both are positive $\Theta^{\pm} > 0$, the surface is anti-trapped. In our case, we have that,

$$\Theta^{\pm} = \frac{2}{R}(U \pm \Gamma). \tag{54}$$

In spherical symmetry, we can consider that any point (r, t) is a closed surface Σ with a proper radius R. These points can be classified as normal, trapped, and anti-trapped. Specifically, the transition from a normal to a trapped surface should satisfy $\Theta^{-} < 0$ and $\Theta^{+} = 0$, which corresponds to a marginally trapped surface, usually called the "apparent horizon". Taking into account that $\Theta^{+}\Theta^{-} = \frac{4}{R^2}(U^2 - \Gamma^2)$, the condition for the apparent horizon is given by $U^2 = \Gamma^2 \Rightarrow 2M = R$.

In Figure 2, $2M/R$ and the congruences $\Theta^{\pm}$ for a supercritical (with an amplitude $\delta_m > \delta_c$) Gaussian curvature fluctuation can be seen, after the first horizon has been formed. The same qualitative behaviour can be applied for any perturbation of type I PBHs (with $w \neq 0$).

Looking at Figure 2, we can see three points where $2M = R$, and therefore three different surfaces that we can characterise:

- Surface (A–B): In this region, we have a transition between $\Theta^{+} > 0$ and $\Theta^{-} < 0$ (normal region) to $\Theta^{+} < 0$ and $\Theta^{-} < 0$. This horizon, moves inwards of the computational domain, which eventually will encounter a singularity at $r = 0$;
- Surface (B–C): In this region, we have a transition between $\Theta^{+} < 0$ and $\Theta^{-} < 0$ to $\Theta^{+} > 0$ and $\Theta^{-} < 0$ (normal region). This horizon, typically called "the apparent horizon", moves outwards of the computational domain. Actually, this horizon and the previous one emerge from a single marginally trapped surface. During the evolution, the motion of the horizon can be followed to obtain the PBH mass $M(r_*, t)$, where $r_*(t)$ is the location of the outer horizon in time. At the final stage of the gravitational collapse for very late times, this horizon will become static, i.e., $r_* =$ const;
- Surface (C–D): In the last region, we have a transition between $\Theta^{+} > 0$ and $\Theta^{-} < 0$ (normal region) to $\Theta^{+} > 0$ and $\Theta^{-} > 0$. This case corresponds to an anti-trapped surface, which corresponds to the cosmological horizon, moving outwards. We should mention that this horizon does not correspond to the cosmological horizon R_H from the

FLRW background, since the perturbed medium (deviating from the FLRW solution) affects the evolution of the cosmological horizon.

For a more formal discussion about horizons, we suggest the reader check [131–140].

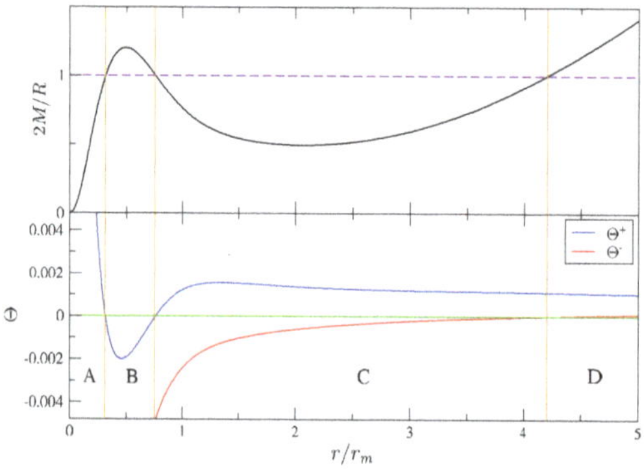

Figure 2. (**Top-panel**): Profile of $2M/R$ once the trapped surface has been formed. The dashed line corresponds to $2M/R = 1$. (**Bottom-panel**): Profiles of the two expansions $\Theta^{\pm}$ at the same time t as the top panel. The vertical orange lines correspond to points where $\Theta^{\pm} = 0$. The initial fluctuation considered corresponds to a Gaussian profile.

4.4. PBH Mass and Accretion

Once the apparent horizon has formed, the initial PBH mass, i.e., the mass of the PBH $M_{\mathrm{PBH},i}$ at the moment of formation of the first apparent horizon t_{AH}, will start to grow to arrive at a stationary value $M_{\mathrm{PBH},f}$. This situation is different from the case of dust collapse $w = 0$, where the mass would increase forever since no pressure gradients would avoid the accretion of energy–matter.

The process of accretion of energy–matter from the FLRW background has been studied in the past [9,11,86,141,142] in different scenarios. To estimate the final mass of the PBH, one could follow the motion of the apparent horizon until very late times when the horizon remains static. However, since it could be computationally very expensive, we can use the Zeldovich–Novikov formula Equation (55) [11], which considers Bondi accretion [11,143,144], and in our case, assume that the energy density right outside the apparent horizon decreases as in an FLRW Universe. It is essential to indicate that this is not valid at the time t_{AH}, since it neglects the cosmological expansion of the spacetime [145], but we can apply from sufficiently late time for $t \gg t_{AH}$ considering an effective constant accretion efficiency rate F [143,144], which depends on the non-linear process of the accretion flow and the equation of state. This approximation was already used numerically in the context of PBH formation from domain walls in [146] and from curvature fluctuations in [80] with great success.

In particular, for very late times $t \gg t_{AH}$, the mass accretion follows,

$$\frac{dM}{dt} = 4\pi F R_{\mathrm{PBH}}^{2} \rho_b(t) \,, \tag{55}$$

where F is commonly numerically found to be of order $O(1)$. In the previous equation, the cosmological expansion is neglected, and also, it assumes a quasi-stationary flow onto a PBH. The inaccuracies of these assumptions are absorbed in F. We should also mention that this formula assumes that PBHs are at rest relative to the FLRW background or have negligible peculiar velocities, which is a good approximation in our case. We refer the reader to [142] for a discussion about the scenario with relative velocities.

Taking into account the condition of the apparent horizon in spherical symmetry $R_{\text{PBH}} = 2M_{\text{PBH}}$, the previous equation is solved analytically as:

$$M_{\text{PBH}}(t) = \frac{1}{\frac{1}{M_a} + \frac{3}{2}F\left(\frac{1}{t} - \frac{1}{t_a}\right)}, \tag{56}$$

where M_a is the initial mass when the asymptotic approximation is used at the time t_a. The value of F has been found to weakly depend on the specific shape of the perturbation and its amplitude, but strongly on the equation of state w. Specifically, F depends on the high non-linear process of accretion and the counterpart of pressure gradients. F can be obtained numerically by fitting the evolution of $M_{\text{PBH}}(t)$ (at sufficiently late times after the formation of the apparent horizon, when it is valid). Using that, the PBH mass is estimated as the asymptotic mass value at $t \to \infty$, i.e.,

$$M_{\text{PBH},f} = M_{\text{PBH}}(t \to \infty) = \left(\frac{1}{M_a} - \frac{3F}{2t_a}\right)^{-1}. \tag{57}$$

The final PBH mass will be dependent on the specific profile of the fluctuation, its initial amplitude δ_m, and w. As was already shown in the past, in the critical regime where δ_m is very close to the critical value δ_c, the black hole mass follows a scaling law [63–65],

$$M_{\text{PBH},f} = \mathcal{K} M_H (\delta_m - \delta_c)^{\gamma}, \tag{58}$$

where $\gamma \approx 0.356$ in radiation $w = 1/3$. The values of γ in terms of the equation of state w were found semi-analytically in [67] and are plotted in Figure 3. These results were numerically confirmed in [94].

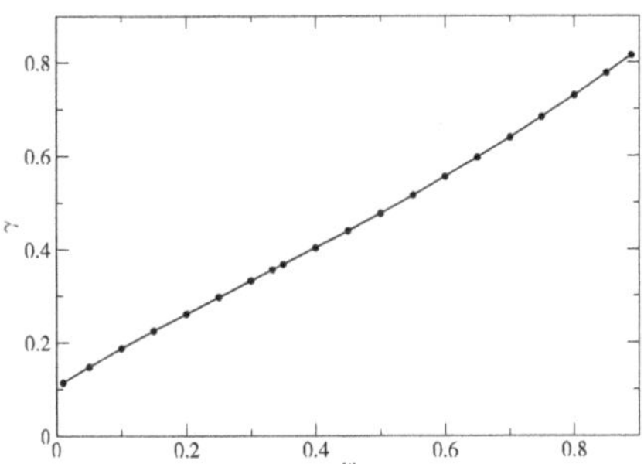

Figure 3. Values of $\gamma(w)$ plotted from the data of Table 1 in [67].

In Equation (58), the constant $\mathcal{K}$ depends on the specific shape of the profile considered and $M_H \equiv 1/2H(t_m)$ is the Hubble mass, which is calculated at the time t_m. Specifically, M_H in terms of the initial values of the perturbation can be written as:

$$M_H = \frac{3(1+w)}{4} t_0 (a_0 \Lambda)^{\frac{3(1+w)}{1+3w}}, \tag{59}$$

which will depend on the initial length scale r_m and the equation of state w (we can consider gauge values $a_0 = 1$ and $t_0 = 1$). In the case of using the $\tilde{r}$ coordinate, we just need to replace $\Lambda \to \Lambda e^{\zeta(\Lambda R_H(t_0))}$ in Equation (59). The scaling law starts to deviate at $(\delta_m - \delta_c) \gtrsim 2 \times 10^{-2}$ (for $w = 1/3$), as was indicated in [94]. It was shown explicitly in [80] that for large δ_m beyond the critical case, the scaling law seems to deviate by order $O(10\%)$ (at least for $w = 1/3$), but it can be smaller or larger depending on the profile considered. The deviation from the critical regime is expected to be also found for other w's, but the quantification still is missing in the literature. As pointed out in [147], it is important to consider the scaling law behaviour Equation (58) for the predictions of the PBH mass spectra.

5. Numerical Techniques

Most of the times, Partial Differential Equations (PDEs) cannot be solved analytically, and numerical methods are needed. This is especially common in general relativity and is the case we have here. Currently, several methods and tools can be employed in numerical relativity [148–156]. In the case of PBH formation, there is the extra difficulty that the gravitational collapse happens on an FLRW background, so not on asymptotically flat spacetime.

Different authors have employed different numerical procedures to simulate PBH formation from the collapse of curvature fluctuations under spherical symmetry, as we mentioned in Section 1. From a historical perspective, several works have used the standard and very well-known finite differences approach to solve Hernandez–Misner–Sharp equations in the comoving gauge [62,63,65,78,79,82,94,122,157]. Alternatively, simulations were also performed at earlier times using a different gauge, in particular the mean curvature gauge [77]. The numerical results of both procedures with different gauges were verified to be consistent with each other in [76] at leading order in gradient expansion. A new numerical method came up recently using pseudospectral methods [80] to solve Misner–Sharp equations together with the use of an excision technique to obtain the final mass of the black hole.

In this review, we describe the pseudospectral collocation technique used in [80]. The basics of the numerical code are available here [158] (currently, the only one in the literature publicly available), and therefore, the reader can reproduce or extend some of the results presented in this review.

5.1. Pseudospectral Methods

Pseudospectral methods have been extensively used in computer science for solving complex physics phenomena [159–161]. The application to the field of numerical relativity has been also very popular to study different problems, from astrophysics to more oriented toward high energy physics [162–165,165–176].

In what follows, we outline how to use the basics of the pseudospectral technique applied to our problem, which allows solving partially algebraically the differential equations describing the gravitational collapse. See also [177,178] for much more details and mathematical foundations. In particular, we use the Chebyshev collocation method (but there are also other formulations).

Let us consider a function $f(x)$ and fit it with $N_{\rm cheb}$ Chebyshev polynomials (any other orthonormal function could be used as well). The function $f(x)$ can be composed by a sum of Chebyshev polynomials as,

$$f_{N_{\rm cheb}}(x) = \sum_{k=0}^{N_{\rm cheb}} c_k T_k(x), \tag{60}$$

where $T_k(x)$ are the Chebyshev polynomials of order k. The coefficients c_k with $k = 0, 1, \ldots, N_{\rm cheb}$ are then obtained by solving $f_{N_{\rm cheb}}(x_k) = f(x_k)$ where $x_k = \cos(k\pi/N_{\rm cheb})$. The points x_k, which satisfy $T'_k(x_k) = 0$, are called Chebyshev collocation points.

Therefore,

$$f_{N_{\rm cheb}}(x) = \sum_{k=0}^{N_{\rm cheb}} L_k(x) f(x_k), \tag{61}$$

$$L_k(x) = \frac{(-1)^{k+1}(1-x^2)T'_{N_{\rm cheb}}(x)}{\bar{c}_k N^2_{\rm cheb}(x - x_k)}, \tag{62}$$

where $\bar{c}_k = 2$ if $k = 0, N$ and $\bar{c}_k = 1$ in the other cases. The functions L_k are the Lagrange interpolation polynomials.

The derivate of p-order of the function will be given by,

$$f^{(p)}_{N_{\rm cheb}}(x_i) = \sum_{k=0}^{N_{\rm cheb}} L^{(p)}_k(x_i) f_{N_{\rm cheb}}(x_k). \tag{63}$$

We define the Chebyshev differentiation matrix as $D^{(p)} = \{L^{(p)}_k(x_i)\}$. The components of the matrix are given by:

$$D^{(1)}_{i,j} = \frac{\bar{c}_i}{\bar{c}_j}\frac{(-1)^{i+j}}{(x_i - x_j)}, (i \neq j), i, j = 1, \ldots, N_{\rm cheb} - 1, \tag{64}$$

$$D^{(1)}_{i,i} = -\frac{x_i}{2(1 - x_i^2)}, i = 1, \ldots, N_{\rm cheb} - 1, \tag{65}$$

$$D^{(1)}_{0,0} = -D^{(1)}_{N_{\rm cheb},N_{\rm cheb}} = \frac{2N^2_{\rm cheb} + 1}{6}. \tag{66}$$

To improve round-off errors in the numerical computations [178], the following identity of the diagonal components of the matrix D is used:

$$D^{(1)}_{i,i} = -\sum_{j=0, j\neq i}^{N_{\rm cheb}} D^{(1)}_{i,j}. \tag{67}$$

In spectral methods, the error decays exponentially with $N_{\rm cheb}$. This represents a significant improvement in comparison with finite differences, where the error decays algebraically as $1/N^v$, where $v > 0$ and N is the number of points of the grid. The exponential convergence in spectral methods is obtained thanks to the fact that the derivate at a given point is computed taking into account all the other points of the grid, instead of finite differences, which only consider the neighbours.

The radial domain Ω in our case corresponds to $\Omega = [r_{\rm min}, r_{\rm max}]$ where $r_{\rm max} = N_H R_H(t_0)$ and $r_{\rm min} = 0$. N_H is the number of initial cosmological horizons where we put the final point of the grid, for which it usually is enough to take $N_H \sim 10^2$. The domain of the Chebyshev polynomials is $[-1, 1]$; therefore, we need to make a mapping between the physical domain and the spectral one. Different options are possible, but a linear mapping is commonly used:

$$\tilde{x}_k = \frac{r_{\rm max} + r_{\rm min}}{2} + \frac{r_{\rm max} - r_{\rm min}}{2} x_k. \tag{68}$$

$\tilde{x}_k$ are the new Chebyshev points rescaled to our physical domain Ω. Furthermore, the Chebyshev matrix can be rescaled using the chain rule:

$$\tilde{D} = \frac{2}{r_{\rm max} - r_{\rm min}} D. \tag{69}$$

The imposition of boundary conditions in spectral methods is straightforward. In the case of the Dirichlet boundary condition at x_k, such that $f(x = x_k) = u_{D,bc}$, we just need to satisfy $f_{N_{\rm cheb}}(x = x_k) = u_{D,bc}$. In the case of the Neumann boundary condition such that $f^{(1)}(x = x_k) = u_{N,bc}$, we should satisfy that $(D \cdot f_{N_{\rm cheb}})(x = x_k) = u_{N,bc}$. The numerical stability will depend on the values chosen for dt and $N_{\rm cheb}$. We do not have complete "freedom" to choose those values. If the number of points of the grid is increased, it will require a reduction of the time step dt, in order to avoid instabilities during the evolution. The Courant–Friedrichs–Lax (CFL) condition for hyperbolic PDEs already indicates that the maximum allowed time step should fulfil $dt \propto 1/N^2$.

There are situations where pressure gradients are substantially large during the numerical evolution. This, for instance, can happen when the curvature fluctuation has very sharp profiles or when w becomes large. In this situation, it is necessary to increase the numerical accuracy of the simulation. One possibility is to increase the number of points of the grid N_{cheb}, but of course, this would imply reducing the time step dt to ensure stability. The other one (which is actually better and more clever) is to use a composite Chebyshev grid: split the full domain into several Chebyshev grids depending on the density of points needed in specific regions of the domain. In particular, the full domain Ω is split into M subdomains as $\Omega_l = [r_l, r_{l+1}]$ with $l = 0, 1 \ldots, M$. A mapping between the spectral and physical domain for each is required. We choose a linear mapping again as,

$$\tilde{x}_{k,l} = \frac{r_{l+1} + r_l}{2} + \frac{r_{l+1} - r_l}{2} x_{k,l}, \tag{70}$$

where $\tilde{x}_{k,l}$ are the new Chebyshev points re-scaled to the subdomain Ω_l. The Chebyshev differentiation matrix is re-scaled again using the chain rule:

$$\tilde{D}_l = \frac{2}{r_{l+1} - r_l} D_l. \tag{71}$$

The numerical evolution is performed independently in each subdomain, i.e., the spatial derivative is computed using the Chebyshev differentiation matrix $\tilde{D}_l$ associated with each subdomain, and the time integration is performed applying the Runge–Kutta 4 method on each subdomain.

To evolve across the different subdomains Ω_l, it is necessary to impose boundary conditions. The approach followed is the one of [179], which is based on performing an analysis of the characteristics of the field variables. The prescription is that the incoming fields' derivative is replaced by the time derivative of the outgoing fields from the neighbouring domain at the boundaries. It was checked in [84] that all the fields of the Misner–Sharp equations are incoming except for the density field. Therefore, the boundary conditions that must be applied between each subdomain are:

$$\begin{aligned} \dot{M}(t, r_{l+1,i}) &= \dot{M}(t, r_{l,f}), \\ \dot{U}(t, r_{l+1,i}) &= \dot{U}(t, r_{l,f}), \\ \dot{R}(t, r_{l+1,i}) &= \dot{R}(t, r_{l,f}), \\ \dot{\rho}(t, r_{l,f}) &= \dot{\rho}(t, r_{l+1,i}). \end{aligned} \tag{72}$$

where $r_{l,f}$ and $r_{l+1,i}$ mean the last grid point in the subdomain Ω_l and the first grid point in the subdomain Ω_{l+1}, respectively. A diagram of the imposition of the boundary conditions between the subdomains can be found in Figure 4.

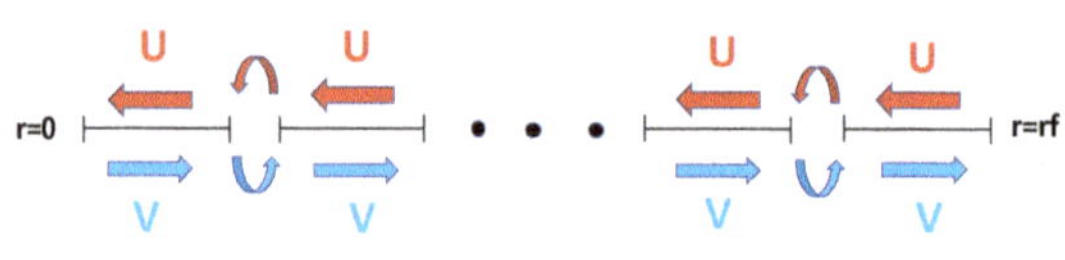

Figure 4. Sketch of the application of the boundary conditions between the different subdomain grids. Through an analysis of the characteristics, the flow of the field variables u and v can be identified (in our case u and v would correspond to ρ, M, R, U), and we apply the prescription of Equation (72).

The use of multigrids in spectral methods has a similar idea to the use of an Adaptive Mesh Refinement (AMR) in finite differences, to increase the resolution of the grid in some regions. Actually, the use of AMR was essential in [94] to verify the scaling law behaviour for the PBH mass in the critical regime. The procedure of the AMR is based on splitting the cell grids on a particular desired region and making subsequent subdivisions of the cells. A summarised and basic procedure is the following: (1) During the evolution, individual grid cells are tagged for refinement using some criteria (high-density regions have to be highly resolved, for instance). (2) All tagged cells are then refined, which means that a finer grid (inside each initial cell) is overlaid on the big one. (3) Finally, after refinement has been performed, individual grid patches on a single fixed level of refinement are evolved in time using the time-discretisation scheme, implemented from the differential equations of the problem. If the level of refinement implemented in a cell is greater than what is needed, the refinement of the cell can be reduced. The disadvantage as in spectral methods is that the stability of the method is modified, and a smaller time step will be required. Therefore, the AMR (and in general, any non-uniform grid procedure) allows resolving problems that are intractable due to the high amount of resolution needed for some regions of the domain.

A known weak point of spectral methods is when they have to handle discontinuous solutions, in particular with the presence of shocks, since the derivative at each point is computed globally, and therefore, the discontinuity is quickly propagated to the other parts of the grids. In this situation, it is better to use a finite differences approach, although usually, a modification of the numerical implementation is still needed to be able to capture shocks, such as the introduction of artificial viscosity.

However, as mentioned in Section 1, in our case, shocks are not formed in the simulations, since we should consider initial conditions at super-horizon scales [94].

5.2. Numerical Procedure

In all the simulations, we fixed the value of w, $t_0 = 1$ and $a_0 = 1$. From that, we automatically obtained $H_0 = \alpha$, $R_H(t_0) = 1/H_0$, and recall that $\alpha = 2/3(1+w)$. For the initial length scale of the perpetuation, $r_m = 10R_H(t_0)$ $(\Lambda = 10)$ is usually considered, which is sufficient to ensure the long wavelength approximation, as was indicated in [79].

First, we focus on describing the procedure to obtain the threshold δ_c, and we leave the method to obtain the final mass $M_{\mathrm{PBH},f}$ for the following subsection. To obtain the threshold, a bisection method was used, which compares different regimes of δ_m until finding the range in which the formation of the apparent horizon appears. The threshold δ_c corresponds to the midpoint of this range.

An example of a successful procedure is the following:

- Create the different grid points $\tilde{x}_{k,l}$ for the different subdomains Ω_l with a given number of points $N_{\mathrm{cheb,l}}$, and compute the different Chebyshev differentiation matrices D_l. Introduce also the initial time step dt_0, which usually $dt_0 \leq 10^{-3}$ is enough to ensure stability;
- Specify a lower and an upper limit in δ_m of the bisection. The maximum value $\delta_{\mathrm{m,max}} = f(w)$ can be chosen for the upper limit, and for the lower bound, $\delta_{\mathrm{m,min}} \approx w$ (which corresponds to the threshold estimation using a Jeans length criteria). To reduce the computational time, a closer domain to δ_c can be chosen if such a domain is known;
- A value δ_m is taken once the bisection starts. From that, the curvature profiles $K(r)$ or $\zeta(\tilde{r})$ can be built, and the location of the compaction function peak r_m can be found, $\tilde{r}_m$, using Equations (51) and (53) respectively. Subsequently, compute the perturbations of the hydrodynamical variables following Equations (40) and (41);
- Integrate Equations (29) and (32) at each time step dt using the fourth-order Runge–Kutta method, imposing at each internal time step the boundary conditions at $r = 0$ and $r = r_f$. Include the internal boundary conditions between the different Chebyshev grids Equation (72);
- To distinguish numerically the formation of an apparent horizon or not, we used the peak value of the compaction function in time. When the peak approaches $\mathcal{C}_{\max} \approx 1$, this implies the formation of the apparent horizon [3]. In this case, the value of δ_m chosen would imply the formation of a black hole, i.e., $\delta_{\mathrm{m,yes}}$, and therefore, a lower value δ_m closer to δ_c must be found modifying the bound of the bisection in such way that $\delta_c \in [\delta_{c,\mathrm{min}}, \delta_{m,\mathrm{yes}}]$. In the opposite case, when $\mathcal{C}_{\max}$ decreases continuously for some time, this will imply the dispersion of the perturbation, and the bisection bound should be modified as $\delta_c \in [\delta_{m,\mathrm{no}}, \delta_{c,\mathrm{max}}]$;
- The previous steps are iterated until the difference between $\delta_{m,\mathrm{yes}}$ and $\delta_{m,\mathrm{no}}$ is lower than the desired resolution, i.e., $\delta_{m,\mathrm{yes}} - \delta_{m,\mathrm{no}} \lesssim \delta(\delta_c)$. From that, the threshold value is defined as $\delta_c = (\delta_{m,\mathrm{yes}} + \delta_{m,\mathrm{no}})/2 \pm \delta(\delta_c)$. When δ_m approaches the critical value, in

some cases, the grid resolution initially set up is not enough to follow the evolution; if so, then δ_m is shifted according to $\delta(\delta_c)$ to avoid being close to δ_c. Another alternative would be the dynamical inclusion of more subgrids in the desired region where gradients start to be large.

During the time evolution, a conformal time step $dt = dt_0(t/t_0)^\alpha$ is used, which substantially improves the performance of the simulation. The two-norm of the Hamiltonian constraint Equation (26) is computed at each time step. It is expected that the constraint should remain constant from $t = t_0$ if the Einstein equations are correctly solved. In particular,

$$\mathcal{H} = D_r M - 4\pi\Gamma\rho R^2, \tag{73}$$

$$|| \mathcal{H} ||_2 \equiv \sum_l \frac{1}{N_{\text{cheb,l}}} \sqrt{\sum_k \left| \frac{M'_{k,l}/R'_{k,l}}{4\pi\rho_{k,l}R^2_{k,l}} - 1 \right|^2}, \tag{74}$$

where the label l refers to the sum of the different Chebyshev grids and k to the different points of the grid.

5.3. Numerical Estimation of the Final PBH Mass

To obtain the final mass of the PBH $M_{\text{PBH},f}$, we used an excision technique, and we describe it with detail in this section. Saying that, other alternatives in the literature have used null coordinates with the Hernandez–Misner equations [63,71,73] to obtain the final mass. We refer the reader to these references, but the primary approach followed can be summarised as follows: (i) The simulation was set up with some initial data using the Misner–Sharp equations. (ii) Starting the evolution, an outgoing null radial ray is sent from the centre to the outer point of the grid, and the values of the field variables are stored once the ray crosses each grid point. (iii) The Misner–Hernandez equations are evolved using the initial conditions obtained in the previous step, which are basically the values stored.

Let us focus now on excision. The main idea of an excision technique is based on the fact that nothing inside the event horizon can affect the physics outside. Therefore, the aim is to dynamically remove the segment of the computational domain inside the apparent horizon, to avoid large gradients appearing and breaking down the simulation.

There are different procedures to implement excision, but the one we used here is the following: We define two parameters, Δr and dr (we always consider $\Delta r > dr$), where Δr is the distance separation between the apparent horizon and the excision surface that we establish after each redefinition of the excision surface, and dr is the maximum allowed displacement of the apparent horizon before we create a new excision boundary. At each time step, we locate the position of the apparent horizon, taking into account that $2M(r_*, t)/R(r_*, t) = 1$, where $r_*(t)$ is the location of the apparent horizon in time. Since we were handling a discretised grid, we used a cubic spline interpolation to find the correct value of $r_*(t)$ (the difference of taking a quadratic spline interpolation is $O(0.01\%)$ for the PBH mass values).

In particular, the method used is the following:

For a supercritical perturbation, when the peak of the compaction function $\mathcal{C}_{\max} \approx 1.2$ (the final result is not affected by the exact value that we take, as long as we consider $\mathcal{C}_{\max} \approx O(1)$), part of the computation domain is removed, creating an excision surface close to the location of the apparent horizon, whose separation from the excision boundary is basically given by Δr. Once having done this, we created a dense, but small Chebyshev grid starting from the excised boundary. Another single Chebyshev grid can cover the other domain since large gradients are only developed near the excision surface.

We evolved in time the Chebyshev grids in the usual form independently, and we tracked the apparent horizon motion at each dt. Once the apparent horizon had been moved a distance greater than dr, we created a new excision surface separated a distance Δr from the apparent horizon. During the evolution, we needed to continuously reduce the values Δr,dr, since the velocity of the apparent horizon decreases in time. This is especially important for very small $\delta_m - \delta_c$. We decided to reduce Δr,dr when the Hamiltonian constraint starts to be continually violated. In particular, it was found that considering $\Delta r \approx 2dr \approx O(10^{-2})$ worked nicely for our purposes.

There is no physical boundary condition to apply at the excision surface. Still, it was found that freezing (leave it constant) the gradient of ρ (i.e., ρ') at the excision surface, after each redefinition of the excision boundary, allowed increasing the stability of the method without changing the numerical results.

6. Numerical Results

In this section, we review some numerical results using the numerical procedure pointed out in the previous section. First of all, we introduce a set of curvature profiles that allows us to span all different possible thresholds for PBH formation. Later on, we will see the dynamical evolution of the collapse for three different cases in terms of the initial amplitude δ_m and with a specific profile chosen. Finally, we show the results regarding the final PBH mass in terms of the different profiles, as well as the effect of the accretion from the FLRW background.

The numerical results were obtained using the procedure of Section 5 with a computer, i-7 core, 16 GB of RAM, and without parallelisation procedures. As an example of the efficiency, in the case of $w = 1/3$ and with a Gaussian profile, to obtain the threshold δ_c with a resolution $\delta_m - \delta_c \approx 10^{-3}$, the running time was ~3–5 min using the bisection method in Section 5.2. To estimate the final mass $M_{\mathrm{PBH},f}$, we needed to run the excision method in Section 5.3 for ~3 h for the largest PBHs. These values can differ when changing the profiles or the equation of state w. The typical size of the grid chosen was two or three grids with ~100 points in each grid. However, more points could be needed for sharp initial profiles. With these configurations, the maximum resolution obtained for the threshold in the case of a Gaussian profile was $\delta_m - \delta_c \approx 10^{-5}$ (in the case of $w = 1/3$). With sharper profiles or larger w, this resolution slightly reduced.

6.1. Curvature Fluctuations

The seed of the generation of the curvature fluctuation is the power spectrum, namely $P(k)$, where k is the wavelength. We briefly review the procedure assuming the Gaussian statistics of [116] applied to our purposes. We considered that the amplitudes of the curvature perturbations are statistically Gaussian distributed accordingly to the specific inflationary model considered, and therefore, under this assumption, the rare over-densities leading to PBH formation are, with a very good approximation, spherically symmetric [116].

From the power spectrum $P_\zeta(k)$, we can relate it to the curvature fluctuation ζ Equation (37) as,

$$< \zeta_k \zeta_{k'} >= (2\pi)^3 \frac{2\pi^3 P_\zeta(k)}{k^3} \delta_D^3(k+k'), \tag{75}$$

where ζ_k is the Fourier mode k of the field ζ,

$$\zeta(\vec{x}) = \int \frac{d\vec{k}}{(2\pi)^3} e^{i\vec{k}\cdot\vec{x}} \zeta(\vec{k}). \tag{76}$$

Notice that in linear theory and at super-horizon scales, the density contrast $\delta\rho/\rho_b(\tilde{r})$ is related to $\zeta(\tilde{r})$ as,

$$\frac{\delta\rho}{\rho_b}(\tilde{r}) \approx -\frac{2(1+w)}{5+3w} \frac{1}{a^2H^2} \nabla^2 \zeta(\tilde{r}), \tag{77}$$

and therefore:

$$\left(\frac{\delta\rho}{\rho_b}\right)_k \approx -\frac{2(1+w)}{5+3w} \frac{k^2}{a^2H^2} \zeta_k. \tag{78}$$

From that, we can build the normalised two-point correlation function by,

$$\psi(\tilde{r}) = \frac{1}{\sigma_0^2} \int W(k) P_\zeta(k) \operatorname{sinc}(k\tilde{r}) \frac{dk}{k}, \tag{79}$$

where $W(k)$ is the window function, and the variance of the field ζ is given by,

$$< \zeta^2 >= \sigma_0^2 = \int \frac{dk}{k} W(k) P_\zeta(k). \tag{80}$$

Therefore, the curvature fluctuation $\zeta(r)$ assuming peak theory can be built as,

$$\zeta(\tilde{r}) = \mu\psi(\tilde{r}) = \frac{\mu}{\sigma_0} \int W(k) P_\zeta(k) \operatorname{sinc}(k\tilde{r}) \frac{dk}{k}, \tag{81}$$

where μ is the peak value of ζ.

In conclusion, using Equation (81), we can relate the power spectrum to our curvature fluctuation $\zeta(\tilde{r})$ and, from that, set up the initial conditions following Equations (41) and (42), or alternatively also build $K(r)$ using the transformations of Equation (39). Although it goes beyond the scope of the review, we should mention that one could also consider the effect of the inclusion of non-Gaussianities in Equation (81), which can have important consequences for the PBH abundance estimation. There are vast and extensive works that consider its impact on the thresholds and PBH abundances [103,107,108,180–195], including some of them that have used numerical simulations following the basics shown in this review [196,197].

Saying that, it is much more practical to already span a set of curvature profiles instead of considering different power spectrums when we are interested in studying the physical process of PBH formation. The reason is that setting up directly the curvature fluctuation allows us to modulate different shapes and therefore obtain all possible thresholds and estimate their effects on the PBH mass. In any case, some typical power spectrum templates can be found in [60,180].

In the literature, exponential families of profiles have been commonly used to modulate $K(r)$ [77,79,81,82]. Another possibility is a polynomial family [84]. Both can be written in a convenient way as:

$$K_{\text{pol}}(r) = \frac{\delta_m}{f(w) r_m^2} \frac{1+1/q}{1+\frac{1}{q}\left(\frac{r}{r_m}\right)^{2(q+1)}}, \tag{82}$$

$$K_{\text{exp}}(r) = \frac{\delta_m}{f(w) r_m^2} \left(\frac{r}{r_m}\right)^{2\lambda} e^{\frac{(1+\lambda)^2}{q}\left(1-\left(\frac{r}{r_m}\right)^{\frac{2q}{1+\lambda}}\right)}. \tag{83}$$

The profiles of Equations (82) and (83) correspond to polynomial and exponential (for $\lambda = 0$) non-centrally peaked (for $\lambda \neq 0$) profiles, respectively. Notice that the profiles depend on the ratio r/r_m. The profiles are written in a convenient way with the amplitude δ_m, in such away that when computing $\mathcal{C}(r)$ using Equation (50), this automatically gives $\mathcal{C}(r_m) = \delta_m$ when evaluating $r = r_m$.

On the other hand, the parameter q is a dimensionless parameter, which relates the shape around the peak of the compaction function, which is defined as,

$$q = -\frac{\mathcal{C}''(r_m) r_m^2}{4\mathcal{C}(r_m)}. \tag{84}$$

In terms of the $\tilde{r}$ coordinate, the transformation in Equation (39) can be used, as was done in [198], to obtain,

$$q = -\frac{\mathcal{C}''(\tilde{r}_m)\tilde{r}_m^2}{4\mathcal{C}(\tilde{r}_m)\left[1-\frac{3}{2}\mathcal{C}(\tilde{r}_m)\right]}. \tag{85}$$

Interestingly, in [83], it was found numerically that profiles with the same q have the same threshold δ_c upon a deviation of O(3–4%) in the case $1/3 \leq w \leq 1$ (and O(1–2%) in the particular case of a radiation-dominated Universe), but we will discuss this in more detail in Section 7, since this result will be used to obtain analytical estimations on δ_c. The benefit of Equation (82) in comparison with Equation (83) is that at $r \to 0$, this fulfils the regularity conditions, whereas Equation (83) does not for some small values of q. Actually, the profile Equation (82) can be considered as a basis profile, since it allows spanning all possible threshold values, as we will see in more detail in Section 6.3. The profiles of $\mathcal{C}(r)$ for the polynomial profile can be found in Figure 5. When $q \gg 1$, the peak of the compaction function is sharp, while when $q \ll 1$, the peak is broad (which would be a homogeneous sphere). The opposite behaviour can be found for the density contrast (bottom panel of Figure 5). In this case, for small q, the density contrast profile is sharper, whereas for large q, the shape becomes broad up to r_m (a homogeneous sphere).

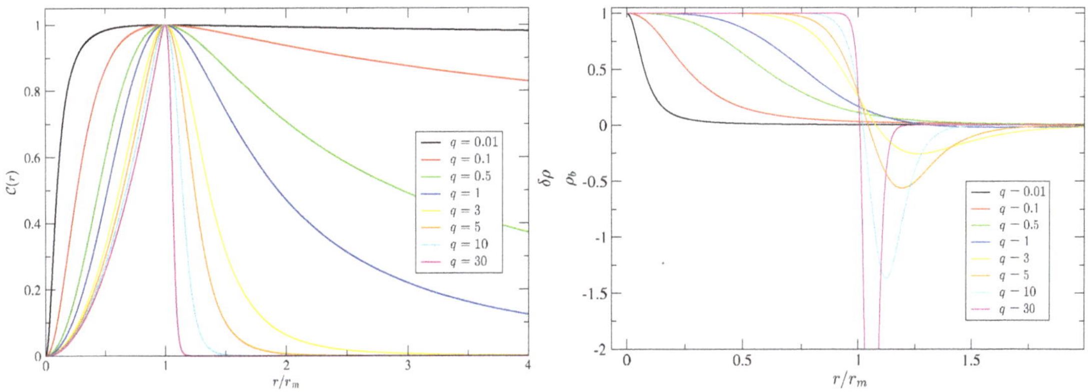

Figure 5. (**Left-panel**): Profiles of $\mathcal{C}(r)$ of Equation (82) for several values of q. The values δ_m are normalised to $\delta_m = 1$. (**Right-panel**): Density contrast $\delta\rho/\rho_b$ for the same profiles as the top panel. The profiles are normalised to the peak $\delta\rho(r=0)/\rho_b$. The negative peak value for $q = 30$ around $r = r_m$ goes up to ~ -4.

A profile with two modulating peaks in $\mathcal{C}$ was also considered in [87] with the aim to compare the other profiles. It is similar to the one considered in [78], where double-PBH formation was studied. This kind of profile would represent a perturbed region superposed on another one at much larger scales, as discussed in [78]. The compaction function profile is given by,

$$\mathcal{C}_{tt}(r) = \mathcal{C}_b(r, \delta_1, q_1, r_{m1}) + \theta(r - r_j)\mathcal{C}_b(r - r_j, \delta_2, q_2, r_{m2}), \tag{86}$$

where $\mathcal{C}_b$ is equal to:

$$\mathcal{C}_b(r, \delta_j, q_j, r_{mj}) = \delta_j \left(\frac{r}{r_{mj}}\right)^2 \frac{1 + 1/q_j}{1 + \frac{1}{q_j}\left(\frac{r}{r_{mj}}\right)^{2(q_j+1)}}; \tag{87}$$

the value of the second peak of $\mathcal{C}$ can be modulated as,

$$\mathcal{C}_{tt(\text{peak},2)} = \frac{(1+q_1)(r_j + r_{m2})^2 \delta_1}{q_1 r_{m1}^2 + (r_j + r_{m2})^2 (r_j + r_{m2}/r_{m1})^{2q_1}} + \delta_2. \tag{88}$$

Different profiles of the three families are plotted in Figure 6. In particular, we plot the profiles of Equations (82), (83), and (86), and from them, we also plot the corresponding compaction function from Equation (50), as well as the density contrast Equation (41). The conversion from $K(r)$ to $\zeta(\tilde{r})$ can be made using Equation (39).

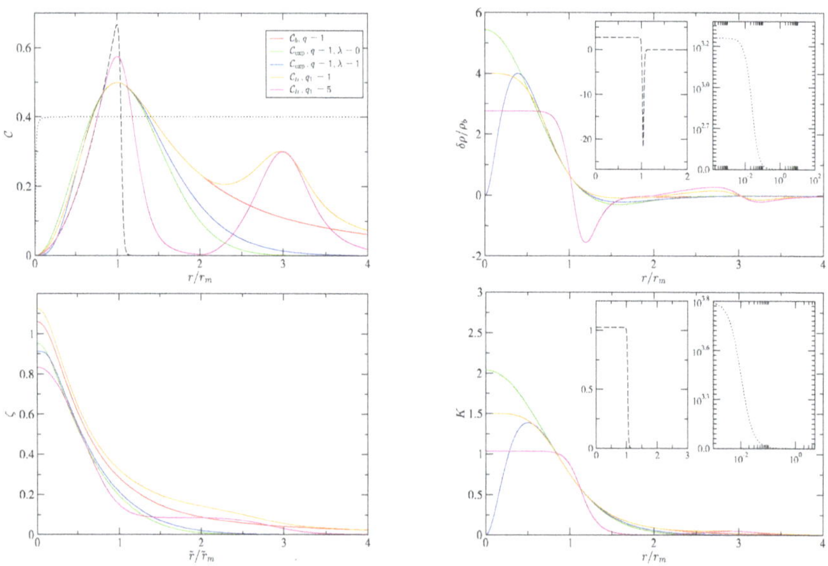

Figure 6. Plots for the profiles with Equations (82), (83), and (86) for $\mathcal{C}(r)$ (**top-left**), $\delta\rho(r)/\rho_b$ (**top-right**), $\zeta(r)$ (**bottom-left**), and K (**bottom-right**) as a function of r/r_m for $\delta_m \sim \delta_c$. The dotted black line corresponds to $q = 0$ and the dashed black line to $q \to \infty$. The values taken for $\mathcal{C}_{tt}(r)$ are $\delta_1 = \delta_c(q_1)$, $q_2 = 3$, $r_{m1} = r_{m2} = 1$, $r_j = 2r_{m1}$, $\mathcal{C}_{tt(\text{peak},2)} = 0.3$, with the corresponding δ_2 obtained from Equation (88) using the previous parameters, $q_1 = 5$ (violet) and $q_1 = 1$ (orange).

6.2. Non-Linear Behaviour of the Gravitational Collapse

In this subsection, we show the behaviour of the gravitational collapse of the curvature fluctuations for three different regimes. As an example, a Gaussian profile given by Equation (83) with $q = 1$ ($\lambda = 0$) and for different values of δ_m and $w = 1/3$ is considered. However, the same qualitative behaviour can be found for other profiles and other $w \neq 0$. In particular, three different regimes can be considered: subcritical ($\delta_m < \delta_c$), supercritical ($\delta_m > \delta_c$) with $\mid \delta_m - \delta_c \mid \gg O(10^{-3})$ in both previous cases, and critical $\mid \delta_m - \delta_c \mid \leqslant O(10^{-3})$. In Figure 7 (supercritical), Figure 8 (subcritical), and Figure 9 (critical), we see the evolution of the variables ρ, Γ, U, and $\mathcal{C}$.

In the case of a supercritical evolution, which is the case of Figure 7, we can observe that the compaction function grows during the evolution. The gravitational collapse increases the energy density on the central regions, and therefore, the mass excess grows. The formation of the two horizons is also obvious, as discussed in Section 4.3. The outer horizon moves outwards, and the inner one moves inwards, faster than the outwards one. The simulation breaks down when the inner horizon approaches $r = 0$. This is when the excision technique has to be applied to remove the singularity.

In Figure 8 (the subcritical case), the compaction function decreases continuously as the perturbation is diluted away due to the superiority of the pressure gradients against gravity. In general, for late times, the peak $\mathcal{C}_{\max}$ is pushed outwards due to the pressure gradients. This behaviour can also be seen with the peak value of ρ/ρ_b. In the supercritical case, the peak value is continuously increasing. In contrast, in the subcritical case, the peak value starts to decrease at some moment (which means the perturbation starts to disperse). In addition, from Figures 7–9, it can be observed that Γ is not constant during the evolution. As expected, the gradient expansion approximation fails for sufficiently late times $t \gg t_0$.

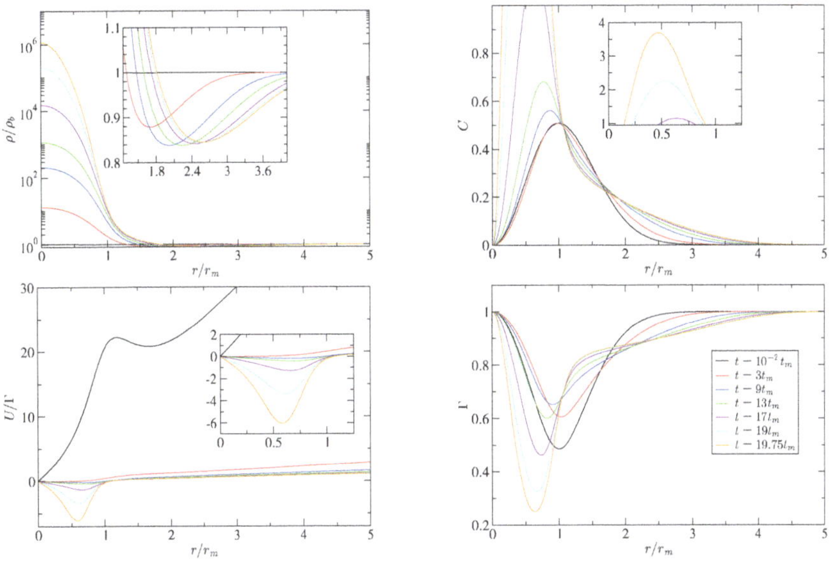

Figure 7. Dynamics of different magnitudes in the case of a supercritical fluctuation (Gaussian profile) at given times t. In particular, $\delta_m = 0.51$ and $\delta_c = 0.49774 \pm 2 \times 10^{-5}$.

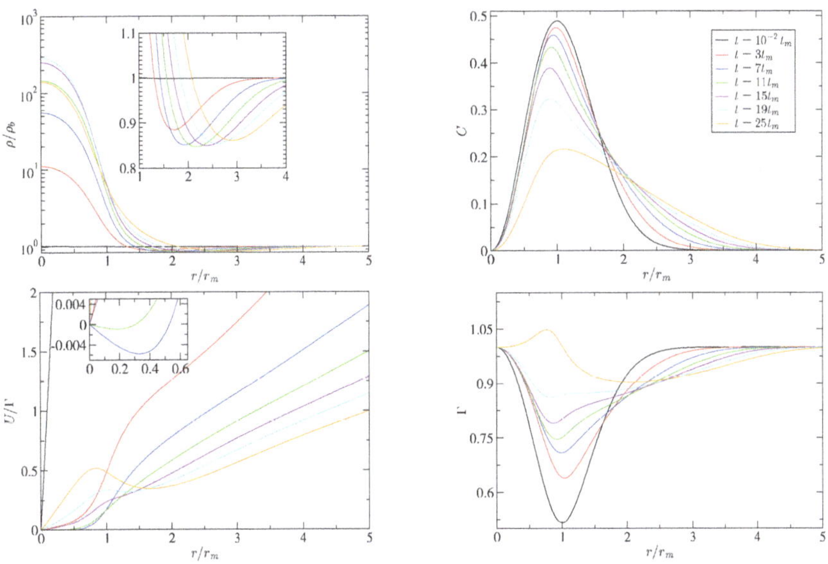

Figure 8. Dynamics of different magnitudes in the case of a subcritical fluctuation with $\delta_m = 0.49$ (Gaussian profile) at given times t. In particular, $\delta_m = 0.49$ and $\delta_c = 0.49774 \pm 2 \times 10^{-5}$.

In Figure 7 (supercritical case), we observe that U/Γ decreases quickly in time, and a negative velocity mainly dominates the profile before r_m for late times, which indicates the collapse of the fluid. Instead, in Figure 8 (subcritical case), only a small negative value U/Γ is reached for early times, and beyond that, the values are positive, which indicates that the perturbation is dispersing, avoiding the gravitational collapse.

The most relevant behaviour is found in the critical regime, in Figure 9. Here, the fluid is divided into two parts, one going outwards (positive U) and one inwards (negative U), which generates an under-dense region. This under-dense region re-attracts the fluid with the net effect of a compression and rarefaction process, which increases its velocity in time. In this situation, the gradients become very large, and eventually, the simulation breaks down if a sufficient resolution has not been set up for the simulation.

It is interesting to mention that in the presence of secondary peaks in $\mathcal{C}(r)$ (not only one isolated one), with an amplitude roughly equal to or less than the first one, it was verified numerically in [197] that those secondary peaks decrease and disperse on the FLRW background, even in the case of supercritical perturbations. A different situation could happen when a secondary peak is sufficiently high. In such a case, the secondary peak could also collapse even if the first peak is not greater than the threshold value to collapse itself.

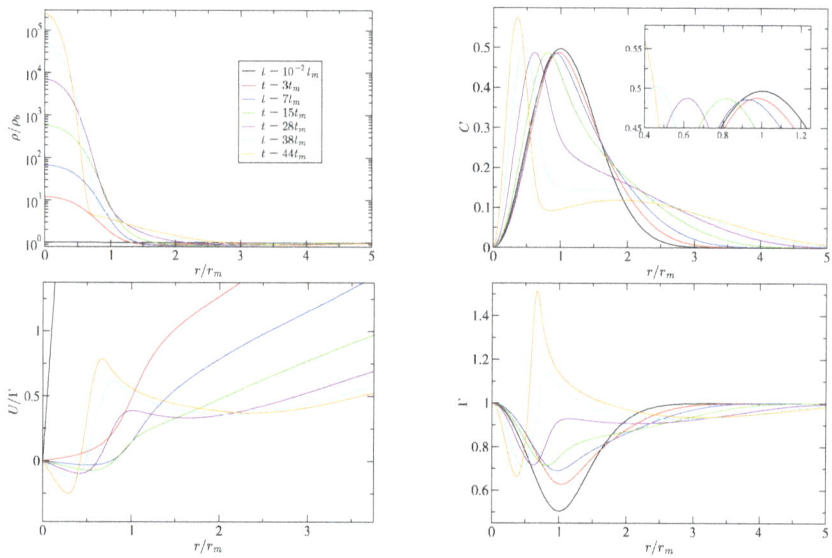

Figure 9. Dynamics of different magnitudes in the case of a perturbation close to the critical threshold $\delta_m \approx \delta_c$ (Gaussian profile) at given times t. In particular, $\delta_m = 0.49775$ and $\delta_c = 0.49774 \pm 2 \times 10^{-5}$.

In Figure 10 is shown the correctness of the numerical evolution using the Hamiltonian constraint of Equation (74) at each step. It corresponds to different evolutions of the gravitational collapse, both subcritical and supercritical. In the figure, it can be seen that the constraint starts to be violated for late times for $(\delta_m - \delta_c) \approx O(10^{-5})$, which gives a bound on the maximal resolution for δ_c (in this particular example)

The convergence with the use of spectral methods can be found in Figure 11. It is clear that the convergence is exponential taking into account the fits performed (see the caption of the figure). In particular, the use of a multidomain grid improves the performance and accuracy of the simulations substantially, as is shown in the right panel of Figure 11, where the spectral accuracy is obtained with fewer points (in comparison with the left panel). Making the first grid Ω_1 from $r = 0$ up to $r = 2r_m$ allows capturing the region where large gradients are developed. Then, for the second grid, Ω_2 goes from $r = 2r_m$ up to $r = r_f$, and only a few points are needed.

An interesting behaviour is found for t_{AH} when w is changed, as noticed in [84]. Although the pressure gradients work against the gravitational collapse, since they are also a form of gravitational energy, it will mostly be favoured when the collapse is active [84]. Consequently, this implies a shorter time of formation for larger w, as can be observed in Figure 12.

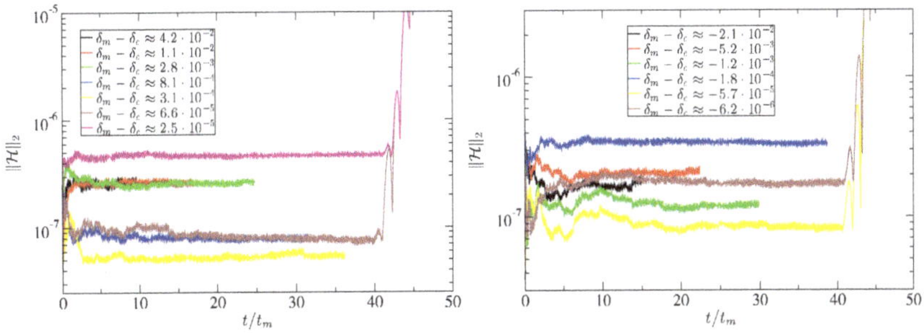

Figure 10. (**Left-panel**): Hamiltonian constraint for the numerical iterations of the bisection procedure in the case of a profile given by a Gaussian fluctuation for the supercritical case with Equation (83) and $q = 1$, $\lambda = 0$. (**Right-panel**): The same as the left panel, but for subcritical fluctuations. The initial Hamiltonian constraint is subtracted for each evolution of δ_m in both cases.

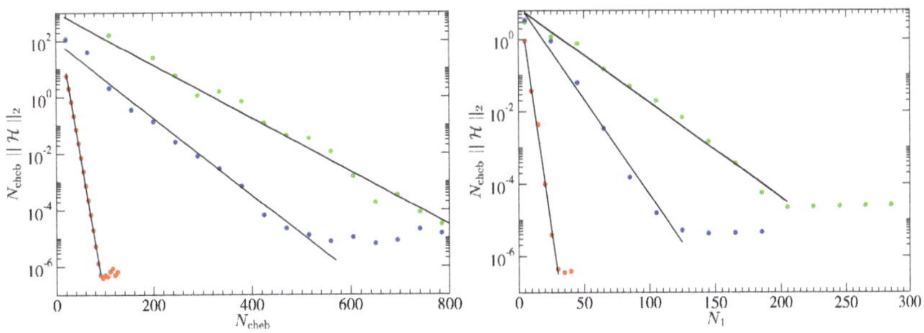

Figure 11. Spectral convergence for different profiles. Red points correspond to the profile of Equation (83) with $q = 1$, blue points to $q = 5$, and green points to $q = 10$ ($\lambda = 0$ in all cases). The black solid line is the exponential fit $\sim e^{-\alpha N_{\rm cheb}}$. (**Left-panel**): Case with a single grid $N_{\rm cheb}$ and $\alpha \approx 0.23, 0.032, 0.022$ respectively for the previous cases. (**Right-panel**): Case with two grids N_1 and $N_2 = 50$, with the second grid starting at $r = 2r_m$, and $\alpha \approx 0.59, 0.12, 0.06$, with the same profiles of the left panel. The simulation corresponds to $\delta_m = 0.5$ and $t = t_H$ for $w = 1/3$.

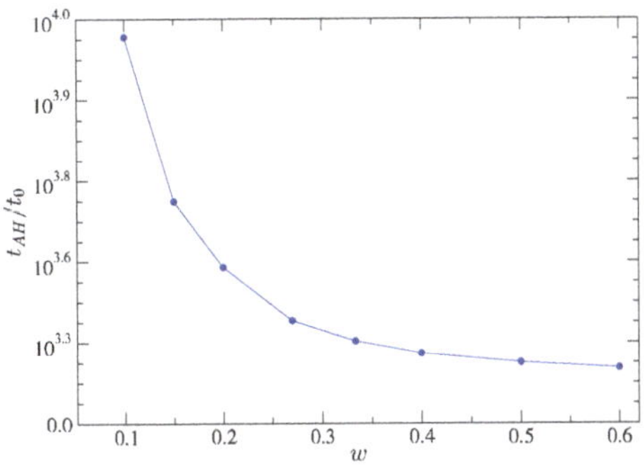

Figure 12. Time that it takes the fluctuation to form the apparent horizon t_{AH} in terms of w. The fluctuation chosen corresponds to Equation (82) with $q = 1$ and $\delta_m = \delta_c + 10^{-2}$.

6.3. Thresholds for PBH Formation

As we have mentioned, the threshold for PBH formation depends on the shape of the curvature fluctuation. Following the numerical procedure in Section 5.2, all the possible thresholds for PBH formation can be obtained using the basis polynomial profile of Equation (82). The results are shown in Figure 13 for different profiles. In the particular case of a radiation fluid, it was found in [79,83] that the range of all possible thresholds is

$\delta_c \in [0.4, 2/3]$. The minimum threshold $\delta_{c,\min} = 0.4$ corresponds to the limit case of $q \to 0$ (broad profile in $\mathcal{C}$). Instead, the maximum threshold $\delta_{c,\max} = 2/3$ corresponds to $q \to \infty$ (peaked profile in $\mathcal{C}$). Increasing q, the pressure gradients are larger, and therefore, the threshold value is higher.

Remarkably, the relative deviation between the thresholds for the different profiles in the left panel of Figure 13 with the same q is within 2%. This was found for the first time in [83], and it was shown that different profiles with the same q (shape around the peak of the compaction function) have the same thresholds within a small deviation.

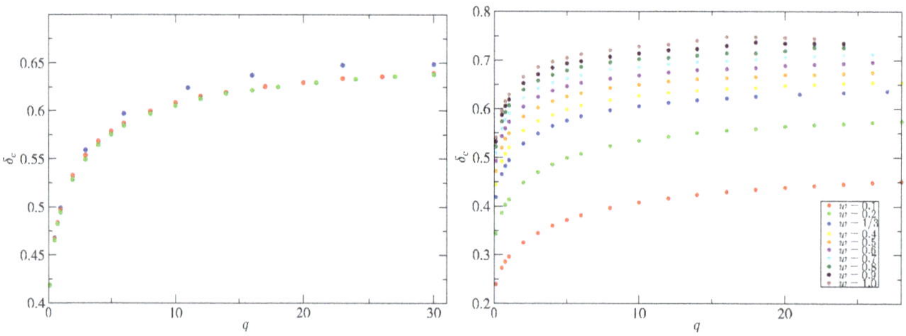

Figure 13. (**Left-panel**): Threshold values δ_c in terms of q for the case of $w = 1/3$. Green points correspond to Equation (82), red points to Equation (83) with $\lambda = 0$, and blue to Equation (83) with $\lambda = 1$. (**Right-panel**): Threshold values δ_c in terms of q for different values of w corresponding to the polynomial profile of Equation (82).

In [84], simulations of PBH formation in the case of a perfect fluid with $w \in (0, 1]$ were performed. The functional behaviour of δ_c in terms of q for different w's is roughly similar, as is shown in the left panel of Figure 13 (the case with fixed $w = 1/3$), but quantitatively, the values are different. It is clear that increasing w, the gradients will be larger, and therefore, the threshold increases. The limits $q \to 0$ and $q \to \infty$ for the general case in terms of w are both amenable to further analysis:

- The case $q \to 0$ corresponds to a $\mathcal{C}$ that becomes approximately constant over a wide range of scales. Simulations, in this case, become difficult due to the presence of conical singularities [84]. To solve this issue, in [84], a modified profile of Equation (82) was used to subtract the mass excess beyond $r \gg r_m$ and fulfil that at the boundaries, the FLRW background is recovered. In Figure 14 are shown the numerical values of $\delta_c(q \to 0)$ in terms of w, which show a strong dependence on w. As expected, for $w \to 0$, the threshold $\delta_c \to 0$;
- The case of large $q \to \infty$ corresponds to the case of a sharply peaked profile for the compaction function. For this kind of profile, the pressure gradients acting against the gravitational collapse are maximal, and hence, the compaction function should be maximal as well. Analytically, this is shown by Equation (50) that $\mathcal{C}(r_m) < f(w)$ (at super-horizon scales). As was proven numerically in [79], $\delta_{c,\max} \to f(w)$ in the case of radiation $w = 1/3$. The intuition is that the saturation should persist to larger values of w because larger values of w also imply larger pressures that fight against the collapse. Therefore, for $w \geq 1/3$, the compaction function of a peaked profile must also saturate the bound $f(w)$. This was verified numerically in [80]. This behaviour is no longer true for $w < 1/3$, something obvious in the case of dust $w = 0$, since $f(0) = 0.6$, but $\delta_c(w = 0) = 0$ always for any profile. This is shown in the left panel of Figure 14. It should be taken into account that beyond $q > 40$, the simulations do not obtain the threshold with enough resolution. This is the reason why the numerical value of the threshold for $w = 1/3$ does not exactly coincide with $f(1/3)$. One expects that in the asymptotic limit, the threshold for $q \to \infty$ should match with $f(1/3)$, as was pointed out in [79] with the same argument.

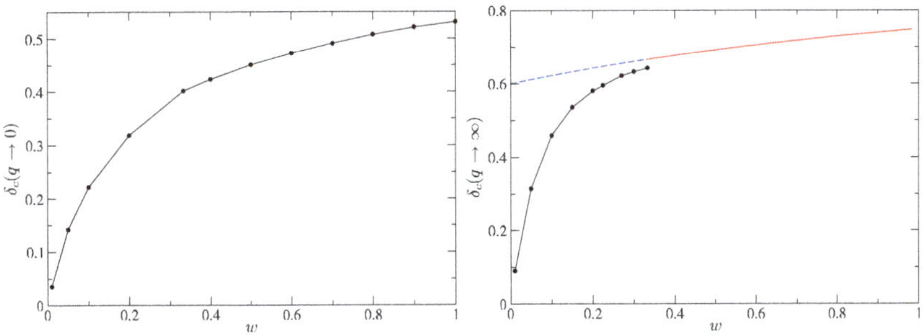

Figure 14. (**Left-panel**): Threshold δ_c in terms of w for $q \to 0$. (**Right-panel**): Threshold δ_c in terms of w for $q \to \infty$; in particular, we are able to perform simulation up to $q = 40$. The red line corresponds to the analytical case $f(w)$, where the black dots correspond to the numerical simulations in the region where $\delta_c(q \to \infty) \neq f(w)$, as shown in [84]. The blue dashed line corresponds to $f(w)$ on that region.

It is interesting also to notice the difference between considering $K(r)$ and $\zeta(\tilde{r})$ for the threshold and PBH mass. As we have seen in Section 4, there is a non-linear relation between ζ and $\mathcal{C}$, which leads to some differences between the thresholds obtained for both profiles. Consider for instance a Gaussian profile in both cases, i.e., $K(r) = \mathcal{A}\, e^{-(r/r_m)^2}$ and $\zeta(r) = \mu\, e^{-(r/r_m)^2}$. Due to the non-linear relation between $\mathcal{C}$ and ζ at super-horizon scales, the $\delta_{m,\zeta}$ (for this example, let us call it $\delta_{m,K}$ and $\delta_{m,\zeta}$ for both profiles) depends non-linearly on μ as,

$$
\begin{aligned}
\delta_{m,K} &= f(w)\mathcal{A} r_m^2 e^{-1}, \\
\delta_{m,\zeta} &= 4 f(w)(e-\mu)\mu e^{-2}.
\end{aligned}
\tag{89}
$$

Due to this, the q_ζ value in the case of the fluctuation ζ is dependent also on its amplitude μ and w,

$$
q_\zeta = \frac{e^2(e-2\mu)}{(e-\mu)(e^2 - 6f(w)\mu + 6f(w)\mu^2)}, \tag{90}
$$

when for the $K(r)$ in this example, it is just $q = 1$. A plot of the thresholds δ_c for both profiles can be found in Figure 15 for different w's. Although both profiles are Gaussian, they have a different δ_c due to the non-linear relation with $\zeta(\tilde{r})$. The threshold for $\zeta(\tilde{r})$ is higher than $K(r)$ since the shape around the compaction function is sharper, which means a larger q, as can be observed in the bottom panel of the subplot in Figure 15. It also has an implication for the PBH mass, as we will see in the next subsection.

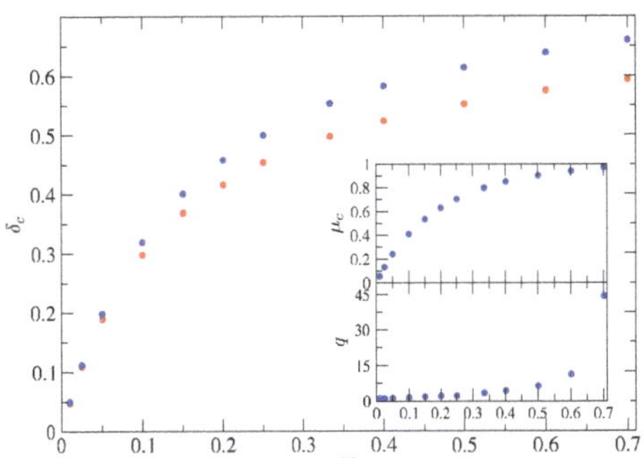

Figure 15. Thresholds δ_c in terms of w for two profiles. Red dots correspond to the Gaussian profile with $K(r)$ and blue dots to $\zeta(\tilde{r})$ instead. The (**top-subplot**) is the critical amplitude μ_c of the profile in $\zeta(\tilde{r})$. The (**bottom-subplot**) is the value q of the critical profile $\zeta(\tilde{r})$ for each w.

A Gaussian family of profiles with $\zeta(\tilde{r}) = \mu e^{-(r/r_m)^{2p}}$ was considered in [106] for the estimation of PBH abundances taking into account the non-linearity of $\zeta(\tilde{r})$ for $w = 1/3$. There, a range of thresholds was found, $0.442 < \delta_c < 0.656$ for $0.34 \lesssim p \lesssim 2$.

6.4. PBH Masses

Once the apparent horizon is formed, an excision technique following Section 5.3 can be applied to remove the singularity. An example of the evolution of the PBH mass $M_{\rm PBH}(t)$ for the profile of Equation (83) with $q = 1$, $\lambda = 0$, and different amplitudes δ_m can be found in Figure 16. The initial time of the curves corresponds to the initial mass of the PBH $M_{\rm PBH,i}$ at the time of formation of the apparent horizon t_{AH}. As can be seen in Figure 16, when the amplitudes δ_m increase, the PBH mass will be higher.

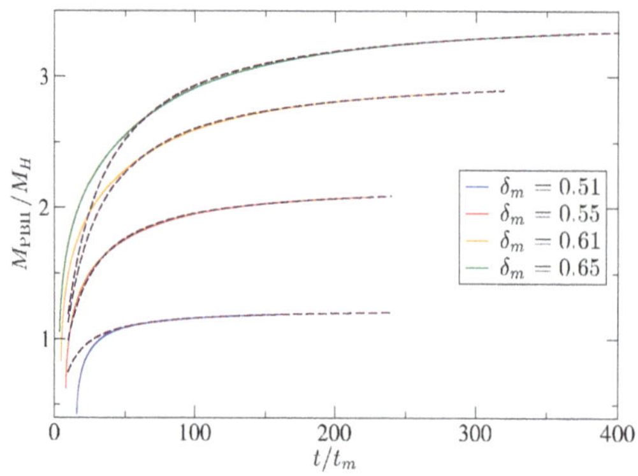

Figure 16. Evolution of the mass of the PBH for $w = 1/3$ after the formation of the apparent horizon for the profile of Equation (83) with $q = 1$ and $\lambda = 0$. The colour lines correspond to different values of initial amplitudes δ_m. The dashed line corresponds to the analytical fit using Equation (56). Threshold value $\delta_c = 0.49774 \pm 2 \times 10^{-5}$.

The amount of time needed until reaching the stationary regime where the PBH mass remains constant could be substantial and therefore computationally expensive. As we have mentioned, to avoid that, Equation (56) can be used to estimate the final mass of the PBH after the accretion process accurately.

In order to find the domain in time where the approximation of Equation (56) is valid, in [80], the ratio of the increment in time of the black hole mass with respect to the Hubble scale $\Psi = \dot{M}/HM$ was computed, which is predicted to be $\Psi < 1$ when the evolution of the PBH mass satisfies this regime. During the application of the excision technique, the Hamiltonian constraint is computed using Equation (74) to check the correctness of the numerical evolution. In the simulations in [80,87], the constraint was fulfilled until very late times, as shown in Figure 17. Nevertheless, it seems that the evolution of the mass is not affected substantially by the violation of the constraint. Interestingly, in the bottom panel of Figure 17, a crossing for different evolutions of Ψ at a given time t^* is observed. In Figure 18 is shown the evolution of $M_{\rm PBH}(t)$ in time for different values of q and two different profiles. It is clear that the larger is q, the smaller is the final PBH mass $M_{\rm PBH,f}$, since the pressure gradients are stronger and prevent accretion from the FLRW background. Notice the differences of $M_{\rm PBH}(t)$ between the two profiles considered in the figure. Even with the same q, the values for the mass are different since the full shape of the profile will contribute to the final mass.

To obtain the parameters t_a, M_a, and F, a non-linear fit using the function of Equation (56) with the data $M_{\rm PBH}(t)$ coming from the simulations can be performed. The range of values to make the fit are those that fulfil $\Psi \lesssim 0.1$. The values of F that were obtained in the literature [80] making the non-linear fit were in the range $F \in [3.2, 3.8]$ (for the case of $w = 1/3$) in terms of the different profiles and amplitudes δ_m. Some indicative results regarding the accuracy of the fits performed are: $\sigma_{\max} \approx 10^{-2}$ (variance) $s_d(M_a) = 10^{-4.2}$, $s_d(t_a) = 10^{-4.1}$, and $s_d(F) = 10^{-3.6}$, where s_d is the standard deviation of the parameters. This is consistent with what was already stated in [11,143,144]: the numerical coefficient F must be $O(1)$. In particular, in the case of the collapse of the domain walls, $F \approx 3.8$ was found for large black holes [146].

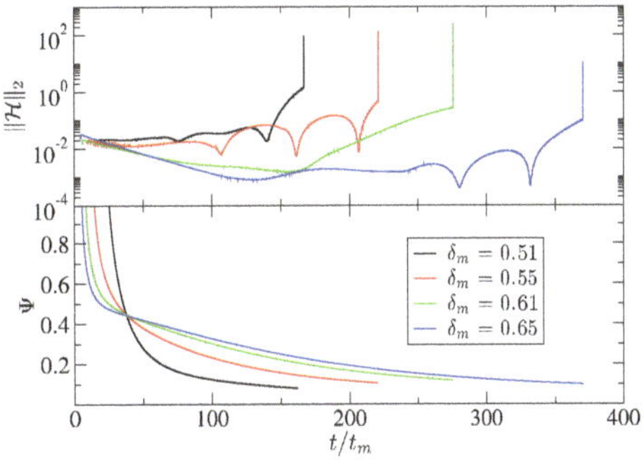

Figure 17. (**Top-panel**): Time evolution of the Hamiltonian constraint for different amplitude values δ_m during the excision procedure. (**Bottom-panel**): Numerical evolution of Ψ in time. The crossing point is around $t/t_m \approx 37.5$. The profile used is the same as in Figure 16. The excision starts at very early times, at $t \approx 10t_m$.

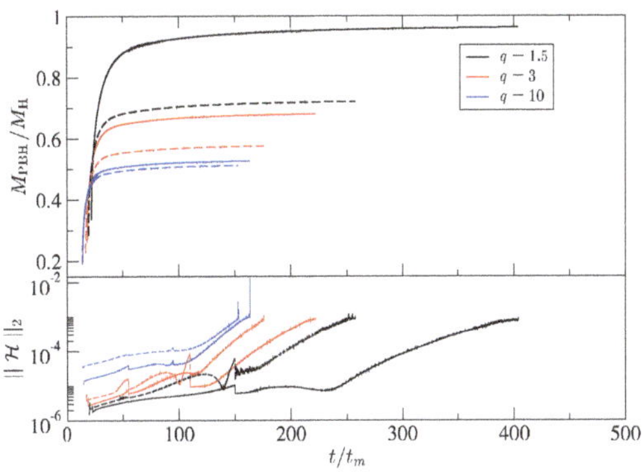

Figure 18. (**Top-panel**): Evolution of $M_{PBH}(t)$ for the profiles $q = 1.5$, $q = 3$, and $q = 10$. The dashed line corresponds to Equation (83) with $\lambda = 0$ and the solid line to Equation (82). (**Bottom-panel**): Time evolution of the Hamiltonian constraint. In all cases, $\delta_m - \delta_c = 5 \times 10^{-3}$.

Interestingly, in [87], the effect of the accretion (from the FLRW background) to the growth of the PBH until reaching the stationary regime was quantified, namely $M_{\mathrm{PBH},f}/M_{\mathrm{PBH},i}$, for the case $w = 1/3$. As was pointed out, the accretion should not be important for small PBHs, i.e., with δ_m near δ_c. However, the situation differs for relatively large PBHs. This can be precisely seen in Figure 19. The accretion can be significantly important for large δ_m.

Sharp profiles, corresponding to large q, have larger pressure gradients, and therefore, the ratio $M_{\mathrm{PBH},f}/M_{\mathrm{PBH},i}$ is smaller, even for large δ_m, since the gradients prevent the accretion. For low $M_{\mathrm{PBH},f}$, the ratio $M_{\mathrm{PBH},f}/M_{\mathrm{PBH},i}$ should be small, as expected [9,86]. When $M_{\mathrm{PBH},f} \simeq M_H$, i.e., for PBHs with a higher probability to form, $M_{\mathrm{PBH},f} \simeq 3M_H$ was obtained. In comparison, a similar result was also found in the case of the collapse of a massless scalar field [141] and in [146,199] from BHs formed from the collapse of domain walls or vacuum bubbles, with a factor $M_{\mathrm{PBH},f}/M_{\mathrm{PBH},i} \leq 2$.

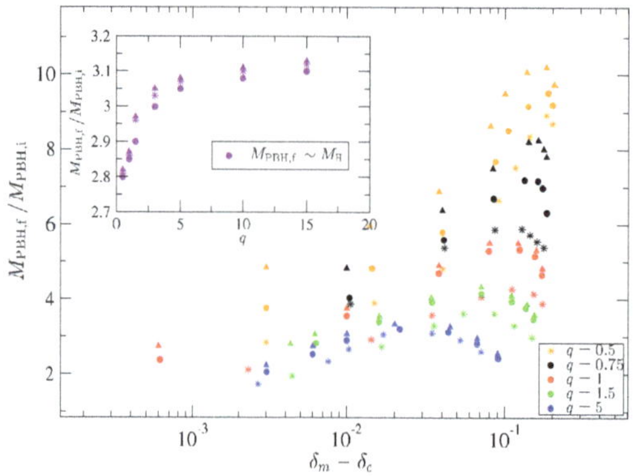

Figure 19. Accretion effect on the final PBH mass measured as the ratio $M_{\mathrm{PBH},f}/M_{\mathrm{PBH},i}$ in terms of $\delta_m - \delta_c(q)$ for different profiles with $w = 1/3$. Stars correspond to Equation (83) with $\lambda = 0$, circles to Equation (82), and triangles to Equation (86). The ratio $M_{\mathrm{PBH},f}/M_{\mathrm{PBH},i}$ for PBHs with $M_{\mathrm{PBH},f} \simeq M_H$ is shown in the subplot.

An example of the values of PBH mass for other values of w can be found in Figure 20, where the ratio $M_{\mathrm{PBH},f}/M_H$ is shown using the Gaussian profile Equation (83) with $q = 1$ ($\lambda = 0$) for those PBHs $M_{\mathrm{PBH},f} \approx M_H$. The PBH mass $M_{\mathrm{PBH},f}$ is very sensitive to w, as expected. When w increases, the gradients are larger, preventing the accretion and the increase of the PBH mass. The ratio $M_{\mathrm{PBH},f}/M_H$ is larger for smaller w at the same $\delta_m - \delta_c$, since the accretion is substantially higher when the pressure gradients are decreased, as can be seen in the bottom panel of Figure 20. In this case, the values of F differ substantially from $w = 1/3$. In particular, for the set of values $w \in [0.2, 1/3, 0.5, 0.7]$, this corresponds to $F \approx [7.2, 3.8, 2.0, 1.37]$, respectively, which already clearly indicates the substantial accretion for small w. Moreover, the time needed to reach the asymptotic regime of Equation (56) is much higher the smaller w is. For example, in the case of $\delta_m - \delta_c = 10^{-2}$ for $w = 0.2$ and $w = 0.5$, a time of order $t/t_0 \backsim 10^4$ and $t/t_0 \backsim 10^3$ was needed, respectively, to reach the asymptotic regime.

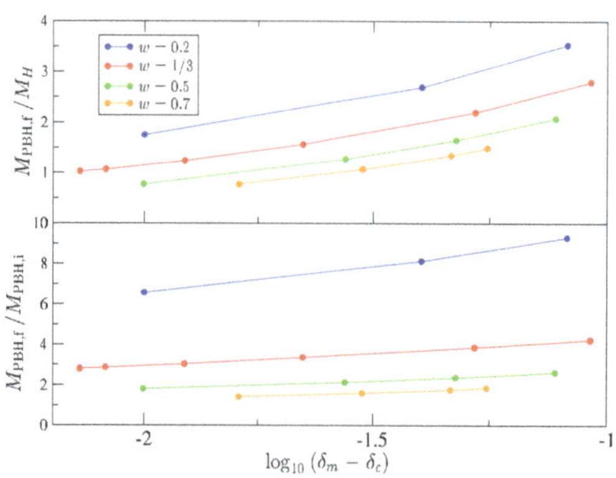

Figure 20. (Top-panel) : Values of the final PBH mass in terms of M_H for some values of w as a function of $\delta_m - \delta_c$. **(Bottom-panel)**: Accretion effect on the final PBH mass for some values of w as a function of $\delta_m - \delta_c$. In both cases, the profile of Equation (83) was used with $q = 1$ and $\lambda = 0$.

On the other hand, in [86], some analytical approximations were made regarding the maximum size of the PBHs formed from hydrodynamical perturbations at the time of the formation of the apparent horizon t_{AH}; in particular, it was found that:

$$\left(\frac{R_{\mathrm{PBH},i}}{R_{H,i}}\right)_{\max} = \left(\frac{2}{1+3w}\right)^3 \left[\frac{3(1+w)}{2(1+\sqrt{w})}\right]^{\frac{3(1+w)}{1+3w}} w^{3/2}. \quad (91)$$

This analytical result (derived for $0 < w \leq 1$) was verified numerically in the case of a radiation fluid in [87], where an extensive study on the values $M_{\mathrm{PBH},i}$, $R_{\mathrm{PBH},i}$, and t_{AH} was also performed for different profiles and amplitudes δ_m.

To compute the value of the constant $\mathcal{K}$ in front, the scaling law Equation (58) is needed to estimate the final mass of the black hole $M_{\mathrm{PBH},f}$ using perturbations with an amplitude near the critical regime, i.e., $\delta_m \to \delta_c$. A range of $10^{-3} \lesssim \delta_m - \delta_c \lesssim 10^{-2}$ is sufficient to estimate it, since the gravitational collapse of the fluctuations happens in the critical regime. Beyond that, the scaling law deviates. In particular, for a radiation fluid, the value $\gamma = 0.357$ [66,68] can be taken directly to obtain $\mathcal{K}$ from a given $\delta_m - \delta_c$ (for other values of w, check Figure 3 and the caption).

In Figure 21 are shown the values of $\mathcal{K}$ for the profiles of Equations (82), (83), and (86) as an example of $\mathcal{K}$ in terms of the profile. The value of the PBH mass $M_{\mathrm{PBH},f}$ (and therefore, also $\mathcal{K}$) is affected by the shape of the profile beyond the peak of $\mathcal{C}$ since the accretion process depends on the full shape of the fluctuation, in comparison with the threshold δ_c, as was pointed out in Section 6.2, which mainly depends on the shape around the compaction function peak.

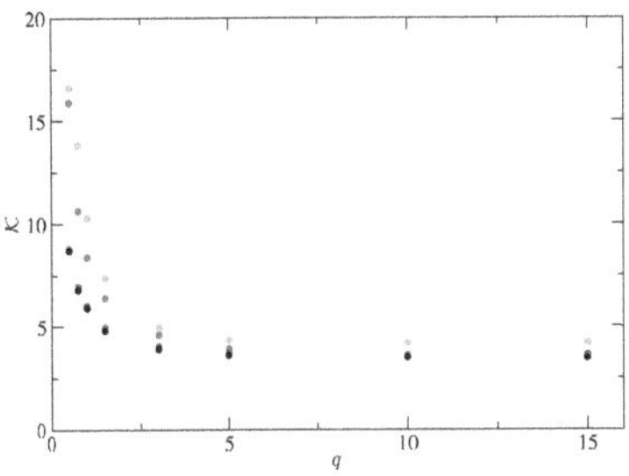

Figure 21. Dependence of $\mathcal{K}$ as a function q for the profiles in Equation (82) (red), Equation (83) with $\lambda = 0$ (black), Equation (83) with $\lambda = 1$ (blue), and Equation (86) (green). The parameters used for the profile $C_{tt}(r)$ are $q_2 = 3$, $r_j = 2r_{m1}$, $C_{tt(\mathrm{peak},2)} = 0.3$, and δ_2, obtained from Equation (88) using the previous values with and $q_1 = q$.

In the case of the profile Equation (82), $\mathcal{K}$ tends to ≈ 3.5 for large values of q. The values of $\mathcal{K}$ tend to increase as q decreases, as shown Figure 21. Numerically, in [87], it was not possible to obtain the final mass $M_{\mathrm{PBH},f}$ for profiles $q \lesssim 0.5$, due to conic singularities, as was already found in [84].

On the other hand, in Figure 22, an explicit example of $M_{\mathrm{PBH},f}$ can be seen for a Gaussian profile (in both $K(r)$ and $\zeta(r)$) in terms of M_H for large values of δ_m beyond the critical regime $\delta_m \gtrsim 10^{-2}$ with $w = 1/3$. As can be observed in the subplot of Figure 22, the scaling law deviates for the highest allowed values of δ_m to O(15–20%) for both cases. For these particular cases, the maximum allowed mass of the PBH formed is $M_{\mathrm{PBH},f} \approx 3.7 M_H$ and $M_{\mathrm{PBH},f} \approx 2.3 M_H$, respectively. It is expected in principle that this deviation from the scaling regime will not significantly modify the estimation of the PBH abundance, due to the unusualness of such perturbations since those PBHs with a higher probability to form have $M_{\mathrm{PBH},f} \approx M_H$. It is important to also notice that the masses generated with the profile $\zeta(\tilde{r})$ are lower than those with $K(r)$, since although they are both Gaussian, they have $q > 1$ and $q = 1$, respectively, and sharper profiles have less mass due to stronger pressure gradients, as we see in Figure 18.

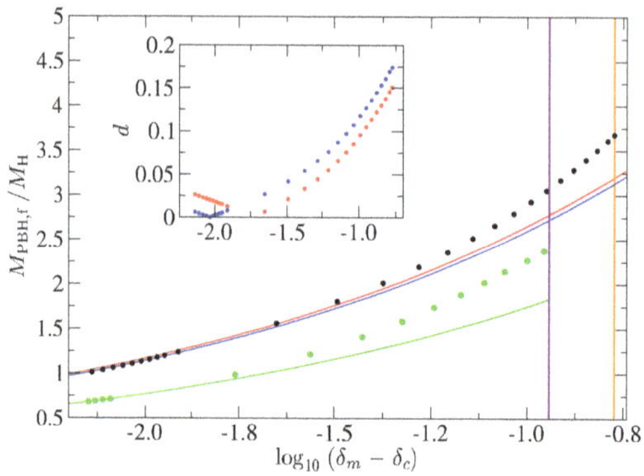

Figure 22. Values of $M_{\mathrm{PBH},f}/M_H$ in terms of $(\delta_m - \delta_c)$ for two Gaussian profiles in $K(r)$ and $\zeta(\tilde{r})$. Dark points correspond to the numerical values for the profile in $K(r)$. The solid red line corresponds to the scaling law behaviour with $\gamma = 0.357$, $\delta_c = 0.49774$ for $\mathcal{K} = 6.03$ (pointed out in [94]) and the blue solid line with the numerical value obtained in the review with $\mathcal{K} = 5.91$. The absolute value of the relative deviation d with respect to the numerical values and the ones coming from the scaling law are shown in the subplot. The orange vertical line is the value $\delta_{\max} = 2/3$. On the other hand, the green dots correspond to the case of the Gaussian profile in ζ. The solid green line is the scaling law with $\gamma = 0.357$, $\delta_c = 0.55257$, and $\mathcal{K} = 3.98$ (in [106], for this profile, $\mathcal{K} \simeq 4.0$ was given). The vertical magenta line corresponds to the case $\delta_{\max} = 2/3$ for this profile. In both cases, $w = 1/3$.

7. Analytical Estimation for the Threshold of PBH Formation

In this section, we review in detail the current analytical estimations used in the literature for the threshold of PBH formation from curvature fluctuations with an equation of state w. The first analytical estimation came from [74], where using a Jeans length approximation, it was shown that,

$$\delta_{\mathrm{Carr}} = w\,. \tag{92}$$

Although this approximation is simple, it fulfils some properties: (i) at $w = 0$ (dust), any curvature fluctuation should collapse, i.e., $\delta_c(w = 0) = 0$; (ii) when w increases, the pressure gradients are larger, so the threshold should be higher. This approximation was obtained assuming the argument that the size of an over-density (that leads to PBH formation), at the maximum expansion, should be smaller than the particle horizon and larger than the Jean radius (Jeans criterion). In particular,

$$R_{\mathrm{J}} \lesssim R_{\max} \lesssim R_{\mathrm{PH}}, \tag{93}$$

where $R_{\mathrm{PH}} \backsim \sqrt{3/8\pi\rho_{\max}}$, $R_{\mathrm{J}} \backsim c_s R_{\mathrm{PH}}$. See also Section 2 for a simple derivation.

Later on, in [75], the previous estimation of [74] was improved by using a sophisticated and interesting model (a three-zone model) of the collapse of a homogeneous over-dense sphere surrounded by a thin under-dense shell. We suggest the reader check [75] for more details on the model and the assumptions made. The threshold estimation that they obtained was,

$$\delta_{\mathrm{HYK}} = \frac{3(1+w)}{5+3w}\sin^2\left(\frac{\pi\sqrt{w}}{1+3w}\right). \tag{94}$$

It was also provided with an upper bound $\delta^{+}_{\mathrm{HYK}}$ and a lower bound $\delta^{-}_{\mathrm{HYK}}$ on δ_c, to take into account the different ways of applying the relativistic Jeans criterion for this model.

$$\delta^{+}_{\mathrm{HYK}} = \frac{3(1+w)}{5+3w}\sin\left(\frac{2\pi\sqrt{w}}{1+2\sqrt{w}+3w}\right). \tag{95}$$

$$\delta^{-}_{\mathrm{HYK}} = \frac{3(1+w)}{5+3w}\sin\left(\frac{\pi\sqrt{w}}{1+\sqrt{w}+3w}\right). \tag{96}$$

These previous estimations (in particular, Equation (94)) have been extensively used in the literature, but as has already been noticed [74,75], these estimations are not profile dependent and therefore do not take into account the specific shape of the curvature fluctuation.

7.1. Shape-Dependent Analytical Estimation of the Threshold for Radiation Fluid

In [83], it was proven that, to a very good approximation, the threshold for the $w = 1/3$ case only depends on the curvature of the compaction function at its maximum, under the assumption of a central over-dense peak in the density distribution. It was used to build an analytical formula for the threshold with a dependence on the profile considered. The formula was accurate enough and only deviated 2% in comparison with the numerical simulations.

The two main ingredients of the procedure used in [83] were: (i) parametrise the shape around the compaction function peak with a dimensionless parameter q; (ii) use the average of the critical compaction function. As was shown in [83] and as we have already mentioned, to a very good approximation, the threshold only depends on,

$$q \equiv -\frac{r_m^2 \mathcal{C}''(r_m)}{4\mathcal{C}(r_m)}, \tag{97}$$

which is a dimensionless measure of the curvature of $\mathcal{C}(r)$ at its maximum. This means that different profiles, but with the same q value, will have the same threshold upon a small deviation of <2%.

The next point is to define a "basis" (or fiducial set of curvature profiles) such that, by varying q, this set covers the whole range of interesting thresholds and shapes with $q \in (0, \infty)$ while also being regular at $r = 0$ and having $\rho'(r = 0, t) = 0$. In [83], this basis was given in terms of the exponential functions Equation (83) (with $\lambda = 0$). Although it strictly does not satisfy regular conditions for $q \to 0$ at $r = 0$, this limit was tested numerically with the polynomial basis of Equation (82).

Taking this into account, the averaged critical compaction function integrated from $r = 0$ up to the peak $r = r_m$ is defined as,

$$\bar{\mathcal{C}}_c \equiv \frac{3}{r_m^3}\int_0^{r_m} \mathcal{C}_c(r) r^2 dr\,. \tag{98}$$

Introducing the profile considered into Equation (98), we can solve the integral analytically to obtain:

$$\bar{\mathcal{C}}_c = \frac{3}{2} e^{\frac{1}{q}} q^{-1+\frac{5}{2q}} \left[\Gamma\left(\frac{5}{2q}\right) - \Gamma\left(\frac{5}{2q}, \frac{1}{q}\right)\right] \delta_c, \tag{99}$$

where $\Gamma(x)$ [4] is the gamma function, $\Gamma(x, y)$ [5] is the incomplete gamma function, and we have used $\mathcal{C}_c(r_m) \equiv \delta_c$. In the case of a radiation fluid, when $q \to \infty$, it was confirmed numerically that $\delta_c(q \to \infty) = f(w = 1/3) = 2/3$ [79]. Taking into account this analytical limit into Equation (99), we obtain:

$$\bar{\mathcal{C}}_c(q \to \infty) = \frac{3}{5}\delta_c(q \to \infty) = \frac{2}{5}. \tag{100}$$

The assumption made in [83] was to consider that $\bar{\mathcal{C}}_c$ is a profile-independent quantity, and therefore, $\bar{\mathcal{C}}_c = 2/5$ for any value of q. The assumption was tested extensively and numerically for different profiles and was found to be consistent with numerical simulations up to a small deviation of $O(2\%)$ in δ_c. Inverting Equation (99) in δ_c with $\bar{\mathcal{C}}_c = 2/5$, we finally obtain the analytical threshold formula δ_c in terms of the specific profile q, in the case of a radiation fluid,

$$\delta_c(q) = \frac{4}{15} e^{-\frac{1}{q}} \frac{q^{1-\frac{5}{2q}}}{\Gamma\left(\frac{5}{2q}\right) - \Gamma\left(\frac{5}{2q}, \frac{1}{q}\right)}. \tag{101}$$

This remarkable result in [83] was the first analytical estimation shown in the literature for the threshold of PBH formation, taking into account the shape of the fluctuation, with enough accuracy to be used for estimations of PBH abundances [95]. A plot of Equation (101) in terms of q can be found in Figure 23. Interestingly, a study of the thresholds for different power spectrums (leading to perturbations in ζ with the coordinate $\tilde{r}$) was performed in [198] using Equation (101), with the corresponding q value obtained from Equation (85) and connecting with the density contrast of Equation (43).

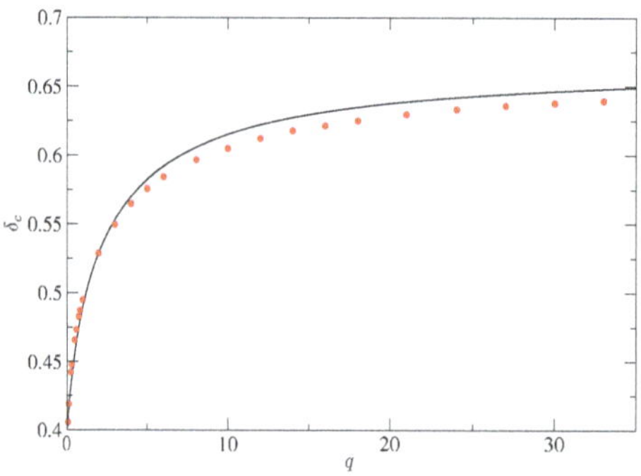

Figure 23. Plot of $\delta_c(q)$ for $w = 1/3$. The solid black line corresponds to the analytical estimation of Equation (101). Red points correspond to the numerical thresholds of Equation (82) with $w = 1/3$.

7.2. Generalisation of the Analytical Threshold beyond the Radiation Fluid

Later on, the procedure used in [83] was generalised considering PBH formation within a perfect fluid with $w \in (0, 1]$ on an FLRW background, and it was found that the threshold averaged compaction function within a concentric sphere of radius $r_m(1 - \alpha(w)) \leq r \leq r_m$ is, to a very good approximation, again universal in the case $1/3 \leq w \leq 1$, where $\alpha(w)$ is the location where we should start to integrate the compaction function, and it should depend on w. This was used to provide a new and improved analytical formula for the threshold that only depends on the normalised second derivative of the compaction function at its maximum and the equation of state w. This formula was proven to be accurate at order 6% compared to numerical simulations and reproducing the same result for the case of radiation $w = 1/3$. It was also again found that the δ_c for different profiles with the same q are equal within a deviation less than 4%.

In [83], the basis profiles used were given in terms of the exponential functions; despite that, since the boundary conditions at $r = 0$ are violated for $q < 1/2$, the basis profile of Equation (82) was used instead. This set of profiles satisfies the appropriate boundary and regularity conditions.

The critical compaction function, averaged within a spherical shell extending from radius $[1 - \alpha(w)]\, r_m$ to r_m (the peak), is defined to be:

$$\bar{\mathcal{C}}_c(w, \text{profile}) \equiv \frac{3}{r_m^3 V[\alpha(w)]} \int_{r_m[1-\alpha(w)]}^{r_m} \mathcal{C}_c(r) r^2 dr\,, \tag{102}$$

where $V[\alpha(w)] = \alpha(w)\,[3 + (\alpha(w) - 3)\alpha(w)]$ and $\mathcal{C}_c(r) = \mathcal{C}(r)\Big|_{\mathcal{C}(r_m)=\delta_c}$. Inserting Equation (82) into Equation (102) yields:

$$\bar{\mathcal{C}}_c(w, \text{basis}) = \delta_c(w, q)\, g(q, w) \left[-F_1(q) + (1 - \alpha)^{3-2q} F_2(q, \alpha)\right], \tag{103}$$

with:

$$g(q,w) = \frac{3(1+q)}{\alpha(2q-3)[3+\alpha(\alpha-3)]}, \tag{104}$$

$$F_1(q) = {}_2F_1\left[1, 1-\frac{5}{2(1+q)}, 2-\frac{5}{2(1+q)}, -q\right], \tag{105}$$

and:

$$F_2(q,w) = {}_2F_1\left[1, 1-\frac{5}{2(1+q)}, 2-\frac{5}{2(1+q)}, -q(1-\alpha)^{-2(1+q)}\right], \tag{106}$$

where ${}_2F_1$ is the hypergeometric function[6].

The key point is that as was shown numerically in [84], the dependence of the averaged critical compaction function on the profile shape is weak enough to be ignored, and therefore universal. Then,

$$\bar{\mathcal{C}}_c(w, \text{profile}) \simeq \bar{\mathcal{C}}_c(w)\,. \tag{107}$$

Therefore, the analytic expression for the critical threshold value reads as:

$$\delta_c(w,q) = \frac{\bar{\mathcal{C}}_c(w)}{g(q,w)}\frac{1}{\left[-F_1(q) + (1-\alpha)^{3-2q}F_2(q,\alpha)\right]}\,. \tag{108}$$

The functions $\alpha(w)$ and $\bar{\mathcal{C}}_c(w)$ were found using two different procedures: (I) a numerical double fit minimisation using the numerical results of the thresholds for different profiles and (II) making a single numerical fit minimisation already using the analytical result that $\delta_c(q \to \infty)$ for $w \geq 1/3$. Both procedures gave equivalent results. Therefore, the assumption of $\bar{\mathcal{C}}_c(w)$ being universal was found good enough and verified numerically for $w \gtrsim 1/3$. This result verified the previous one obtained for a radiation fluid $w = 1/3$ in [83], where $\alpha = 1$ and $\bar{\mathcal{C}}_c = 4$. In particular, using the procedure (I), it was found,

$$\bar{\mathcal{C}}_c(w) = a + b\,\text{Arctan}(c\,w^d) \tag{109}$$

$$\alpha(w) = e + f\,\text{Arctan}(g\,w^h), \tag{110}$$

with $a = -0.140381$, $b = 0.79538$, $c = 1.23593$, $d = 0.357491$, $e = 2.00804$, $f = -1.10936$, $g = 10.2801$, and $h = 1.113$. Instead, using (II),

$$\bar{\mathcal{C}}_c(w) = i + j\,\text{Arctan}(p\,w^l), \tag{111}$$

$$\alpha(w) = m + t\,\text{Arctan}(r\,w^s), \tag{112}$$

with $i = -0.140381$, $j = 0.79538$, $p = 1.23593$, $l = 0.357491$, $m = 2.00804$, $t = -1.10936$, $r = 10.2801$, and $s = 1.113$.

7.3. A Comparison between Analytical Estimates

In [84], the analytical estimation obtained with Equation (108) was compared with the previous ones in the literature, i.e., Equations (92) and (94). In Figure 24 is shown the comparison. The solid lines in Figure 24 show these approximations; the symbols with error bars show $\delta_c(w)$ from the numerical simulations of the profiles of Equation (82) with $q = 0.015$, $q = 0.1$, $q = 1$, and $q = 30$. The dotted and dashed curves, which significantly better describe the result of the numerical simulations, show the result of inserting Equations (109) and (110) or Equations (111) and (112), into Equation (108). In both cases, Equation (108) (as well as the simulations) exceeds even the upper bound claimed by [75] at even lower w as q increases, which are Equations (95) and (96).

As can be observed, the discrepancy of the estimation Equations (92) and (94) for the case of small w is as large as 50% for the case $q = 0.1$. This is the limit that was considered successful and optimal for the approximations on which Equation (94) is based. This deviation is even bigger than the one found earlier because Equation (94) was compared with a Gaussian profile (Equation (83) with $q = 1$ and $\lambda = 0$). Actually, in the case $w < 0.15$, the solid black curve in Figure 24 give a satisfactory description of the simulations with $q = 1$. However, the top-hat profile with $q \to 0$ (top-hat for the profile in the compaction function), which was the one used in the analytic estimations of [75], is quite better approximated by $q \ll 1$. For $q = 0.1$, the formula Equation (94) does not describe the simulations particularly well, and the discrepancy at $w < 1/3$ is even worse when $q = 0.015$ is taken. It was claimed in [84] that this disagreement seems to suggest that the supposed agreement shown in Figure 3 of [75] is a result of numerical coincidences, with no physics implications.

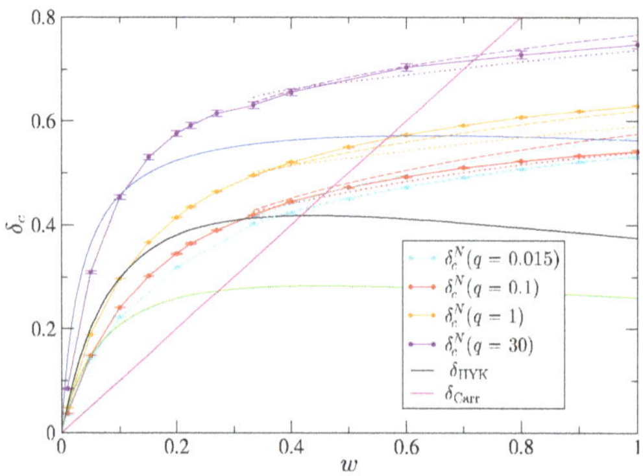

Figure 24. Dependence of threshold δ_c on w when the initial profile is given by Equation (83) with $q = 30, q = 1, q = 0.1, q = 0.015$, and $\lambda = 0$ in all the cases ($q = 0$ would be a homogeneous sphere in the compaction function). The solid lines with dots and error bars show the results of the numerical simulations. The green and blue curves show the minimal and maximal bounds of δ_c from [75], in particular Equations (95) and (96), respectively. The magenta line shows the approximation for $\delta_{\rm Carr}$ of Equation (92). The black curve shows $\delta_{\rm HYK}$ from Equation (94). The other curves show the approximation (Equation (108)) in which δ_c depends both on w and q. Finally, the dotted curves use Equations (111) and (112) in Equation (108), whereas the dashed curves use Equations (109) and (110) in Equation (108).

Finally, the polynomial basis profile of Equation (82) fulfils the regularity conditions for $q \to 0$, as we have pointed out before, something not fulfilled with the exponential basis Equation (83) (with $\lambda = 0$). Precisely, this led in previous literature [79] to inferring that the minimum threshold for PBH formation (in the case of a radiation fluid) must coincide with the estimation of [75], i.e., $\delta_{c,\rm min}(w = 1/3) = \delta_{\rm HYK}(w = 1/3) \approx 0.41$. In [79], the profile leading to the minimum numerical threshold of PBH formation was given by Equation (83) with $q = 0.1$ and $\lambda = 0$, since as was indicated there, the profiles were too sharp to take even smaller values of q and produce a numerical simulation. This actually looks related to the fact that the exponential profile Equation (83) does not satisfy the regularity conditions at $r \to 0$ for $q < 1/2$. Instead, in the case of using the polynomial profile Equation (82), it allows performing simulations even with smaller values in q, up to $q = 0.01$, as done in [83]. Therefore, this led to the conclusion in [83] that, for a radiation fluid, the correct range of values for the threshold is $0.4 \leq \delta_c \leq 2/3$, and therefore, the minimum value was lower than $\delta_{\rm HYK}$. Indeed, the generalisation of δ_c in w already indicates that the minimum threshold should not correspond necessarily to the one of $\delta_{\rm HYK}$.

8. Other Scenarios of PBH Formation

In this review paper, we mainly focused on the study of PBH formation from the spherical collapse of primordial density perturbations, which is currently the case most studied in the literature and with more results and applications [39]. Despite that, many other mechanisms and scenarios could have led to black holes formed in the very early

Universe [12]. In this last section, we try to briefly enumerate other scenarios beyond the ones considered in this review.

The case $w = 0$ is very special as the threshold of PBH formation, for an infinitely long matter-dominated era; it is zero if non-spherical effects are not considered. However, as was pointed out in [200], the dust phase could induce an angular momentum on the collapsing perturbation, therefore effectively increasing the threshold value for the black hole formation, making it non-vanishing. Thus, the study of abundances, where the Universe is matter-dominated for a finite time, differs greatly from the case of radiation. Some works have considered this scenario with scalar field types [201–206], specifically taking into account the oscillatory behaviour of the scalar field during the reheating. Recently, in [207], simulations were performed in full GR using scalar fields to modulate the perturbation and the expanding dust background of the Universe, showing that the PBH mass grows beyond the self-similar limit $M_{\rm PBH} \propto H^{-1}$, at least initially. The dust collapse has also been studied in a more oriented perfect fluid scenario [208]. The PBH formation has also been considered in the context of the Lemaître–Tolman–Bondi (LTB) model [209,210]. In [211], the effect of inhomogeneity on PBH formation in the matter-dominated era and its consequences on the PBH production were studied. On the other hand, in [200,212], the non-spherical effects and spins for PBH formation in the case of a dust-dominated Universe were studied, showing this could have a significant effect.

A non-spherical effect, in principle, could dramatically change the threshold value and PBH mass [213]. Some analytical estimations taking into account an ellipsoidal collapse were performed in the past, showing that the effect could be important [209,214]. However, in [215], for the first time in the literature, the first numerical simulation in $3+1$ spacetime (beyond spherical symmetry) for PBH formation from the collapse of curvature fluctuations was performed, with a small (perturbative) non-sphericity initial condition (with a radiation fluid). The main conclusion was that the effect of ellipticity on the threshold for PBH formation is negligible in the standard scenario of PBH formation in the radiation-dominated Universe. This situation could dramatically change if an equation of state different from radiation is considered, as indicated in [215]. In particular, for a matter-dominated Universe, the effect of any non-sphericity could be important [200,208,212]. Another possibility is the inclusion of angular momentum to analyse the spin generation to PBHs; some analytical estimations have been performed [216], but still, numerical simulations to test those arguments are necessary. Recently, PBH formation with an anisotropic perfect fluid has also been considered analytically [217].

The usual scenario of the collapse of cosmological density perturbations considers a constant equation of state w (mainly $w = 1/3$ in the literature). However, the Universe has a thermal history where the equation of state is not necessarily constant, i.e., $w(t)$. A particular (and very popular) example applied to the PBH scenario is the QCD epoch [218–221]. During the QCD phase transition, the equation of state w could have been softened. Looking at the result of the simulations in Figure 24, it is already clear that the threshold δ_c will be reduced, and therefore, this makes the formation probability of PBHs exponentially more likely. Several works have addressed this scenario with progressive refinements [218,222–227], specifically with the new calculations regarding lattice QCD [228,229]. However, numerical simulations are still needed to fully verify the effect on how much the threshold, PBH mass, and mass function modifying the equation of state in time could change.

A bit more different mechanism of PBH formation comes from the collapse of Q-balls and oscillons, which are features of supersymmetric extensions of the standard model [230]. Specific realisations of those ideas have been made through theory-motivated scalar field potentials, e.g., the axion–monodromy potential [231]. The mechanism for PBH formation of all these cases is naively similar: small number densities of defects lead to large fluctuations relative to the background density. These fluctuations become gravitationally bound and collapse to form black holes once the relic density has come to dominate. Finally, the relics decay due to some instabilities. Precisely, it has been argued that some solitonic type solutions such as Q-balls and oscillons could produce a significant fraction of dark matter in the form of PBHs [232–236], without relying on any spectrum of density perturbations. Some recent numerical works on oscillons and Q-balls have obtained results in this direction [237–239].

On the other hand, scalar force instability can lead to the growth of structures and the formation of halos of interacting particles (even in a radiation-dominated Universe) [240–244]. Actually, in [243], it was theorised that these structures could have collapsed and formed PBHs, but later on, it was realised that they should remain as virialised dark matter [244]. Precisely, in [245], it was shown that the radiative cooling due to the scalar radiation allows those halos to collapse into black holes. This was shown with a simple model with fermions interacting via the Yukawa interaction. The general relativistic formulation of the interaction between a Fermi gas and a scalar field in cosmology was carefully studied recently in detail in [246].

All these mechanisms could induce small or large spins to the PBHs in terms of their formation evolution and the process of radiative cooling, as was pointed out in [213]. Another proposal is the one in [247], where the dilution of heavy quarks produced during inflation, which become confined by QCD flux tubes after horizon re-entry, could have led to PBH production.

Another mechanism is through the collapse of domain walls. Domain walls are topological defects that may form when a discrete symmetry is spontaneously broken in the very early Universe [248,249]. Several works have addressed this scenario and have shown the possibility of forming PBHs, with astrophysical consequences [146,250–253]. Another possibility comes from the collapse of cosmic strings [254–256], which are $1+1$ topological defects that are predicted beyond the Standard Model and could have led to the production of PBHs [257–263].

Similarly, it was also shown that vacuum bubbles could lead to a successful channel for PBH production or also be involved in related mechanisms [199,264–271]. They could have been formed during the inflationary epoch through a nucleation process. After inflation and depending on their size, these bubbles could collapse or even collide, forming black holes [146,181,197,272–274].

Other results have also been pointed out in the context of modified gravity, in particular in [275], where the PBH production in G-inflation was studied, as well as in [276], where this was studied with the inclusion of a Gauss–Bonnet term. In [277], the effects on the threshold in Eddington-inspired Born–Infeld gravity were considered.

Finally, a new mechanism appeared in [278], where PBHs could have been formed from the collapse of primordial CDM isocurvature fluctuations. Related to that, a new work has numerically shown the formation of PBHs [279] from isocurvature fluctuations generated by a massless scalar field.

9. Conclusions

In this review, we focused on explaining in detail the process of PBH formation from the collapse of curvature fluctuations under spherical symmetry on an FLRW background and its numerical implementation through a particular numerical technique. Specifically, we saw that numerical simulations are essential to capture the correct and full-non-linear behaviour of the collapse and formation of PBHs. This leads to a better and accurate determination of the threshold of PBH formation and their masses compared to analytical estimations, something essential for the precise analysis of PBH abundances. In this review, we saw a simple and efficient numerical procedure based on pseudospectral methods that allows us to fully solve the non-linear gravitational collapse of those perturbations and obtain the needed values of the threshold and PBH mass in the range of practical interest.

Two aspects in the literature have been remarkably important for the development of the field: (I) the results found from numerical simulations about the critical scaling regime for the PBH mass and (II) the analytical estimations for the threshold of PBH formation. Both allow us to estimate the abundances of PBHs and constrain them. In particular, we saw that the threshold and PBH mass are both dependent on the specific shape of the curvature fluctuation considered, and it is crucial to take this into account for the precise analysis of PBH abundances instead of using a constant threshold or PBH mass value. Moreover, we saw that the gravitational collapse is highly dependent on the equation of state of the fluid w. The larger w is, the smaller is the mass of the PBHs and the accretion from the FLRW background, but the higher the threshold is.

It is interesting to mention that the compaction function, firstly used in [77] years ago, has become currently essential, not only for the definition of the threshold of PBH formation, but also to obtain an analytical estimation of it, using its average.

There is still much effort to be made in other directions and different scenarios of PBH formation to consider, as the ones already pointed out in Section 8. Future theoretical developments and the implementation of numerical simulations to test and check those scenarios could lead to new possibilities for PBHs to account for dark matter and update the abundance estimations. In particular, it would be interesting to compare the results already pointed out in this review in other scenarios.

We hope that in the next few years, several scientific results in this and other directions will help to fully understand and constrain the PBH scenario and its consequences in our Universe, especially taking into account future gravitational wave observations.

Funding: A.E. is currently supported by "bourse de post-doctorat" from Université Libre de Bruxelles (ULB) and was supported by Contract PID2019-106515GB-I00/AEI/10.13039/501100011033 (Spanish Ministry of Science and Innovation).

Institutional Review Board Statement: Not applicable.

Informed Consent Statement: Not applicable.

Data Availability Statement: The numerical results shown in this review have been obtained using the code that can be downloaded in [158].

Acknowledgments: I want to thank Vicente Atal, Guillem Domenech, Cristiano Germani, Claudia Gonzalez-Boquera, and Marc Oncins for extremely useful comments about the manuscript. I would also like to appreciate the discussions on the PBH topic with Vicente Atal, Sebastien Clesse, Guillem Domenech, Jaume Garriga, Cristiano Germani, Carsten Gundlach, Ilia Musco, Marc Oncins, Javier G. Subils, Ravi K. Sheth, Antonio Enea Romano, Yuichiro Tada, Shuichiro Yokoyama, and Chulmoon-Yoo. Finally, I also want to thank the anonymous Referees for the constructive and valuable comments that have helped to improve this work.

Conflicts of Interest: The author declares no conflict of interest.

Abbreviations

PBHs	Primordial Black Holes
DM	Dark Matter
FLRW	Friedman–Lemaître–Robertson–Walker
GR	General Relativity
CMB	Cosmic Microwave Background
LIGO	Laser Interferometer Gravitational-Wave Observatory
BH	Black Hole
PDE	Partial Differential Equation
AMR	Adaptive Mesh Refinement
NANOGrav	North American Nanohertz Observatory for Gravitational Waves

Notes

1. The confluent hypergeometric function is defined as $\Gamma_U(a,b,z) = \frac{1}{\Gamma(a)} \int_0^\infty e^{-zt} t^{a-1} (1+t)^{b-a-1} dt$.
2. The erf function is defined as $\mathrm{erf}(x) = \frac{2}{\sqrt{\pi}} \int_0^z e^{-t^2} dt$.
3. Technically, the condition of the formation of the apparent horizon is given when $2M(r,t)/R(r,t) = 1$. However, for computationally efficiency, ot is easier to check whatever $\mathcal{C} \approx 1$, since $\mathcal{C} = 2M(r,t)/R(r,t) - 2M_b(r,t)/R(r,t)$, which avoids finding the cosmological horizon.

4 The gamma function $\Gamma(x)$ is defined as $\Gamma(x) = \int_0^\infty t^{z-1}e^{-t}dt$.

5 The incomplete gamma function is defined as $\Gamma(x,y) = \int_y^\infty t^{x-1}e^{-t}dt$.

6 The hypergeometric function is defined as ${}_2F_1(a,b,c,z) = \sum_{k=0}^\infty \frac{(a)_k(b)_k}{(c)_k}\frac{z^k}{k!}$.

References

1. Particle Data Group. Review of Particle Physics. *Prog. Theor. Exp. Phys.* **2020**, *2020*, 083C01. [CrossRef]
2. Bertone, G.; Hooper, D. History of dark matter. *Rev. Mod. Phys.* **2018**, *90*, 045002. [CrossRef]
3. Carr, B.; Kohri, K.; Sendouda, Y.; Yokoyama, J. Constraints on Primordial Black Holes. *Rep. Prog. Phys.* **2021**, *84*, 116902. [CrossRef] [PubMed]
4. Carr, B.; Kuhnel, F. Primordial Black Holes as Dark Matter Candidates. *arXiv* **2021**, arXiv:2110.02821.
5. Carr, B.; Kuhnel, F.; Visinelli, L. Black holes and WIMPs: All or nothing or something else. *Mon. Not. R. Astron. Soc.* **2021**, *506*, 3648–3661. [CrossRef]
6. Carr, B.; Kuhnel, F.; Visinelli, L. Constraints on Stupendously Large Black Holes. *Mon. Not. R. Astron. Soc.* **2021**, *501*, 2029–2043. [CrossRef]
7. Carr, B.J.; Kohri, K.; Sendouda, Y.; Yokoyama, J. New cosmological constraints on primordial black holes. *Phys. Rev. D* **2010**, *81*, 104019. [CrossRef]
8. Keith, C.; Hooper, D.; Blinov, N.; McDermott, S.D. Constraints on Primordial Black Holes From Big Bang Nucleosynthesis Revisited. *Phys. Rev. D* **2020**, *102*, 103512. [CrossRef]
9. Carr, B.J.; Hawking, S.W. Black Holes in the Early Universe. *Mon. Not. R. Astron. Soc.* **1974**, *168*, 399–415. [CrossRef]
10. Hawking, S. Gravitationally Collapsed Objects of Very Low Mass. *Mon. Not. R. Astron. Soc.* **1971**, *152*, 75–78. [CrossRef]
11. Zel'dovich, Y.B.; Novikov, I.D. The Hypothesis of Cores Retarded during Expansion and the Hot Cosmological Model. *Sov. Astron.* **1967**, *10*, 602.
12. Carr, B.; Kühnel, F.; Sandstad, M. Primordial black holes as dark matter. *Phys. Rev. D* **2016**, *94*, 083504. [CrossRef]
13. García-Bellido, J.; Linde, A.; Wands, D. Density perturbations and black hole formation in hybrid inflation. *Phys. Rev. D* **1996**, *54*, 6040–6058. [CrossRef]
14. Khlopov, M.Y. Primordial black holes. *Res. Astron. Astrophys.* **2010**, *10*, 495–528. [CrossRef]
15. Sasaki, M.; Suyama, T.; Tanaka, T.; Yokoyama, S. Primordial black holes—perspectives in gravitational wave astronomy. *Class. Quantum Gravity* **2018**, *35*, 063001. [CrossRef]
16. Inomata, K.; Kawasaki, M.; Mukaida, K.; Tada, Y.; Yanagida, T.T. Inflationary primordial black holes as all dark matter. *Phys. Rev. D* **2017**, *96*, 043504. [CrossRef]
17. Georg, J.; Watson, S. A preferred mass range for primordial black hole formation and black holes as dark matter revisited. *J. High Energy Phys.* **2017**, *2017*, 138. [CrossRef]
18. Carr, B.; Silk, J. Primordial black holes as generators of cosmic structures. *Mon. Not. R. Astron. Soc.* **2018**, *478*, 3756–3775. [CrossRef]
19. Bird, S.; Cholis, I.; Muñoz, J.B.; Ali-Haïmoud, Y.; Kamionkowski, M.; Kovetz, E.D.; Raccanelli, A.; Riess, A.G. Did LIGO Detect Dark Matter? *Phys. Rev. Lett.* **2016**, *116*, 201301. [CrossRef]
20. Ali-Haimoud, Y.E. Electromagnetic probes of primordial black holes as dark matter. *arXiv* **2019**, arXiv:1903.04424.
21. Clesse, S.; García-Bellido, J. The clustering of massive Primordial Black Holes as Dark Matter: Measuring their mass distribution with Advanced LIGO. *Phys. Dark Univ.* **2017**, *15*, 142–147. [CrossRef]
22. Luis Bernal, J.; Bellomo, N.; Raccanelli, A.; Verde, L. Cosmological implications of Primordial Black Holes. *J. Cosmol. Astropart. Phys.* **2017**, *10*, 052. [CrossRef]
23. Clesse, S.; García-Bellido, J. Massive Primordial Black Holes from Hybrid Inflation as Dark Matter and the seeds of Galaxies. *Phys. Rev. D* **2015**, *92*, 023524. [CrossRef]
24. Clesse, S.; García-Bellido, J.; Orani, S. Detecting the Stochastic Gravitational Wave Background from Primordial Black Hole Formation. *arXiv* **2018**, arXiv:1812.11011.
25. Tada, Y.; Yokoyama, S. Primordial black hole tower: Dark matter, earth-mass, and LIGO black holes. *Phys. Rev. D* **2019**, *100*, 023537. [CrossRef]
26. Atal, V.; Sanglas, A.; Triantafyllou, N. NANOGrav signal as mergers of Stupendously Large Primordial Black Holes. *J. Cosmol. Astropart. Phys.* **2021**, *06*, 022. [CrossRef]
27. Atal, V.; Sanglas, A.; Triantafyllou, N. LIGO/Virgo black holes and dark matter: The effect of spatial clustering. *J. Cosmol. Astropart. Phys.* **2020**, *11*, 036. [CrossRef]
28. De Luca, V.; Franciolini, G.; Pani, P.; Riotto, A. Primordial Black Holes Confront LIGO/Virgo data: Current situation. *J. Cosmol. Astropart. Phys.* **2020**, *06*, 044. [CrossRef]
29. Bartolo, N.; De Luca, V.; Franciolini, G.; Peloso, M.; Racco, D.; Riotto, A. Testing primordial black holes as dark matter with LISA. *Phys. Rev. D* **2019**, *99*, 103521. [CrossRef]
30. Clesse, S.; García-Bellido, J. Seven Hints for Primordial Black Hole Dark Matter. *Phys. Dark Univ.* **2018**, *22*, 137–146. [CrossRef]
31. Ezquiaga, J.M.; García-Bellido, J.; Vennin, V. The exponential tail of inflationary fluctuations: Consequences for primordial black holes. *J. Cosmol. Astropart. Phys.* **2020**, *03*, 029. [CrossRef]

32. Takhistov, V. Primordial Black Hole Dark Matter and Ways to Find It. In Proceedings of the 40th International Conference on High Energy Physics, Prague, Czech Republic, 30 July–5 August 2020; p. 610. [CrossRef]
33. Jedamzik, K. Primordial Black Hole Dark Matter and the LIGO/Virgo observations. *J. Cosmol. Astropart. Phys.* **2020**, *09*, 022. [CrossRef]
34. Vallejo-Peña, S.A.; Romano, A.E. Are primordial black holes produced by entropy perturbations in single field inflationary models? *J. Cosmol. Astropart. Phys.* **2019**, *11*, 015. [CrossRef]
35. Kashlinsky, A. LIGO gravitational wave detection, primordial black holes and the near-IR cosmic infrared background anisotropies. *Astrophys. J. Lett.* **2016**, *823*, L25. [CrossRef]
36. Kashlinsky, A.; Arendt, R.G.; Atrio-Barandela, F.; Cappelluti, N.; Ferrara, A.; Hasinger, G. Looking at cosmic near-infrared background radiation anisotropies. *Rev. Mod. Phys.* **2018**, *90*, 025006. [CrossRef]
37. Kashlinsky, A. Cosmological Advection Flows in the Presence of Primordial Black Holes asDark Matter and Formation of First Sources. *Phys. Rev. Lett.* **2021**, *126*, 011101. [CrossRef]
38. Pi, S.; Zhang, Y.l.; Huang, Q.G.; Sasaki, M. Scalaron from R^2-gravity as a heavy field. *J. Cosmol. Astropart. Phys.* **2018**, *05*, 042. [CrossRef]
39. Carr, B.; Kuhnel, F. Primordial Black Holes as Dark Matter: Recent Developments. *Ann. Rev. Nucl. Part. Sci.* **2020**, *70*, 355–394. [CrossRef]
40. Bean, R.; Magueijo, J.A. Could supermassive black holes be quintessential primordial black holes? *Phys. Rev. D* **2002**, *66*, 063505. [CrossRef]
41. Bernal, J.L.; Raccanelli, A.; Verde, L.; Silk, J. Signatures of primordial black holes as seeds of supermassive black holes. *J. Cosmol. Astropart. Phys.* **2018**, *05*, 017; Erratum in *J. Cosmol. Astropart. Phys.* **2020**, *01*, E01. [CrossRef]
42. Meszaros, P. Primeval black holes and galaxy formation. *Astron. Astrophys.* **1975**, *38*, 5–13.
43. Carr, B.J. Pregalactic black hole accretion and the thermal history of the Universe. *Mon. Not. R. Astron. Soc.* **1981**, *194*, 639–668. [CrossRef]
44. Abbott, B.P.E.A. Observation of Gravitational Waves from a Binary Black Hole Merger. *Phys. Rev. Lett.* **2016**, *116*, 061102. [CrossRef]
45. Sasaki, M.; Suyama, T.; Tanaka, T.; Yokoyama, S. Primordial Black Hole Scenario for the Gravitational-Wave Event GW150914. *Phys. Rev. Lett.* **2016**, *117*, 061101; Erratum in *Phys. Rev. Lett.* **2018**, *121*, 059901. [CrossRef] [PubMed]
46. Franciolini, G.; Baibhav, V.; De Luca, V.; Ng, K.K.Y.; Wong, K.W.K.; Berti, E.; Pani, P.; Riotto, A.; Vitale, S. Quantifying the evidence for primordial black holes in LIGO/Virgo gravitational-wave data. *arXiv* **2021**, arXiv:2111.12693.
47. Sasaki, M.; Takhistov, V.; Vardanyan, V.; Zhang, Y.l. Establishing the Non-Primordial Origin of Black Hole-Neutron Star Mergers. *arXiv* **2021**, arXiv:2110.09509.
48. Abbott, R.; Abbott, T.D.; Acernese, F.; Ackley, K.; Adams, C.; Adhikari, N.; Bouffanais, Y. Search for subsolar-mass binaries in the first half of Advanced LIGO and Virgo's third observing run. *arXiv* **2021**, arXiv:2109.12197.
49. Chen, Z.C.; Yuan, C.; Huang, Q.G. Confronting the primordial black hole scenario with the gravitational-wave events detected by LIGO-Virgo. *arXiv* **2021**, arXiv:2108.11740.
50. De Luca, V.; Desjacques, V.; Franciolini, G.; Pani, P.; Riotto, A. GW190521 Mass Gap Event and the Primordial Black Hole Scenario. *Phys. Rev. Lett.* **2021**, *126*, 051101. [CrossRef]
51. Arzoumanian, Z.; Baker, P.T.; Blumer, H.; Bécsy, B.; Brazier, A.; Brook, P.R.; NANOGrav Collaboration. The NANOGrav 12.5 yr Data Set: Search for an Isotropic Stochastic Gravitational-wave Background. *Astrophys. J. Lett.* **2020**, *905*, L34. [CrossRef]
52. Vaskonen, V.; Veermäe, H. Did NANOGrav see a signal from primordial black hole formation? *Phys. Rev. Lett.* **2021**, *126*, 051303. [CrossRef]
53. De Luca, V.; Franciolini, G.; Riotto, A. NANOGrav Data Hints at Primordial Black Holes as Dark Matter. *Phys. Rev. Lett.* **2021**, *126*, 041303. [CrossRef]
54. Kohri, K.; Terada, T. Solar-Mass Primordial Black Holes Explain NANOGrav Hint of Gravitational Waves. *Phys. Lett. B* **2021**, *813*, 136040. [CrossRef]
55. Bian, L.; Cai, R.G.; Liu, J.; Yang, X.Y.; Zhou, R. Evidence for different gravitational-wave sources in the NANOGrav dataset. *Phys. Rev. D* **2021**, *103*, L081301. [CrossRef]
56. Sugiyama, S.; Takhistov, V.; Vitagliano, E.; Kusenko, A.; Sasaki, M.; Takada, M. Testing Stochastic Gravitational Wave Signals from Primordial Black Holes with Optical Telescopes. *Phys. Lett. B* **2021**, *814*, 136097. [CrossRef]
57. Domènech, G.; Pi, S. NANOGrav Hints on Planet-Mass Primordial Black Holes. *arXiv* **2020**, arXiv:2010.03976.
58. Bhattacharya, S.; Mohanty, S.; Parashari, P. Implications of the NANOGrav result on primordial gravitational waves in nonstandard cosmologies. *Phys. Rev. D* **2021**, *103*, 063532. [CrossRef]
59. Inomata, K.; Kawasaki, M.; Mukaida, K.; Yanagida, T.T. NANOGrav Results and LIGO-Virgo Primordial Black Holes in Axionlike Curvaton Models. *Phys. Rev. Lett.* **2021**, *126*, 131301. [CrossRef]
60. Germani, C.; Musco, I. Abundance of Primordial Black Holes Depends on the Shape of the Inflationary Power Spectrum. *Phys. Rev. Lett.* **2019**, *122*, 141302. [CrossRef]
61. Nadezhin, D.K.; Novikov, I.D.; Polnarev, A.G. The hydrodynamics of primordial black hole formation. *Sov. Astron.* **1978**, *22*, 129–138.

62. Niemeyer, J.C.; Jedamzik, K. Near-Critical Gravitational Collapse and the Initial Mass Function of Primordial Black Holes. *Phys. Rev. Lett.* **1998**, *80*, 5481–5484. [CrossRef]
63. Niemeyer, J.C.; Jedamzik, K. Dynamics of primordial black hole formation. *Phys. Rev. D* **1999**, *59*, 124013. [CrossRef]
64. Hawke, I.; Stewart, J.M. The dynamics of primordial black-hole formation. *Class. Quantum Gravity* **2002**, *19*, 3687–3707. [CrossRef]
65. Musco, I.; Miller, J.C.; Polnarev, A.G. Primordial black hole formation in the radiative era: Investigation of the critical nature of the collapse. *Class. Quantum Gravity* **2009**, *26*, 235001. [CrossRef]
66. Koike, T.; Hara, T.; Adachi, S. Critical Behaviour in Gravitational Collapse of Radiation Fluid: A Renormalization Group (Linear Perturbation) Analysis. *Phys. Rev. Lett.* **1995**, *74*, 5170–5173. [CrossRef] [PubMed]
67. Maison, D. Nonuniversality of critical behaviour in spherically symmetric gravitational collapse. *Phys. Lett. B* **1996**, *366*, 82–84. [CrossRef]
68. Evans, C.R.; Coleman, J.S. Critical phenomena and self-similarity in the gravitational collapse of radiation fluid. *Phys. Rev. Lett.* **1994**, *72*, 1782–1785. [CrossRef]
69. May, M.M.; White, R.H. Stellar Dynamics and Gravitational Collapse. *Methods Comput. Phys.* **1967**, *7*, 219–258.
70. Misner, C.W.; Sharp, D.H. Relativistic Equations for Adiabatic, Spherically Symmetric Gravitational Collapse. *Phys. Rev.* **1964**, *136*, B571–B576. [CrossRef]
71. Hernandez, W.C., Jr.; Misner, C.W. Observer Time as a Coordinate in Relativistic Spherical Hydrodynamics. *Astrophys. J.* **1966**, *143*, 452. [CrossRef]
72. Baumgarte, T.W.; Shapiro, S.L.; Teukolsky, S.A. Computing supernova collapse to neutron stars and black holes. *Astrophys. J.* **1995**, *443*, 717–734. [CrossRef]
73. Musco, I.; Miller, J.C.; Rezzolla, L. Computations of primordial black-hole formation. *Class. Quantum Gravity* **2005**, *22*, 1405–1424. [CrossRef]
74. Carr, B.J. The primordial black hole mass spectrum. *Astrophys. J.* **1975**, *201*, 1–19. [CrossRef]
75. Harada, T.; Yoo, C.M.; Kohri, K. Threshold of primordial black hole formation. *Phys. Rev. D* **2013**, *88*, 084051. [CrossRef]
76. Harada, T.; Yoo, C.M.; Nakama, T.; Koga, Y. Cosmological long-wavelength solutions and primordial black hole formation. *Phys. Rev. D* **2015**, *91*, 084057. [CrossRef]
77. Shibata, M.; Sasaki, M. Black hole formation in the Friedmann universe: Formulation and computation in numerical relativity. *Phys. Rev. D* **1999**, *60*, 084002. [CrossRef]
78. Nakama, T. The double formation of primordial black holes. *J. Cosmol. Astropart. Phys.* **2014**, *10*, 040. [CrossRef]
79. Musco, I. Threshold for primordial black holes: Dependence on the shape of the cosmological perturbations. *Phys. Rev. D* **2019**, *100*, 123524. [CrossRef]
80. Escrivà, A. Simulation of primordial black hole formation using pseudospectral methods. *Phys. Dark Universe* **2020**, *27*, 100466. [CrossRef]
81. Hidalgo, J.C.; Polnarev, A.G. Probability of primordial black hole formation and its dependence on the radial profile of initial configurations. *Phys. Rev. D* **2009**, *79*, 044006. [CrossRef]
82. Nakama, T.; Harada, T.; Polnarev, A.; Yokoyama, J. Identifying the most crucial parameters of the initial curvature profile for primordial black hole formation. *J. Cosmol. Astropart. Phys.* **2014**, *2014*, 037. [CrossRef]
83. Escrivà, A.; Germani, C.; Sheth, R.K. Universal threshold for primordial black hole formation. *Phys. Rev. D* **2020**, *101*, 044022. [CrossRef]
84. Escrivà, A.; Germani, C.; Sheth, R.K. Analytical thresholds for black hole formation in general cosmological backgrounds. *J. Cosmol. Astropart. Phys.* **2021**, *01*, 030. [CrossRef]
85. Gundlach, C. Critical phenomena in gravitational collapse. *Living Rev. Rel.* **1999**, *2*, 4. [CrossRef]
86. Harada, T.; Carr, B.J. Upper limits on the size of a primordial black hole. *Phys. Rev. D* **2005**, *71*, 104009. [CrossRef]
87. Escrivà, A.; Romano, A.E. Effects of the shape of curvature peaks on the size of primordial black holes. *J. Cosmol. Astropart. Phys.* **2021**, *05*, 066. [CrossRef]
88. Green, A.M.; Kavanagh, B.J. Primordial Black Holes as a dark matter candidate. *J. Phys. Nucl. Part. Phys.* **2021**, *48*, 043001. [CrossRef]
89. Yuan, C.; Huang, Q.G. A topic review on probing primordial black hole dark matter with scalar induced gravitational waves. *arXiv* **2021**, arXiv:2103.04739.
90. Villanueva-Domingo, P.; Mena, O.; Palomares-Ruiz, S. A brief review on primordial black holes as dark matter. *Front. Astron. Space Sci.* **2021**, *8*, 87. [CrossRef]
91. Domènech, G. Scalar induced gravitational waves review. *Universe* **2021**, *7*, 398. [CrossRef]
92. Allahverdi, R. The First Three Seconds: A Review of Possible Expansion Histories of the Early Universe. *arXiv* **2020**, arXiv:2006.16182. [CrossRef]
93. Neilsen, D.W.; Choptuik, M.W. Critical phenomena in perfect fluids. *Class. Quant. Grav.* **2000**, *17*, 761–782. [CrossRef]
94. Musco, I.; Miller, J.C. Primordial black hole formation in the early universe: Critical behaviour and self-similarity. *Class. Quantum Gravity* **2013**, *30*, 145009. [CrossRef]
95. Germani, C.; Sheth, R.K. Nonlinear statistics of primordial black holes from Gaussian curvature perturbations. *Phys. Rev. D* **2020**, *101*, 063520. [CrossRef]

96. De Luca, V.; Franciolini, G.; Kehagias, A.; Peloso, M.; Riotto, A.; Ünal, C. The Ineludible non-Gaussianity of the Primordial Black Hole Abundance. *J. Cosmol. Astropart. Phys.* **2019**, *07*, 048. [CrossRef]
97. Kalaja, A.; Bellomo, N.; Bartolo, N.; Bertacca, D.; Matarrese, S.; Musco, I.; Raccanelli, A.; Verde, L. From Primordial Black Holes Abundance to Primordial Curvature Power Spectrum (and back). *J. Cosmol. Astropart. Phys.* **2019**, *10*, 031. [CrossRef]
98. Erfani, E.; Kameli, H.; Baghram, S. Primordial Black Holes in the Excursion Set Theory. *Mon. Not. R. Astron. Soc.* **2021**, *505*, 1787–1793. [CrossRef]
99. Wu, Y.P. Peak statistics for the primordial black hole abundance. *Phys. Dark Univ.* **2020**, *30*, 100654. [CrossRef]
100. De Luca, V.; Franciolini, G.; Riotto, A. On the Primordial Black Hole Mass Function for Broad Spectra. *Phys. Lett. B* **2020**, *807*, 135550. [CrossRef]
101. Young, S.; Musso, M. Application of peaks theory to the abundance of primordial black holes. *J. Cosmol. Astropart. Phys.* **2020**, *11*, 022. [CrossRef]
102. Yoo, C.M.; Harada, T.; Hirano, S.; Kohri, K. Abundance of Primordial Black Holes in Peak Theory for an Arbitrary Power Spectrum. *Prog. Theor. Exp. Phys.* **2021**, *2021*, 013E02. [CrossRef]
103. Yoo, C.M.; Gong, J.O.; Yokoyama, S. Abundance of primordial black holes with local non-Gaussianity in peak theory. *J. Cosmol. Astropart. Phys.* **2019**, *09*, 033. [CrossRef]
104. Gow, A.D.; Byrnes, C.T.; Cole, P.S.; Young, S. The power spectrum on small scales: Robust constraints and comparing PBH methodologies. *J. Cosmol. Astropart. Phys.* **2021**, *02*, 002. [CrossRef]
105. Young, S. The primordial black hole formation criterion re-examined: Parametrisation, timing and the choice of window function. *Int. J. Mod. Phys. D* **2019**, *29*, 2030002. [CrossRef]
106. Young, S.; Musco, I.; Byrnes, C.T. Primordial black hole formation and abundance: Contribution from the non-linear relation between the density and curvature perturbation. *J. Cosmol. Astropart. Phys.* **2019**, *11*, 012. [CrossRef]
107. Young, S.; Byrnes, C.T. Primordial black holes in non-Gaussian regimes. *J. Cosmol. Astropart. Phys.* **2013**, *08*, 052. [CrossRef]
108. Young, S.; Byrnes, C.T.; Sasaki, M. Calculating the mass fraction of primordial black holes. *J. Cosmol. Astropart. Phys.* **2014**, *07*, 045. [CrossRef]
109. Yoo, C.M.; Harada, T.; Garriga, J.; Kohri, K. PBH abundance from random Gaussian curvature perturbations and a local density threshold. *arXiv* **2018**, arXiv:1805.03946.
110. Suyama, T.; Yokoyama, S. A novel formulation of the primordial black hole mass function. *Prog. Theor. Exp. Phys.* **2020**, *2020*, 023E03. [CrossRef]
111. Ando, K.; Inomata, K.; Kawasaki, M. Primordial black holes and uncertainties in the choice of the window function. *Phys. Rev. D* **2018**, *97*, 103528. [CrossRef]
112. Zaballa, I.; Green, A.M.; Malik, K.A.; Sasaki, M. Constraints on the primordial curvature perturbation from primordial black holes. *J. Cosmol. Astropart. Phys.* **2007**, *03*, 010. [CrossRef]
113. Yokoyama, J. Cosmological constraints on primordial black holes produced in the near critical gravitational collapse. *Phys. Rev. D* **1998**, *58*, 107502. [CrossRef]
114. Tada, Y.; Vennin, V. Statistics of coarse-grained cosmological fields in stochastic inflation. *arXiv* **2021**, arXiv:2111.15280.
115. Press, W.H.; Schechter, P. Formation of galaxies and clusters of galaxies by selfsimilar gravitational condensation. *Astrophys. J.* **1974**, *187*, 425–438. [CrossRef]
116. Bardeen, J.M.; Bond, J.R.; Kaiser, N.; Szalay, A.S. The Statistics of Peaks of Gaussian Random Fields. *Astrophys. J.* **1986**, *304*, 15. [CrossRef]
117. Green, A.M.; Liddle, A.R.; Malik, K.A.; Sasaki, M. A New calculation of the mass fraction of primordial black holes. *Phys. Rev. D* **2004**, *70*, 041502. [CrossRef]
118. Yoo, C.M.; Harada, T.; Garriga, J.; Kohri, K. Primordial black hole abundance from random Gaussian curvature perturbations and a local density threshold. *Prog. Theor. Exp. Phys.* **2018**, *2018*, 123E01. [CrossRef]
119. Carr, B.; Raidal, M.; Tenkanen, T.; Vaskonen, V.; Veermäe, H. Primordial black hole constraints for extended mass functions. *Phys. Rev. D* **2017**, *96*, 023514. [CrossRef]
120. Tokeshi, K.; Inomata, K.; Yokoyama, J. Window function dependence of the novel mass function of primordial black holes. *J. Cosmol. Astropart. Phys.* **2020**, *12*, 038. [CrossRef]
121. Salopek, D.S.; Bond, J.R. Nonlinear evolution of long-wavelength metric fluctuations in inflationary models. *Phys. Rev. D* **1990**, *42*, 3936–3962. [CrossRef]
122. Polnarev, A.G.; Musco, I. Curvature profiles as initial conditions for primordial black hole formation. *Class. Quantum Gravity* **2007**, *24*, 1405–1431. [CrossRef]
123. Arnowitt, R.L.; Deser, S.; Misner, C.W. The Dynamics of general relativity. *Gen. Rel. Grav.* **2008**, *40*, 1997–2027. [CrossRef]
124. Arnowitt, R.; Deser, S.; Misner, C.W. Dynamical Structure and Definition of Energy in General Relativity. *Phys. Rev.* **1959**, *116*, 1322–1330. [CrossRef]
125. Tanaka, Y.; Sasaki, M. Gradient expansion approach to nonlinear superhorizon perturbations. II. A Single scalar field. *Prog. Theor. Phys.* **2007**, *118*, 455–473. [CrossRef]
126. Lyth, D.H.; Malik, K.A.; Sasaki, M. A General proof of the conservation of the curvature perturbation. *J. Cosmol. Astropart. Phys.* **2005**, *05*, 004. [CrossRef]
127. Sugiyama, N.S.; Komatsu, E.; Futamase, T. δN formalism. *Phys. Rev. D* **2013**, *87*, 023530. [CrossRef]

128. Romano, A.E.; Sasaki, M.; Starobinsky, A.A. Effects of inhomogeneities on apparent cosmological observables: 'Fake' evolving dark energy. *Eur. Phys. J. C* **2012**, *72*, 2242. [CrossRef]
129. Polnarev, A.G.; Nakama, T.; Yokoyama, J. Self-consistent initial conditions for primordial black hole formation. *J. Cosmol. Astropart. Phys.* **2012**, *09*, 027. [CrossRef]
130. Kopp, M.; Hofmann, S.; Weller, J. Separate Universes Do Not Constrain Primordial Black Hole Formation. *Phys. Rev. D* **2011**, *83*, 124025. [CrossRef]
131. Mello, M.M.C.; Maciel, A.; Zanchin, V.T. Evolving black holes from conformal transformations of static solutions. *Phys. Rev. D* **2017**, *95*, 084031. [CrossRef]
132. Helou, A.; Musco, I.; Miller, J.C. Causal Nature and Dynamics of Trapping Horizons in Black Hole Collapse. *Class. Quant. Grav.* **2017**, *34*, 135012. [CrossRef]
133. Dafermos, M. Spherically symmetric spacetimes with a trapped surface. *Class. Quantum Gravity* **2005**, *22*, 2221–2232. [CrossRef]
134. Williams, C. Asymptotic Behaviour of Spherically Symmetric Marginally Trapped Tubes. *Ann. Henri Poincare* **2008**, *9*, 1029–1067. [CrossRef]
135. Booth, I.; Brits, L.; Gonzalez, J.A.; Van Den Broeck, C. Marginally trapped tubes and dynamical horizons. *Class. Quant. Grav.* **2006**, *23*, 413–440. [CrossRef]
136. Ashtekar, A.; Krishnan, B. Isolated and dynamical horizons and their applications. *Living Rev. Rel.* **2004**, *7*, 10. [CrossRef]
137. Faraoni, V. Evolving Black Hole Horizons in General Relativity and Alternative Gravity. *Galaxies* **2013**, *1*, 114–179. [CrossRef]
138. Hayward, S.A. General laws of black-hole dynamics. *Phys. Rev. D* **1994**, *49*, 6467–6474. [CrossRef]
139. Jaramillo, J.L.; Macedo, R.P.; Moesta, P.; Rezzolla, L. Black-hole horizons as probes of black-hole dynamics. II. Geometrical insights. *Phys. Rev. D* **2012**, *85*, 084031. [CrossRef]
140. Yoo, C.M.; Okawa, H. Black hole universe with a cosmological constant. *Phys. Rev. D* **2014**, *89*, 123502. [CrossRef]
141. Harada, T.; Carr, B.J. Growth of primordial black holes in a universe containing a massless scalar field. *Phys. Rev. D* **2005**, *71*, 104010. [CrossRef]
142. Custodio, P.S.; Horvath, J.E. Evolution of a primordial black hole population. *Phys. Rev. D* **1998**, *58*, 023504. [CrossRef]
143. Guedens, R.; Clancy, D.; Liddle, A.R. Primordial black holes in braneworld cosmologies: Accretion after formation. *Phys. Rev. D* **2002**, *66*, 083509. [CrossRef]
144. Nayak, B.; Singh, L.P. Accretion, primordial black holes and standard cosmology. *Pramana* **2011**, *76*, 173–181. [CrossRef]
145. Carr, B.; Harada, T.; Maeda, H. Can a primordial black hole or wormhole grow as fast as the universe? *Class. Quant. Grav.* **2010**, *27*, 183101. [CrossRef]
146. Deng, H.; Garriga, J.; Vilenkin, A. Primordial black hole and wormhole formation by domain walls. *J. Cosmol. Astropart. Phys.* **2017**, *2017*, 050. [CrossRef]
147. Kühnel, F.; Rampf, C.; Sandstad, M. Effects of Critical Collapse on Primordial Black-Hole Mass Spectra. *Eur. Phys. J. C* **2016**, *76*, 93. [CrossRef]
148. Baumgarte, T.W.; Shapiro, S.L. *Numerical Relativity: Solving Einstein's Equations on the Computer*; Cambridge University Press: Cambridge, UK, 2010.
149. Lehner, L. Numerical relativity: A Review. *Class. Quant. Grav.* **2001**, *18*, R25–R86. [CrossRef]
150. Palenzuela, C. Introduction to Numerical Relativity. *Front. Astron. Space Sci.* **2020**, *7*, 58. [CrossRef]
151. Gourgoulhon, E. 3 + 1 formalism and bases of numerical relativity. *arXiv* **2007**, arXiv:gr-qc/0703035.
152. Grandclément, P.; Novak, J. Spectral Methods for Numerical Relativity. *Living Rev. Relativ.* **2009**, *12*, 1. [CrossRef] [PubMed]
153. Font, J.A. Numerical hydrodynamics in general relativity. *Living Rev. Rel.* **2000**, *3*, 2. [CrossRef]
154. Clough, K.; Figueras, P.; Finkel, H.; Kunesch, M.; Lim, E.A.; Tunyasuvunakool, S. GRChombo: Numerical Relativity with Adaptive Mesh Refinement. *Class. Quant. Grav.* **2015**, *32*, 245011. [CrossRef]
155. Löffler, F.; Faber, J.; Bentivegna, E.; Bode, T.; Diener, P.; Haas, R.; Laguna, P. The Einstein Toolkit: A Community Computational Infrastructure for Relativistic Astrophysics. *Class. Quant. Grav.* **2012**, *29*, 115001. [CrossRef]
156. Ruchlin, I.; Etienne, Z.B.; Baumgarte, T.W. SENR/NRPy+: Numerical Relativity in Singular Curvilinear Coordinate Systems. *Phys. Rev. D* **2018**, *97*, 064036. [CrossRef]
157. Bloomfield, J.; Bulhosa, D.; Face, S. Formalism for Primordial Black Hole Formation in Spherical Symmetry. *arXiv* **2015**, arXiv:1504.02071.
158. Albert Escrivà Webpage. 2021. Available online: https://aescriva.wixsite.com/albertescriva (accessed on 15 November 2021).
159. Guo, B. *Spectral Methods and Their Applications*; World Scientific: Shanghai, China, 1998. [CrossRef]
160. Hussaini, M.Y.; Zang, T.A. Spectral Methods in Fluid Dynamics. *Annu. Rev. Fluid Mech.* **1987**, *19*, 339–367. [CrossRef]
161. Hesthaven, J.S.; Gottlieb, S.; Gottlieb, D.I. *Spectral Methods for Time-Dependent Problems*; Cambridge University Press: Cambridge, UK, 2007.
162. Santos-Oliván, D.; Sopuerta, C.F. New Features of Gravitational Collapse in Anti–de Sitter Spacetimes. *Phys. Rev. Lett.* **2016**, *116*, 041101. [CrossRef]
163. Canizares, P.; Sopuerta, C.F.; Jaramillo, J.L. Pseudospectral collocation methods for the computation of the self-force on a charged particle: Generic orbits around a Schwarzschild black hole. *Phys. Rev. D* **2010**, *82*, 044023. [CrossRef]
164. Santos-Oliván, D.; Sopuerta, C.F. Numerical Relativity with Arbitrary Precision Arithmetic: Applications to Gravitational Collapse. *arXiv* **2018**, arXiv:1803.00858.

165. Santos-Oliván, D.; Sopuerta, C.F. Moving closer to the collapse of a massless scalar field in spherically symmetric anti–de Sitter spacetimes. *Phys. Rev. D* **2016**, *93*, 104002. [CrossRef]
166. Meringolo, C.; Servidio, S.; Veltri, P. A spectral method algorithm for numerical simulations of gravitational fields. *Class. Quant. Grav.* **2021**, *38*, 075027. [CrossRef]
167. Bonazzola, S.; Gourgoulhon, E.; Marck, J.A. Spectral methods in general relativistic astrophysics. *J. Comput. Appl. Math.* **1999**, *109*, 433–473. [CrossRef]
168. Sengupta, T.K.; Sundaram, P.; Sengupta, A. Analysis of pseudospectral methods used for numerical simulation of turbulence. *arXiv* **2021**, arXiv:2109.00255.
169. Moxon, J.; Scheel, M.A.; Teukolsky, S.A.; Deppe, N.; Fischer, N.; Hébert, F.; Kidder, L.E.; Throwe, W. The SpECTRE Cauchy-characteristic evolution system for rapid, precise waveform extraction. *arXiv* **2021**, arXiv:2110.08635.
170. Alcoforado, M.A.; Aranha, R.F.; Barreto, W.O.; de Oliveira, H.P. Multidomain Galerkin-Collocation method: Spherical collapse of scalar fields II. *Class. Quantum Gravity* **2021** , *38*, 225004. [CrossRef]
171. Alcoforado, M.A.; Barreto, W.O.; de Oliveira, H.P. Multidomain Galerkin-Collocation method: Characteristic spherical collapse of scalar fields. *Gen. Rel. Grav.* **2021**, *53*, 42. [CrossRef]
172. Oltean, M.; Sopuerta, C.F.; Spallicci, A.D.A.M. Particle-without-Particle: A practical pseudospectral collocation method for linear partial differential equations with distributional sources. *J. Sci. Comput.* **2019**, *79*, 827. [CrossRef]
173. Frauendiener, J.; Hennig, J. Fully pseudospectral solution of the conformally invariant wave equation near the cylinder at spacelike infinity. II: Schwarzschild background. *Class. Quant. Grav.* **2017**, *34*, 045005. [CrossRef]
174. Schwabe, B.; Gosenca, M.; Behrens, C.; Niemeyer, J.C.; Easther, R. Simulating mixed fuzzy and cold dark matter. *Phys. Rev. D* **2020**, *102*, 083518. [CrossRef]
175. Edwards, F.; Kendall, E.; Hotchkiss, S.; Easther, R. PyUltraLight: A Pseudo-Spectral Solver for Ultralight Dark Matter Dynamics. *J. Cosmol. Astropart. Phys.* **2018**, *10*, 027. [CrossRef]
176. Musoke, N.; Hotchkiss, S.; Easther, R. Lighting the Dark: Evolution of the Postinflationary Universe. *Phys. Rev. Lett.* **2020**, *124*, 061301. [CrossRef]
177. Boyd, J.P. *Chebyshev and Fourier Spectral Methods*; DOVER: Mineola, NY, USA, 2000.
178. Trefethen, L.N. *Spectral Methods in MATLAB*; SIAM: Mineola, NY, USA, 2000.
179. Kidder, L.E.; Scheel, M.A.; Teukolsky, S.A.; Carlson, E.D.; Cook, G.B. Black hole evolution by spectral methods. *Phys. Rev. D* **2000**, *62*, 084032. [CrossRef]
180. Atal, V.; Germani, C. The role of non-Gaussianities in Primordial Black Hole formation. *Phys. Dark Univ.* **2019**, *24*, 100275. [CrossRef]
181. Atal, V.; Garriga, J.; Marcos-Caballero, A. Primordial black hole formation with non-Gaussian curvature perturbations. *J. Cosmol. Astropart. Phys.* **2019**, *09*, 073. [CrossRef]
182. Passaglia, S.; Hu, W.; Motohashi, H. Primordial black holes and local non-Gaussianity in canonical inflation. *Phys. Rev. D* **2019**, *99*, 043536. [CrossRef]
183. Cai, Y.F.; Chen, X.; Namjoo, M.H.; Sasaki, M.; Wang, D.G.; Wang, Z. Revisiting non-Gaussianity from non-attractor inflation models. *J. Cosmol. Astropart. Phys.* **2018**, *05*, 012. [CrossRef]
184. Bullock, J.S.; Primack, J.R. NonGaussian fluctuations and primordial black holes from inflation. *Phys. Rev. D* **1997**, *55*, 7423–7439. [CrossRef]
185. Pattison, C.; Vennin, V.; Assadullahi, H.; Wands, D. Quantum diffusion during inflation and primordial black holes. *J. Cosmol. Astropart. Phys.* **2017**, *2017*, 046. [CrossRef]
186. Pina Avelino, P. Primordial black hole constraints on non-Gaussian inflation models. *Phys. Rev. D* **2005**, *72*, 124004, [CrossRef]
187. Riccardi, F.; Taoso, M.; Urbano, A. Solving peak theory in the presence of local non-Gaussianities. *J. Cosmol. Astropart. Phys.* **2021**, *08*, 060. [CrossRef]
188. Young, S.; Byrnes, C.T. Long-short wavelength mode coupling tightens primordial black hole constraints. *Phys. Rev. D* **2015**, *91*, 083521. [CrossRef]
189. Young, S.; Regan, D.; Byrnes, C.T. Influence of large local and non-local bispectra on primordial black hole abundance. *J. Cosmol. Astropart. Phys.* **2016**, *02*, 029. [CrossRef]
190. Hidalgo, J.C. The effect of non-Gaussian curvature perturbations on the formation of primordial black holes. *arXiv* **2007**, arXiv:0708.3875.
191. Atal, V.; Domènech, G. Probing non-Gaussianities with the high frequency tail of induced gravitational waves. *J. Cosmol. Astropart. Phys.* **2021**, *06*, 001. [CrossRef]
192. Kitajima, N.; Tada, Y.; Yokoyama, S.; Yoo, C.M. Primordial black holes in peak theory with a non-Gaussian tail. *J. Cosmol. Astropart. Phys.* **2021**, *10*, 053. [CrossRef]
193. Davies, M.W.; Carrilho, P.; Mulryne, D.J. Non-Gaussianity in inflationary scenarios for primordial black holes. *arXiv* **2021**, arXiv:2110.08189.
194. Taoso, M.; Urbano, A. Non-Gaussianities for primordial black hole formation. *J. Cosmol. Astropart. Phys.* **2021**, *08*, 016. [CrossRef]
195. Cai, Y.F.; Ma, X.H.; Sasaki, M.; Wang, D.G.; Zhou, Z. One Small Step for an Inflaton, One Giant Leap for Inflation: A novel non-Gaussian tail and primordial black holes. *arXiv* **2021**, arXiv:2112.13836.

196. Kehagias, A.; Musco, I.; Riotto, A. Non-Gaussian Formation of Primordial Black Holes: Effects on the Threshold. *arXiv* **2019**, arXiv:1906.07135.
197. Atal, V.; Cid, J.; Escrivà, A.; Garriga, J. PBH in single field inflation: The effect of shape dispersion and non-Gaussianities. *J. Cosmol. Astropart. Phys.* **2020**, *05*, 022. [CrossRef]
198. Musco, I.; De Luca, V.; Franciolini, G.; Riotto, A. Threshold for primordial black holes. II. A simple analytic prescription. *Phys. Rev. D* **2021**, *103*, 063538. [CrossRef]
199. Deng, H.; Vilenkin, A. Primordial black hole formation by vacuum bubbles. *J. Cosmol. Astropart. Phys.* **2017**, *12*, 044. [CrossRef]
200. Harada, T.; Yoo, C.M.; Kohri, K.; Nakao, K.I. Spins of primordial black holes formed in the matter-dominated phase of the Universe. *Phys. Rev. D* **2017**, *96*, 083517; Erratum in *Phys. Rev. D* **2019**, *99*, 069904. [CrossRef]
201. Khlopov, M.Y.; Malomed, B.A.; Zeldovich, Y.B. Gravitational instability of scalar fields and formation of primordial black holes. *Mon. Not. R. Astron. Soc.* **1985**, *215*, 575–589. [CrossRef]
202. Hidalgo, J.C.; De Santiago, J.; German, G.; Barbosa-Cendejas, N.; Ruiz-Luna, W. Collapse threshold for a cosmological Klein Gordon field. *Phys. Rev. D* **2017**, *96*, 063504. [CrossRef]
203. Carr, B.; Tenkanen, T.; Vaskonen, V. Primordial black holes from inflaton and spectator field perturbations in a matter-dominated era. *Phys. Rev. D* **2017**, *96*, 063507. [CrossRef]
204. Carrion, K.; Hidalgo, J.C.; Montiel, A.; Padilla, L.E. Complex Scalar Field Reheating and Primordial Black Hole production. *J. Cosmol. Astropart. Phys.* **2021**, *07*, 001. [CrossRef]
205. Martin, J.; Papanikolaou, T.; Pinol, L.; Vennin, V. Metric preheating and radiative decay in single-field inflation. *J. Cosmol. Astropart. Phys.* **2020**, *05*, 003. [CrossRef]
206. Padilla, L.E.; Hidalgo, J.C.; Malik, K.A. A new mechanism for primordial black hole formation during reheating. *arXiv* **2021**, arXiv:2110.14584.
207. de Jong, E.; Aurrekoetxea, J.C.; Lim, E.A. Primordial black hole formation with full numerical relativity. *arXiv* **2021**, arXiv:2109.04896.
208. Khlopov, M.; Polnarev, A. Primordial black holes as a cosmological test of grand unification. *Phys. Lett. B* **1980**, *97*, 383–387. [CrossRef]
209. Harada, T.; Jhingan, S. Spherical and nonspherical models of primordial black hole formation: Exact solutions. *Prog. Theor. Exp. Phys.* **2016**, *2016*, 093E04. [CrossRef]
210. Harada, T.; Goymer, C.; Carr, B.J. Tolman-Bondi collapse in scalar tensor theories as a probe of gravitational memory. *Phys. Rev. D* **2002**, *66*, 104023. [CrossRef]
211. Kokubu, T.; Kyutoku, K.; Kohri, K.; Harada, T. Effect of Inhomogeneity on Primordial Black Hole Formation in the Matter Dominated Era. *Phys. Rev. D* **2018**, *98*, 123024. [CrossRef]
212. Harada, T.; Yoo, C.M.; Kohri, K.; Nakao, K.I.; Jhingan, S. Primordial black hole formation in the matter-dominated phase of the Universe. *Astrophys. J.* **2016**, *833*, 61. [CrossRef]
213. Flores, M.M.; Kusenko, A. Spins of primordial black holes formed in different cosmological scenarios. *Phys. Rev. D* **2021**, *104*, 063008. [CrossRef]
214. Kühnel, F.; Sandstad, M. Ellipsoidal collapse and primordial black hole formation. *Phys. Rev. D* **2016**, *94*, 063514. [CrossRef]
215. Yoo, C.M.; Harada, T.; Okawa, H. Threshold of primordial black hole formation in nonspherical collapse. *Phys. Rev. D* **2020**, *102*, 043526. [CrossRef]
216. He, M.; Suyama, T. Formation threshold of rotating primordial black holes. *Phys. Rev. D* **2019**, *100*, 063520. [CrossRef]
217. Musco, I.; Papanikolaou, T. Primordial black hole formation for an anisotropic perfect fluid: Initial conditions and estimation of the threshold. *arXiv* **2021**, arXiv:2110.05982.
218. Carr, B.; Clesse, S.; García-Bellido, J. Primordial black holes from the QCD epoch: Linking dark matter, baryogenesis and anthropic selection. *Mon. Not. R. Astron. Soc.* **2021**, *501*, 1426–1439. [CrossRef]
219. Widerin, P.; Schmid, C. Primordial black holes from the QCD transition? *arXiv* **1998**, arXiv:astro-ph/9808142.
220. Boeckel, T.; Schettler, S.; Schaffner-Bielich, J. The Cosmological QCD Phase Transition Revisited. *Prog. Part. Nucl. Phys.* **2011**, *66*, 266–270. [CrossRef]
221. Jedamzik, K. Primordial black hole formation during the QCD epoch. *Phys. Rev. D* **1997**, *55*, 5871–5875. [CrossRef]
222. Byrnes, C.T.; Hindmarsh, M.; Young, S.; Hawkins, M.R.S. Primordial black holes with an accurate QCD equation of state. *J. Cosmol. Astropart. Phys.* **2018**, *08*, 041. [CrossRef]
223. Gao, F.; Oldengott, I.M. Cosmology meets functional QCD: First-order cosmic QCD transition induced by large lepton asymmetries. *arXiv* **2021**, arXiv:2106.11991.
224. Abe, K.T.; Tada, Y.; Ueda, I. Induced gravitational waves as a cosmological probe of the sound speed during the QCD phase transition. *J. Cosmol. Astropart. Phys.* **2021**, *06*, 048. [CrossRef]
225. Clesse, S.; Garcia-Bellido, J. GW190425, GW190521 and GW190814: Three candidate mergers of primordial black holes from the QCD epoch. *arXiv* **2020**, arXiv:2007.06481.
226. Carr, B.; Clesse, S.; García-Bellido, J.; Kühnel, F. Cosmic conundra explained by thermal history and primordial black holes. *Phys. Dark Univ.* **2021**, *31*, 100755. [CrossRef]
227. Sobrinho, J.L.G.; Augusto, P.; Gonçalves, A.L. New thresholds for Primordial Black Hole formation during the QCD phase transition. *Mon. Not. R. Astron. Soc.* **2016**, *463*, 2348–2357. [CrossRef]

228. Borsányi, S.; Fodor, Z.; Guenther, J.; Kampert, K.H.; Katz, S.D.; Kawanai, T.; Kovacs, T.G.; Mages, S.W.; Pasztor, A.; Pittler, F.; et al. Calculation of the axion mass based on high-temperature lattice quantum chromodynamics. *Nature* **2016**, *539*, 69–71. [CrossRef] [PubMed]
229. Bhattacharya, T.; Buchoff, M.I.; Christ, N.H.; Ding, H.T.; Gupta, R.; Jung, C.; Karsch, F.; Lin, Z.; Mawhinney, R.D.; McGlynn, G.; et al. QCD Phase Transition with Chiral Quarks and Physical Quark Masses. *Phys. Rev. Lett.* **2014**, *113*, 082001. [CrossRef] [PubMed]
230. Kusenko, A.; Shaposhnikov, M.E. Supersymmetric Q balls as dark matter. *Phys. Lett. B* **1998**, *418*, 46–54. [CrossRef]
231. Ballesteros, G.; Rey, J.; Rompineve, F. Detuning primordial black hole dark matter with early matter domination and axion monodromy. *J. Cosmol. Astropart. Phys.* **2020**, *06*, 014. [CrossRef]
232. Cotner, E.; Kusenko, A. Primordial Black Holes from Supersymmetry in the Early Universe. *Phys. Rev. Lett.* **2017**, *119*, 031103. [CrossRef]
233. Cotner, E.; Kusenko, A. Primordial black holes from scalar field evolution in the early universe. *Phys. Rev. D* **2017**, *96*, 103002. [CrossRef]
234. Cotner, E.; Kusenko, A.; Takhistov, V. Primordial Black Holes from Inflaton Fragmentation into Oscillons. *Phys. Rev. D* **2018**, *98*, 083513. [CrossRef]
235. Cotner, E.; Kusenko, A.; Sasaki, M.; Takhistov, V. Analytic Description of Primordial Black Hole Formation from Scalar Field Fragmentation. *J. Cosmol. Astropart. Phys.* **2019**, *10*, 077. [CrossRef]
236. Flores, M.M.; Kusenko, A. Primordial black holes as a dark matter candidate in theories with supersymmetry and inflation. *arXiv* **2021**, arXiv:2108.08416.
237. Kou, X.X.; Tian, C.; Zhou, S.Y. Oscillon Preheating in Full General Relativity. *Class. Quant. Grav.* **2021**, *38*, 045005. [CrossRef]
238. Nazari, Z.; Cicoli, M.; Clough, K.; Muia, F. Oscillon collapse to black holes. *J. Cosmol. Astropart. Phys.* **2021**, *05*, 027. [CrossRef]
239. Kim, J.; McDonald, J. General Analytical Conditions for Inflaton Fragmentation: Quick and Easy Tests for its Occurrence. *arXiv* **2021**, arXiv:2111.12474.
240. Gradwohl, B.A.; Frieman, J.A. Dark Matter, Long-Range Forces, and Large-Scale Structure. *Astrophys. J.* **1992**, *398*, 407. [CrossRef]
241. Gubser, S.S.; Peebles, P.J.E. Structure formation in a string inspired modification of the cold dark matter model. *Phys. Rev. D* **2004**, *70*, 123510. [CrossRef]
242. Nusser, A.; Gubser, S.S.; Peebles, P.J.E. Structure formation with a long-range scalar dark matter interaction. *Phys. Rev. D* **2005**, *71*, 083505. [CrossRef]
243. Amendola, L.; Rubio, J.; Wetterich, C. Primordial black holes from fifth forces. *Phys. Rev. D* **2018**, *97*, 081302. [CrossRef]
244. Savastano, S.; Amendola, L.; Rubio, J.; Wetterich, C. Primordial dark matter halos from fifth forces. *Phys. Rev. D* **2019**, *100*, 083518. [CrossRef]
245. Flores, M.M.; Kusenko, A. Primordial Black Holes from Long-Range Scalar Forces and Scalar Radiative Cooling. *Phys. Rev. Lett.* **2021**, *126*, 041101. [CrossRef]
246. Domènech, G.; Sasaki, M. Cosmology of strongly interacting fermions in the early universe. *J. Cosmol. Astropart. Phys.* **2021**, *06*, 030. [CrossRef]
247. Dvali, G.; Kühnel, F.; Zantedeschi, M. Primordial Black Holes from Confinement. *Phys. Rev. D* **2021**, *104*, 123507. [CrossRef]
248. Basu, R.; Guth, A.H.; Vilenkin, A. Quantum creation of topological defects during inflation. *Phys. Rev. D* **1991**, *44*, 340–351. [CrossRef]
249. Ipser, J.; Sikivie, P. Gravitationally repulsive domain wall. *Phys. Rev. D* **1984**, *30*, 712–719. [CrossRef]
250. Liu, J.; Guo, Z.K.; Cai, R.G. Primordial Black Holes from Cosmic Domain Walls. *arXiv* **2019**, arXiv:1908.02662.
251. Rubin, S.G.; Khlopov, M.Y.; Sakharov, A.S. Primordial black holes from nonequilibrium second order phase transition. *Grav. Cosmol.* **2000**, *6*, 51–58.
252. Khlopov, M.Y.; Rubin, S.G.; Sakharov, A.S. Primordial structure of massive black hole clusters. *Astropart. Phys.* **2005**, *23*, 265. [CrossRef]
253. Tanahashi, N.; Yoo, C.M. Spherical Domain Wall Collapse in a Dust Universe. *Class. Quant. Grav.* **2015**, *32*, 155003. [CrossRef]
254. Kibble, T.W.B. Topology of cosmic domains and strings. *J. Phys. A Math. Gen.* **1976**, *9*, 1387–1398. [CrossRef]
255. Vilenkin, A. Cosmic strings and domain walls. *Phys. Rep.* **1985**, *121*, 263–315. [CrossRef]
256. Hindmarsh, M.B.; Kibble, T.W.B. Cosmic strings. *Rep. Prog. Phys.* **1995**, *58*, 477–562. [CrossRef]
257. Jenkins, A.C.; Sakellariadou, M. Primordial black holes from cusp collapse on cosmic strings. *arXiv* **2020**, arXiv:2006.16249.
258. Vilenkin, A.; Levin, Y.; Gruzinov, A. Cosmic strings and primordial black holes. *J. Cosmol. Astropart. Phys.* **2018**, *11*, 008. [CrossRef]
259. James-Turner, C.; Weil, D.P.B.; Green, A.M.; Copeland, E.J. Constraints on the cosmic string loop collapse fraction from primordial black holes. *Phys. Rev. D* **2020**, *101*, 123526. [CrossRef]
260. Özsoy, O.; Parameswaran, S.; Tasinato, G.; Zavala, I. Mechanisms for Primordial Black Hole Production in String Theory. *J. Cosmol. Astropart. Phys.* **2018**, *07*, 005, [CrossRef]
261. Garriga, J.; Vilenkin, A. Black holes from nucleating strings. *Phys. Rev. D* **1993**, *47*, 3265–3274. [CrossRef]
262. Caldwell, R.R.; Casper, P. Formation of black holes from collapsed cosmic string loops. *Phys. Rev. D* **1996**, *53*, 3002–3010. [CrossRef]
263. Polnarev, A.; Zembowicz, R. Formation of primordial black holes by cosmic strings. *Phys. Rev. D* **1991**, *43*, 1106–1109. [CrossRef]

264. Deng, H. Primordial black hole formation by vacuum bubbles. Part II. *J. Cosmol. Astropart. Phys.* **2020**, *09*, 023. [CrossRef]
265. Dymnikova, I.; Koziel, L.; Khlopov, M.; Rubin, S. Quasilumps from first order phase transitions. *Grav. Cosmol.* **2000**, *6*, 311–318.
266. Hawking, S.W.; Moss, I.G.; Stewart, J.M. Bubble collisions in the very early universe. *Phys. Rev. D* **1982**, *26*, 2681–2693. [CrossRef]
267. Jung, T.H.; Okui, T. Primordial black holes from bubble collisions during a first-order phase transition. *arXiv* **2021**, arXiv:2110.04271.
268. De Luca, V.; Franciolini, G.; Riotto, A. Bubble Correlation in First-Order Phase Transitions. *Phys. Rev. D* **2021**, *104*, 123539. [CrossRef]
269. Lewicki, M.; Vaskonen, V. On bubble collisions in strongly supercooled phase transitions. *Phys. Dark Univ.* **2020**, *30*, 100672. [CrossRef]
270. Baker, M.J.; Breitbach, M.; Kopp, J.; Mittnacht, L. Primordial Black Holes from First-Order Cosmological Phase Transitions. *arXiv* **2021**, arXiv:2105.07481.
271. Gross, C.; Landini, G.; Strumia, A.; Teresi, D. Dark Matter as dark dwarfs and other macroscopic objects: Multiverse relics? *J. High Energy Phys.* **2021**, *09*, 033. [CrossRef]
272. Garriga, J.; Vilenkin, A.; Zhang, J. Black holes and the multiverse. *J. Cosmol. Astropart. Phys.* **2016**, *02*, 064. [CrossRef]
273. Kusenko, A.; Sasaki, M.; Sugiyama, S.; Takada, M.; Takhistov, V.; Vitagliano, E. Exploring Primordial Black Holes from the Multiverse with Optical Telescopes. *Phys. Rev. Lett.* **2020**, *125*, 181304. [CrossRef]
274. Maeso, D.N.; Marzola, L.; Raidal, M.; Vaskonen, V.; Veermäe, H. Primordial black holes from spectator field bubbles. *arXiv* **2021**, arXiv:2112.01505.
275. Vallejo-Peña, S.A.; Romano, A.E. Effects of the modification of gravity on the production of primordial black holes. *Phys. Lett. B* **2021**, *817*, 136289. [CrossRef]
276. Kawai, S.; Kim, J. Primordial black holes from Gauss-Bonnet-corrected single field inflation. *Phys. Rev. D* **2021**, *104*, 083545. [CrossRef]
277. Chen, C.Y. Threshold of primordial black hole formation in Eddington-inspired-Born–Infeld gravity. *Int. J. Mod. Phys. D* **2021**, *30*, 02. [CrossRef]
278. Passaglia, S.; Sasaki, M. Primordial Black Holes from CDM Isocurvature. *arXiv* **2021**, arXiv:2109.12824.
279. Yoo, C.M.; Harada, T.; Hirano, S.; Okawa, H.; Sasaki, M. Primordial black hole formation from massless scalar isocurvature. *arXiv* **2021**, arXiv:2112.12335.

universe MDPI

Review

Scalar Induced Gravitational Waves Review

Guillem Domènech

INFN Sezione di Padova, I-35131 Padova, Italy; domenech@pd.infn.it

Abstract: We provide a review on the state-of-the-art of gravitational waves induced by primordial fluctuations, so-called induced gravitational waves. We present the intuitive physics behind induced gravitational waves and we revisit and unify the general analytical formulation. We then present general formulas in a compact form, ready to be applied. This review places emphasis on the open possibility that the primordial universe experienced a different expansion history than the often assumed radiation dominated cosmology. We hope that anyone interested in the topic will become aware of current advances in the cosmology of induced gravitational waves, as well as becoming familiar with the calculations behind.

Keywords: induced gravitational waves; primordial black holes; early universe; inflation

1. Introduction

Cosmology with Gravitational Waves (GWs) is a new and unique doorway to the primordial universe. The furthest in time we can observe with electromagnetic waves is around neutrino decoupling, roughly one second after the Big Bang and some time before Big Bang Nucleosynthesis (BBN). Prior to that, we have scarce evidence of the evolution of the primordial universe. See Ref. [1] for a review on possible expansion histories of the early universe. Nevertheless, we learned a lot from studies of the Cosmic Microwave Background[1] (CMB). For instance, we know that there are Gaussian primordial fluctuations with an almost scale invariant spectrum [2,3], which by causality must have been generated before the Big Bang. This fact provides strong evidence that there was a period of accelerated expansion in the primordial universe, so-called inflation [4–7]. During inflation, quantum vacuum fluctuations are stretched out to the largest scales and become primordial fluctuations [8–12], which we later see as CMB anisotropies and galaxies. Thus, through the observation of primordial fluctuations we have access to inflation. However, the CMB only probes the largest scales and, therefore, only a small fraction of inflation. Any information on much smaller scales,[2] and the last stages of inflation, is basically erased by complicated astrophysical processes. In contrast, GWs barely interact with intervening matter. This means that with GWs we can explore the universe before neutrino decoupling and even before the Big Bang towards the last stages of inflation. GWs provide a unique opportunity to complete our picture of the very early universe. A quantitative picture of the potential of cosmology with GWs can be found in Figure 1.

GWs generated by a cosmological process in the early universe will today appear as randomly distributed in all directions and with a very large number of unresolved sources.[3] The superposition of all incoming unresolved GWs leads to the Stochastic GW Background (SGWB). Encouragingly, the typical frequency[4] of such cosmic GWs falls right into the frequency range of future GWs detectors. For example, Pulsar Timing Arrays (PTA) [13–18], cover around $10^{-9}\sim10^{-7}$ Hz, LIGO, VIRGO, KAGRA and ET [19], are sensitive roughly for $10\sim10^3$ Hz. In between, we will have LISA [20,21], DECIGO [22–24], AION/MAGIS [25], Taiji [26] and Tainqin [27]. For an illustration see Figure 1.

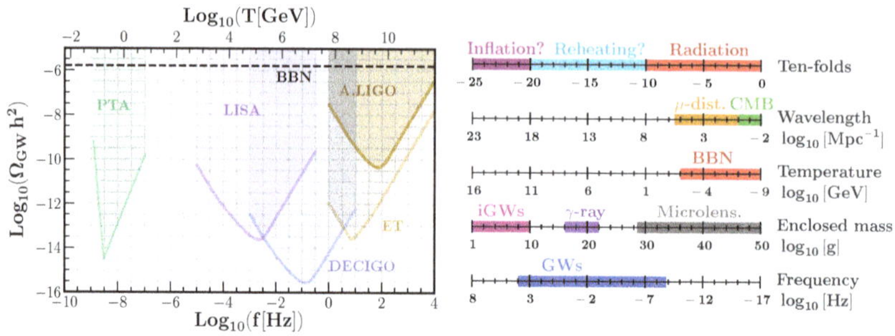

Figure 1. On the left we show the power-law integrated sensitivity curves [28] in terms of the GW spectral density for some GW detectors. In the lower and upper horizontal axis, we respectively show the GW frequency and the associated temperature of the universe at the time of GW generation, assuming radiation domination. On the right, we show the typical quantities associated to the cosmological horizon at a particular time, given in terms of 10-folds from matter-radiation equality. We also show the ranges covered by CMB anisotropies and μ spectral distortions, BBN constraints and GWs. If PBH form, they can be probed by induced GWs (iGW), emission of γ rays and microlensing.

Typical sources of cosmological GWs in the very early universe include (for more details see [29]): phase transitions, which may lead to collisions of bubbles or a universe filled with cosmic strings, resonances during reheating, quantum (gravity) fluctuations during inflation (so-called primordial GWs) and GWs induced from large primordial fluctuations. Such large primordial fluctuations may also collapse to form primordial black holes (PBHs). Out of these cosmological sources, GWs induced by primordial fluctuations, and to some extend PBHs, are our direct window to the latest stages of inflation. The information we could gain about inflation by the so-called induced GWs potentially covers scales around $k \sim 10^7$–$10^{18}\,\mathrm{Mpc}^{-1}$, otherwise inaccessible by any other probe. Even the absence of induced GWs will place new constraints on the primordial spectrum in unexplored regimes, potentially down to $\mathcal{P}_\mathcal{R} \sim 10^{-4}$–$10^{-5}$ [30]. On top of these very promising prospects, induced GWs are generated when primordial fluctuations re-enter the horizon sometime between the end of inflation and BBN. Since we have no evidence of the content and expansion history of the universe around that time, induced GWs not only provide access to the last stages of inflation but they also contain information on the content of the primordial universe. In this review, we will be opening to the possibility that the universe much before BBN was not dominated by radiation. In the future, information on the primordial spectrum obtained by induced GWs will complement those from other probes such as spectral distortions [31,32] in the multimessenger cosmology era [33].

1.1. *Induced GWs History*

The possibility of having GWs induced by density fluctuations was first noticed, as far as the author is aware, by K. Tomita in 1967 [34]. They were later rediscovered in the 1990s by Matarrese, Pantano and Saez [35,36] when studying second order cosmological perturbations in a dust dominated universe. Quite interestingly, in 1997 Matarrese, Mollerach and Bruni [37] noticed that induced GWs in a dust dominated universe suffer from gauge ambiguities. They proceeded to argue that only the oscillating part of the induced tensor modes were identified as true gravitational waves. We will discuss more about the gauge ambiguities in Section 7. However, Refs. [35–37] concluded that induced GWs were too small to be practically observed.

It was not until 20 years later that Ananda, Clarkson, and Wands [38] started to uncover the potential of induced GWs. In Ref. [38] they proposed to use induced GWs, generated in a radiation dominated universe, to constrain the spectral tilt of primordial fluctuations. A blue tilted primordial spectrum even with the CMB normalization might end up yielding large enough induced GWs. Some months later, Baumann, Steinhardt, Takahashi and Ichiki [39] looked into more detail at the transfer function of induced GWs taking into account the radiation-matter equality at around $z \sim 3400$ and the anisotropic stress due to neutrinos (see also [40] on the latter). They found that assuming the CMB normalisation of an almost scale invariant spectrum, modes with $k \sim k_{\rm eq} \sim 0.01\ {\rm Mpc}^{-1}$ were enhanced due to the matter dominated stage. Notably, the peak in the induced GW background can be larger than the primordial one if the tensor to scalar ratio is $r < 0.1$. However, the frequency of the peak of such GWs is extremely low, around $f_{\rm eq} \sim 10^{-17}$ Hz. This is only observable perhaps by CMB B-mode polarization experiments or cosmic shear [41]. Nevertheless, this was an indication that an early epoch of dust domination might yield interesting results. In the same period, Martineau and Brandenberger [42] derived a lower bound to induced GWs for various inflationary models and alternatives, although they referred to it as lower bound from "backreaction" of scalar fluctuations. Another application of GWs induced by primordial fluctuations in the curvaton scenario, sourced between the end of inflation and the curvaton decay, was studied by Bartolo, Matarrese, Riotto and Väihkönnen [43]. An action formalism for a scalar field acting as a perfect fluid and its induced GWs was studied by Boubekeur, Creminelli, Noreña and Vernizzi [44].

An important realization was done by Saito and Yokoyama in 2008 [45,46]: if the primordial spectrum of fluctuations is large enough on small scales, it does not only induce GWs but might also collapse to form PBHs. This means that the induced GW spectrum can be used in the future to place bounds on the PBH abundance. Even more, if PBHs were found, one should also find the induced GW counterpart. These ideas were further pursued by Bugaev and Klimai [47–49]. Extending the work of Saito and Yokoyama a bit further, Assadullahi and Wands proposed induced GWs as probes of the primordial spectrum [50]. They also considered the early dust dominated case [51] which confirmed the large enhancement of induced GWs. The same group, now including Arroja, Assadullahi, Koyama and Wands [52], studied the matching conditions on superhorizon scales for such GWs induced at second order. Later, in 2012 the induced GWs from particular models of inflation were studied by Alabidi, Kohri, Sasaki and Sendouda in radiation [53] and dust domination [54]. Around the same time, a curvaton scenario with a blue tilted primordial spectrum yielding large induced GWs was proposed by Kawasaki, Kitajima and Yokoyama [55]. More investigations on the upper bounds of induced GWs by overproduction of PBHs were done by Nakama and Suyama [56,57]. Other works related to induced GWs include the decay of two curvatons by Suyama and Yokoyama [58], the impact of anisotropic stress due to free streaming particles by Saga, Ichiki and Sugiyama [59] and the B-modes due to induced GWs in the CMB by Fidler et al. [60].

In 2016, LIGO reported the detection of GWs from the merger of a binary black hole [61] and the amount of new works on induced GWs and their PBH counterpart exploded. The amount of works is so large that we will not attempt to go through all of them in chronological order but instead we will classify them into seven main directions below. We give a more detailed discussion in the corresponding sections.

- General semi-analytical formulation: Since induced GWs are a second order effect one needs to integrate in time and momenta over the linear evolution of primordial fluctuations. The analytical transfer functions for radiation domination are derived in Refs. [62,63] and later generalized to constant equation of state parameter in Ref. [64];
- Induced GWs for different expansion histories and different contents of the universe: Induced GWs may have been generated in a non-radiation dominated universe. This leaves characteristic signatures in the induced GW spectrum. Studies in early matter era can be found in Refs. [65–67]. The extension to an early PBH dominated epoch is investigated in Refs. [68–71]. More general thermal histories are studied

in Refs. [64,72–76]. The impact of additional free streaming particles is studied in Ref. [77];

- Induced GW spectral features: There are cases where the induced GW spectrum may be investigated semi-analytically. These are for example, the low frequency tail [74,78,79], the UV tail [80,81] and the log-normal peak in the primordial spectrum [82]. Furthermore, the primordial spectrum may also present oscillatory features which are captured into the induced GW spectrum [83–87]. On top of that, large primordial non-Gaussianities may have a non-trivial impact on the induced GW spectrum [81,88–93]. Other effects include: anisotropic non-gaussianities, which may be a source of superhorizon tensor modes [94], resonances that may occur during inflation, which enhance the induced GW spectrum [95–98], and non-Bunch Davies initial conditions in inflation [99], although the latter does not yield an observable signature;
- Explanations of current observations: Induced GWs have been extensively used as counterpart of the PBH scenario as a totality or a fraction of dark matter. For example, the induced GWs from various inflationary models can be found in Refs. [100–118] and [62,116,119,120] in the context of Higgs inflation. In particular, the large induced GW counterpart to PBH as the totality of dark matter is studied in Refs. [88,110,121–125]. Other possibilities include an explanation to the LIGO observations [121,126–128] and the NANOGrav results [129–135];
- Current and future GW constraints: It is important to place constraints on the current absence of induced GWs and also asses future capabilities to constraint/find different models. A study using current PTA and LIGO data on SGWBs can be found in Refs. [136–139] and an analysis of future GW prospects in Refs. [33,75,140–146];
- SGWB anisotropies: In the far future we may not only observe the GW spectrum but also its anisotropies. The anisotropies due to induced GWs associated to PBH formation in the case of a monochromatic mass function are investigated in Refs. [122,123,147,148];
- The gauge issue of induced GWs: Tensor modes are subject to gauge ambiguities at second order due to mode mixing. Since the work of Hwang, Jeong and Noh [149] in 2017 there has been an extensive discussion on the gauge issue of induced GWs [150–162]. The source of the problem and the applicability of predictions is by now well understood, although the gauge issue persists in the strictest sense.

1.2. Structure and Scope of the Review

Induced GWs are the most promising probe of the primordial universe. In this review, we will revisit and unify the current analytical formulation and main predictions for induced GWs. Note that the scope of this review is basically limited to the author's works on induced GWs [64,70,71,74,81,132,150,161], although we discuss and properly cite the relevant literature. As interesting as it is, we will not enter in the details of PBH formation nor on the GW data analysis. Nevertheless, we briefly discuss these relevant topics and give appropriate credit, when needed.

For convenience, we proceed to briefly explain the organization of the review. In this way, the reader interested in certain aspects may jump directly to the relevant section. In Section 2 we present an intuitive picture of the induced GW generation. We estimate their typical frequency, the spectral shape and we briefly comment on their PBH counterpart for different expansion histories in the primordial universe. The details of the calculations and derivation of analytical approximations are reviewed in Sections 3 and 4. The analytical approximations for the induced GW spectrum for typical primordial curvature spectra are presented in Section 5. The special case of the dust dominated universe is discussed in some detail in Section 6. We also include the case where PBH dominate the universe in Section 6.2. The gauge issue of induced GWs is explained in Section 7. We list other GW counterparts associated to PBHs in Section 8 and discuss the future observational prospects in Section 9. We also provide a summary of the main formulas in Section 10.

These formulas are ready for calculating the induced GW spectrum for general primordial spectra. Thus, the reader interested solely in the final formulas of the induced GW may jump directly to Section 5 and/or Section 10. We end with a discussion on future directions in the conclusions Section 11. We present many details of the numerical factors, formulas and calculations used in the main text in the Appendices.

We work in natural units where $\hbar = c = 1$. We keep the reduced Planck mass $M_{\rm pl} = (8\pi G)^{-1/2}$ in the equations, except in Sections 2.2 and 4 where for simplicity in the calculations we set it to $M_{\rm pl} = 1$. At the beginning of each section we list the literature in which the section is based, except for Sections 8 and 9 which are relatively short. We provide a list of useful related reviews below.

A list of other useful reviews. The context of induced GWs is very broad. It covers cosmological perturbation theory up to second order, stochastic GW backgrounds, primordial black holes and primordial non-Gaussianity. Unfortunately, all the details cannot be covered in this review. Instead, we list below existing reviews on topics directly related to induced GWs, which have been useful in writing this manuscript.

- *Cosmological perturbation theory.* The classical reviews on cosmological perturbation theory at linear order are the one by Kodama and Sasaki [12] and by Mukhanov, Feldman and Brandenberger [163]. A typical reference for second order cosmological perturbation theory is the review by Malik and Wands [164]. Since then there have been many other reviews on cosmological perturbation theory. Some that we found particularly up to date and useful are Refs. [165–168]. For a take on multi-field inflation we suggest Ref. [169]. For primordial features in the primordial spectrum we refer the reader to Ref. [170];
- *Stochastic GW backgrounds.* Induced GWs are not the only source of GW backgrounds. Reviews on the different cosmic and astrophysical sources can be found in the review by Caprini and Figueroa [29] and by Christensen [171]. A useful collection of cosmic GW spectra can be found in Kuroyanagi, Chiba and Takahashi [172], although it is technically not a review. A review focused on GWs from inflation is given by Guzzetti, Bartolo, Liguori and Matarrese [173];
- *Primordial black holes.* The literature on primordial black holes is very vast and currently under refinement. A review that has been used in particular is the one by Sasaki, Suyama, Tanaka and Yokoyama [174]. Other interesting reviews are Refs. [175–178]. A complementary review on induced GWs with more focus on the PBH counterpart is given by Yuan and Huang [179];
- *Primordial non-Gaussianity.* Although quantum fluctuations during inflation are drawn from a Gaussian distribution, they can develop small departures from such Gaussian distribution due to gravitational or general interactions. The reader interested in primordial non-Gaussianities may check the reviews in the context of inflation and CMB observations, e.g., [180–183];
- *Alternative expansion histories.* A recent review encompassing many of the new physics of a primordial universe which is not filled with radiation is given in Ref. [1].

2. Estimates and Intuitive Picture

In this section we present the intuitive physical picture behind induced GWs. We do not intend to do a rigorous derivation but to lay out the basic assumptions behind the calculations. Before we go into the details of the induced GWs themselves, we first review in Section 2.1 what is considered to be the "observable" quantity for cosmic GWs. This will also be useful when dealing with the gauge issue in Section 7. We later in Section 2.2 derive intuitively, based on heuristic calculations and order of magnitude estimations, the main features of the induced GW spectrum in general cosmological backgrounds. Lastly in Section 2.3, we give the relation between the typical frequencies of the induced GWs with the corresponding PBH masses also in general cosmological backgrounds.

Main references: Section 2.1 is mainly based on Misner, Thorne and Wheeler's [184] and Maggiore's [185] books, while the original work is essentially due to Isaacson [186,

187]. We follow the notation of, e.g., Ref. [78,161]. Section 2.2 essentially follows from [74] although it builds up from previous works in the literature, e.g., [39,63,64,78]. Section 2.3 is mainly extracted from [64,132,174]. All relevant equations can be found in Appendix E.

2.1. The Spectral Density of GWs in Cosmology

A GW detector, such as an interferometer, measures the GW strain at a given frequency and time. In the case of the SGWB, one is measuring the time and/or ensemble average of the superposition of all incoming GWs. For convenience, cosmologists often use the energy density fraction of GWs as the "observable". However, it is not an easy task to theoretically associate an energy density to GWs in general situations. This is because in general relativity there is no local notion of energy of the gravitational field. By the equivalence principle, one may always go to a locally flat frame and erase any sign of gravity. Of course, we know that GWs carry momentum and energy (after all, they have been detected). Thus, indeed one can define the energy of GWs in some limits of interest. For instance, when the frequency of the GW is much higher than that of a slowly varying background [185].

Relevant to the later discussion, let us briefly review the case of GWs propagating in a Ricci flat spacetime, i.e., a spacetime with no matter sources. We can then split the total metric as $g_{\mu\nu} = \bar{g}_{\mu\nu} + h_{\mu\nu}$ where $\bar{g}_{\mu\nu}$ is the background metric and $h_{\mu\nu} \ll 1$ are the GWs, treated as a high frequency perturbation. Then, by the non-linear nature of gravity, these GW backreact onto the background metric. Mathematically speaking, if we expand the Einstein tensor up to second order we will find terms quadratic in $h_{\mu\nu}$ which can be thought of as a "matter" source. This means that, if we somehow "integrate out" the high frequency perturbation, we can write[5]

$$\bar{G}_{\mu\nu}[\bar{g}_{\mu\nu}] = M_{\rm pl}^{-2} t^{\rm GW}_{\mu\nu}[h_{\mu\nu}]\,, \tag{1}$$

where $G_{\mu\nu}$ is the Einstein tensor and $t^{\rm GW}_{\mu\nu}$ is the (pseudo)-energy momentum tensor of GWs. A way to "integrate out" the high frequency modes is by taking an average. For instance, we may focus on a small box and take the volume (and/or time) average. The box shall not be too small, so that several GW wavelengths fit in. In this case, we have that the (pseudo)-energy momentum tensor of GWs reads [186,187]

$$t^{\rm GW}_{\mu\nu} = \frac{M_{pl}^2}{4}\left\langle \partial_\mu h^{\alpha\beta}\partial_\nu h_{\alpha\beta} - \frac{1}{2}\bar{g}_{\mu\nu}\partial_\sigma h^{\alpha\beta}\partial^\sigma h_{\alpha\beta}\right\rangle, \tag{2}$$

where all contractions of indices are done with the background metric.

In cosmology we face the extra difficulty of having a matter source to the Einstein equations which drives the expansion of the universe. Because of the expansion, we also have a cosmological horizon roughly given by $1/H$ where H is the Hubble parameter and quantifies the expansion rate. For these reasons, the definition of GWs in a cosmological background is more subtle than in Ricci flat space times. Nevertheless, with some tweaks we shall write in analogy

$$G_{\mu\nu} = M_{\rm pl}^{-2}\left(T_{\mu\nu} + t^{\rm GW}_{\mu\nu}\right), \tag{3}$$

where $T_{\mu\nu}$ is the matter energy-momentum tensor. This time, however, we must be more careful on the averaging procedure. Furthermore, we must introduce a slightly different, but important, notation. First, in the helicity decomposition of the metric, the transverse-traceless component of the spatial metric is referred to as tensor mode. A tensor mode with a co-moving wavenumber k stays constant if its physical size is larger than the cosmological horizon, i.e., $k/a \ll H$. In the opposite limit, a tensor mode behaves as a GW if its physical size is much smaller than the cosmological horizon, that is $k/a \gg H$. Thus, Equation (3) only makes sense for the high frequency tensor modes deep inside the horizon. This also means that the small boxes we used to take the average have to be much smaller than the horizon size but still larger than several GW wavelengths. In addition to that, in cosmology we often deal with a SGWB and, by ergodicity, we may take an ensemble average over all small boxes. Under these assumptions we may use Equation (2) to compute the energy density of GWs in cosmology. Be aware though that these assumptions are strictly speaking not enough as we shall argue in Section 7.

Proceeding with the cosmological SGWB, we consider GWs propagating in a Friedmann-Lemaître-Robertson-Walker (FLRW) metric, namely[6]

$$ds^2 = -dt^2 + a^2(t)(\delta_{ij} + h_{ij})dx^i dx^j\,, \tag{4}$$

where $a(t)$ is the scale factor. The evolution of the scale factor is dictated by the Friedmann equations, which we provide in Appendix E. Using Equation (2) under the ensemble average procedure, we find that the energy density of GWs (the 00 component of $t^{\rm GW}_{\mu\nu}$) in Fourier space (see Appendix D for notations) reads

$$\rho_{\rm GW} = M_{\rm pl}^2 \int d\ln k \frac{k^3}{16\pi^2}\left\{\sum_\lambda \langle \dot h_{\mathbf{k},\lambda}\dot h_{-\mathbf{k},\lambda}\rangle^! + \frac{k^2}{a^2}\langle h_{\mathbf{k},\lambda}h_{-\mathbf{k},\lambda}\rangle^!\right\}, \tag{5}$$

where $\dot{} \equiv d/dt$, λ is the GW polarization and we already used that the coincident ensemble average of two tensor modes in a homogeneous and isotropic universe is proportional to a Dirac delta,[7] namely

$$\langle h_{\mathbf{k},\lambda}h_{\mathbf{k}',\lambda}\rangle = \langle h_{\mathbf{k},\lambda}h_{-\mathbf{k},\lambda}\rangle^!(2\pi)^3\delta^{(3)}(\mathbf{k}+\mathbf{k}')\,. \tag{6}$$

The notation "!" in Equations (5) and (6) is to indicate that the Dirac delta has been factored out, so that $\langle h_{\mathbf{k},\lambda}h_{-\mathbf{k},\lambda}\rangle^!$ is directly related to the power spectrum, which by isotropy is a function of $k = |\mathbf{k}|$ only. Equation (5) is the total energy density of the GWs filling the universe. It is often more convenient to use the spectral density fraction, i.e., the GW power per logarithm of a given wavenumber k (or frequency) over the critical density, which is defined as

$$\Omega_{\rm GW}(k) \equiv \frac{1}{3M_{\rm pl}^2 H^2}\frac{d\rho_{\rm GW}}{d\ln k} = \frac{k^2}{12a^2H^2}\sum_\lambda \mathcal{P}_{h,\lambda}(k)\,, \tag{7}$$

where $H = \dot a/a$. This is often what is used to compare theoretical predictions with current constraints and future observational prospects. Note that in Equation (7) we used the definition of the dimensionless power spectrum, concretely

$$\langle h_{\mathbf{k},\lambda}h_{-\mathbf{k},\lambda}\rangle^! = \frac{2\pi^2}{k^3}\mathcal{P}_{h,\lambda}(k)\,. \tag{8}$$

We also used the approximation that for a freely propagating wave we have that

$$\dot h_{\mathbf{k},\lambda} \approx \frac{k}{a}h_{\mathbf{k},\lambda}\,. \tag{9}$$

With these ingredients we can use Equation (7) to estimate the amplitude and spectral shape of induced GWs.

2.2. GWs Induced by Primordial Fluctuations

Let us derive some estimates by considering a rough, yet instructive, toy model for induced GWs. Take a tensor mode h_{ij} in a FLRW background (4) sourced by scalar field fluctuations $\delta\varphi$. Its equations of motion, in terms of conformal time $d\tau = dt/a$, read

$$h''_{ij} + 2\mathcal{H}h'_{ij} - \partial_k\partial^k h_{ij} = \mathcal{P}^{ab}{}_{ij}\{T_{ab}\} \approx \mathcal{P}^{ab}{}_{ij}\{\partial_a\delta\varphi\partial_b\delta\varphi\}\,, \tag{10}$$

where $' \equiv d/d\tau$, $\mathcal{H} = H/a = a'/a$, T_{ij} is the spatial components of the energy-momentum tensor of a scalar field (see Equation (45)) and $\mathcal{P}^{ab}{}_{ij}$ is the transverse-traceless projector defined in Appendix E. Without understanding much about the evolution of these fluctuations, we can already see that

$$\mathcal{P}_h \equiv \sum_\lambda \mathcal{P}_{h,\lambda} \propto \mathcal{P}^2_{\delta\varphi} . \tag{11}$$

If the scalar field fluctuations were sourced during inflation, we can relate them to the curvature perturbation $\mathcal{R}$ so that $\mathcal{P}_{\delta\varphi} \sim \mathcal{P}_\mathcal{R}$. Furthermore, the spectral density of GWs in a radiation dominated universe remains constant, as GWs behave as radiation themselves. Thus, we can boldly estimate that the amplitude of induced GW spectral density measured today is given by

$$\Omega^{\rm induced}_{\rm GW} h^2 \sim \frac{1}{12}\Omega_{r,0}h^2 \times \mathcal{P}^2_\mathcal{R} , \tag{12}$$

where $\Omega_{r,0} \equiv \rho_{r,0}/(3H_0^2M_{\rm pl}^2)$ is the density fraction of radiation, H_0 is the Hubble rate today and $h \equiv H_0/(100\ {\rm km/Mpc/s})$. We have introduced $\Omega_{r,0}$ to take into account the dilution of the GWs (as radiation) as the universe expands and goes through the cold dark matter and dark energy dominated stages. We give more details on this factor in Section 5. h considers the uncertainty in the exact value of H_0. From the latest Planck 2018 results [2] we have that $\Omega_{r,0}h^2 \approx 4 \times 10^{-5}$. With all the above simplifications, we arrive at

$$\Omega^{\rm induced}_{\rm GW} h^2 \sim 10^{-6}\mathcal{P}^2_\mathcal{R} . \tag{13}$$

The estimate (13) can be now contrasted with future observational prospects. The sensitivity of future GW detectors after collecting data over several years might reach $\Omega^{\rm induced}_{\rm GW} h^2 \sim 10^{-16}$, if we use the power-law sensitivity curves of Thrane and Romano [28]. This optimistic value is for the best DECIGO sensitivity [23,188]. Then, in the absence of an induced GW detection we can use Equation (13) to place an optimistic upper bound to the primordial spectrum on the corresponding scales (roughly at 10^{16}–10^{12} Mpc^{-1}) of about $\mathcal{P}_\mathcal{R} < 10^{-5}$. This is already three orders of magnitude better than current constraints from the absence of PBHs (around $\mathcal{P}_\mathcal{R} < 10^{-2}$ [174]), although it is still four orders of magnitude above the CMB extrapolation which would be around $\mathcal{P}_\mathcal{R} \sim 10^{-9}$ [3]. The most interesting point though is that PBHs which formed by the collapse of large primordial fluctuations, which requires $\mathcal{P}_\mathcal{R} \gtrsim 10^{-2}$, must have an observable large induced GW counterpart. Thus, the absence of induced GWs completely rules out the PBH scenario from large primordial fluctuations [189,190]. Note that PBH could have formed by other mechanisms though. For example, PBH could form in first order phase transitions [191,192], tunneling during inflation [193], the collapse of Q-balls [194,195] and might also be the result of long range interactions stronger than gravity [196–198].

In the above estimations we have completely neglected the evolution of the scalar fluctuations since the end of inflation onwards, which is by no means irrelevant. As shown in Refs. [51,54,65,66], a period of dust domination might substantially enhance the amplitude of induced GWs. The amplification is such that GWs induced by tiny primordial fluctuations might observable [54,66]. This case exemplifies the importance of the expansion history of the primordial universe. At this point, we would like to emphasise that contrary to studies of the CMB we have scarce evidence of the content of the primordial universe. Therefore, when deriving theoretical predictions, and eventually contrasting them with the data, it would be a good idea to be open and acknowledge our ignorance by allowing a free equation of state parameter $w \equiv \rho/p$ and a free propagation speed of fluctuations c_s, which describe the unknown content of the primordial universe. A nice review on the implications of various expansion histories can be found in Ref. [1]. For these reasons, let us estimate the general impact of the equation of state w and sound speed c_s onto the induced GW spectrum. To do that, we continue with the toy model of the scalar field. This time, however, we will consider the evolution of $\delta\varphi$. Fluctuations of a K-essence massless scalar field roughly follow the Klein-Gordon equation

$$\delta\varphi_k'' + 2\mathcal{H}\delta\varphi_k' + c_s^2 k^2 \delta\varphi_k = 0\,, \tag{14}$$

where for simplicity we considered a constant w and c_s. The former implies constant expansion rate, i.e.,

$$\epsilon \equiv -\frac{\dot{H}}{H^2} = \frac{3}{2}(1+w) = \text{constant}\,. \tag{15}$$

Such a constant expansion rate implies a scale factor which goes as a power-law of conformal time, explicitly from the equations in Appendix E we have

$$a \propto \tau^{1+b} \quad \text{where} \quad b = \frac{1-3w}{1+3w}\,. \tag{16}$$

We introduced the parameter b for later convenience. $b = 0$ corresponds to a universe filled with radiation ($w = 1/3$). Then, $b < 0$ and $b > 0$ respectively correspond to a stiffer and softer fluid. Note that if we allow $w \gg 1$ the range of b goes as $-1 < b < \infty$ for $\infty > w > -1/3$.

For the sake of simplicity, we solve the Klein-Gordon Equation (14) in the super-(sound)horizon, $c_s k \ll \mathcal{H}$, and sub-(sound)horizon, $c_s k \gg \mathcal{H}$, limits. This leads us to

$$\delta\varphi_k \approx \delta\varphi_i \begin{cases} \text{constant} & c_s k\tau \ll 1 \\ a^{-1} e^{i c_s k \tau} & c_s k\tau \gg 1 \end{cases}, \tag{17}$$

where $\delta\varphi_i$ is an initial value. The above solutions are all we need to solve for the induced GWs. For analytical viability, let us take that there are fluctuations of the scalar field only in a single scale k_p. Namely the scalar fluctuations are Dirac-delta like in Fourier space, i.e.,

$$\delta\varphi_i \sim \delta\varphi_p \times \delta(\ln(k/k_p))\,, \tag{18}$$

where $\delta\varphi_p$ is the primordial amplitude, e.g., set by inflation. All the above simplifications lead to a simple differential equation for the tensor modes given by

$$h_k'' + 2\mathcal{H}h_k' + k^2 h_k \approx k_p^2 \delta\varphi_p^2\,. \tag{19}$$

We readily see that, if we focus on $\delta\varphi_p = $ constant and we evaluate h_k at horizon crossing, that is when the mode k is of the size of the comoving horizon, i.e., $k = aH$, then the amplitude of induced GWs is proportional to $h_k \sim \delta\varphi_p^2$ for $k \sim k_p$, as used in our previous estimate (12). Now, to derive the spectral features we can focus on two interesting limits: (i) the infra-red (IR) tail for modes with $k \ll k_p$ and (ii) the near peak behaviour for modes with $k \sim k_p$.

2.2.1. The IR Tail

If we focus on modes with $k \ll k_p$ there will be no competition between the k^2 and k_p^2 terms nor any resonance. Thus, we can peacefully solve the tensor modes in the superhorizon ($k\tau \ll 1$) and subhorizon ($k\tau \gg 1$) regimes. First, in the superhorizon regime we find that

$$h(k\tau \ll 1) \approx \begin{cases} \left(k_p\tau\right)^2 & c_s k_p \tau \ll 1 \\ \text{constant} + \left(k_p\tau\right)^{-2b} & c_s k_p \tau \gg 1 \end{cases}. \tag{20}$$

What Equation (20) tells us is that tensor modes initially grow as τ^2 due to a constant source. Then, after the scalar source enters the sub(sound)-horizon regime, i.e., at $c_s k_p \tau_p = 1$, the tensors developed a constant amplitude. Now, if $b > 0$ ($w < 1/3$) the scalar source decays much faster than the cosmological background, that is $\delta\varphi_k^2/\mathcal{H}^2 \propto (k\tau)^{-b} \to 0$, and the tensor growth stops. However, for $b < 0$ ($w > 1/3$) the scalar source decays slower than the background expansion, which yields a second superhorizon growth of the tensor modes [74]. To follow tensor mode evolution, the second line of Equation (20) has to be matched at horizon crossing for the tensor mode ($k\tau = 1$) with the subhorizon solution. By doing so, we arrive at

$$h(k\tau \gg 1) \approx (k\tau)^{-1-b} e^{ik\tau}\left(\text{constant} + (k_p/k)^{-2b}\right). \tag{21}$$

The final ingredient is that somehow the universe transitions to a radiation dominated universe to recover the standard cosmology. For simplicity, we take an instantaneous reheating at $\tau = \tau_{\rm rh}$ with the corresponding reheating scale $k_{\rm rh} = \mathcal{H}_{\rm rh}$. From that moment on the spectral density of induced GWs stays constant and we can derive our predictions. Inserting Equation (21) evaluated at $\tau = \tau_{\rm rh}$ into Equation (5) we find that the IR slope of the induced GW spectrum is in general given by

$$\frac{d\ln\Omega_{\rm GW}^{\rm induced}}{d\ln k}(k_{\rm rh} \ll k \ll k_p) \approx 3 - 2|b|\,. \tag{22}$$

We later find that for $\beta \in \mathbb{Z}$ there is a logarithmic correction [74]. Additionally, if the scalar spectrum of fluctuations is really a Dirac delta, the spectrum slope is given by $2 - 2|b|$ due to the unphysical divergence of $\delta\varphi_p$ at $k = k_p$. Note that for tensor modes which entered during the radiation domination era we have to match the first line of (20) to the subhorizon solution. This yields a spectral index given by

$$\frac{d\ln\Omega_{\rm GW}^{\rm induced}}{d\ln k}(k \ll k_{\rm rh}) \approx 3\,, \tag{23}$$

which is in agreement with the general results of Ref. [78]. These naive derivations will be later recovered from the analytical formulas in Section 4. It is interesting to see that for $b > 3/2$ ($w < -1/15$) the induced GW spectrum peaks at $k \sim k_{\rm rh}$ which is clearly different than the peak in the primordial spectrum at $k \sim k_p$.

The intuitive physical picture behind (22) is the following. In a universe with constant b, modes of a massless field (e.g., the tensor modes h_{ij} or the scalar field $\delta\varphi$) which enter the horizon early get more diluted than modes which enter later. For example, consider that right after horizon crossing, i.e., when $k = \mathcal{H}_k$, we have that the energy density of the massless field starts to dilute as $\rho \propto (a_k/a)^4$. Since we have that $a_k \propto \tau_k^{1+b} \propto k^{-1-b}$ then $\rho \propto a_k^4 \propto k^{-4b}$ and $\rho/H^2 \propto a^{-2b/(1+b)} \propto k^{-2b}$. Thus, if the IR tail of induced GWs in radiation domination goes as k^3, we expect $\Omega_{\rm GWs} \propto k^{3-2b}$ due to dilution [78]. However, as we have shown, this is incomplete for induced GWs from a peaked spectrum. For $b < 0$, the density fraction of the scalar field $\rho_{\delta\varphi}$ grows as $a^{-2b/(1+b)}$. This means that induced tensor modes which are generated later have a larger source and a larger amplitude than those generated earlier. In other words, for $b < 0$ small k have a larger amplitude than large k by a factor $a_k^{-4} \propto k^{4b}$. Then, for $b < 0$ the density fraction of induced GWs goes as $\Omega_{\rm GWs} \propto k^{3-2b+4b} \propto k^{3+2b}$ [74]. This gives the general formula $\Omega_{\rm GWs} \propto k^{3-2|b|}$ that we derived in (22). Whether this absolute value of b is unique to induced GWs is something yet to explore.

Before moving on to the near peak limit, let us comment on the implications of Equation (22). While the far most IR tail of the induced GW spectrum indeed must go as k^3, we may have an intermediate regime where $\Omega_{\rm GWs} \propto k^{3-2|b|}$. This has interesting consequences when comparing with GWs from other sources. Let us model a general GW spectrum by a broken power-law around a characteristic scale k_c, as done in Ref. [172]. Then, for example, the IR spectrum of induced GWs for $b = 2$ ($w = -1/9$) resembles that of GWs generated by a strong first order phase transition which have $\Omega_{\rm GWs} \propto k^{2.8}$ for $k < k_c$ and $\Omega_{\rm GWs} \propto k^{-1}$ for $k > k_c$. This gives another very good reason to be open about a primordial universe with a different equation of state than radiation.

2.2.2. The Near Peak Regime

Looking closely at Equation (19), we might suspect that there could be a resonance if the wavenumber k of a tensor mode approaches that of twice the scalar mode $2k_p$. This resonance may be understood in two ways. First, we know that a harmonic oscillator with an external force has a resonance when the period of the force matches that of the harmonic oscillator. If that occurs, the amplitude of the oscillations diverges in time. In short, the external force is kicking the oscillator at the right times. The second way to see it is that two scalar modes with physical momentum $c_s k_p$ have a preferred window to produce a tensor mode with physical momentum equal to twice the scalar momenta, i.e., $k = 2c_s k_p$, if allowed by momentum conservation. To see this mathematically, let us redefine the tensor modes in Equation (19) to remove the Hubble friction by

$$h_k = v_k / a \,. \tag{24}$$

With this new variable v_k Equation (19) now reads

$$v_k'' + \left(k^2 - \frac{b(1+b)}{\tau^2}\right) v_k = \frac{k_p^2}{a} \delta\varphi_p^2 e^{2ic_s k_p \tau} , \tag{25}$$

where we only focused on the oscillatory behaviour of the scalar field for $c_s k_p \tau \gg 1$. The particular solution to v_k can be found by the Green's function method (see Appendix B for the details). Neglecting the term proportional to $1/\tau^2$ in Equation (25), the homogeneous solutions to v_k are $e^{\pm ik\tau}$. With such homogeneous solutions, the particular solution by the Green's method reads

$$v_k \approx \int_{\tau_i}^{\tau} d\tilde{\tau} \, \sin(k(\tau - \tilde{\tau}))\tilde{\tau}^{-1-b} e^{2ic_s k_p \tilde{\tau}} \,. \tag{26}$$

We clearly see from the integrand in Equation (26) that there is a resonance when the frequency of the sine equals that of the exponential. Then the oscillations interfere and leave the integral of a power-law. Since we are only interested in how the amplitude diverges, let us only focus on the divergent part of (26) which is given by

$$v_k \propto \int d\tilde{\tau} \, \tilde{\tau}^{-1-b} e^{-i\tilde{\tau}(k-2c_s k_p)} \propto \Gamma[-b](k - 2c_s k_p)^b \,. \tag{27}$$

We see that when $b \leq 0$ the resonance is effective and the amplitude diverges. Note that for $b = 0$ the divergence is logarithmic. When $b > 0$ the resonance is not effective enough to cause a divergence as the amplitude of the integrand decays faster than τ^{-1}. However, since we so far used a Taylor expansion around $k \sim 2c_s k_p$ we expect that for $1 > b > 0$ Equation (27) still yields the leading contribution after the constant term. Thus, for $1 > b > 0$ we also expect a peak at around $k \sim 2c_s k_p$, albeit not divergent. For $b > 1$ the situation is less clear and a detailed analysis is necessary. Nevertheless for $b < 1$, we can roughly write that the tensor modes in the near peak regime have an amplitude given by

$$h_k(k \sim 2c_s k_p, b < 1) \propto \text{constant} - (k - 2c_s k_p)^b \,. \tag{28}$$

This also implies that the induced GW spectrum has a peak near $k \sim 2c_s k_p$ with a growth rate proportional to

$$\Omega_{\rm GW}^{\rm induced}(k \sim 2c_s k_p) \propto \begin{cases} |k - 2c_s k_p|^{-2|b|} & b < 0 \\ \text{constant} - |k - 2c_s k_p|^b & 1 > b > 0 \end{cases} . \tag{29}$$

These estimates will be later checked by the general analytical solutions in Section 4. Thus, the resonant peak in the induced GW spectrum has information on the sound horizon $c_s k_p$ by the position of the peak and on the expansion history b by the growth rate around the peak. This type of resonant structure is characteristic of induced GWs. Thus, the observation of such resonant peak could give strong evidence for induced GWs. Then, the sound speed of fluctuations c_s and the peak k_p could be disentangled by the simultaneous observation of the PBH counterpart, which we proceed to discuss. Note that in (29) the radiation domination case $b = 0$ is "special" because it is exactly when the amplitude of the source term is proportional to the background density.

2.3. Primordial Black Hole Counterpart

In this section we briefly introduce the PBH counterpart to the induced GWs. It is by no means a review on PBH formation and it does not intend to be so. A detailed review on PBHs can be found in Ref. [174]. Other interesting reviews are listed at the Section 1.2. PBHs from large primordial fluctuations will form in those Hubble patches where the density contrast $\delta\rho/\rho$ is above a certain threshold[8] δ_c [190,203–208], which depends on the EoS parameter w at horizon re-entry [209–216]. Since fluctuations generated during inflation closely follow a Gaussian distribution centred around zero, only the tail of the distribution has a large enough amplitude to collapse to PBH. This means that PBH formation is an extremely rare event. The fraction of Hubble patches where PBH forms is exponentially suppressed and sensitive to the root mean squared of the distribution, which is related to the power spectrum. This is why we generally need $\mathcal{P}_\mathcal{R} \sim 10^{-2}$ to form a substantial fraction of PBHs. Precise estimations of the fraction of PBH is an active and ongoing research field, e.g., see Refs. [202,214–221] and references therein. For our purposes, it is enough to know that only large induced GWs have a large associated fraction of PBH and, inversely, the absence of large induced GWs rules out PBHs from large primordial fluctuations.

For simplicity, we continue with the assumptions of the previous section and take a primordial spectrum with a Dirac delta peak at k_p. This essentially leads to an almost monochromatic[9] PBH mass function. When the peak in the scalar spectrum enters the horizon, those fluctuations which are above δ_c collapse into a PBH. By the end of the collapse, only a fraction γ of the total mass enclosed inside the horizon at horizon crossing ($H_\circledast a_\circledast = k_p$) will end up as a PBH. We use the symbol $\circledast$ to refer to evaluation at PBH formation time. In mathematical terms, we have that PBHs mainly form with a mass equal to the mass enclosed inside the Hubble sphere, that is

$$M_{\rm PBH,\circledast} = \gamma \frac{4\pi}{3} \frac{\rho_\circledast}{H_\circledast^3} = \frac{4\pi\gamma M_{\rm pl}^2}{H_\circledast}\,. \tag{30}$$

To grasp the magnitude of the actual PBH mass, let us write it in terms of a solar mass $M_\odot$ by referring all quantities in (30) in terms of the size of the horizon at matter-radiation equality, which is very well measured by Planck [2]. The numerical factors and formulas used to derive this and the following estimates can be found Appendix A. After some numerical manipulations we arrive at

$$M_{\rm PBH,\circledast} \approx 3.97 \times 10^{-12} M_\odot \left(\frac{k_p}{k_{\rm rh}}\right)^{-2-b} \left(\frac{k_{\rm rh}}{10^{12}{\rm Mpc}^{-1}}\right)^{-2} \times \left(\frac{\gamma}{0.2}\right) \left(\frac{g_*(T_{\rm rh})}{106.75}\right)^{1/2} \left(\frac{g_{*s}(T_{\rm rh})}{106.75}\right)^{-2/3}, \tag{31}$$

where we kept our ignorance on the primordial universe by leaving a free b (or w) (16). We took $\gamma \sim 0.2$ in a radiation dominated universe [206] but it may differ for different expansion histories [209–212]. $g_*(T_{\rm rh})$ and $g_{*s}(T_{\rm rh})$ are the effective degrees of freedom in the energy density and entropy evaluated at reheating. It should be noted that we also assumed an instantaneous reheating. In that case, the horizon scale at reheating $k_{\rm rh}$ can be related to the reheating temperature by

$$k_{\rm rh} = 1.2 \times 10^{12}\,{\rm Mpc}^{-1}\left(\frac{T_{\rm rh}}{5\times 10^{4}\,{\rm GeV}}\right)\left(\frac{g_*(T_{\rm rh})}{106.75}\right)^{1/2}\left(\frac{g_{*s}(T_{\rm rh})}{106.75}\right)^{-1/3}. \tag{32}$$

As we have seen in Section 2.2, the peak in the primordial curvature power spectrum induces GWs. These GWs have a typical frequency f_p associated to the wavenumber k_p. Since the typical mass and the typical frequency are related by the same typical scale k_p, we can write one in terms of the other. In this way, the typical frequency evaluated today reads

$$f_p = \frac{k_p}{2\pi a_0} \approx 1.54\times 10^{-3}\,{\rm Hz}\left(\frac{M_{\rm PBH,\circledast}}{3.97\times 10^{-12}M_\odot}\right)^{\frac{-1}{2+b}}\left(\frac{f_{\rm rh}}{1.54\times 10^{-3}\,{\rm Hz}}\right)^{\frac{b}{2+b}}. \tag{33}$$

For a radiation dominated universe ($b = 0$) we have a one-to-one correspondence between the frequency and the mass. If we fix the reheating scale, the relation (33) roughly tells us that PBH with a given mass have an induced GW counterpart with a peak frequency at around f_p. The exact relation between f_p and $M_{\rm PBH,\circledast}$ depends on the value of b. For instance, if $b < 0$ ($w > 1/3$) the expansion rate was faster and for a fixed k_p corresponds to smaller PBH mass at formation. The opposite holds for $b > 0$ ($w < 1/3$). It should also be noted that for $b > 3/2$ ($w < -1/15$) there is another peak in the induced GWs at $k \sim k_{\rm rh}$. This may break the correspondence between the PBH mass and the peak of the induced GWs if not all of the GW spectrum is seen.

3. General Formalism

In this section we present the general formulation of induced GWs. This includes the derivation of the equations of motion in Section 3.1 and the general form of the solutions in Section 3.2. Then in Section 3.3 we include the leading terms due to local-type non-Gaussianity. In Section 3.1, we derive the equations of motion using the action formalism and so we do not follow the conventional derivation, e.g., of Refs. [35,38]. The final result is obviously the same, but the derivation is rather quick and clear. Throughout this section and on, we shall neglect the effect of vector perturbations as they typically decay. Anyone interested in the approximations and applications may jump directly to Section 4.

Main references: In Section 3.1 we closely follow Ref. [150] although we work at the level of the Lagrangian. For alternative derivations starting from the Einstein equations see for example Refs. [35,38,39]. Then, Section 3.2 is extracted from [64] which is built upon and uses the notation of Ref. [63] except for the numerical factor 2 involving h_{ij}. Lastly, Section 3.3 on the primordial non-Gaussianity is mainly based on Ref. [81] updated and corrected by Ref. [92].

3.1. Derivation from the Action

We need to arrive at the equations of motion for the transverse-traceless component of the spatial metric at the second order in cosmological perturbation theory. Although it is a second order calculation, we shall derive it quickly from the action formalism; if we work in a particular gauge and we only focus on the interaction of tensor with scalars. It is particularly convenient to work in a gauge similar to the Newtonian gauge[10] with the exponential notation typical of Misner, Throne and Wheeler [184]. With this choice and assuming a perturbed FLRW universe, the metric reads

$$ds^2 = g_{\mu\nu}dx^\mu dx^\nu = -e^{2\Psi}dt^2 + a^2(t)e^{2\Phi}Y_{ij}dx^i dx^j\,, \tag{34}$$

where $g_{\mu\nu}$ is the metric of our space-time, $i = \{1, 2, 3\}$ are the spatial components, $a(t)$ is the scale factor and we have further used the conformal decomposition of the spatial metric such that

$$\frac{\partial}{\partial t}\det Y = Y^{ij}\frac{\partial}{\partial t}Y_{ij} = 0\,. \tag{35}$$

Note that for a flat FLRW background in Cartesian coordinates we have $Y_{ij} = \delta_{ij}$. The conformal decomposition of the spatial metric is very convenient and simplifies calculations. It is also used, e.g., in Numerical Relativity [222]. In short, the conformal decomposition completely splits the trace (volume changing) from the trace-free (volume preserving) degrees of freedom. In the Newtonian gauge this decomposition directly splits the scalar and tensor modes of the spatial metric. This means that Y_{ij} only contains the transverse-traceless degrees of freedom. As we shall see, it also leaves a clean splitting between Y_{ij} and Φ at the level of the action.

To consider a sensible cosmological set up, we include a canonical scalar field φ in the perturbed FLRW metric. We can later generalize it to a perfect fluid quite straightforwardly. The action, without any decomposition, is given by

$$S = \int d^4x\sqrt{-g}\left\{\frac{1}{2}R - \frac{1}{2}g^{\mu\nu}\partial_\mu\varphi\partial_\nu\varphi - V(\varphi)\right\}, \tag{36}$$

where g is the determinant of $g_{\mu\nu}$, R is the 4D Ricci scalar, $\partial_\mu \equiv \partial/\partial x^\mu$ and $V(\varphi)$ is the potential of φ. In the (3 + 1) conformal decomposition (34), after some algebra and integration by parts, the action becomes

$$\begin{aligned} S = \int d^3x\, dt\Big\{ & a\, e^{\Psi+\Phi}\left(\frac{1}{2}R^{(3)}[Y_{ij}] - 2Y^{ij}D_iD_j\Phi - Y^{ij}D_i\Phi D_j\Phi - \frac{1}{2}Y^{ij}D_i\varphi D_j\varphi\right. \\ & \left. + a^3e^{3\Phi-\Psi}\left(\frac{1}{8}Y^{ij}Y^{kl}\dot{Y}_{ik}\dot{Y}_{jl} - 3(H+\dot{\Phi})^2 + \frac{1}{2}\dot{\varphi}^2\right) - a^3e^{3\Phi+\Psi}V(\varphi)\right\}, \end{aligned} \tag{37}$$

where $R^{(3)}[Y_{ij}]$ and D_i are respectively the 3D Ricci scalar and the covariant derivative associated to Y_{ij}. Since we work in Cartesian coordinates, we already used that $\det Y = 1$. In going from (36) to (37) we took a big leap in the (3 + 1) and conformal decomposition of the 4D Ricci scalar. Some steps can be found in Appendix C. For more details, the interested reader is referred to E. Poisson's book [223] for the (3 + 1) or ADM decomposition [224], and to the appendix of R. Wald's book [225] for the conformal transformation rules.

Now, we could take the variation of (37) with respect to Y_{ij}, having in mind that the result should still be transverse and traceless, and obtain the transverse-traceless spatial component of Einstein Equations. However, this would not be very illuminating. Before taking the variation, let us use cosmological perturbation theory and expand the action. A smart way to decompose the spatial metric and to keep the requirement (35) is to take the exponential matrix, i.e.,

$$Y_{ij} = (e^h)_{ij} = \delta_{ij} + h_{ij} + \frac{1}{2}\delta^{kl}h_{ik}h_{jl} + O(h^3), \tag{38}$$

where $h_{ij} \ll 1$ are the transverse-traceless (tensor) modes and satisfy $\delta^{ij}h_{ij} = \partial^i h_{ij} = 0$. This is used, e.g., in Maldacena's work on non-gaussianities [226]. For the details of the expansion in general gauges see [150]. From now on, spatial indices are contracted with the background spatial metric δ_{ij}, for example $h^{ij}h_{ij} \equiv \delta^{ij}\delta^{kl}h_{ik}h_{jl}$. The inverse metric is then $Y^{ij} = \delta^{ik}\delta^{jl}(e^{-h})_{kl}$. With this expansion, the 3D Ricci scalar up to second order is given by

$$R^{(3)}[e^h] = -\frac{1}{4}\partial_i h_{kl}\partial^i h^{kl} + O(h^3). \tag{39}$$

We stopped at second order since we are only interested in scalar-scalar-tensor interactions. We split the other variables as

$$\Psi = \Psi(t,\mathbf{x}) \quad , \quad \Phi = \Phi(t,\mathbf{x}) \quad , \quad \varphi = \bar{\varphi}(t) + \delta\varphi(t,\mathbf{x}), \tag{40}$$

where $\bar{\varphi}(t)$ is the background solution and Φ, Ψ and $\delta\varphi$ are the perturbations. Using the perturbative expansions (38) and (40) and only selecting the terms with two scalars and one tensor, we arrive at

$$S = \int d^3x\, dt \left\{ \frac{a^3}{8}\dot{h}^{ij}\dot{h}_{ij} - \frac{a}{8}\partial_i h_{kl}\partial^i h^{kl} - 2ah^{ij}\partial_i(\Phi+\Psi)\partial_j\Phi + ah^{ij}\partial_i\Phi\partial_j\Phi + \frac{a}{2}h^{ij}\partial_i\delta\varphi\partial_j\delta\varphi \right\}, \quad (41)$$

where the first two terms correspond to the second order Lagrangian of the tensor modes. By taking the variation with respect to h_{ij} we obtain the equations of motion of induced GWs, namely

$$\ddot{h}_{ij} + 3H\dot{h}_{ij} - a^{-2}\Delta h_{ij} = a^{-2}\mathcal{P}^{ab}{}_{ij}\{-8\partial_a(\Phi+\Psi)\partial_b\Phi + 4\partial_a\Phi\partial_b\Phi + 2\partial_a\delta\varphi\partial_b\delta\varphi\}, \quad (42)$$

where $H = \dot{a}/a$ is the Hubble parameter and $\mathcal{P}^{ab}{}_{ij}$ is the transverse-traceless projector, so that the equation is consistent with the transverse-traceless degrees of freedom. For the moment, it is enough to know that it satisfies the requirements of a transverse-traceless object for both pairs of indices. Also note that in the current case of study, we have no source of anisotropic stress. Thus, for the sake of simplicity and consistency, we momentarily cheat and use the traceless part of the spatial component of the linear Einstein equations in Appendix E, which in the absence of anisotropic stress yields

$$\Psi + \Phi = 0. \quad (43)$$

Note that this is not strictly true in the early universe where neutrinos have a large mean free path and give a tiny contribution to the anisotropic stress [39,40].

Before going to the general solutions of the induced GWs, let us translate the current calculation to the perfect fluid picture. A perfect fluid with energy density ρ and pressure P is described by the following energy-momentum tensor:

$$T_{\mu\nu} = (\rho + P)u_\mu u_\nu + P g_{\mu\nu}, \quad (44)$$

where u^μ is the fluid 4-velocity. The perfect fluid is specified once the equation of state $w = P/\rho$ is given. In the perturbative expansion, one takes $\rho = \bar{\rho} + \delta\rho$ and $u_i = \partial_i v$ where v is the velocity perturbation and we neglected vector modes. The energy momentum tensor (44) has to be compared with the one for the scalar field, which reads

$$T^{\varphi}_{\mu\nu} = \partial_\mu\varphi\partial_\nu\varphi - g_{\mu\nu}\left(\frac{1}{2}\partial_\alpha\varphi\partial^\alpha\varphi + V(\varphi)\right). \quad (45)$$

The perfect fluid description of the scalar field follows from the identification

$$u_\mu = \frac{\partial_\mu\varphi}{\sqrt{-\partial_\alpha\varphi\partial^\alpha\varphi}}. \quad (46)$$

If we simply focus on the spatial component u_i we find that

$$\delta\varphi \leftrightarrow v\sqrt{\rho + P}. \quad (47)$$

Thus, in terms of the perfect fluid description we have that

$$\ddot{h}_{ij} + 3H\dot{h}_{ij} - a^{-2}\Delta h_{ij} = a^{-2}\mathcal{P}^{ab}{}_{ij}\{4\partial_a\Phi\partial_b\Phi + 2(\rho+P)\partial_a v\partial_b v\}. \quad (48)$$

This result coincides with the equations derived in Ref. [38,39]. Note that Equation (48) will slightly change in the case of a multi-perfect fluid system. We expect contributions from the relative velocities of the fluids. For a radiation-matter dominated universe, the source term to induced GWs can be found, e.g., in Ref. [70].

3.2. General Solutions

To solve for the induced GWs, we must solve beforehand the first order equations of motion. Continuing with the scalar field fluctuations $\delta\varphi$, we find that they are related by the momentum constraint to Φ by [183]

$$\delta\varphi = -\sqrt{\frac{2}{\epsilon}}\left(\Phi + \frac{\Phi'}{\mathcal{H}}\right). \tag{49}$$

Thus, we must only solve for the only scalar degree of freedom Φ, which we will do briefly. For the moment, we take a general approach and split Φ into an initial value $\Phi_{\mathbf{k}}$ and a transfer function T_Φ as

$$\Phi(\mathbf{k},\tau) = T_\Phi(k,\tau)\Phi_{\mathbf{k}}\,. \tag{50}$$

If we are considering primordial fluctuations, the initial spectrum for $\Phi_{\mathbf{k}}$ is set on superhorizon scales ($k\tau \ll 1$) by quantum fluctuations during inflation. However, in general, any source of fluctuations to the curvature perturbation Φ will lead to induced GWs. We will discuss this possibility later in Section 6. With Equations (49) and (50) we can formally solve the equations of motion for induced GWs [38,39], which in Fourier space read

$$h''_{\mathbf{k},\lambda} + 2\mathcal{H}h'_{\mathbf{k},\lambda} + k^2 h_{\mathbf{k},\lambda} = \mathcal{S}_{\mathbf{k},\lambda}\,, \tag{51}$$

with

$$\mathcal{S}_{\mathbf{k},\lambda} = 4\int\frac{d^3q}{(2\pi)^3}e^{ij}_\lambda(k)q_iq_j\Phi_{\mathbf{q}}\Phi_{|\mathbf{k}-\mathbf{q}|}f(\tau,q,|\mathbf{k}-\mathbf{q}|)\,, \tag{52}$$

and

$$\begin{aligned} f(\tau,q,|\mathbf{k}-\mathbf{q}|) =& T_\Phi(q\tau)T_\Phi(|\mathbf{k}-\mathbf{q}|\tau) \\ &+ \frac{1+b}{2+b}\left(T_\Phi(q\tau) + \frac{T'_\Phi(q\tau)}{\mathcal{H}}\right)\left(T_\Phi(|\mathbf{k}-\mathbf{q}|\tau) + \frac{T'_\Phi(|\mathbf{k}-\mathbf{q}|\tau)}{\mathcal{H}}\right), \end{aligned} \tag{53}$$

where we wrote ϵ in terms of b using Equations (15) and (16). To derive Equation (51) from Equation (48), we used the Fourier transform for h_{ij} in terms of the polarization tensors $e_{ij}(k)$ which are defined in Appendix D. Then, we used that the projection of e_{ij} with the Fourier transform of the transverse-traceless projector $\tilde{\mathcal{P}}^{ab}{}_{ij}$ is trivial, i.e., $e^{ij}\tilde{\mathcal{P}}^{ab}{}_{ij} = e^{ab}$. Now, with the Green's method we can find a formal solution to (51) given by

$$h_{\mathbf{k},\lambda}(\tau) = \int_{\tau_i}^{\tau} d\tilde{\tau}\, G_h(\tau,\tilde{\tau})\mathcal{S}_{\mathbf{k},\lambda}(\tilde{\tau})\,, \tag{54}$$

where $G_h(\tau,\tilde{\tau})$ is the Green's function of the homogeneous solutions to (51) given in Appendix B. The formal solution (54) is defined such that at the initial time τ_i we have $h_{\mathbf{k},\lambda}(\tau_i) = h'_{\mathbf{k},\lambda}(\tau_i) = 0$. Any GW with primordial origin, i.e., generated during inflation, may be simply added to the solution (54).

Since we are interested in the main observable of the SGWB, that is the GW power, let us compute the 2-point correlation function of the induced GWs which is given by

$$\langle h_\lambda(k,\tau)h_\lambda(k',\tau)\rangle = \int_0^\tau d\tau_1\int_0^\tau d\tau_2 G(\tau,\tau_1)G(\tau,\tau_2)\langle S_\lambda(k,\tau_1)S_\lambda(k',\tau_2)\rangle\,. \tag{55}$$

Here we have neglected any primordial signal but we could add one without further issue. In fact, this was considered in Ref. [88] where they may have a non-trivial cross-correlated spectrum by tensor-scalar-scalar couplings during inflation. We will not pursue this possibility further in this review. Continuing with Equation (55), we can compute the two-point function of the source term by

$$\langle S_\lambda(k,\tau_1)S_\lambda(k',\tau_2)\rangle = 4^2\int\frac{d^3q}{(2\pi)^3}\int\frac{d^3q'}{(2\pi)^3}e^{ij}_\lambda(k)q_iq_je^{ij}_\lambda(k')q'_iq'_j \times f(\tau_1,q,|\mathbf{k}-\mathbf{q}|)f(\tau_2,q',|\mathbf{k}'-\mathbf{q}'|)\langle\Phi_q\Phi_{|\mathbf{k}-\mathbf{q}|}\Phi_{q'}\Phi_{|\mathbf{k}'-\mathbf{q}'|}\rangle\,. \quad (56)$$

We see that the induced GW spectrum depends on the 4-point function of the scalar fluctuations. Such a 4-point function may be decomposed in general into a disconnected, i.e., the product of 2-point functions, and a connected piece [89]. If we consider that the stochastic fluctuations are drawn from a Gaussian distribution, the connected 4-point function vanishes and we are led to[11]

$$\langle\Phi_q\Phi_{|\mathbf{k}-\mathbf{q}|}\Phi_{q'}\Phi_{|\mathbf{k}'-\mathbf{q}'|}\rangle = \frac{2\pi^2}{q^3}\mathcal{P}_\Phi(q)\frac{2\pi^2}{|\mathbf{k}-\mathbf{q}|^3}\mathcal{P}_\Phi(|\mathbf{k}-\mathbf{q}|) \times(2\pi)^6\delta^{(3)}(\mathbf{q}+\mathbf{q}')\delta^{(3)}(\mathbf{k}+\mathbf{k}'-\mathbf{q}-\mathbf{q}')+(\mathbf{q}\leftrightarrow\mathbf{k}-\mathbf{q})\,. \quad (57)$$

We discuss the case of mildly non-Gaussian fluctuations in Section 3.3. With further manipulations of Equations (55) and (8), integrating one internal momentum using a Dirac delta and writing the remaining integral in spherical (momentum) coordinates, we arrive at

$$\overline{\mathcal{P}_h} = 8\int_0^\infty dv\int_{|1-v|}^{1+v}du\left(\frac{4v^2-(1-u^2+v^2)^2}{4uv}\right)^2\overline{I^2(\tau,k,v,u)}\mathcal{P}_\Phi(ku)\mathcal{P}_\Phi(kv)\,, \quad (58)$$

where we followed the notation of Kohri and Terada [63]. In Equation (58) we already summed over polarizations (see Appendix D for the details) and we have introduced for convenience

$$v\equiv\frac{q}{k}\quad,\quad u\equiv\frac{|\mathbf{k}-\mathbf{q}|}{k}\,. \quad (59)$$

We also took the oscillation average, i.e., we integrated over half period and divided by π, to take into account that observations of the SGWB actually measure an average over many wavelengths. Furthermore, we have inserted all the time dependence into a "kernel" or "transfer function" defined by

$$I(\tau,k,u,v)\equiv\int_{\tau_i}^{\tau}d\tilde{\tau}\,G(\tau,\tilde{\tau})f(\tilde{\tau},k,u,v)\,. \quad (60)$$

The (averaged) power spectrum (58) is the main quantity needed for the calculations of induced GWs. Knowing (58) we can compute the spectral density of GWs by Equation (7). Note that the induced tensor power spectrum (58) is valid for any curvature perturbation regardless of its origin as long as it is Gaussian. In the next section, we discuss in more detail the case when the fluctuations Φ are generated during inflation.

3.3. Inclusion of Primordial Non-Gaussianity

Primordial fluctuations generated out of quantum fluctuations during inflation are very close to being Gaussian, although not exactly. Small departures of a Gaussian distribution are expected due to gravitational interactions [183,226]. These perturbative departures from the Gaussian distribution are often referred to as Non-Gaussianity (NG).[12] Depending on the formation mechanism, e.g., due to large interactions among fields, we may obtain some significant level of NG. The possibility of observing NG of primordial fluctuations is very exciting, as it might provide information on the particle content during inflation. For this reason, the study of NG is sometimes referred to as cosmological collider physics [227].

For practical convenience, the predictions of primordial fluctuations generated during inflation are often given in terms of the curvature perturbation $\mathcal{R}$. This is because $\mathcal{R}$ is a non-linearly conserved quantity on superhorizon scales [228]. This makes it easy to set well-defined initial conditions on superhorizon scales for whatever scalar variable one might want to consider, independently of what happened previously (e.g., how inflation ended). For instance, since we are working in the Newtonian gauge, we have to relate $\mathcal{R}$ with Φ. It can be shown that on superhorizon scales the linear relation is given by [229]

$$\mathcal{R} = \frac{5+3w}{3(1+w)}\Phi = \frac{2b+3}{b+2}\Phi \,. \tag{61}$$

In the case at hand, we are interested on NGs generated during inflation, i.e., primordial NG. These primordial NGs are often parametrized by the amplitude and shape of 3-point function or the bispectrum [180,181]. For simplicity, we usually consider only the local type NG which can be expressed as a local perturbative expansion around the Gaussian curvature perturbation $\mathcal{R}^g$ by

$$\mathcal{R}(x) = \mathcal{R}^g(x) + \frac{3}{5} f_{NL} (\mathcal{R}^g(x))^2 \,, \tag{62}$$

where the factor 3/5 is by convention.[13] Note, however, that this is just a particular shape of the NG. Currently though, there is no study on the impact of other shapes of NG on the induced GW spectrum. Now, if we go to Fourier space the squared term in (62) becomes a convolution. Then, we can write that for the Φ_q modes the perturbative expansion is given by

$$\Phi_q = \Phi_q^g + F_{NL} \int \frac{d^3 l}{(2\pi)^3} \Phi_l^g \Phi_{|\mathbf{q}-\mathbf{l}|}^g \,, \tag{63}$$

where to avoid unnecessary numerical factors we have introduced

$$F_{NL} \equiv \frac{3}{5}\frac{2b+3}{b+2} f_{NL} \,. \tag{64}$$

We can then expand perturbatively the 4-point function in (56) as

$$\begin{aligned}
&\langle \Phi_q \Phi_{|\mathbf{k}-\mathbf{q}|} \Phi_{q'} \Phi_{|\mathbf{k}'-\mathbf{q}'|} \rangle = \langle \Phi_q^g \Phi_{|\mathbf{k}-\mathbf{q}|}^g \Phi_{q'}^g \Phi_{|\mathbf{k}'-\mathbf{q}'|}^g \rangle \\
&+ F_{NL}^2 \int \frac{d^3 l}{(2\pi)^3} \int \frac{d^3 l'}{(2\pi)^3} \langle \Phi_q^g \Phi_l^g \Phi_{|\mathbf{k}-\mathbf{q}-\mathbf{l}|}^g \Phi_{q'}^g \Phi_{l'}^g \Phi_{|\mathbf{k}'-\mathbf{q}'-\mathbf{l}'|}^g \rangle \\
&+ (|\mathbf{k}-\mathbf{q}| \leftrightarrow q) + (|\mathbf{k}'-\mathbf{q}'| \leftrightarrow q') + (q \leftrightarrow q'; |\mathbf{k}-\mathbf{q}| \leftrightarrow |\mathbf{k}'-\mathbf{q}'|) \\
&+ F_{NL}^2 \int \frac{d^3 l}{(2\pi)^3} \int \frac{d^3 l'}{(2\pi)^3} \langle \Phi_l^g \Phi_{|\mathbf{q}-\mathbf{l}|}^g \Phi_{l'}^g \Phi_{|\mathbf{k}-\mathbf{q}-\mathbf{l}'|}^g \Phi_{q'}^g \Phi_{|\mathbf{k}'-\mathbf{q}'|}^g \rangle \\
&+ (q \leftrightarrow q'; |\mathbf{k}-\mathbf{q}| \leftrightarrow |\mathbf{k}'-\mathbf{q}'|) + O(F_{NL}^4) \,.
\end{aligned} \tag{65}$$

The first term on the right hand side of (65) is already computed in Equation (57). If we focus on the second line in the NG contribution of (65), we see that there are 6 possible non-vanishing Wick contractions in the first 6 point function of Φ^g which, together with the 4 remaining possibilities of the third line, makes 24 terms for the first NG contribution. In the fourth line of (65) there are 8 possible non-vanishing contractions, which make 16 possibilities counting the fifth line. The total is then 40 non-vanishing contractions.[14] The NG contribution can be further classified into three different terms following the notation of Ref. [92]: the "H", the "C", the "Z". Using Equation (65) in (8) and (55), these NG contributions to the induced tensor spectrum can be respectively written as

$$\overline{\mathcal{P}_h^{\rm H}} = 2^5\, F_{NL}^2 k^3 \sum_\lambda \int \frac{d^3q}{2\pi} \left(e_\lambda^{ij}(k) q_i q_j\right)^2 \overline{I^2(\tau, q, |\mathbf{k}-\mathbf{q}|)} \\ \times \frac{\mathcal{P}_\Phi(q)}{q^3} \int \frac{d^3l}{2\pi} \frac{\mathcal{P}_\Phi(l)}{l^3} \frac{\mathcal{P}_\Phi(|\mathbf{k}-\mathbf{q}-\mathbf{l}|)}{|\mathbf{k}-\mathbf{q}-\mathbf{l}|^3}, \tag{66}$$

$$\overline{\mathcal{P}_h^{\rm C}} = 2^6 F_{NL}^2 k^3 \sum_\lambda \int \frac{d^3q}{2\pi} \int \frac{d^3l}{2\pi} e_\lambda^{ij}(k) q_i q_j e_\lambda^{ij}(k) l_i l_j \overline{I(\tau, q, |\mathbf{k}-\mathbf{q}|) I(\tau, l, |\mathbf{k}-\mathbf{l}|)} \\ \times \frac{\mathcal{P}_\Phi(|\mathbf{k}-\mathbf{l}|)}{|\mathbf{k}-\mathbf{l}|^3} \frac{\mathcal{P}_\Phi(l)}{l^3} \frac{\mathcal{P}_\Phi(|\mathbf{q}-\mathbf{l}|)}{|\mathbf{q}-\mathbf{l}|^3}, \tag{67}$$

and

$$\overline{\mathcal{P}_h^{\rm Z}} = 2^6\, F_{NL}^2 k^3 \sum_\lambda \int \frac{d^3q}{2\pi} \int \frac{d^3l}{2\pi} e_\lambda^{ij}(k) q_i q_j e_\lambda^{ij}(k) l_i l_j \overline{I(\tau, q, |\mathbf{k}-\mathbf{q}|) I(\tau, l, |\mathbf{k}-\mathbf{l}|)} \\ \times \frac{\mathcal{P}_\Phi(q)}{q^3} \frac{\mathcal{P}_\Phi(l)}{l^3} \frac{\mathcal{P}_\Phi(|\mathbf{k}-\mathbf{q}-\mathbf{l}|)}{|\mathbf{k}-\mathbf{q}-\mathbf{l}|^3}. \tag{68}$$

These derived formulas agree with those of Ref. [92]. We will assume that we are in the perturbative regime, so that the NG contributions (66) and (67) should be thought of as a correction to the Gaussian contribution (58). The first contribution (66) is also referred to as "hybrid" [90]. Additionally, note that the second line in (66) can be reabsorbed into a redefinition of the primordial power spectrum to include non-Gaussianities. For instance, we could write

$$\mathcal{P}_\mathcal{R}^{NL}(q) = \mathcal{P}_\mathcal{R}(q) + F_{NL}^2 q^3 \int \frac{d^3l}{2\pi} \frac{\mathcal{P}_\mathcal{R}(l)}{l^3} \frac{\mathcal{P}_\mathcal{R}(|\mathbf{q}-\mathbf{l}|)}{|\mathbf{q}-\mathbf{l}|^3}, \tag{69}$$

which follows from the computation of $\langle\Phi(k)\Phi(k')\rangle$ using Equation (63). This is the term considered by Cai, Pi and Sasaki [89] and acts as a local redefinition of the primordial spectrum. With the form of Equation (69) it is much easier to compute the induced GW spectrum up to $O(F_{NL}^4)$ since it just accounts for a replacement of $\mathcal{P}_\mathcal{R}$ in (58) for $\mathcal{P}_\mathcal{R}^{NL}$.The second contribution (67) was first considered by Unal [90] and the third (68) by Ref. [81]. Unal also considered the terms of $O(F_{NL}^4)$ [90]. Most recently, Adshead, Lozanov and Weiner [92] presented a detailed and complete analysis of all the contributions up to $O(F_{NL}^4)$. The interested reader is referred to Ref. [92] for the detailed formulas. We discuss the main features of the NG corrections in Section 5. It should be noted that the effect of primordial NG might compete with the terms due to non-linearities, e.g., from studied third order contributions to the source term of induced GWs. The reader is referred to Refs. [143,179] for the details on the third order expansion. Lastly, it should also be noted that for simplicity we assumed a scale invariant f_{NL} in (62). However, in general situations one expects that f_{NL} is scale dependent. For a recent study on the impact of the scale dependence see Ref. [93].

4. Analytical Transfer Functions

In general situations the kernel (60) and the power spectrum (58) have to be computed numerically. The problem is that the time integral in (60) is extremely demanding for $k\tau \gg 1$, as the integrand is the product of three oscillating functions with frequencies that depend on the tensor and scalar mode wavenumbers. In fact, $k\tau \gg 1$ is the range of scales we are interested in. If we extend the integral until today, say at $\tau = \tau_0 \propto H_0^{-1}$, and we look at scales much smaller than the horizon, we trivially have $k\tau_0 \gg 1$. However, there is more we can learn. For instance, scales accessible to future GW detectors range from $k \sim 10^7$ to $10^{18}\,\mathrm{Mpc}^{-1}$. These very small scales entered the horizon much before

BBN, which roughly correspond to the time when modes with $k \sim 10^3\,\mathrm{Mpc}^{-1}$ entered the horizon.[15] Luckily, this implies that we can evaluate the kernel (60) some time prior to BBN when induced GWs are really inside the horizon and in a radiation dominated universe. Then, from that moment on, we may simply treat these induced GWs as a radiation fluid with $w = 1/3$. Note that we will also consider the case that the modes of interest enter the horizon in an epoch where $w \neq 1/3$. In that case, we must follow induced GWs until the universe transitions to the standard radiation dominated era. We will provide more details later.

Surprisingly, there are relevant regimes where the integrals in (60) can be done analytically. This is the case of a constant equation of state parameter w and constant propagation speed of fluctuations c_s. For example, this includes a perfect fluid whether it is made of a scalar field[16] or an adiabatic perfect fluid, which are commonly used in cosmology. Even when there are transitions between one perfect fluid domination to another, there will be periods in which w and c_s are almost constant. Thus, the constant w and c_s limit is a good approximation in a cosmological set up if the modes of interest enter the horizon outside the transition periods. With these assumptions, we present the solutions to first order cosmological perturbations in the Newtonian gauge in Section 4.1 and derive a very simple form of the source term (53) for the induced GWs. In the general situation where $w \neq 1/3$, the integral (60) can be analytically done for modes which either enter the horizon before or after the transition. In this way, in Sections 4.2 and 4.3 we respectively derive the kernel for modes which are subhorizon and superhorizon before the transition. When then follow our solutions to the radiation dominated epoch.

Main references: The whole section is a generalisation of Refs. [64,74] to account for not only general constant w but general constant c_s. The notation used here improves and unifies those of [64,74]. The main base of Refs. [64,74] was started in [39,63] in the radiation dominated universe.

4.1. First Order Solutions

At the moment, the formal solution (54) for the induced tensor modes does not tell us much. To extract some meaningful information, we have to solve the equations of motion for first order perturbations, which are given in detail in Appendix E. After some simplifications, the equation for the Newtonian potential for general w and c_s and in the absence of isocurvature[17] fluctuations reads[18]

$$\Phi'' + (2\epsilon - \eta)\mathcal{H}\Phi' - \left(\eta + 2s\left(1 + \epsilon - \eta - 2s + \frac{\dot{s}}{Hs}\right)\right)\mathcal{H}^2\Phi + c_s^2k^2\Phi = 0\,, \qquad (70)$$

where we have also introduced ϵ as in Equation (15) as well as

$$\eta \equiv \frac{\dot{\epsilon}}{H\epsilon} \quad \text{and} \quad s \equiv \frac{\dot{c}_s}{Hc_s}\,. \qquad (71)$$

The notation ϵ, η and s is typical of inflationary models and simplifies the form of the equations. One may of course write Equation (70) in a more conventional notation, e.g., that in Mukhanov's book [229] which for constant c_s reads

$$\Phi'' + 3\mathcal{H}\left(1 + c_w^2\right)\Phi' + \left(2\mathcal{H}' + (1 + 3c_w^2)\mathcal{H}^2\right)\Phi + c_s^2k^2\Phi = 0\,, \qquad (72)$$

where $c_w^2 = \dot{P}/\dot{\rho}$ and $c_s^2 = \delta P/\delta\rho$. In the case of an adiabatic perfect fluid one has $c_s^2 = c_w^2 = w$. In the case of a canonical scalar field we have $c_w^2 = w$ and $c_s^2 = 1$. As we already mentioned, for analytical viability we are mostly interested in the case of constant w and c_s. This requirement in turn implies $\eta = s = 0$, so that the Newtonian potential in Equation (70) behaves as a massless field. The solution to (70) for general c_s and w is given in terms of Bessel functions of the first and second kind, explicitly

$$\Phi(k\tau) = (c_s k\tau)^{-b-3/2}(C_1 J_{b+3/2}(c_s k\tau) + C_2 Y_{b+3/2}(c_s k\tau)), \tag{73}$$

where b in terms of w is defined in Equation (16). Now, we must give some kind of initial conditions to Φ. If the primordial spectrum was generated by quantum fluctuations during inflation, the initial conditions are well-defined and constant on super(sound)horizon scales, that is when $c_s k\tau \ll 1$. By picking up the asymptotically constant term in (73), we find that $C_2 = 0$ and

$$\Phi(k\tau) = \Phi_{\mathbf{k}}\, 2^{b+3/2}\Gamma[b+5/2](c_s k\tau)^{-b-3/2} J_{b+3/2}(c_s k\tau), \tag{74}$$

where $\Phi_{\mathbf{k}}$ is the primordial (stochastic) value set by inflation. Note that the general solution (73) is not valid for $c_s = 0$ where the gradient term in (70) is absent. Nevertheless, the solution for constant w is just $\Phi = \Phi_{\mathbf{k}}$ on all scales. The case of $c_s = 0$ is quite particular as the fact that Φ does not decay implies a constant source to induced GWs. This makes the detailed dynamics of the transition to radiation domination very important for the final GW spectrum [65,66], when the source term experiences drastic changes. We dedicate Section 6 to this particular case.

With an exact analytical formula for the first order perturbations, we can attempt to compute the resulting induced GWs. For that, we first need to know the two independent homogeneous solutions to the tensor mode Equation (51), say h_1 and h_2. For constant w they are given as well in terms of Bessel functions with one order less than in Φ, concretely

$$h_1(k\tau) = (k\tau)^{-b-1/2} J_{b+1/2}(k\tau) \quad \text{and} \quad h_2(k\tau) = (k\tau)^{-b-1/2} Y_{b+1/2}(k\tau). \tag{75}$$

In Equation (75), h_1 and h_2 respectively are the growing and decaying modes. Note that there is no c_s since in general relativity tensor modes propagate at the speed of light. This yields a Green's function (see Appendix B) given by

$$G(\tau, \tilde{\tau}) = \frac{\pi}{2k}\frac{(k\tilde{\tau})^{b+3/2}}{(k\tau)^{b+1/2}}(J_{b+1/2}(k\tilde{\tau})Y_{b+1/2}(k\tau) - J_{b+1/2}(k\tau)Y_{b+1/2}(k\tilde{\tau})). \tag{76}$$

This finishes the list of necessary ingredients for our induced GWs calculations. In passing, since there always appears the combination $k\tau$ or $k\tilde{\tau}$ we introduce the useful notation

$$x \equiv k\tau, \tag{77}$$

which we will use from now on. Additionally, a tilde on x implies a tilde on τ, i.e., $\tilde{x} = k\tilde{\tau}$.

Thus far it seems that we may have to deal with several integrals containing at least three Bessel functions, which is challenging to do analytically in general. However, the source term to the induced GWs ends up acquiring a simple enough form after several simplifications. For instance, using the Bessel functions properties (see Appendix F), we find that Equation (53) reduces to

$$\begin{aligned} f(x,u,v) =& \frac{2^{2b+3}\Gamma^2[b+5/2]}{(2b+3)(b+2)}(c_s x)^{-2b-1}(uv)^{-b-1/2} \\ &\times \left(J_{b+1/2}(c_s vx)J_{b+1/2}(c_s ux) + \frac{b+2}{b+1}J_{b+5/2}(c_s vx)J_{b+5/2}(c_s ux)\right), \end{aligned} \tag{78}$$

where we used x from (77) and u and v from (59). To arrive at (78) one needs to write the Bessel function $J_{b+3/2}$ that appears in Φ in a combination of $J_{b+1/2}$ and $J_{b+5/2}$ which can be found in Appendix F. So far, it is not entirely clear whether there is any physical reason behind the simplification (78) which turns out to render the source term as a sum of the squared of Bessel functions of the same order. This simple form of the source term is crucial to obtain analytical formulas for general b and c_s that we derive below.

Replacing (77) and (78) into Equation (60) we find that the kernel can be written as

$$I(x,u,v) = \pi 4^b \Gamma^2[b+3/2]\frac{2b+3}{b+2}(c_s^2 uvx)^{-b-1/2}\left(J_{b+1/2}(x)\mathcal{I}_Y - Y_{b+1/2}(x)\mathcal{I}_J\right), \quad (79)$$

where we defined for compactness

$$\mathcal{I}_{J/Y} \equiv \int_0^x d\tilde{x}\,\tilde{x}^{1/2-b}\left\{\begin{matrix} J_{b+1/2}(\tilde{x}) \\ Y_{b+1/2}(\tilde{x}) \end{matrix}\right\}$$
$$\times\left(J_{b+1/2}(c_s v\tilde{x})J_{b+1/2}(c_s u\tilde{x}) + \frac{b+2}{b+1}J_{b+5/2}(c_s v\tilde{x})J_{b+5/2}(c_s u\tilde{x})\right). \quad (80)$$

Unfortunately, we are not aware of a general analytical formula for the time integrals (80) unless $b \in \mathbb{Z}$, in which case the Bessel functions reduce to spherical Bessel functions and can be written in terms of transcendental functions. This is for example the case of radiation domination $b = 0$ ($w = 1/3$), where the general kernel can be found in Ref. [63]. Nevertheless, we can integrate (80) in two relevant limits, $x \gg 1$ (subhorizon) and $x \ll 1$ (superhorizon).

Let us explain the assumptions needed for the approximations before jumping into the actual calculations. We are considering that right after inflation there is a period when the universe has a free b (or w) and free c_s. However, to recover the standard hot big bang cosmology, such an epoch must end, and the universe must enter a radiation domination stage. Thus, whatever solution we find in this section for the induced GWs must be followed until we reach the radiation dominated stage. For simplicity, we will assume an instantaneous transition. Then, we will refer to the time of transition as "reheating" time $\tau_{\rm rh}$ and the comoving size of the horizon at that time as "reheating" scale $k_{\rm rh}$. The value of $k_{\rm rh}$ in terms of the "reheating" temperature $T_{\rm rh}$ can be found in Equation (32). The assumption of an instantaneous transition might be important for those tensor modes with wavenumber $k \sim k_{\rm rh}$. Namely, those modes which entered the horizon around the transition. If the transition is gradual, then our approximations will not be very good for modes which entered during the last stages of the transition. In any case, as we shall see, the instantaneous approximation gives a very good idea of the shape of the induced GW spectrum even for $k \sim k_{\rm rh}$. On top of that, we will assume that the scale corresponding to the peak of the primordial spectrum enters the horizon well before reheating so that during and after reheating there is no significant source of induced GWs. In the case of a top-hat primordial spectrum, this means that the scale corresponding to the IR cut-off enters the horizon much before reheating. Under these two assumptions, i.e., sudden transition and "localized" primordial spectrum, we can proceed to derive our analytical formulas. Note that this does not apply to the case when $c_s = 0$ which we discuss in Section 6.

4.2. General Subhorizon Kernel

Let us start with the simplest situation. This corresponds to the case when all modes of interest entered the horizon much before reheating, that is $k \gg k_{\rm rh}$. Then, we can safely send the upper integration limit in (80) to infinity, i.e., $x = k\tau \gg 1$. This is because when we eventually match our solution at reheating to a free GW propagating in a radiation dominated universe, the upper limit would be $x_{\rm rh} = k\tau_{\rm rh} \sim k/k_{\rm rh} \gg 1$. The integrals (80) for $x \to \infty$ are derived by Gervois and Navelet [233] and given in Appendix G. Here we just give the form of (79) after integration, which reads

$$
\begin{aligned}
I(x \gg 1, u, v) =& x^{-b-1} 4^{b} \Gamma^{2}[b+3/2] \frac{2b+3}{b+2} \frac{|1-y^{2}|^{b/2}}{c_s^{2} uv} \\
& \times \left\{ \cos\left(x - \tfrac{b\pi}{2}\right) \left(\mathsf{P}_b^{-b}(y) + \frac{b+2}{b+1} \mathsf{P}_{b+2}^{-b}(y) \right) \Theta[c_s(u+v) - 1] \right. \\
& + \frac{2}{\pi} \sin\left(x - \tfrac{b\pi}{2}\right) \left(\mathsf{Q}_b^{-b}(y) + \frac{b+2}{b+1} \mathsf{Q}_{b+2}^{-b}(y) \right) \Theta[c_s(u+v) - 1] \\
& \left. - \frac{2}{\pi} \sin\left(x - \tfrac{b\pi}{2}\right) \left(\mathcal{Q}_b^{-b}(-y) + 2\frac{b+2}{b+1} \mathcal{Q}_{b+2}^{-b}(-y) \right) \Theta[1 - c_s(u+v)] \right\}, \quad (81)
\end{aligned}
$$

where $\Theta[x]$ is the Heaviside function, $\mathsf{P}_b^{-b}(y)$ and $\mathsf{Q}_b^{-b}(y)$ are the Ferrer's function of the first and second kind, valid for $|y| < 1$, and $\mathcal{Q}_b^{-b}(y)$ is the associated Legendre function of the second kind, valid for $|y| > 1$. Their explicit expressions in terms of Hypergeometric functions can be found in Appendix H. We also defined for simplicity

$$
y = 1 - \frac{1 - c_s^2(u-v)^2}{2c_s^2 uv} = -1 - \frac{1 - c_s^2(u+v)^2}{2c_s^2 uv}, \quad (82)
$$

and we Taylor expanded the Bessel functions for large argument. Note that y is related to the area of the triangle made by the three momenta, $c_s|\mathbf{k} - \mathbf{q}|$, $c_s|\mathbf{q}|$ and k. The presence of the Heaviside function in (81) signals the possible resonance we anticipated in Section 2.2 when the wavenumber of a tensor modes equals the sum of two typical scalar modes $2c_s k_p \sim k$. In more practical terms, when $c_s(u+v) \sim 1$ the three Bessel functions in the definite integral (80) for $x \to \infty$ conspire (by rendering their product as coherent oscillations) and yield a divergent integral. Additionally, note that Equation (81) has the same time dependence as a subhorizon tensor mode propagating in a constant b and c_s universe, e.g., see Equation (75). In other words, it decays as a^{-2} and oscillates with a frequency proportional to x. This was expected as the source term for induced GWs decays once inside the horizon and, therefore, induced GWs behave as a freely propagating GW. This does not apply in the case of a dust dominated universe though.

In the current form Equation (81) is general enough so that it can also be used in the C (67) and Z (68) NG terms. However, in this review we are more interested with the Gaussian contribution (58). Thus, we are more concerned about the averaged square kernel. After some simple algebra, that is squaring (81) and calculating the oscillation average, we arrive at

$$
\begin{aligned}
\overline{I^2(x,u,v)} =& x^{-2(b+1)} 4^{2b} \Gamma^{4}[b+3/2] \left(\frac{2b+3}{b+2}\right)^{2} \frac{|1-y^{2}|^{b}}{2c_s^{4} u^{2} v^{2}} \\
& \times \left\{ \left(\mathsf{P}_b^{-b}(y) + \frac{b+2}{b+1} \mathsf{P}_{b+2}^{-b}(y) \right)^{2} \Theta[c_s(u+v) - 1] \right. \\
& + \frac{4}{\pi^2} \left(\mathsf{Q}_b^{-b}(y) + \frac{b+2}{b+1} \mathsf{Q}_{b+2}^{-b}(y) \right)^{2} \Theta[c_s(u+v) - 1] \\
& \left. + \frac{4}{\pi^2} \left(\mathcal{Q}_b^{-b}(-y) + 2\frac{b+2}{b+1} \mathcal{Q}_{b+2}^{-b}(-y) \right)^{2} \Theta[1 - c_s(u+v)] \right\}. \quad (83)
\end{aligned}
$$

With the averaged kernel (83) we are ready to compute the induced GWs in quite general situations. We are only left with the integral over momenta which can be easily computed numerically. Analytical formulas for the final GW spectrum are discussed in Section 5. Let us note that although (83) might look technically complicated to implement, it is by far more useful than attempting a numerical integration of three Bessel functions. Furthermore, the superhorizon approximation is very good for modes with $k \gg k_{\rm rh}$ regardless of the shape of the primordial spectrum. This is because induced GWs with

$k \gg k_{\rm rh}$ are essentially already free GWs. Thus, soon after horizon re-entry and thereon they will behave as relativistic particles in an expanding universe. This also means that we can evaluate (7) and (83) at the instantaneous reheating time, i.e., $\tau = \tau_{\rm rh}$, to estimate the observable spectral density of induced GWs. We will do this shortly. If the transition is gradual, the kernel (83) is a good approximation for modes which enter during a period with almost constant b.

In the next subsections, we will have a closer inspection to the kernel (83). After matching to radiation domination, we will look at the behaviour of the kernel near the resonant point and in the IR tail.

4.2.1. Matching to Radiation Domination

Let us discuss the matching to radiation domination in more technical terms. After matching the background quantities at $\tau = \tau_{\rm rh}$, that is the scale factor a and the Hubble parameter $\mathcal{H}$ we find that a and $\mathcal{H}$ in the radiation dominated period follows the typical solution (found in (16) with $w = 1/3$) with a shifted conformal time, which reads

$$\bar{\tau} \equiv \tau - \frac{b}{1+b}\tau_{\rm rh}\,, \tag{84}$$

where b is related to the equation of state of the previous epoch by (16). Now, for the matching of tensor modes it is important to realize that the continuity of h_{ij} and its first derivative imply the continuity of the kernel (60) and its first derivative. Thus, from now on we simply deal with the matching of the kernel (60). Since we are dealing with subhorizon scales we have that the kernel before the transition is

$$I(x \gg 1, u, v) \approx x^{-b-1}\left(A_{1,b}\sin\left(x - \frac{b\pi}{2}\right) + A_{2,b}\cos\left(x - \frac{b\pi}{2}\right)\right), \tag{85}$$

where the coefficients, not important at the moment, can be extracted from Equation (81). The kernel after the transition is also for subhorizon modes, which must have the form of a freely propagating GW in a radiation dominated universe, namely

$$I_{RD}(x \gg 1, u, v) \approx A_{1,RD}\frac{\sin\bar{x}}{\bar{x}} + A_{2,RD}\frac{\cos\bar{x}}{\bar{x}}\,. \tag{86}$$

It is easy to convince oneself that after matching (85) and (86) at $\tau = \tau_{\rm rh}$ one finds that

$$A^2_{1,RD} + A^2_{2,RD} = \left(\frac{k\tau_{\rm rh}}{1+b}\right)^2\left(A^2_{1,b} + A^2_{2,b}\right). \tag{87}$$

This implies that although we should be calculating

$$\Omega_{\rm GW}(k \gg k_{\rm rh}, \tau \gg \tau_{\rm rh}) = \frac{k^2}{12a^2H^2}\overline{\mathcal{P}_h^{RD}}\,, \tag{88}$$

with

$$\overline{\mathcal{P}_h^{RD}} = 8\int_0^\infty dv \int_{|1-v|}^{1+v} du \left(\frac{4v^2 - (1-u^2+v^2)^2}{4uv}\right)^2 \overline{I^2_{RD}(x \gg 1, v, u)}\mathcal{P}_\Phi(ku)\mathcal{P}_\Phi(kv)\,, \tag{89}$$

it is equivalent to evaluate

$$\Omega_{\rm GW}(k \gg k_{\rm rh}, \tau \gg \tau_{\rm rh}) = \frac{k^2}{12a^2H^2}\overline{\mathcal{P}_h}\Big|_{\tau=\tau_{\rm rh}}\,, \tag{90}$$

with

$$\overline{\mathcal{P}_h} = 8\int_0^\infty dv \int_{|1-v|}^{1+v} du \left(\frac{4v^2-(1-u^2+v^2)^2}{4uv}\right)^2 \overline{I^2(x \gg 1, v, u)}\mathcal{P}_\Phi(ku)\mathcal{P}_\Phi(kv)\,, \quad (91)$$

where the kernel is that given in (83).

It is important to note that our subhorizon approximation is valid for modes with $k\tau_{\rm rh} \gg 1$. Later, we may try to join the subhorizon approximation with the superhorizon approximation valid for $k\tau_{\rm rh} \ll 1$ by extrapolating up to $k\tau_{\rm rh} \sim 1$. Let us call such scale k_m. However, there is an important subtlety in joining the two approximations. If we look at the scale that last crossed the horizon at reheating, namely

$$k_{\rm rh} = \mathcal{H}_{\rm rh} = \frac{1+b}{\tau_{\rm rh}}\,, \quad (92)$$

we see that if we compare with k_m we have that

$$\frac{k_m}{k_{\rm rh}} = \frac{1}{1+b}\,. \quad (93)$$

For $b > 0$ we have that $k_m < k_{\rm rh}$. This means that our subhorizon approximation is even valid for $k \sim k_{\rm rh}$ and, therefore, we expect a good matching with the superhorizon approximation at $k \sim k_{\rm rh}$ (since for $k < k_{\rm rh}$ the modes are superhorizon before the transition). However, for $b < 0$ we find $k_m > k_{\rm rh}$. This time the subhorizon approximation breaks down for $k > k_{\rm rh}$ and so we expect the matching with the superhorizon approximation to be at $k \sim k_m$. This is what we eventually find in Section 4. In addition to the above, we have that the subhorizon approximation has corrections of $O(x^{1+b})$ [74]. So for $b < 0$ the corrections quickly become important for $x \sim 1$. A detailed studied for $b < 0$, perhaps with numerical methods, is needed to clarify the exact shape of the GW spectrum. Nevertheless, our results should be a good order of magnitude estimate.

Let us remind the reader that when implementing (83) inside the tensor spectrum (58), we must bear in mind that if we are considering primordial curvature fluctuations from inflation, we have to rescale the spectrum of Φ accordingly, namely

$$\mathcal{P}_\Phi = \left(\frac{b+2}{2b+3}\right)^2 \mathcal{P}_\mathcal{R}\,. \quad (94)$$

4.2.2. Resonances

Here we focus on the general behaviour near the resonant point for the averaged kernel (83) that we discussed in Section 2.2. We start by noting that we have three vectors, the one from the tensor mode $\mathbf{k}$ and the two scalar modes, $\mathbf{q}$ and $\mathbf{k}-\mathbf{q}$, which of course satisfy momentum conservation. Recall that from Equation (59) q corresponds to v and $|\mathbf{k}-\mathbf{q}|$ to u. This means that our three variables satisfy a triangle inequality, that is

$$|k-q| < |\mathbf{k}-\mathbf{q}| < k+q \quad \text{or} \quad ||\mathbf{k}-\mathbf{q}|-q| < k < |\mathbf{k}-\mathbf{q}|+q\,, \quad (95)$$

which translates into

$$|1-v| < u < 1+v \quad \text{or} \quad |u-v| < 1 < u+v\,. \quad (96)$$

The area A of the triangle made by k, q and $|\mathbf{k}-\mathbf{q}|$ is related to the angle between $\mathbf{k}$ and $\mathbf{q}$ by

$$\sin\theta_k = \frac{2A}{kq}\,. \quad (97)$$

This means that the projection with the polarization tensor (see Appendix D) is proportional to the area, namely

$$e_{\lambda}^{ij} q_i q_j \propto \sin^2 \theta_k \propto A^2 \,. \tag{98}$$

This implies that the saturation of the triangle inequality at $u + v = 1$ corresponds to zero area and, therefore, does not contribute to the integral (58). Now, from the arguments in Section 2.2 we expect a resonance when

$$c_s(u+v) = 1 \,. \tag{99}$$

Thus, if we have $c_s^2 = 1$ the resonant point coincides with the saturation of the triangle inequality and, thus, has vanishing contribution. In this way we see that, by momentum conservation, there is no resonance for $c_s^2 = 1$ in the induced GW spectrum. Nevertheless, for $c_s^2 < 1$ we do have a resonance inside the integration region (see Figure 2).

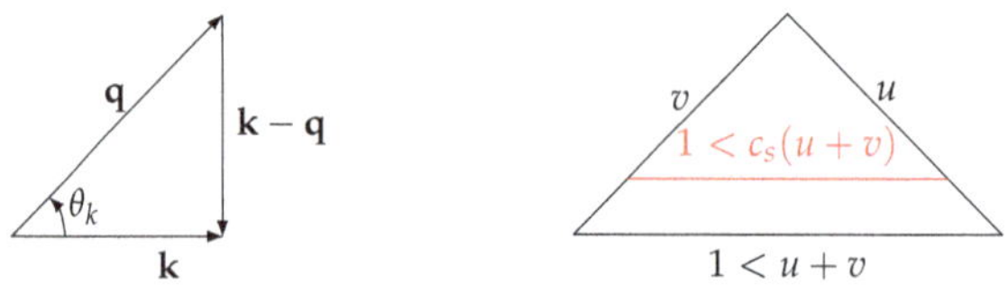

Figure 2. On the left, relation between the three vectors involved in the integrals of induced GWs for the Gaussian part. On the right, illustration of the triangle inequality. We see that when $1 < c_s(u+v)$ and $c_s < 1$ we are below the triangle inequality and the resonance is allowed by momentum conservation.

We can confirm these expectations by expanding the kernel (83) around $c_s(u+v) \sim 1$ which corresponds to $y \sim -1$, where the associated Legendre functions might have a divergence. The asymptotic behaviours around $y \sim -1$ are given in Appendix H. Focusing first on the case $b < 0$, we find that the kernel diverges as

$$\overline{I^2_{\rm res}(x,u,v,b<0)} \approx x^{-2(b+1)} 4^{2b} \Gamma^4[b+5/2] \frac{8}{c_s^4 u^2 v^2} \frac{\csc^2(b\pi)}{\Gamma^2[3+b]} |1-|y||^{2b} \,. \tag{100}$$

This recovers Equation (29) which showed that a very peaked primordial spectrum, so that $u \sim v \sim k_p/k$, leads to

$$\Omega^{\rm res}_{\rm GWs}(b<0) \propto |1-|y||^{2b} \propto |k - 2c_s k_p|^{2b} \,. \tag{101}$$

The case $b = 0$ is explicitly given in Equation (209). In the case when $b > 0$, we arrive at

$$\overline{I^2_{\rm res}(x,u,v,b>0)} \approx x^{-2(b+1)} 4^{3b} \Gamma^4[b+5/2] \frac{32}{\pi^2 c_s^4 u^2 v^2} \left(\frac{(1+b+b^2)}{(b+1)(b+2)} \frac{\Gamma[b]}{\Gamma[2b+3]} \right)^2 , \tag{102}$$

which is independent of y. Thus, as expected there is no resonance for $b > 0$. Nevertheless, as shown in [64] for $1 > b > 0$ there is still a peak at $k = 2c_s k_p$ in the induced GWs spectrum for a peaked primordial spectrum. We can roughly understand the peak by noting that since $\mathcal{Q}_b^{-b}(y) \sim (1-y)^{-b}$, the next leading order must be $\mathcal{Q}_b^{-b}(y) \sim (1-y)^{-b} + \text{constant}$ since $b < 1$. Then, inserting this into the kernel (83) we find that

$$\Omega^{\rm res}_{\rm GWs}(1<b<0) \propto \text{constant} - |1-|y||^{b} \propto \text{constant} - |k - 2c_s k_p|^{b} \,. \tag{103}$$

For $b > 1$ and $c_s \neq 1$ there is no detailed study of the kernel (83). Nevertheless, by looking at the explicit expression for $b = 2$ (213) we see that there is nothing special at $c_s(u+v) = 1$. Before ending this section, note that the kernel is symmetric around the resonant point $c_s(u+v) = 1$. It displays the same behaviour when approaching from both directions, $c_s(u+v) \to 1^{\pm}$.

4.2.3. Infrared Regime

Here we show the behaviour of the kernel when $u \sim v \gg 1$ which is the relevant regime for the IR tail of the GW spectrum as $v \sim k_p/k$ and $k \ll k_p$. This limit means that, if there is any peak in the spectrum of scalar fluctuations, we are looking at scales much larger than the scale of the peak. Thus, such limit of the kernel (83) is relevant for wavenumbers $k_{\rm rh} \ll k \ll k_p$, where we included $k_{\rm rh}$ since it is the limit of the subhorizon approximation. The cases $b < 0$ and $b > 0$ exhibit different behaviours as anticipated in Section 2.2. In the limit $y \to 1$, the average kernel (83) is respectively given by

$$\overline{I^2_{\rm IR}(x \gg 1, u, v, b < 0)} \approx x^{-2(b+1)} 4^{2b} \Gamma^4[b+5/2] \frac{8}{c_s^4 u^2 v^2} \frac{\csc^2(b\pi)}{\Gamma^2[3+b]} |1-y|^{2b}, \tag{104}$$

for $b < 0$ and

$$\overline{I^2_{\rm IR}(x \gg 1, u, v, b > 0)} \approx x^{-2(b+1)} 4^{3b} \Gamma^4[b+5/2] \frac{32}{\pi^2 c_s^4 u^2 v^2} \left(\frac{(1+b+b^2)}{(b+1)(b+2)} \frac{\Gamma[b]}{\Gamma[2b+3]} \right)^2, \tag{105}$$

for $b > 0$. They are actually very similar to (100) and (102), except for the fact that we are now looking at $u \sim v \gg 1$, so that $1 - y \sim v^{-2}$. Evaluating (104) and (105) at the time of reheating and taking into account the factor k^2 in (58), we see that whatever the IR scaling in the radiation dominated universe (which for localized sources is k^3 [78]) we have to correct the exponent by a factor $-2|b|$, i.e., to $k^{3-2|b|}$ for a localized sourced. This is the same result as in Section 2.2. The absolute value is due to an extended superhorizon growth of tensor modes for $b < 0$, as explained in Section 2.2.

4.3. Superhorizon Kernel Approximation

In the previous subsections we have studied in detail the subhorizon kernel. However, there are modes which are still superhorizon at the time of reheating. Thus, for such modes we cannot take the upper limit in (80) to $x \to \infty$ as $x \ll 1$ at reheating. In addition to the approximation $x \ll 1$, we will use the assumption that the peak in the scalar spectrum $\mathcal{P}_\Phi$ is on scales $k_p \gg k_{\rm rh}$. This means that we are only concerned with the region of integration of (80) around the peak in $\mathcal{P}_\Phi$, where we have that $v \sim k_p/k \gg 1$ and therefore $u \sim v \gg 1$. Using this fact we can expand the first Bessel function for a small argument in (80) and integrate in the limit of $u \sim v$. By doing so we arrive at Equation (20) but for the kernel, namely

$$I(x \ll 1, u, v) \approx B_{1,b} + B_{2,b} x^{-2b}, \tag{106}$$

where the exact coefficients are given by

$$B_{1,b} = -\frac{(3+2b)^2(1+b+b^2)}{4b(1+b)^2(2+b)}(c_s v)^{-2} \quad , \quad B_{2,b} = \frac{4^{1+b}\Gamma^2[b+5/2]}{b(1+b)(2+b)\pi}(c_s v)^{-2(1+b)}. \tag{107}$$

The details on the integrations can be found in Appendix G. Thus, we have found the kernel for modes which are superhorizon before reheating. However, these induced tensor modes are not yet GWs so we must follow them after they enter the horizon in the radiation domination era.

Matching to Radiation Domination

The kernel (106) is valid for superhorizon modes during an arbitrary $b =$ constant period. After this period ends with a sudden reheating, we shall assume that the source term is shut off and tensor modes evolve as freely propagating massless tensor mode. As explained in Section 4.3 the continuity of h_{ij} implies continuity of the kernel. This means that the kernel goes from (106) to the kernel of a superhorizon tensor mode in radiation domination, which reads

$$I_{RD}(\bar{x} \ll 1, u, v) \approx B_{1,RD} + B_{2,RD}(k\bar{\tau})^{-1} \quad \text{with} \quad \bar{\tau} \equiv \tau - \frac{b}{1+b}\tau_{\rm rh}\,, \tag{108}$$

where we introduced $\bar{\tau}$ from the continuity of the background FLRW metric, i.e., we imposed continuity of a and $\mathcal{H}$ at $\tau = \tau_{\rm rh}$. Matching (108) and its first derivatives with (106), we find that

$$B_{1,RD} = B_{1,b} + \frac{1-b}{1+b}B_{2,b}(k\tau_{\rm rh})^{-2b} \quad , \quad B_{2,RD} = \frac{2b}{(1+b)^2}B_{2,b}(k\tau_{\rm rh})^{1-2b}\,. \tag{109}$$

Thus, we have effectively continued the superhorizon kernel from the $b =$ constant era to the radiation era. We can now follow the kernel down to subhorizon scales, which must behave as a tensor mode in a radiation dominated universe as there is no source, that is Equation (75) with $b = 0$. Then, taking the averaged square kernel yields

$$\overline{I^2_{RD}(k \ll k_{\rm rh}, \tau \gg \tau_{\rm rh})} \approx \frac{1}{2\bar{x}^2}\left(B^2_{1,RD} + B^2_{2,RD}\right). \tag{110}$$

After picking up the most relevant contributions, that is the leading terms for $v \gg 1$, we arrive at

$$\begin{aligned}\overline{I^2_{RD}(k \ll k_{\rm rh}, \tau \gg \tau_{\rm rh})} \approx \frac{1}{2\bar{x}^2}\left(\frac{(3+2b)^2}{4b(1+b)^2(2+b)}\right)^2 \\ \times \left((1+b+b^2)(c_s v)^{-2} - 4^{1+b}\Gamma^2[b+3/2]\frac{(1-b)}{\pi}(c_s v)^{-2(1+b)}\left((1+b)\frac{k}{k_{\rm rh}}\right)^{-2b}\right)^2.\end{aligned} \tag{111}$$

This is the kernel squared that should be used in the induced GW spectrum (58) for scales $k \ll k_{\rm rh}$. In this way, the kernel (111) yields the right IR scaling for localized sources in a radiation dominated universe, i.e., $\Omega^{\rm IR}_{\rm GWs} \propto k^3$ [78]. The matching between the superhozion (111) and subhorizon (83) approximations at $k \sim k_{\rm rh}$ has shown to be very good in the instantaneous reheating case [74]. Note that for a gradual transition, the superhorizon approximation should still be good enough except for those modes which enter during the transition. We can then calculate the spectral density of induced GWs by

$$\Omega_{\rm GW}(k \gg k_{\rm rh}, \tau \gg \tau_{\rm rh}) = \frac{k^2}{12a^2H^2}\overline{\mathcal{P}^{RD}_h}\,, \tag{112}$$

with

$$\overline{\mathcal{P}^{RD}_h} = 8\int_0^\infty dv \int_{|1-v|}^{1+v} du \left(\frac{4v^2 - (1-u^2+v^2)^2}{4uv}\right)^2 \overline{I^2_{RD}(x \gg 1, v, u)}\mathcal{P}_\Phi(ku)\mathcal{P}_\Phi(kv)\,, \tag{113}$$

and using Equation (111). Nevertheless, it is worth noting that for a Dirac delta or a fairly sharp peak and $b > 0$ a good approximation to the full induced GW spectrum is to stop the subhorizon spectral density (90) at $k \sim 3/4k_{\rm rh}$ and match with a power-law $\Omega_{\rm GW} = \Omega_{\rm GW,rh}(4k/3k_{\rm rh})^2$ [74]. The amplitude of $\Omega_{\rm GW,rh}$ can be found by matching the two.

It should be noted that Equation (111) does not recover the logarithmic correction typical of induced GWs in a radiation dominated universe [79]. The main reason is that we assumed the source term to stop at reheating and, therefore, that superhorizon tensor modes at that time behave as free tensor modes. This is a good approximation taken for simplicity. However, the precise matching would require to first match the Newtonian potential Φ from a b = constant universe to radiation domination and then continue the kernel (60) after reheating with the matched Φ. This would give a more accurate calculation and would most likely recover the logarithmic correction in the IR tail.

5. Typical Induced GW Spectra

In Section 4 we derived analytical approximations for the kernel (or transfer functions) of induced GWs. However, we still have the integral over the scalar momenta. While the kernel was not so sensitive to the shape of the primordial spectrum, i.e., the approximation is valid as long as all the relevant primordial spectrum enters the horizon before reheating, the integral over momenta crucially depends on the shape of the primordial spectrum. In this section, we will present the final form of the induced GW spectrum and discuss their main features. We will not be so concerned in the series of approximations to achieve analytical formulas of the GW spectrum, which can be found in the papers. Before doing so, we will carefully derive in Section 5.1 what is the spectral density of induced GWs measured today. Then we will focus on the simplest types of motivated primordial spectra. These are a very sharp peak Section 5.2, a log-normal peak Section 5.3, a broken power-law Section 5.4 and a scale invariant spectrum Section 5.4.2. We will also comment on possible oscillatory features Section 5.5 on top of the basic shapes and the impact of primordial NG on the final GW spectrum Section 5.6.

Main references: Although there are many studies on the induced GW spectrum for various primordial spectra, we focus on the cases where analytical formulas have been derived. In this way, Section 5.1 can be derived using the formulas in Appendix A and can also be found in [64,74]. Section 5.2 is mainly based on [74]. Section 5.3 follows [74,82]. Then, Section 5.4 is based on [81]. Section 5.5 briefly describes the results of Refs. [84,87]. Lastly, Section 5.6 reports the main findings from Refs. [89,90,92].

5.1. The GW Spectral Density Today

To find the power of GWs that we would observe we have to evaluate (7) today, namely

$$\Omega_{\rm GW,0} = \frac{1}{3M_{\rm pl}^2 H_0^2}\frac{d\rho_{\rm GW,0}}{d\ln k}\,. \tag{114}$$

However, our computations in Section 4 were done up to and including the transition to a radiation dominated universe. Since deep inside the horizon we have that $\rho_{\rm GW} \propto \rho_r \propto a^{-4}$ then after reheating we have that $\Omega_{\rm GW} = \Omega_{\rm GW,rh}$ = constant. To relate $\Omega_{\rm GW,rh}$ with $\Omega_{\rm GW,0}$ we can make use of the fact that GWs behave as radiation and write

$$\begin{aligned}\Omega_{\rm GW,0}h^2 &= \Omega_{r,0}h^2\frac{1}{\rho_{r,0}}\frac{d\rho_{\rm GW,0}}{d\ln k}\\ &= 1.62\times 10^{-5}\left(\frac{\Omega_{r,0}h^2}{4.18\times 10^{-5}}\right)\left(\frac{g_*(T_{\rm rh})}{106.75}\right)\left(\frac{g_{*s}(T_{\rm rh})}{106.75}\right)^{-4/3}\Omega_{\rm GW,rh}\,.\end{aligned} \tag{115}$$

where $\Omega_{r,0}h^2 \approx 4.18\times 10^{-5}$ is the density fraction of radiation today given by Planck [2]. We also traced $\rho_{r,0}$ and $\rho_{\rm GW,0}$ back to the reheating time taking into account that the effective degrees of freedom change with temperature. In the case that the induced GWs are generated during a radiation domination era we should replace the subscript "rh" for the evaluation at a time when the induced tensor modes start to behave as a GW. This is denoted with a subscript "c" by Inomata et al. [126].

5.2. Dirac Delta Peak

The simplest case is that of a very sharp peak in the primordial spectrum. While this type of spectrum is not possible to generate during single field inflation [103,109,234], it may be achieved in multi-field models of inflation [95,235–241,243]. If the peak is extremely sharp, it is often modelled by a Dirac delta as

$$\mathcal{P}_{\mathcal{R}}(k) = \mathcal{A}_{\mathcal{R}}\delta(\ln(k/k_p)) \,. \tag{116}$$

Although this might be an unrealistic situation, it captures the essence of sharp peaks and gives insight on the kernel. We discuss a more realistic situation in Section 5.3. Now, for the power spectrum inside the integral (58) we have

$$\mathcal{P}_{\mathcal{R}}(vk) = \mathcal{A}_{\mathcal{R}}\delta(\ln(vk/k_p)) = \mathcal{A}_{\mathcal{R}}\frac{k_p}{k}\delta\left(v - \frac{k_p}{k}\right), \tag{117}$$

and similarly for u. Recall, however, that the integral over u is bounded by momentum conservation to $|1-v| < u < 1+v$. So, when we integrate over v and u using the Dirac delta, the range is now $|k-k_p| < k_p < k+k_p$. This translates to a range in k given by $0 < k < 2k_p$. After taking this finite range of k into consideration, we may simply evaluate at $v = u = k_p/k$ the integrand (89) (and (113)) and insert it into (88) (and (112)) to arrive at

$$\Omega_{\rm GW,rh}(k) = \frac{2}{3}\left(\frac{k_p}{k_{\rm rh}}\right)^2\left(1 - \frac{k^2}{4k_p^2}\right)^2 \overline{I^2_{RD}(k/k_{\rm rh}, k/k_p)}\Theta(2k_p - k) \,. \tag{118}$$

Note that we are being general by allowing a $b =$ constant phase before the standard radiation domination. Since we are assuming an instantaneous transition at $\tau = \tau_{\rm rh}$ from the $b =$ constant to radiation domination, we have to include the scale $k_{\rm rh}$ corresponding to the last scale that entered the horizon at the transition. In more detail, $k_{\rm rh}$ is related to $\tau_{\rm rh}$ by

$$k_{\rm rh} = \mathcal{H}_{\rm rh} = \frac{1+b}{\tau_{\rm rh}} \,. \tag{119}$$

Thus, we have that $x_{\rm rh} = k\tau_{\rm rh} = (1+b)k/k_{\rm rh}$. If we are considering the induced GWs in a radiation dominated universe we may simply replace $k_{\rm rh} \to k_p$ and recover the standard formula [63]. In Figures 3 and 4 we see the GW spectrum for different values of b and respectively for $c_s^2 = w$ and $c_s^2 = 1$, so that the resonance is evident.

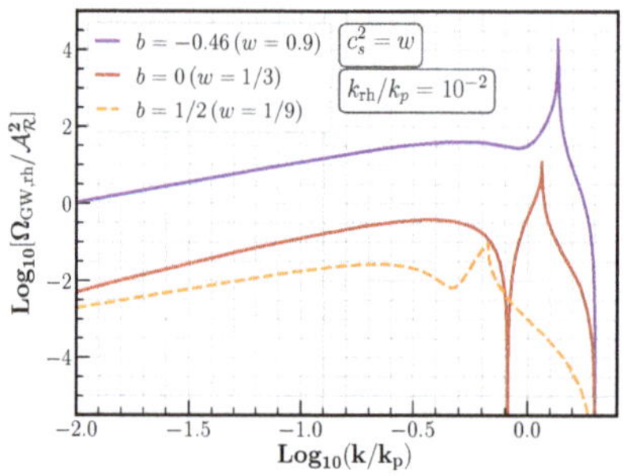

Figure 3. Induced GW spectral density for a Dirac delta primordial spectrum, in terms of wavenumber k/k_p. We restricted the plot to $k > k_{\rm rh}$. We show the induced GW generated in a radiation dominated universe (red line), in a stiff fluid domination with $w \sim 0.9$ (purple line) and in a soft fluid domination with $w = 1/9$ (orange line). All cases have $c_s^2 = w$ and, thus, the spectrum presents the resonant peak at $k \sim 2c_s^2 k_p$. We see that for softer w, the position of the peak moves to the left. We also see that stiff w enhances the overall amplitude of the induced GW spectrum and has a sharper resonant peak.

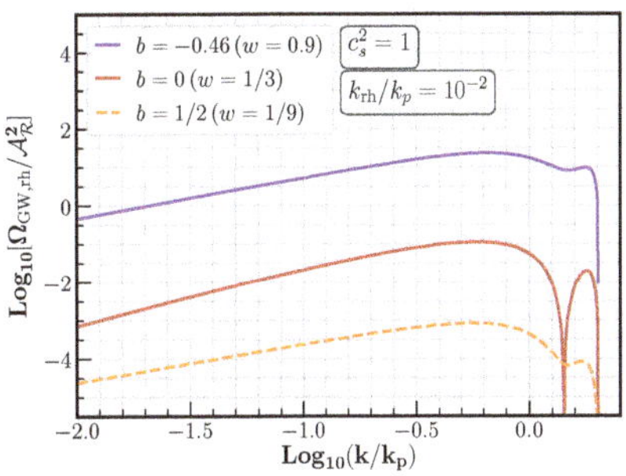

Figure 4. Induced GW spectral density for a Dirac delta primordial spectra, in terms of wavenumber k/k_p. This time we chose $c_s^2 = 1$ and, therefore, the resonant peaks of Figure 3 are absent. We show in red $w = 1/3$, in purple $w \sim 0.9$ and in orange $w = 1/9$. We see how the stiffer the w, the larger the overall amplitude of the induced GW spectrum.

Despite the fact that (118) is already a completely analytical formula, the functional complication does not provide much insight yet. Nevertheless, we can check one interesting limit, which is the IR tail of the spectrum. Expanding (118) for $k \ll k_p$ we arrive at a general analytical approximation for the low frequency tail of the GW spectrum. First, for $b < 0$ we have that

$$\Omega_{\rm GW,rh}(b<0, k \ll k_p) = \frac{\mathcal{A}_{\mathcal{R}}^2}{12}\left(\frac{2^{1+b}(2+b)\Gamma^2[3/2+b]}{\pi c_s^{2(1+b)}(1+b)^{1+b}}\right)^2\left(\frac{k_{\rm rh}}{k_p}\right)^{2+4b} \times \begin{cases} \dfrac{2^{3+2b}}{\pi(1+b)^{2b}}\left(\dfrac{k}{k_{\rm rh}}\right)^2 & (k \lesssim k_{\rm rh}/(1+b)) \\ \left(\dfrac{\pi}{\sin(b\pi)\Gamma[2+b]}\right)^2\left(\dfrac{k}{k_{\rm rh}}\right)^{2+2b} & (k \gtrsim k_{\rm rh}/(1+b)) \end{cases}, \quad (120)$$

where we set the broken power-law knee at $k \sim k_{\rm rh}/(1+b)$ since it is when the subhorizon approximation breaks down for $b < 0$. A more detailed discussion can be found in Section 4.2. Second, for $b > 0$ we find

$$\Omega_{\rm GW,rh}(b>0, k \ll k_p) = \frac{\mathcal{A}_{\mathcal{R}}^2}{24\pi}\left(\frac{(2+b)(1+b+b^2)}{c_s^2 b(1+b)^2}\right)^2\left(\frac{k_{\rm rh}}{k_p}\right)^{2} \times \begin{cases} \left(\dfrac{k}{k_{\rm rh}}\right)^2 & (k \lesssim k_{\rm rh}) \\ \dfrac{1}{2}\left(\dfrac{2^{1+b}\Gamma[b+3/2]}{(1+b)^{1+b}}\right)^2\left(\dfrac{k}{k_{\rm rh}}\right)^{2-2b} & (k \gtrsim k_{\rm rh}) \end{cases}, \quad (121)$$

where this time the matching is roughly at $k \sim k_{\rm rh}$. The above formulas agree with those in Ref. [74]. The difference between $b > 0$ and $b < 0$ is the continued superhorizon growth in the case of $b < 0$ which is explained in Section 2. The most important aspect of Equations (120) and (121) is that it clearly shows how the IR spectral tilt of the induced GW spectrum is directly related to the equation of state parameter at the time of induced GW generation. Thus, the observation of the IR tail gives insight on the expansion history of the primordial universe. Interestingly, for $b > 1$ ($w < 0$) the spectrum has a peak at $k \sim k_{\rm rh}$ which is larger than that at $k \sim k_p$. This means that induced GWs for $b < 1$ peaks at a different scale than the peak of the primordial spectrum. This has been used, e.g., in [132] to fit the NANOGrav results [18] and at the same time predict PBH with a mass smaller than the solar mass. If instead one assumes induced GWs in radiation domination to fit the NANOGrav frequency range, the associated PBH counterpart must be of the order of solar mass. We discuss more about these possibilities in Section 9. Another important point is that the amplitude of the spectrum at $k \sim k_p$ strongly depends on the ratio $k_{\rm rh}/k_p$ and often mildly on b. We thus see that the smaller the ratio $k_{\rm rh}/k_p$, the more suppressed is the spectrum at $k \sim k_{\rm rh}$. This is because the main induced GW generation occurs when the scalar mode k_p enters the horizon. Then, since the modes with $k \sim k_{\rm rh}$ have been superhorizon until the transition they can only grow by a factor $(k_{\rm rh}/k_p)^2$. For $b < 0$ the argument is slightly different because of the second superhorizon growth and so the suppression is minor compared to $b > 0$ with a power index that depends on b.

5.3. Log-Normal Peak

We now turn to a more realistic situation where the peak in the primordial spectrum has a finite width. Inspired by several models of multi-field inflation the peak of the primordial spectrum may be parametrized by a log-normal [82], namely

$$\mathcal{P}_{\mathcal{R}}(k) = \frac{\mathcal{A}_{\mathcal{R}}}{\sqrt{2\pi}\Delta} \exp\left[-\frac{\ln^2(k/k_p)}{2\Delta^2}\right], \tag{122}$$

where Δ is the dimensionless width of the peak. Note that the power spectrum (122) is normalized, i.e., $\int_{-\infty}^{\infty} d\ln k\, \mathcal{P}_{\mathcal{R}}(k) = \mathcal{A}_{\mathcal{R}}$. The log-normal parametrization includes sharp peaks where $\Delta \ll 1$ [95,235–243] and broad peaks where $\Delta \gtrsim 1$ [62,85,102,128,238,244–255]. For $\Delta \to 0$ we recover the Dirac delta case.

In this review we are mainly interested in the case of a narrow peak ($\Delta \ll 1$). The reasons are its simplicity and that we can use our results for general b of Section 5.2. The interested reader for analytical approximation in the broad peak case ($\Delta \gtrsim 1$) in radiation domination is referred to the work of Pi and Sasaki [82]. Here we may directly borrow their result for $\Delta \ll 1$ which reads

$$\Omega_{\rm GW,\Delta}(k) = {\rm erf}\left(\frac{1}{\Delta}\sinh^{-1}\frac{k}{2k_p}\right)\Omega_{{\rm GW},\delta}(k), \tag{123}$$

where ${\rm erf}(x)$ is the error function and $\Omega_{{\rm GW},\delta}(k)$ is the GW spectrum induced by a δ-function peak given in Section 5.2. Then, Equation (123) provides the correction to the Dirac delta spectrum of Section 5.2 due to the small but finite width of the spectrum which leads to the k^3 IR scaling in radiation domination [78]. The most important effect then appears when the ratio of scales k/k_p is smaller than the width Δ. Indeed, when $k \ll k_p$ we have that

$$\Omega_{\rm GW,\Delta}(k) = {\rm erf}\left(\frac{k}{2k_p\Delta}\right)\Omega_{{\rm GW},\delta}(k), \tag{124}$$

which gives a correction of ${\rm erf}(k/(2k_p\Delta)) \sim k/(2k_p\Delta)$ when $\Delta \gg k/(2k_p)$. Thus, due to the presence of Δ our IR tail of Section 5.2 changes respectively for $2k_p\Delta < k_{\rm rh}$ and $2k_p\Delta > k_{\rm rh}$ to

$$\Omega_{\rm GWs}(k \ll k_p, 2k_p\Delta < k_{\rm rh}) \propto \mathcal{A}_{\mathcal{R}}^2 \begin{cases} \left(\dfrac{k}{k_p}\right)^3 & k \ll 2k_p\Delta \ll k_{\rm rh} \\ \left(\dfrac{k}{k_p}\right)^2 & 2k_p\Delta \ll k \ll k_{\rm rh}\,, \\ \left(\dfrac{k}{k_p}\right)^{2-2|b|} & k \ll 2k_p\Delta \ll k_p \end{cases} \tag{125}$$

or

$$\Omega_{\rm GWs}(k \ll k_p, 2k_p\Delta > k_{\rm rh}) \propto \mathcal{A}_{\mathcal{R}}^2 \begin{cases} \left(\frac{k}{k_p}\right)^3 & k \ll k_{\rm rh} \\ \left(\frac{k}{k_p}\right)^{3-2|b|} & k_{\rm rh} \ll k \ll 2k_p\Delta\,. \\ \left(\frac{k}{k_p}\right)^{2-2|b|} & 2k_p\Delta \ll k \ll k_p \end{cases} \tag{126}$$

Note that since $k_{\rm rh}/k_p \ll 1$ the case $2k_p\Delta > k_{\rm rh}$ still allows for a very sharp peak with $k_{\rm rh}/(2k_p) < \Delta \ll 1$. See how the inclusion of the finite width increases the richness of the induced GWs spectrum. We illustrate this in Figure 5. Additionally, note that if $b = 0$ then the induced GW spectrum only has two different slopes and $k_{\rm rh}$ becomes meaningless.

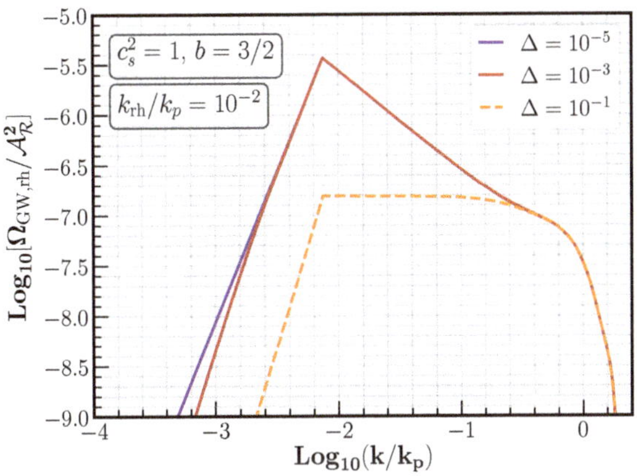

Figure 5. Induced GW spectral density for the log-normal primordial spectrum. We used $c_s^2 = 1$, so there is no resonant peak. We also chose $b = 3/2$ ($w = -1/15$) to show that for $b < 1$ the largest peak of the induced GW spectrum for a Dirac delta is at $k \sim k_{\rm rh}$. In purple we show $\Delta = 10^{-5}$ which for the range of k shown in the plot is effectively the induced GW spectrum of a Dirac delta. For $k < k_{\rm rh}$ the GW spectrum goes as k^2. In red we show $\Delta = 10^{-3}$ so that the effects of the finite width occur within the plot range. Around $k/k_p \sim 5 \times 10^{-2}$ the GW spectrum goes from k^2 to k^3. Lastly, we show in orange $\Delta = 10^{-1}$. The finite width becomes important now for $k_{\rm rh} < k < k_p$ and the GW spectrum transition from $k^{2-2b} = k^{-1}$ to $k^{3-2b} = k^0$.

Before ending this subsection, we briefly discuss the case when $\Delta > 1$. It is shown in Ref. [82] that in the induced GW spectrum in radiation domination for $\Delta > 1$ and near $k \sim k_p$ is well approximated by

$$\Omega^{\rm peak}_{\rm GWs}(\Delta > 1) \approx 0.125\frac{\mathcal{A}_{\mathcal{R}}^2}{\Delta^2}e^{-\frac{\ln^2(k/k_p)}{\Delta^2}}\,. \tag{127}$$

This means that the GW spectrum induced by a log-normal peak in the primordial spectrum also follows a log-normal peak but with a shallower width given by $\Delta/\sqrt{2}$. The factor 2 is a reflection of the secondary nature of induced GWs [82].

5.4. Broken Power-Law

In Section 5.3 we have studied the log-normal sharp peak which is inspired in multi-field models of inflation. However, if the curvature perturbation is enhanced during single field inflation it often takes a broken power-law shape [103,109,234,256]. The induced GWs counterpart for a broken power-law primordial spectrum have been studied in Refs. [80,81,105,141,220]. Here we present the results of Ref. [81], which contain analytical approximations for the amplitude and spectral tilt of both the IR and UV tails of the induced GW spectrum. A detailed discussion on the approximations used can be found in [81]. At the end of this subsection, we also briefly discuss the case of a scale invariant primordial spectrum.

Let us parametrize the primordial curvature power spectrum as a strict broken power-law given by

$$\mathcal{P}_{\mathcal{R}} = \mathcal{A}_{\mathcal{R}} \begin{cases} \left(\dfrac{k}{k_p}\right)^{n_{\rm IR}} & k \leq k_p \\ \left(\dfrac{k}{k_p}\right)^{-n_{\rm UV}} & k \geq k_p \end{cases}, \tag{128}$$

where $n_{\rm IR}$, $n_{\rm UV} > 0$ respectively are the IR and UV spectral tilts of the spectrum which peaks at $k = k_p$. In single field models one often finds that $n_{\rm IR} \sim 4$ [103]. If the enhancement of the curvature fluctuations is due to a bump in the inflationary potential, then $n_{\rm UV}$ can be related to the second derivative of the potential at the bump [80,256]. In such models $n_{\rm UV}$ is also related to the local non-gaussianity parameter f_{NL} [256]. In what follows, we will only focus on induced GWs in the radiation dominated universe and later discuss its generalisations to other values of the equation of state parameter. Currently, only analytical approximations for radiation domination are available. Now, using (91) and (128) we can compute the associated induced GW spectrum. Doing so for the IR tail, i.e., for modes with $k \ll k_p$, we find that

$$\Omega_{\rm GW,rh}(k \ll k_p) \approx 12\mathcal{A}_{\mathcal{R}}^2 \left(\frac{1}{2n_{\rm IR} - 3} + \frac{1}{2n_{\rm UV} + 3}\right)\left(\frac{k}{k_p}\right)^3 \ln^2\left(\frac{k}{k_p}\right). \tag{129}$$

Equation (129) is valid for $n_{\rm IR} > 3/2$ and $n_{\rm UV} > 0$ and gives a good approximation of the IR tail. The condition $n_{\rm IR} > 3/2$ comes from requiring convergence of the integral at large internal momenta ($u \sim v \gg 1$).

The UV tail of the induced GW spectrum presents two different behaviours. If $0 < n_{\rm UV} < 4$ the integral converges even in the strict limit $v \to 0$. This means that we can pull out of the integral (91) all the k dependence in $\mathcal{P}_{\mathcal{R}}$ and have that $\Omega_{\rm GW,rh} \sim \mathcal{P}_{\mathcal{R}}^2$. The remaining piece is just a numerical factor. In this way, we find that the induced GW spectrum in the UV reads

$$\Omega_{\rm GW,rh}(k \gg k_p, n_{\rm UV} < 4) \approx \frac{1}{12}\mathcal{A}_{\mathcal{R}}^2 F(n_{\rm UV})\left(\frac{k}{k_p}\right)^{-2n_{\rm UV}}, \tag{130}$$

where $F(n_{\rm UV})$, for $0 < n_{\rm UV} < 4$, is given by

$$F(n_{\rm UV}) = 8\int_0^\infty dv \int_{|1-v|}^{1+v} du \left(\frac{4v^2 - (1 - u^2 + v^2)^2}{4uv}\right)^2 (uv)^{n_{\rm UV}} \overline{I^2(n_{\rm UV}, k, v, u)}, \tag{131}$$

and converges everywhere. The range of $n_{\rm UV}$ for which this approximation is valid depends on the value of b. We discuss this after presenting the results for radiation domination ($b = 0$). Since (131) has to be computed numerically, we present here a numerical fit for $n_{\rm UV} < 4$, which reads

$$F(n_{\rm UV}) \approx 41 + \frac{16 n_{\rm UV}^2}{\sqrt{16 - n_{\rm UV}^2}}. \tag{132}$$

The form of (132) has been found noting that the integral diverges in the exact limit $n_{UV} \to 4$. After several trial and error attempts, a reasonable fit was found to be (132). A detailed study of (131) might yield better analytical insights than (132). For the moment Equation (132) gives a good order of magnitude estimate. Note that when $n_{UV} = 4$ the induced GW spectrum present a logarithmic divergence [80,81]. Lastly, for $n_{UV} > 4$ the primordial spectrum has a sharp decay and, in analogy with the sharp peak, we do not expect that $\Omega_{\rm GW,rh} \sim \mathcal{P}^2_{\mathcal{R}}$. Instead, there should be a fast fall off at around $k \sim 2k_p$ by momentum conservation. Indeed when $n_{UV} > 4$ we find that

$$\Omega_{\rm GW,rh}(k \gg k_p, n_{UV} > 4) \approx \frac{16}{3}\mathcal{A}^2_{\mathcal{R}}\left(\frac{1}{n_{UV}-4} + \frac{1}{n_{IR}+4}\right)\left(\frac{k}{k_p}\right)^{-4-n_{UV}}. \tag{133}$$

We show the analytical estimates against the numerical results in Figure 6. See how even if the approximations are extrapolated to $k \sim k_p$ they still yield a sensible order of magnitude estimate of the induced GW spectrum.

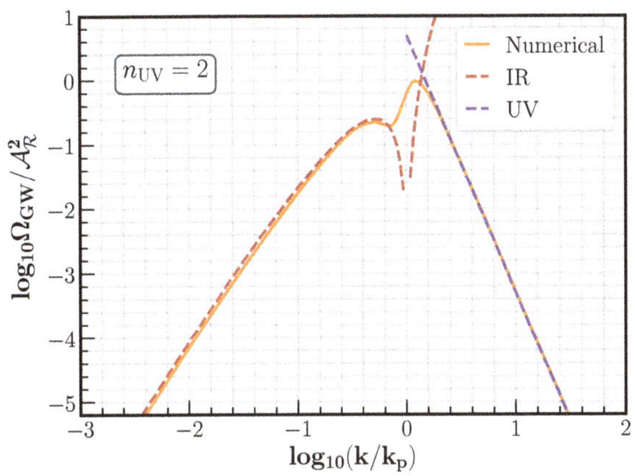

Figure 6. Induced GW spectral density for a broken power-law primordial spectrum with $n_{IR} = 4$ and $n_{UV} = 2$. In orange we show the result of numerical integration. In red we show the IR approximation (129). In purple we show the UV approximation (130). See how the analytical estimates fit well the numerical integration for $k \ll k_p$ and $k \gg k_p$. They even give a good order of magnitude estimate of the amplitude of the GW spectrum around $k \sim k_p$.

With the above analytical estimates for the amplitude and the spectral tilt in the IR and UV limits we conclude that the induced GW spectrum has the following broken power-law shape

$$\Omega_{\rm GWs}(k) \propto \mathcal{A}^2_{\mathcal{R}}\begin{cases}\left(\dfrac{k}{k_p}\right)^3 & k \ll k_p \\ \left(\dfrac{k}{k_p}\right)^{-\Delta} & k \gg k_p\end{cases}, \quad \text{with} \quad \Delta = \begin{cases}2n_{UV} & 0 < n_{UV} < 4 \\ 4 + n_{UV} & n_{UV} > 4\end{cases}. \tag{134}$$

Equation (134) tells us that the UV spectral tilt of the induced GW has information the UV tail of the primordial power spectrum. This is to be contrasted with the IR spectral tilt, which does not tell us much about the inflationary model unless we measure the amplitude very well. Lastly, if we focus on the UV tail of the induced GW spectrum (134), we see that for example it may be degenerate with the GW spectrum from the sound waves in a first order phase transition [172].

5.4.1. Alternative Expansion Histories

Here we briefly present the spectral tilts of the induced GW spectrum if instead of radiation the universe was dominated by another perfect fluid. In this case, we find that (134) changes to [81]

$$\Omega_{\rm GW}(k) \propto \mathcal{A}_{\mathcal{R}}^2 \begin{cases} \left(\frac{k}{k_p}\right)^3 & k \ll k_{\rm rh} \\ \left(\frac{k}{k_p}\right)^{3-2|b|} & k_{\rm rh} \ll k \ll k_p \,, \\ \left(\frac{k}{k_p}\right)^{-\Delta-2b} & k \gg k_p \end{cases} \tag{135}$$

where b is defined in Equation (16) and this time

$$\Delta = \begin{cases} 2n_{\rm UV} & 0 < n_{\rm UV} + b < 4 \\ 4 + n_{\rm UV} & n_{\rm UV} + b > 4 \end{cases} . \tag{136}$$

Previously, in Sections 2.2 and 4, we derived the IR broken power-law behaviour of the induced GW spectrum for general equation of state parameter. Now, in this section we have a power-law UV tail in the induced GW coming from the UV tail of the primordial spectrum. The additional factor $-2b$ with respect to (134) in the UV tail can be understood as follows. All modes with $k \gg k_p$ entered the horizon deep inside the $b =$ constant era and before the scalar peak at $k \sim k_p$. After that the induced tensor modes effectively behave as free GWs irrespective of whether $b < 0$ or $b > 0$. Thus, the free GWs experience the typical relative redshift with respect to the background, which is the factor k^{-2b}. Note that if $b < 0$ the UV tail might have a blue tilt.

5.4.2. Scale Invariant Spectrum

A limiting case of the broken power-law studied (128), setting $n_{\rm UV} = n_{\rm IR} = 0$, is a scale invariant spectrum typical of slow-roll inflation. In this case, it is easy to convince ourselves that the induced GW spectrum for general expansion histories is given by

$$\Omega_{\rm GW,rh} = \mathcal{A}_{\mathcal{R}}^2 \mathcal{F}(b, c_s) \left(\frac{k}{k_{\rm rh}}\right)^{-2b} . \tag{137}$$

where

$$\mathcal{F}(b, c_s) = \frac{2}{3}\int_0^\infty dv \int_{|1-v|}^{1+v} du \left(\frac{4v^2 - (1-u^2+v^2)^2}{4uv}\right)^2 \overline{I^2}(b, c_s, k, v, u) \,, \tag{138}$$

and $\overline{I^2}$ is given by (83). Another explicit formula can be found at the summary Section 10. The integral (138) converges everywhere for any value of b and c_s. However, note that (138) with (83) is only valid for $k \gg k_{\rm rh}$. Nevertheless, since deep inside the radiation domination we expect a scale invariant induced GW spectrum it is reasonable to assume that for scales $k \sim k_{\rm rh}$ we can match the k^{-2b} behaviour to the constant spectral density. A more detailed, perhaps numerical, analysis is needed to determine the spectrum for $k \sim k_{\rm rh}$ as induced GWs are being sourced before and after the transition.

5.5. Oscillatory Features

In the subsections above we have discussed the GWs induced by the simplest shapes of the primordial spectrum, that is a log-normal peak and a broken power-law. However, in some situations, it is expected that the primordial power spectrum presents a series of oscillations usually referred to as "primordial features". For recent reviews on such features see, e.g., [257–259]. These oscillations can be classified according to the

k dependence in their frequency. If the frequency is linear in k, the oscillations resulted from sharp dynamics during inflation. For example, a step in the potential in single field inflation [260–267] or a sharp turn in field space [84,86,254,255,268–270]. This case is also known as a sharp feature. If the frequency is logarithmic in k, the oscillations are called resonant features. They are due to resonances during inflation on subhorizon scales. For instance, resonances occur if the potential in single field inflation has wiggles or a massive scalar field is oscillating at the bottom of the potential [87,170,271–277]. In more general cases, such as general interactions [278,279] or alternative scenarios to inflation [280], the frequency of the oscillations in the primordial spectrum could be a power-law of k.

Remarkably, despite the induced GWs being a second order effect, oscillations are not completely washed out. The main difficulty is to derive analytical approximations to the induced GW spectrum. In this respect, Fumagalli, Renaux-Petel and Witkowski [84,87] and also Braglia, Chen and Hazra [86] did a detailed numerical analysis of the induced GW spectrum and derived motivated semi-analytical templates. Perhaps the most interesting case is when the oscillations in the primordial spectrum are $O(1)$. For example, this is the case when a sharp feature is responsible for the peak in the primordial spectrum [84,255]. Now, if the oscillations are large and separated linearly in k, we can gain intuition by looking at the sum of Dirac delta peaks [83,84]. In this case if we call $\omega_{\rm lin}$ to the frequency of the Dirac delta peaks in the primordial spectrum, then the frequency in the induced GW spectrum is $\omega_{\rm lin}^{\rm GWs} = \omega_{\rm lin}/c_s$ [83,84] (in radiation domination $c_s = 1/\sqrt{3}$). Interestingly, in the more realistic situation studied [84] the induced GW spectrum follows an envelope as if there were no oscillations and, on top of that, oscillations appear near the resonant peak at $k \sim 2c_s k_p$ and reach at most an amplitude of about 20% in radiation domination [84]. We illustrate this possibility in Figure 7. Another example are $O(1)$ oscillations due to a resonant feature in the primordial spectrum. In this case, if we call $\omega_{\rm log}$ the frequency of the resonant feature, the resulting induced GW spectrum exhibits two oscillatory components with frequencies $\omega_{\rm log}$ and $2\omega_{\rm log}$. The relative amplitude of such components depend on both the frequency and the envelope of the primordial spectrum [87]. For the analytical templates and a more detailed discussion, see Refs. [84,87].

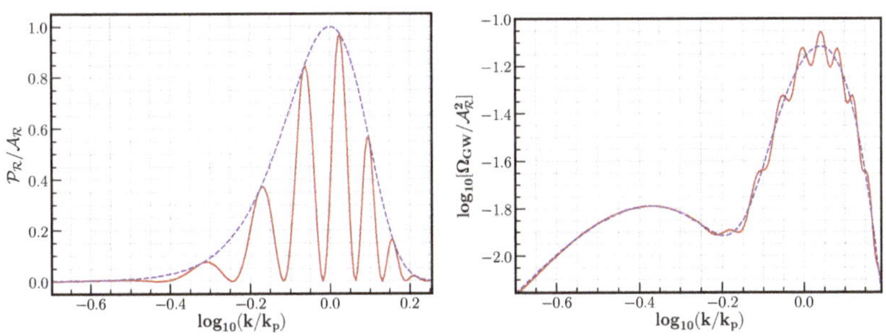

Figure 7. Primordial spectrum and induced GWs from a sharp turn in the inflationary trajectory from the model of Ref. [84], courtesy of L. Witkowski, J. Fumagalli and S. Renaux-Petel. Left: Normalized primordial spectrum in terms of wavenumber k/k_p. The solid line is the power spectrum resulting from the sharp feature. The dashed line is the envelope of the oscillations. They are respectively given by Equations (25) and (26) of Ref. [84] with $\delta = 0.5$ and $\eta_\perp = 20$. Right: Induced GW spectral density normalized by $\mathcal{A}_\mathcal{R}$. The solid line is the induced GW spectrum of the oscillatory power spectrum while the dashed line is the induced GWs from the envelope properly normalised. The oscillations have an amplitude of about 20% larger than the envelope around the induced GW spectrum peak.

5.6. Impact of Non-Gaussianities

Here we briefly discuss the effects that primordial non-gaussianities may have on the induced GW spectrum. We will not dwell on the details, which are well explained in Refs. [89,90,92]. The leading formulas for local-type non-gaussianity can be found in Section 3.3. We describe the main effects of local-type non-gaussianity below.

First, let us focus on the very sharp peaks of Sections 5.2 and 5.3. In this case, the most relevant effect is that the induced GW spectrum does not stop at $k \sim 2k_p$ but continues until $k \sim 3k_p$. This is roughly because when including primordial non-gaussianity, the source term to induced GWs has three scalars. Thus, we have a cut-off at $k \sim 3k_p$ by momentum conservation. Note that this effect is characteristic of the sharp peak in the primordial spectrum. For large non-gaussianity, that is $F_{NL} \gg 1$ in Equation (63), the spectrum does not have the clear resonant peak of Section 5.2 but presents a few peaks with similar amplitude instead [89,92]. Second, in the case where the peak is broad or a broken power-law, the non-gaussian contribution typically has the same IR and UV spectral tilts [81,92]. For large enough $F_{NL} \gtrsim 1$, the induced GW spectrum also shows an additional peak/bump at around $k \sim 3k_p$. If the non-gaussianity is large enough, i.e., $\mathcal{A}_\mathcal{R} F_{NL}^2 > O(1)$, the non-gaussian contribution starts to dominate over the gaussian one [89,92]. However, it is not so clear whether one can achieve $\mathcal{A}_\mathcal{R} F_{NL}^2 > 1$ in inflationary models as in this case the non-gaussian expression (63) cannot be regarded as a perturbative expansion [81]. A more detailed and general analysis is necessary. It should be noted that if we include third order perturbation theory terms, the induced GW spectrum for a Dirac delta peak has three resonant peaks instead of one [143,179].

The primordial non-gaussian contribution to the induced GW spectrum, as we have just presented it, might not yield much information on inflation. For example, one may design a gaussian primordial spectrum such that the resulting induced GW spectrum is exactly as that of a primordial log-normal peak with large non-gaussianities. Although this may be a very complicated task, it is in principle possible. In this sense, there is no telling primordial non-gaussianity apart just from the induced GW spectrum [179]. Nevertheless, the story becomes more interesting when considering the PBH counterpart. Since PBH formation is highly dependent on the tail of the distribution of primordial fluctuations, any tiny deviation from a normal distribution could have an exponential impact on the fraction of PBHs. For the details see the recent discussions in Refs. [217,219–221,256,281–284]. Then, ideally one may use the two observations, the PBH mass function and the induced GW counterpart, to perhaps disentangle the primordial non-gaussianity from the primordial spectrum.

6. The Dust Dominated Universe

In previous sections we have derived intuitive and analytical formulas for the spectrum of induced GWs under the assumption that the universe was not dominated by a pressure-less perfect fluid with vanishing propagation speed of fluctuations, i.e., $w = c_s^2 = 0$, so-called dust. The particularity of a dust dominated universe is that the Newtonian potential is constant on all scales. As can be seen from Equation (70) with $c_s^2 = 0$, $\Phi =$ constant is a solution. This also implies that density fluctuations of the dust fluid grow. In fact, they grow as $\delta\rho/\rho \propto a$. This is the typical behaviour of a CDM dominated universe. This also means that there is a scale below which density fluctuations grow so much that they become non-linear, i.e., larger than $O(1)$. This occurs for modes with [51]

$$k > k_{NL} \sim \sqrt{\frac{3}{2}}\mathcal{P}_\Phi^{-1/4}(\tau_{\rm rh})\mathcal{H}(\tau_{\rm rh})\,. \tag{139}$$

Note that it does not mean that if there are fluctuations on $k > k_{NL}$ there will be no production of induced GWs. It means that we cannot completely trust our calculations since non-linear dynamics might be important, e.g., due to GWs from mergers. Nevertheless, we quite often have $\Phi < 1$ and thus perturbation theory for Φ is sensible.

In Section 6.1 we show and discuss the subtleties of the dust dominated universe and derive analytical estimates for the amplitude of induced GWs in the case of an instantaneous transition. We apply our results to PBH dominated epoch in Section 6.2 and use overproduction of induced GWs to place stringent constraints on the scenario.

Main references: Section 6.1 is based on Inomata et al. [65,66] with the notation of Refs. [70,71]. These works are built upon previous papers by Assadullahi and Wands [51] and Kohri and Terada [63]. Then, Section 6.2 follows the pioneering work by Papanikolau, Vennin and Langlois [69] further extended by [70] to account for the transition to radiation domination. Lastly, Ref. [68] contains useful details on the transition of the PBH dominated universe to radiation domination.

6.1. General Dust Domination

Let us now derive the kernels and some estimates for the induced GWs. First we note that the fact that Φ = constant in an early Matter Dominated (eMD) era seemingly makes naive calculations of induced GWs easier. Just take a transfer function equal to unity in (53), which yields

$$f(\tau, q, |\mathbf{k}-\mathbf{q}|) = \frac{5}{3}, \tag{140}$$

where we took $b = 1$ ($w = 0$). Now we have to simply integrate (60) with a constant source. In the limit where $x \gg 1$, we have that the averaged kernel squared is just a constant [51,63]. In our notation we have

$$\overline{I^2_{eMD}(x \gg 1, q, |\mathbf{k}-\mathbf{q}|)} \approx \left(\frac{20}{3}\right)^2. \tag{141}$$

We can then use (58) to give a naive estimation of the induced GW during the early matter era. However, this is far from being a complete picture. As first realised by Inomata, Kohri, Nakama and Terada, induced GWs from a dust dominated are more subtle and depend on the nature of the transition to radiation domination [65,66]. If the transition is gradual, the induced GW spectrum is actually suppressed [65]. The suppression can be understood from the fact that the initial enhancement in the dust dominated universe found in Ref. [51] is due to the constant Φ. If Φ decays, the enhancement is not as efficient. Instead, if the transition from dust to radiation is instantaneous, the induced GW spectrum gets amplified very much [66]. We will shortly explain the source of the amplification. Let us first note that the reason why we did not find such behaviour in Section 4 comes down to the particular behaviour of Φ in dust domination. In Section 4, due to the fact that we assumed $c_s^2 \neq 0$, the general solution to Φ in a w = constant universe is a decaying and oscillating function. After the transition to radiation domination, Φ is also decaying and oscillating, albeit with different frequency. Thus, the matching between the two periods is well described within the WKB approximation. In this case, only those scales close to $k \sim k_{\rm rh}$ might receive at most a $O(1)$ correction.

What is special about the instantaneous reheating in the dust dominated universe is that we are suddenly going from $\Phi' = 0$ to $\Phi' \neq 0$. Since Φ did not have time to decay and the frequency of the oscillations is proportional to the ratio of the largest wavenumbers k with the reheating scale $k_{\rm rh}$, the amplitude of Φ' right after the sudden reheating will be huge. A more physical picture was presented in Ref. [66] and it goes as follows. Density fluctuations grow as $\delta\rho/\rho \propto a$ and attain a "large" amplitude by the end of the dust dominated era. These large fluctuations suddenly evaporate and are converted into fluctuations in a radiation fluid, which create strong pressure waves. These waves generate spacetime oscillations and are the source to induced GWs.

Let us explicitly show that Φ' becomes very large right after reheating. To do that, we first consider that Φ has an arbitrary amplitude during the eMD of $\Phi = \Phi_{eMD}(k)$. Then, the eMD ends abruptly and we enter a late Radiation Domination (lRD). Matching the general solution (73) and its first derivative for $b = 1$ and $b = 0$ we arrive at

$$\Phi_{IRD} = (c_s k\bar{\tau})^{-3/2}(C_1 J_{3/2}(c_s k\bar{\tau}) + C_2 Y_{3/2}(c_s k\bar{\tau}))\,, \tag{142}$$

where $c_s = 1/\sqrt{3}$ since we are in a radiation dominated universe, $\bar{\tau}$ is defined in (84) and comes from requiring continuity of a and $\mathcal{H}$ and we defined

$$C_1 = -\Phi_{\rm eMD}\frac{\pi}{2}\left(\frac{c_s k\bar{\tau}_{\rm rh}}{2}\right)^{5/2} Y_{5/2}\left(\frac{c_s k\bar{\tau}_{\rm rh}}{2}\right), \tag{143}$$

$$C_2 = \Phi_{\rm eMD}\frac{\pi}{2}\left(\frac{c_s k\bar{\tau}_{\rm rh}}{2}\right)^{5/2} J_{5/2}\left(\frac{c_s k\bar{\tau}_{\rm rh}}{2}\right). \tag{144}$$

If we now compute the variation of $\Phi_{\rm IRD}$ per Hubble time we find that the leading contribution for $x > x_{\rm rh} \gg 1$ is given by

$$\frac{d\Phi_{\rm IRD}}{d\ln\bar{x}} \approx -\Phi_{\rm eMD}\frac{c_s\bar{x}_{\rm rh}^2}{\bar{x}}\sin(c_s(\bar{x}-\bar{x}_{\rm rh}))\,. \tag{145}$$

Thus, we see that the amplitude of the time derivative is roughly

$$\frac{d\Phi_{IRD}}{d\ln\bar{x}} \propto \left(\frac{k}{k_{\rm rh}}\right)^2. \tag{146}$$

This means that if there are fluctuations on scales $k \gg k_{\rm rh}$ the amplitude of $\Phi'_{\rm IRD}$ is very large and so it will produce induced GWs with a large amplitude.

Let us move on to the GWs induced after the sudden transition. For simplicity, we only focus on the largest contribution to the source (53), which comes from the time derivatives and reads

$$f(\tau, q, |\mathbf{k}-\mathbf{q}|) \approx \frac{1}{2}\frac{dT_\Phi(q\tau)}{d\ln\tau}\frac{dT_\Phi(|\mathbf{k}-\mathbf{q}|\tau)}{d\ln\tau}\,. \tag{147}$$

Inserting Equations (145) and (147) into (60) we arrive at the kernel during the IRD, namely

$$I_{IRD}(\bar{x},u,v) \approx \Phi^2_{\rm eMD}\frac{c_s^2\bar{x}_{\rm rh}^4}{2\bar{x}}uv\int_{\bar{x}_{\rm rh}}^{\bar{x}} d\tilde{\bar{x}}\,\frac{\sin(\bar{x}-\tilde{\bar{x}})}{\tilde{\bar{x}}}\sin(c_s v(\tilde{\bar{x}}-\bar{x}_{\rm rh}))\sin(c_s u(\tilde{\bar{x}}-\bar{x}_{\rm rh}))\,. \tag{148}$$

The integral in (148) may be done analytically in terms of sine and cosine integrals but it is not so illuminating. The interested reader is referred to Ref. [66]. Using our experience of Section 4, we know that the integral often diverges when $c_s(u+v) \sim 1$. Thus, within our order of magnitude estimates, we shall focus solely on the neighbourhood of the resonance. In this case, the divergent piece is the cosine integral $\mathrm{Ci}[x]$ and the kernel is approximately given by

$$I_{IRD,\rm res}(\bar{x},u,v) \approx \Phi^2_{\rm eMD}\frac{c_s^2\bar{x}_{\rm rh}^4}{8\bar{x}}uv\,\mathrm{Ci}[|1-c_s(u+v)|\bar{x}_{\rm rh}]\,\sin\bar{x}\,. \tag{149}$$

The calculation of the averaged kernel squared is straightforward and yields

$$\overline{I^2_{IRD,\rm res}(\bar{x},u,v)} \approx \Phi^4_{\rm eMD}\frac{c_s^4\bar{x}_{\rm rh}^8}{128\bar{x}^2}u^2v^2\,\mathrm{Ci}^2[|1-c_s(u+v)|\bar{x}_{\rm rh}]\,. \tag{150}$$

For other approximations regarding the IR tail, such as the large momentum contribution $u \sim v \gg 1$, we refer the reader to Refs. [66,68].

With the analytic approximation for the kernel (150) we shall turn our attention to the dominant contribution to the tensor spectrum (58). For clarity, let us explicitly insert (150) into (58) to obtain the resonant part of the tensor spectrum, which leads us

$$\overline{\mathcal{P}_h^{\rm res}} \approx \frac{c_s^4 \tilde{x}_{\rm rh}^8}{16\tilde{x}^2}\int_0^\infty dv \int_{|1-v|}^{1+v} du \left(\frac{4v^2-(1-u^2+v^2)^2}{4uv}\right)^2 \mathcal{P}_\Phi(ku)\mathcal{P}_\Phi(kv) \\ \times u^2v^2\,{\rm Ci}^2[|1-c_s(u+v)|\tilde{x}_{\rm rh}]\,. \tag{151}$$

Before carrying out the integral, let us analyse some of the assumptions. First, we are assuming that (151) yields the dominant contribution. This will be the case if there is integration support for large k. This means that $\mathcal{P}_\Phi(k)$ should peak at large k. We also argued that there might be a limit to the magnitude of k due to non-linearities given by (139). We should use such a limit if there are no stronger motivations to consider $k > k_{NL}$. However, as we shall see, if PBH dominate the universe, there must be fluctuations above k_{NL} [69]. For these reasons, let us parameterize the scalar power spectrum as power-law with a UV cut-off k_{UV} given by [71]

$$\mathcal{P}_\Phi = \mathcal{A}_\Phi\left(\frac{k}{k_{\rm UV}}\right)^{-n}\Theta(k_{UV}-k)\,, \tag{152}$$

where $\mathcal{A}_\Phi$ is the amplitude and n the spectral tilt. Looking at (151) we see that the integrand peaks at large k if $n < 2$. This includes an almost scale invariant spectrum ($n \sim 0$) and the PBH density fluctuations ($n \sim 1$) [69,70].

With these assumptions we shall proceed to perform the integral in (151). We shall use the fact that ${\rm Ci}[z]$ is divergent for $z \sim 0$ to evaluate the integrand at the resonant point except for ${\rm Ci}[z]$. For simplicity, let us switch integration variables to

$$z \equiv (1-c_s(u+v))\tilde{x}_{\rm rh} \quad \text{and} \quad s \equiv u - v\,, \tag{153}$$

which have a Jacobian given by $|J| = 1/(2c_s\tilde{x}_{\rm rh})$. Replacing (u,v) for (z,s) in (151), evaluating integrand at $z = 0$ except for ${\rm Ci}[z]$ and maintaining the s dependence we arrive at

$$\overline{\mathcal{P}_h^{\rm res}} \approx \frac{\pi\tilde{x}_{\rm rh}^7}{512c_s\tilde{x}^2}\left(1-c_s^2\right)^2\int_{-s_0(k)}^{s_0(k)} ds(1-s^2)^2\mathcal{P}_\Phi\left(k\frac{1+c_s s}{2c_s}\right)\mathcal{P}_\Phi\left(k\frac{1-c_s s}{2c_s}\right), \tag{154}$$

where we already integrated the cosine integral around the divergence[19] and we have defined [66]

$$s_0(k) = \begin{cases} 1 & k_{UV}/k \geq \frac{1+c_s^{-1}}{2} \\ 2\frac{k_{UV}}{k} - c_s^{-1} & \frac{1+c_s^{-1}}{2} \geq \frac{k_{UV}}{k} \geq \frac{c_s^{-1}}{2} \\ 0 & \frac{c_s^{-1}}{2} \geq \frac{k_{UV}}{k} \end{cases}. \tag{155}$$

These bounds on s come from momentum conservation, i.e., $|1-v| < u < 1+v$, evaluated at the resonant point $z = 0$. To have an order of magnitude estimate of the amplitude of the peak of the induced GW spectrum, we can evaluate (154) at $s = 0$. By doing so, the amplitude of the induced GW spectrum at $k \sim k_{\rm UV}$ due to the sudden reheating is roughly given by

$$\Omega_{\rm GWs}^{\rm peak} \approx \frac{\pi}{3\times 2^{12}c_s}\left(1-c_s^2\right)^2\left(4c_s^2\right)^n\left(\frac{k_{\rm UV}}{k_{\rm rh}}\right)^7\mathcal{A}_\Phi^2\,, \tag{156}$$

where we used that $\tilde{x}_{\rm rh} = k\tau_{\rm rh}/2 = k/k_{\rm rh}$ and we used that the numerical evaluation of the integral in s roughly yields an extra factor $1/2$ [70]. Let us remind the reader that

$c_s = 1/\sqrt{3}$ since we are in radiation domination. From (156) we see how if $k_{\rm UV} \gg k_{\rm rh}$ the amplitude of the induced GWs is tremendously large by a power of 7. Thus, not to backreact, we either need $\mathcal{A}_\Phi$ to be extremely small or we restrict the value of $k_{\rm UV}$ not to be too large. Now, knowing the amplitude of the peak (156) we can approximate the induced GW spectrum by

$$\Omega_{\rm GWs,rh}(k \sim k_{\rm UV}) \approx \Omega_{\rm GWs}^{\rm peak}\left(\frac{k}{k_{\rm UV}}\right)^{7-2n}\Theta(k_{\rm UV}-k)\,, \tag{157}$$

where, although we used the cut-off in the induced GWs, which is at $k = 2k_{\rm UV}$, the decay after $k \sim k_{\rm UV}$ is so fast that we can safely neglect it.

The resonant peak is not the only contribution after the sudden reheating induced GWs. There is also the IR tail which corresponds to the $u \sim v \gg 1$ region of integration. Nevertheless, this contribution is always suppressed by a factor $k_{\rm rh}/k_{\rm UV}$. The interested reader can find the corresponding formulas in Refs. [66,68]. In addition to that, there is the contribution of the GWs induced during eMD, i.e., using the kernel (141) into (58). In the case of the sudden transition, we can take the amplitude of the induced GWs as initial conditions for the subsequent radiation domination era. Then the total spectrum is well approximated by the sum of the two contributions [66]. This contribution will be small compared to the resonant part (157) but since it may present a large plateau [51,66,69] it may dominate on the very low frequency band. The case of the gradual transition requires a more careful treatment which can be found in Ref. [65]. We shall proceed with a very interesting application of (157).

6.2. PBH Dominated Era

An early matter dominated period could have been due to PBHs. Once formed, PBHs basically behave like a dust fluid, i.e., a fluid with no pressure and no propagation speed of fluctuations, like non-relativistic matter. Furthermore, their energy density practically redshifts as that of non-relativistic matter. You can see this from the following. Since the number density of PBH, $n_{\rm PBH}$, is conserved (unless they substantially merge or evaporate) we have that $n_{\rm PBH} \propto a^{-3}$. The total energy density contained in these PBHs is $\rho_{\rm PBH} = M_{\rm PBH} n_{\rm PBH}$. Then in periods where PBH evaporation is negligible we have that $\rho_{\rm PBH} \propto a^{-3}$. For concreteness, let us assume that PBH formed in a radiation dominated era, with an initial fraction of PBH given by

$$\beta \equiv \frac{\rho_{{\rm PBH},i}}{\rho_{r,i}} \approx \frac{\rho_{{\rm PBH},i}}{3H_i^2 M_{\rm pl}^2}\,, \tag{158}$$

where in the last step we used the first Friedmann Equation (A33). If β is large enough, then PBH will eventually dominate the universe since we have that $\rho_r \propto a^{-4}$. Now, for simplicity, let us assume that these PBHs formed by the collapse of large primordial fluctuations with a very peaked primordial spectrum at a scale k_p. This has two implications. First, the PBH mass function is almost monochromatic. Then, we can use Equation (30) to estimate the peak mass in terms of the peak scale k_p by using that $H_i \equiv k_p/a_i$. We refer the reader to Section 2.3 for more details. Second, the initial fraction of PBH β is determined by the amplitude of the primordial spectrum. In this section, we take for convenience $M_{\rm PBH}$ and β as the free parameters of the model. One can then relate the scale and the amplitude of the primordial spectrum to $M_{\rm PBH}$ and β, if necessary. Interestingly, in this situation the PBH evaporation occurs almost instantaneously [68] and, therefore, we can use the estimates we derived in Section 10.2 with some corrections that we describe below. It should be noted that the results we will derive here only apply to a sharply peaked PBH mass function. If we considered an extended PBH mass function, induced GWs would be very much suppressed [68].

A remarkable fact is that, as first realised by Papanikolaou, Vennin and Langlois [69], due to the inhomogeneous distribution, PBH themselves create density fluctuations which might later source induced GWs. To see this, note that PBH formation by the collapse of large fluctuations is a rare event [174] and so each PBH form uncorrelated of the others. Additionally, clustering is often negligible. This means that, as a good approximation PBHs will be randomly distributed uniformly in space. In other words, PBHs are distributed according to Poisson statistics. Such Poisson statistics are dictated by the mean inter-PBH co-moving separation. If PBH follow a monochromatic mass function, the mean inter-PBH co-moving separation at formation is given by

$$d_i = \left(\frac{3}{4\pi n_{\mathrm{PBH},i}}\right)^{1/3} \equiv k_{\mathrm{UV}}^{-1}\,. \tag{159}$$

Note that in Equation (159) we identify the co-moving inter-PBH separation as the UV cut-off in Equation (152). This is because the fluid description of the PBH gas is valid only for $k < k_{\mathrm{UV}}$, i.e., at distances much larger than d_i. In this coarse grained regime the initial dimensionless power spectrum of PBH density fluctuations reads [69]

$$\mathcal{P}_{\delta,\mathrm{PBH}} = \frac{2}{3\pi}\left(\frac{k}{k_{\mathrm{UV}}}\right)^{3}\Theta(k_{\mathrm{UV}} - k)\,. \tag{160}$$

We have shown with Equation (160) that PBHs give rise to density fluctuations. However, in the current model, PBHs form in an early Radiation Dominated stage (eRD) and, therefore, such density fluctuations correspond to isocurvature fluctuations. Thus, they do not source induced GWs yet. Induced GWs are mainly generated by curvature fluctuations. If one is interested in the latter aspect, the general source term in a two fluid system can be found in Ref. [70]. The isocurvature nature of the PBH density fluctuations can be understood from the following [70]. PBH formation in the fluid picture may be regarded as a transition of a fraction of the homogeneous radiation fluid into dust. Yet, the total energy density of radiation remains homogeneous, while PBH are distributed randomly. Thus, the inhomogeneity due to PBHs must be isocurvature as the total energy density is homogeneous. Interestingly, such isocurvature is converted into curvature fluctuations when PBH dominate the universe. This type of transition from radiation to dust was studied analytically in detail by Kodama and Sasaki [285,286]. Thus, we can directly borrow their results which provide the transfer function for a vanishing initial curvature perturbation Φ and non-zero isocurvature deep inside the eMD era as

$$T^{\mathrm{iso}\to\mathrm{curv}}_{\mathrm{eRD}\to\mathrm{eMD}}(k; a \gg a_{\mathrm{eq}}) = \begin{cases} \dfrac{1}{5} & k \ll k_{\mathrm{eq}} \\ \dfrac{3}{4}\left(\dfrac{k_{\mathrm{eq}}}{k}\right)^{2} & k \gg k_{\mathrm{eq}} \end{cases}. \tag{161}$$

The subscript "eq" refers to the early radiation-PBH equality. For an intuitive re-derivation, see Ref. [69]. Equation (161) tells us that the initial isocurvature has been transferred to a constant curvature perturbation. Modes which entered the horizon before the early radiation-PBH equality have decayed substantially and, hence, the suppression factor $(k_{\mathrm{eq}}/k)^2$. This means that the curvature power spectrum due to early PBH fluctuations in the eMD on the smallest scales, i.e., $(k_{\mathrm{UV}} > k \gg k_{\mathrm{eq}})$ is given by

$$\mathcal{P}_{\Phi,\mathrm{PBH}} = \frac{3}{8\pi}\left(\frac{k_{\mathrm{eq}}}{k_{\mathrm{UV}}}\right)^{4}\left(\frac{k_{\mathrm{UV}}}{k}\right)\Theta(k_{\mathrm{UV}} - k)\Theta(k - k_{\mathrm{eq}})\,. \tag{162}$$

It is the curvature perturbation spectrum given by (162) which sources induced GWs during the eMD. However, we are most interested in the induced GWs generated right at the start of the lRD. As we shall see, since PBH evaporation has a finite duration even for the monochromatic case, we must take into account an important suppression. Then we shall use our estimate Equation (156). Before that, we need to understand and quantify the PBH domination a bit further.

Let us derive the conditions to have a PBH dominated era in the early universe allowed by current observations. We start with our PBHs given an initial mass $M_{\rm PBH,\it i}$ (30) and an initial fraction β. After formation, these PBHs will evaporate emitting Hawking radiation. This means that β cannot be too small, otherwise PBH evaporate before they dominate. To quantify this requirement it is enough to look at the evaporation time which for a non-spinning black hole reads [68,189,287]

$$t_{\rm eva} \approx \frac{160}{3.8\pi g_H(T_{\rm PBH})}\frac{M^3_{{\rm PBH},i}}{M^4_{\rm pl}}\,, \tag{163}$$

where $g_H(T_{\rm PBH})$ are the spin-weighted degrees of freedom and

$$T_{\rm PBH} \equiv M^2_{\rm pl}/M_{{\rm PBH},i} \approx 1.06\times 10^9{\rm GeV}\left(\frac{M_{\rm PBH,f}}{10^4\,{\rm g}}\right)^{-1}. \tag{164}$$

If we assume only the standard model of particle physics we have that for PBHs which evaporated before BBN then $g_H(T_{\rm PBH}) \approx 108$. If we include a possible PBH spin the evaporation time decreases, reaching a factor $1/2$ for the extremal BH case [288–292]. Thus, we can take into account the PBH spin by replacing $t_{\rm eva} \to t_{\rm eva}/2$. This effect is considered in the induced GWs by Ref. [71]. We will not consider PBH spin in what follows. From Equation (163) we see that the "reheating" temperature, i.e., the temperature of the radiation fluid which dominates the universe after PBH evaporate, is given solely in terms of $M_{{\rm PBH},i}$. Using the current cosmological parameters, which are given in Appendix A, we find that the evaporation temperature can be written as

$$T_{\rm eva} \approx 2.76\times 10^4{\rm GeV}\left(\frac{M_{{\rm PBH},i}}{10^4\,{\rm g}}\right)^{-3/2}\left(\frac{g_H(T_{\rm PBH})}{108}\right)^{1/2}\left(\frac{g_*(T_{\rm eva})}{106.75}\right)^{-1/4}, \tag{165}$$

where g_* are the effective degrees of freedom in the energy density of radiation [293,294]. Here we find our first constraint on the model: PBH cannot evaporate too late if they are to reheat the universe. For a successful BBN the reheating temperature must be $T_{\rm eva} > 4\,{\rm MeV}$ [295–298]. This puts an upper bound on the PBH mass as

$$M_{{\rm PBH},i} < 5\times 10^8\,{\rm g}\,. \tag{166}$$

Using Equation (165), the co-moving scale corresponding to the horizon size at the time of evaporation is given by

$$k_{\rm eva} \approx 4.7\times 10^{11}{\rm Mpc}^{-1}\left(\frac{g_H(T_{\rm PBH})}{108}\right)^{1/2}\left(\frac{g_*(T_{\rm eva})}{106.75}\right)^{1/4}\left(\frac{g_{*,s}(T_{\rm eva})}{106.75}\right)^{-1/3}\left(\frac{M_{{\rm PBH},i}}{10^4\,{\rm g}}\right)^{-3/2}, \tag{167}$$

where $g_{*,s}$ are the effective degrees of freedom in the entropy. With the above formulas we have that the end of the PBH domination era is fixed by the PBH mass. The second constraint on the model comes from requiring that PBHs dominate before they evaporate. This is a constrain on the initial fraction of PBHs in terms of the PBH mass, which is found to be [68,70]

$$\beta > 6.35\times 10^{-10}\left(\frac{g_H(T_{\rm PBH})}{108}\right)^{1/2}\left(\frac{\gamma}{0.2}\right)^{-1/2}\left(\frac{M_{{\rm PBH},f}}{10^4\,{\rm g}}\right)^{-1}. \tag{168}$$

Before we compute the final amplitude of the induced GW spectrum, it is very important to estimate the effects of the finite duration of the evaporation to the perturbations. Although the transition to a radiation dominated universe takes place within less than $1/4$ of e-fold, perturbations in very small scales are very much affected by the finite width of the transition, as was discovered by Inomata et al. [65,66,68]. The main reason is clear if we neglect the expansion of the universe, which is justified as we are interested in very subhorizon scales. As PBHs evaporate their energy density decays as [68]

$$\rho_{\rm PBH} \propto M_{\rm PBH} \propto \left(1 - \frac{t}{t_{\rm eva}}\right)^{1/3}. \tag{169}$$

Now, we are interested in any effect that this decay may have on Φ. The Newtonian potential is related to fluctuations by the Poisson equation which in Fourier space reads

$$2\frac{k^2}{a^2}\Phi \approx \rho_{\rm PBH}\delta_{\rm PBH} + \rho_r\delta_r\,, \tag{170}$$

where $\delta_Q \equiv \delta\rho_Q/\rho_Q$ is the density contrast. We will do a big step here that can be easily confirmed by looking at the formulas in the appendix of Ref. [70] or in Ref. [68]. If we completely neglect the expansion of the universe we find that $\delta_{\rm PBH} \approx$ constant is a solution to the equations of motion of PBH density fluctuations. With this solution, we can estimate the size of the fluctuations in the radiation fluid sourced by the evaporation. With the Green's method to solve the equations of motion of radiation fluctuations, we find that

$$\delta_r \sim \frac{\rho_{\rm PBH}}{\rho_r}\frac{a^2\Gamma^2}{k^2}\delta_{\rm PBH}\,, \tag{171}$$

where Γ is the evaporation rate of PBHs and it is given by

$$\Gamma \equiv -\frac{d\ln M_{\rm PBH}}{dt} \sim \frac{1}{t - t_{\rm eva}}. \tag{172}$$

What (171) tell us is that fluctuations in the radiation fluid δ_r with $k/a \gg \Gamma$ are very suppressed with respect to $\delta_{\rm PBH}$ due to radiation pressure. This means that for such modes the dominant contribution to Φ is given by $\delta_{\rm PBH}$, even when PBH do not dominate the universe. Then by Equation (170) we see that $\Phi \propto \rho_{\rm PBH}$ until $\Gamma \sim k/a$, at which point the fluctuations in radiation dominate and perturbations behave as in a radiation dominated universe. The temporal decay of Φ due to the decay of $\rho_{\rm PBH}$ yields a relative suppression given by [68]

$$S_\Phi(k/k_{\rm eva}) \equiv \frac{\Phi}{\Phi_{\rm instant}} \approx \left(\sqrt{\frac{2}{3}}\frac{k}{k_{\rm eva}}\right)^{-1/3}, \tag{173}$$

where $\Phi_{\rm instant}$ refers to the value of Φ if the transition is instantaneous.

Now, we can use our estimate for an instantaneous transition (156) but taking into account the suppression of Φ until the transition to radiation domination is completely achieved. This yields a spectrum of curvature fluctuations at the start of the lRD given by

$$\begin{aligned}\mathcal{P}_{\Phi,\rm lRD} &= S^2_\Phi(k/k_{\rm eva})\mathcal{P}_{\Phi,\rm PBH}\\ &= \frac{3}{8\pi}\left(\frac{2}{3}\right)^{-1/3}\left(\frac{k_{\rm UV}}{k_{\rm eva}}\right)^{-2/3}\left(\frac{k_{\rm eq}}{k_{\rm UV}}\right)^4\left(\frac{k}{k_{\rm UV}}\right)^{-5/3}\Theta(k_{\rm UV}-k)\Theta(k-k_{\rm eq})\,.\end{aligned} \tag{174}$$

Comparing (174) with (152) we see that $n = 5/3$ and

$$\mathcal{A}_\Phi = \frac{3}{8\pi}\left(\frac{2}{3}\right)^{-1/3}\left(\frac{k_{\rm UV}}{k_{\rm eva}}\right)^{-2/3}\left(\frac{k_{\rm eq}}{k_{\rm UV}}\right)^4. \tag{175}$$

The final ingredient to derive the induced GW spectrum is the hierarchy of the scales involved. First, note that the ratio between the cut-off $k_{\rm UV}$ and $k_{\rm eva}$ is independent of β and it is given by [70]

$$\frac{k_{\rm UV}}{k_{\rm eva}} \approx 2.3\times 10^{6}\left(\frac{g_H(T_{\rm PBH})}{108}\right)^{-1/3}\left(\frac{M_{{\rm PBH},i}}{10^4\,{\rm g}}\right)^{2/3}. \tag{176}$$

The β independence of the ratio is due to the fact that $H_{\rm eva}$ only depends on the PBH mass and $a_{\rm eva}/a_i \propto n_{{\rm PBH},i}^{1/3}$ because it is a dust dominated universe. The hierarchy is closed with the second and third relations respectively given by [70]

$$\frac{k_{\rm eq}}{k_{\rm UV}} = \sqrt{2}\gamma^{1/3}\beta^{2/3} \quad \text{and} \quad \frac{k_{\rm eq}}{k_p} = \sqrt{2}\beta. \tag{177}$$

Now, inserting Equations (174), (176) and (177) into (157) yields an amplitude of induced GWs at the peak of approximately

$$\Omega^{\rm peak}_{\rm GWs,eva} \approx 1.5\times 10^{-2}\left(\frac{\beta}{10^{-6}}\right)^{16/3}\left(\frac{\gamma}{0.2}\right)^{8/3}\left(\frac{g_H(T_{\rm PBH})}{108}\right)^{-17/9}\left(\frac{M_{{\rm PBH},f}}{10^4\,{\rm g}}\right)^{34/9}. \tag{178}$$

Note that the estimate Equation (178) is the amplitude of induced GWs right after evaporation. It should also be noted that (178) is valid for very sharp PBH mass functions. As the mass function widens, the induced GW counterpart gets suppressed [68]. On top of that, we have that scales $k > k_{NL}$ (139), specially for $k \sim k_{UV}$, PBH density fluctuations $\delta_{\rm PBH}$ become $O(1)$ or larger. However, our estimate (178) has been derived in the linear regime and might receive corrections due to non-linear effects. Nevertheless, we expect it to be a crude order of magnitude estimate since the main source of induced GWs, the curvature perturbation, remains always smaller than unity,[20] i.e., $\Phi \ll 1$. Numerical relativity simulations are needed to derive a preciser estimate of the induced GWs generated in the non-linear regime. We show the resulting induced GW spectrum in Figure 8. Using (167) and (176) we find that the peak of induced GWs today lies at a frequency given by

$$f_{\rm UV} \approx 1.7\times 10^{3}{\rm Hz}\left(\frac{g_H(T_{\rm PBH})}{108}\right)^{1/6}\left(\frac{g_*(T_{\rm eva})}{106.75}\right)^{1/4}\left(\frac{g_{*,s}(T_{\rm eva})}{106.75}\right)^{-1/3}\left(\frac{M_{{\rm PBH},i}}{10^4\,{\rm g}}\right)^{-5/6}. \tag{179}$$

Quite surprisingly, the frequency of the peak enters the observational range of LISA, DECIGO and LIGO for $5\times 10^{8}\ {\rm g} > M_{\rm PBH,i} > 2\times 10^{4}$ g. Thus, this scenario is testable in the future.

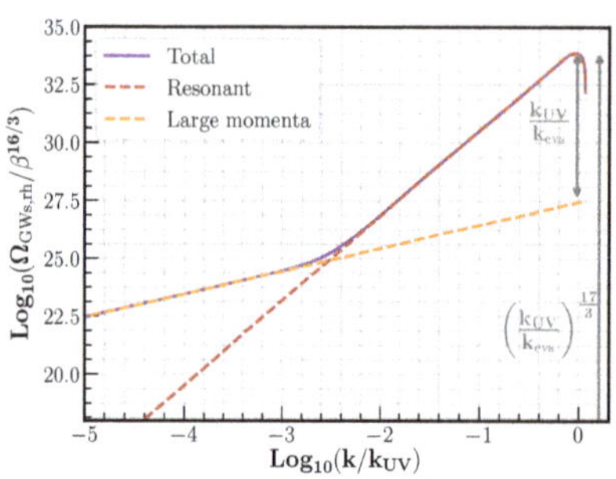

Figure 8. Induced GW spectral density after PBH evaporation normalized by $\beta^{16/3}$ and in terms of co-moving wavenumber $k/k_{\rm UV}$. The dashed red line shows the resonant contribution from Equation (154). The dashed orange line is the large momentum contribution which can be found in [70]. The purple line is the total GW spectrum. Due to the almost sudden transition the GW spectrum is amplified. This implies a very small initial fraction β of PBH not to overproduce induced GWs.

To compute the amount of induced GWs measured today we must consider the redshift of GWs until today. We did this in Section 5 with Equation (115). However, if we want to evaluate the amplitude of induced GWs at BBN, we only need to know the fraction of GWs that compose the total radiation after reheating. This can be done by considering the degrees of freedom of the standard model, which leads us to

$$\Omega_{\rm GW,BBN} = 0.39\left(\frac{g_*(T_{\rm eva})}{106.75}\right)\left(\frac{g_{*s}(T_{\rm eva})}{106.75}\right)^{-4/3}\Omega_{\rm GW,eva}\,. \tag{180}$$

Since these induced GWs act as additional radiation, they contribute to the effective number of relativistic species $N_{\rm eff}$ at BBN. The contribution of GWs to $N_{\rm eff}$ is by convention[21] written as [29]

$$\Omega_{\rm GW,BBN} = \frac{7}{8}\left(\frac{4}{11}\right)^{4/3}\Delta N_{\rm eff} < 0.05\,, \tag{181}$$

where in the last step we used that BBN [299,300] sets an upper bound $\Delta N_{\rm eff} < 0.2$.[22] In this way, we can constrain the amplitude of GWs from existing BBN bounds. Note that to be precise the BBN bound is a constrain on the total energy density of relativistic particles. Thus, we should have actually integrated our induced GW spectrum over $\ln k$ and then compare it with the bound (181). However, since the GW spectrum is very peaked and we are already working with order of magnitude estimates, it is justified to just look at the peak of the GW spectrum. If we use the current constraints from BBN, then we obtain that

$$\beta < 1.5\times 10^{-6}\left(\frac{M_{{\rm PBH},i}}{10^4\,{\rm g}}\right)^{-17/24}\,. \tag{182}$$

This is a new constraint on early epochs of PBH dominations which cannot be obtained by any other means.

Interestingly, PBH evaporation also emits "gravitons", i.e., very high frequency GWs. While these gravitons are not observable by GW detectors, they might be seen as effective relativistic particles at BBN [288–292]. Future CMB probes such as CMB-S4 [303] will be able to improve current bounds down to $\Delta N_{\rm eff} \sim 0.02$. Unfortunately, the graviton contribution of non-spinning PBHs to $N_{\rm eff}$ is out of reach for future CMB probes. Nevertheless, if PBH have a large spin (and dominate the universe at some point), the graviton contribution to $N_{\rm eff}$ will be accessible to future experiments [291,292]. This signature from graviton emission together with induced GWs will be an interesting probe of PBH dominated eras. While the graviton contribution to $N_{\rm eff}$ tells us about PBH spin, the induced GW counterpart has information on the formation mechanism [71]. As a clarification, note that the graviton emission and induced GWs are two very different effects. While the former is emitted by Hawking evaporation, the latter is generated by the time dependence of density fluctuations. The composition of such density fluctuations is not important, e.g., if a fraction are gravitons, as long as it behaves as fluctuations of a radiation fluid.

Before ending this section, we should note that we also have the induced GWs generated in the PBH dominated stage by using the Kernel (141). This is studied in detail in Ref. [69]. However, as we discussed at the beginning of this section this contribution is sensitive to the detailed transition to radiation domination. In the present case, where the transition is almost instantaneous, we can continue the tensor mode amplitude build up during eMD to the lRD. Nevertheless, this contribution is always subdominant compared to the one after reheating [66]. Thus, our estimate (178) provides the dominant peak of the induced GW spectrum.

7. The Gauge Issue

Despite the large amount of research on induced GWs, there is still the open question of what we really observe. This was made explicit in 2017, when Hwang, Jeong and Noh [149] showed that the induced GW spectrum generated in a dust dominated universe depends very much (by orders of magnitude) on the gauge chosen. Although the gauge issue was mentioned earlier by Matarrese, Mollerach and Bruni [37], it was not until [149] that it received more attention. From then and on, there has been a big activity to understand the problem [150–162]. For example in Refs. [151,152] other cosmological backgrounds rather than dust were studied. They found that the gauge dependence is in general present. Later studies [153–162], proposed partial solutions and investigated in which situations the spectrum of induced GW might be well defined. We dedicate Section 7.1 to explain the origin of the gauge issue. Then, we discuss these approaches in detail in Section 7.2.

Main references: The discussion in Section 7.1 is mainly based on [150,161]. Then, in Section 7.2 we revisit the main arguments of [153,154] and explain the improved results of [161].

7.1. The Origin of the Issue

To understand the source of problem, we must distinguish between what we call tensor modes and GWs. On the theoretical side, we define tensor modes in a homogeneous and isotropic universe as the transverse-traceless fluctuations of the spatial metric.[23] However, this definition must depend on how one chooses the spatial hypersurface. Although tensor modes are gauge invariant at linear order, they fail to be so at second order in cosmological perturbation theory. This is reflected by the fact that tensor, vector and scalar modes mix at second order [37]. This may pose a problem when we want to relate the tensor modes in a given slicing to the observational effect, which are the GWs. Note that this is not necessarily an "issue". If the quantity related to observation is clear and calculations are done properly (perhaps in a reasonable coordinate system), one should get proper results. For example, in studies of the CMB the temperature anisotropies and the polarization patterns of the CMB are well defined at second order. The interested reader is referred to Refs. [306–308] and references therein.

The trouble with GW observations is that we only know the GW detector response to linear GWs. Only at first order the relation between the tensor modes and GWs in the linearised theory is clear, since tensor modes are gauge invariant at that order. There is a subtlety, though. A tensor mode whose wavelength is larger than the horizon is frozen, does not evolve. So it acts as a constant anisotropy or shear and, as such, it has nothing to do with actual GWs. Furthermore, since the wavelength of the tensor modes is larger than the curvature scale, the averaging procedure used to derive the (pseudo) energy momentum tensor of GWs in Section 2.2 does not apply. For these reasons, we will be only concerned about tensor modes on subhorizon scales. These tensor modes have a wavelength smaller than the horizon and behave as a wave, which are the GWs we could observe.

In the absence of the GW detector response at second order, we face the problem that we do not know the relation between tensor modes and the true GWs. In most situations though this should not really be a problem, unless one is choosing a very strange coordinate system to describe the observations. For instance, to describe the GWs measured by a GW detector today we usually do not need to consider second order terms as the amplitude of the GWs is already tiny enough. However, this issue is particularly relevant for induced GWs due to their secondary nature. Now, there are two important aspects of the problem. First, we have that the derived tensor spectrum depends on the gauge choice. Second, we expect that once the source term in (48) is absent, i.e., its amplitude decayed enough, such previously induced tensor modes behave as a freely propagating GWs. From that moment on we should be able to relate the induced tensor modes to GWs. Thus, we must reconcile our physical intuition with the fact that tensor modes are gauge dependent. To arrive to that, there are two points that we would like to make as clear as possible. The first one concerns the tensor modes h_{ij} themselves and the second one to the actual observable, which in Section 2.2 we decided it would be the energy density of induced GWs (5). We present them in the following subsections and we later discuss the current state of the issue in Section 7.2.

7.1.1. The Definition of Tensor Modes Is Gauge Dependent

When working at second order in cosmological perturbation theory, it is very important to be clear about our convention. Here we will use the conformal decomposition from Equations (34) and (38) which in a general gauge reads

$$ds^2 = a^2(\tau)\Big(-N^2 d\tau^2 + e^{2\phi}Y_{ij}\Big(dx^i + N^i d\tau\Big)\Big(dx^j + N^j d\tau\Big)\Big)\,, \tag{183}$$

where N and N^i respectively are the lapse and shift vector. We then perturb the variables as follows

$$N = 1 + \alpha \quad , \quad N_i = \partial_i \beta \tag{184}$$

$$[\ln Y]_{ij} = h_{ij} + 2\Big(\partial_i\partial_j - \tfrac{1}{3}\delta_{ij}\Delta\Big)E\,, \tag{185}$$

where we only focused on the tensor and scalar components and ϕ is already a perturbation. Our convention differs from the common expansions, which expands the metric linearly in the perturbation variables instead of exponentially. Nevertheless, our choice considerably simplifies calculations. To relate the two different conventions we refer the reader to the appendix of Ref. [161]. We also use the exponential mapping for a gauge transformation (e.g., see the review by Malik and Wands [164]) which naturally follows from the Hamiltonian flow (see e.g., [150] for the application to second order cosmological perturbations). To be more concrete, let us do an infinitesimal coordinate transformation by

$$\tilde{x}^\mu = x^\mu - \xi^\mu \quad \text{with} \quad \xi^\mu = (T, \partial^i L)\,, \tag{186}$$

where we focused only on the scalar component of the spatial vector ξ^i. Then a quantity Q in one gauge is related to the same quantity evaluated in another gauge $\tilde{Q}$ by the exponential mapping

$$\tilde{Q} = e^{\pounds_\xi} Q\,, \tag{187}$$

where $\pounds_\xi$ is the Lie derivative along ξ. With this convention, we find that the tensor modes at second order change according to

$$\tilde{h}_{ij} = h_{ij} - \widehat{TT}_{ij}{}^{ab}\Big\{ -2\partial_a(\beta - E')\partial_b T + 2\partial_a\partial_k L \partial_k \partial_b E + \partial_a T \partial_b T + \partial_a \partial_k L \partial_b \partial_k L\Big\}\,. \tag{188}$$

Equation (188) is the first cause of the gauge issue for induced GWs. Since the source to induced tensor modes acts at second order in perturbation theory, the actual gauge in which we do the calculations might drastically change the form of the solutions for h_{ij} [149]. The reader interested in the gauge transformations particularly related to induced GWs might go to Ref. [161]. General reviews of cosmological perturbation theory are listed at the Section 1.2.

7.1.2. The Definition of GW Energy Density Is Gauge Dependent

Once we showed that the tensor modes h_{ij} are gauge dependent, let us look at the definition of the "observable" energy density of GWs. Using (2) we have that

$$\rho_{\rm GW} = t_{00}^{\rm GW} = \frac{M_{pl}^2}{8a^2}\left\langle h'^{ij}h'_{ij} + \partial_k h^{ij}\partial^k h_{ij}\right\rangle . \tag{189}$$

Now, looking at (188) and (189) the first question that comes to mind is: h_{ij} in (189) is evaluated in which gauge? Indeed, if we naively apply the gauge transformation (188) to (189) we see that

$$\tilde{\rho}_{\rm GW} \sim \langle \tilde{h}'^{ij}\tilde{h}'_{ij}\rangle = \rho_{\rm GW} + \langle \partial_i T\partial_j T\partial^i T\partial^j T\rangle + \ldots , \tag{190}$$

where we neglected other terms to illustrate the main point, which is that the term quartic in T in (190) does not vanish in general. This shows something that we were expecting in Section 2: the energy density of GWs (189) in analogy of the Ricci flat result (2) is strictly speaking not well-defined on a cosmological background at second order in perturbation theory. Furthermore, it does not matter when you evaluate the energy density, e.g., today. In principle, we can find a coordinate transformation in which $\rho_{\rm GW}$ is very different from the usual estimate. For example, if we go to a frame where the detector is wildly oscillating, Equation (189) might not yield sensible results. This is the second and the most important cause for the gauge dependence of induced GWs.

7.2. Current "Solutions"

In view of the current lack of a description at second order in cosmological perturbation theory of the detector response to passing GWs, all we can do is to try to argue when Equation (189) yields sensible results. Before we start discussing the details, it is important to realize that a gauge invariant formulation of the equations of motion of induced GWs (48) does not help at all. There are infinitely many ways to define gauge invariant variables, each definition corresponding to a particular gauge choice. Thus, the same question remains, although with a slightly different language: which gauge invariant definition of the tensor modes goes into (189)? Similarly, if one tries to find a gauge invariant formulation of (189), which implies going to fourth order in cosmological perturbation theory (and it is clearly not the point), it is not clear what is the relation with the real observable. So, what can we do?

An interesting direction, first explored by [153] and later corrected by [154], is to argue what is the gauge (or coordinate frame) which best describes the GW detection. According to Ref. [153], based on the analogy with asymptotically flat spacetimes, the best gauge is the so-called transverse-traceless (TT) gauge where the metric reads

$$ds^2 = -dt^2 + (\delta_{ij} + h_{ij})dx^i dx^j . \tag{191}$$

In Minkowski spacetime, there are only two degrees of freedom, the ones in h_{ij}, since there is no source to the energy momentum tensor. In this gauge and in the linearised theory, the GW detector is at rest, even for passing GWs, which makes the physical interpretation (and calculations) much easier [184,185]. The closest gauge to the TT gauge in a cosmological background is the synchronous gauge, which is a coordinate system spanned by a family of geodesics. In our convention the metric in the synchronous gauge is given by

$$ds^2 = a^2(\tau)\left(-d\tau^2 + e^{2\phi}Y_{ij}dx^i dx^j\right) , \tag{192}$$

where we set $\alpha = \beta = 0$ in (183). Let us remind the reader that

$$[\ln Y]_{ij} = h_{ij} + 2\left(\partial_i\partial_j - \tfrac{1}{3}\delta_{ij}\Delta\right)E , \tag{193}$$

and therefore we have two more variables, ϕ and E, than in (191). In the end, only one of them is dynamical. Thus, although at linear order (193) resembles the TT gauge (191) it is certainly not the same. At linear order and if we neglect the drift due to the expansion a, the metric (193) does describe a fixed detector. However, at second order the detector will not be fixed and will feel the influence of E. Another "problem" of the synchronous gauge is that it is not a completely fixed gauge. There is a residual gauge ambiguity which corresponds to a change of the reference geodesics. This gauge ambiguity turns out to be very important in the final spectrum of induced tensor modes. If one does not properly fix the synchronous gauge, then induced tensor modes could be very different than, e.g., in the Newtonian gauge [158]. This is because if we start with a family of geodesics defined in the very early universe and we use the same geodesics to describe the GW detection, it is very possible that these geodesics do not match the rest frame of the detector. The resulting induced tensor spectrum might be singular if those geodesics happen to have focusing singularities near the detector [161]. Nevertheless, by properly fixing the gauge one finds that the induced GW spectrum in the synchronous gauge and in the Newtonian gauge yield the same prediction in a radiation dominated universe [153–155,158]. This also holds for any cosmological background with a constant equation of state [161] if it is not dust, i.e., $c_s^2 \neq 0$. This result provides evidence that perhaps the energy density of GWs (189) computed in the Newtonian gauge, which we do for simplicity of the calculations, yields meaningful results. However, it does not explain why these two predictions coincide. Furthermore, the original issue raised in [149] in a dust dominated universe persist.

Here is where the direction explored in Ref. [161] becomes relevant: when and why does the energy density of GWs (189) yield the same predictions and for which gauges? The physical intuition is that the analogy with Ricci flat spacetimes should work well if:

(i) we look at really subhorizon scales ($k \gg \mathcal{H}$), where cosmology should be less relevant;

(ii) there is (barely) no source of induced GWs (48), so that induced tensor modes are freely propagating GWs;

(iii) we use a coordinate system suitable for small distances calculations, so that we are not confused by strange coordinate artefacts.

As shown in [161], if points (i)–(iii) are satisfied then (189) is gauge independent up to corrections of $O(\mathcal{H}^2/k^2)$. Note that conditions (ii) and (iii) yield some restrictions respectively on the type of cosmological background and on the class of gauges. In order to show such approximate gauge independence, Ref. [161] argued that a good gauge choice to start with is the Newtonian gauge. The physical reason for that choice is that on small distances the Newtonian gauge reduces to the well-known Newtonian gravity. The practical reason is that calculations are simpler. Nevertheless, the argument provided in [161] does not strongly depend on the initial choice provided that it is suitable for subhorizon calculations. For example, one could have started with the synchronous gauge or the uniform Hubble gauge.

Starting from the Newtonian gauge we can relate the curvature perturbation Φ in any gauge, say Φ^G, with that of the Newtonian gauge, say Φ^N, by (187) which yields

$$\Phi_G = \Phi_N + \mathcal{H}T_G + \tfrac{1}{3}\Delta L_G\,. \tag{194}$$

Then, the class of suitable gauges for subhorizon physics is defined as those gauges in which

$$\Phi_G(k \gg \mathcal{H}) = O(\Phi_N(k \gg \mathcal{H}))\,. \tag{195}$$

This means that the curvature perturbation need not be the same but have the same qualitative behaviour. We shall see that although this is a requirement on the scalar variable Φ, it yields the gauge independence of h_{ij}. Using (195) into (194) we find conditions on the gauge parameters T_G and L_G. For example, the gauge time parameter must be

$$T_G(k \gg \mathcal{H}) = O(\mathcal{H}^{-1}\Phi_N(k \gg \mathcal{H})) \quad \text{or higher order.} \tag{196}$$

Note that this abstract condition (195) actually includes most of the commonly used gauges such as the synchronous, the flat and the constant Hubble gauges [161]. However, it excludes the comoving slicing gauge [161]. The main reason for such exclusion is that the comoving slicing gauge is a choice where the spatial hypersurface is orthogonal to the flow lines of the fluid. This is a good choice for superhorizon scales where the flow lines are slowly changing with time, but it is a poor choice for subhorizon scales where fluid velocities oscillate. To convince ourselves of the gauge independence of (189), take the solution to Φ^N (74) and h_{ij}^N (83) for constant w, which read

$$\Phi_k^N \propto \left(\frac{k}{\mathcal{H}}\right)^{-2-b} \quad \text{and} \quad h_k^N \propto \left(\frac{k}{\mathcal{H}}\right)^{-1-b}. \tag{197}$$

For such solutions, the condition (196) translates into

$$T_G = O(h_k^N) \quad \text{or higher order}. \tag{198}$$

Then, using (188) with $L_G = 0$ and in Fourier space we have

$$h_k^G = h_k^N - \int \frac{d^3q}{(2\pi)^3} e^{ij} q_i q_j T_G(q) T_G(|\mathbf{k} - \mathbf{q}|)\,, \tag{199}$$

where we used that in the Newtonian gauge we have $E = \beta = 0$. So when substituting (198) in (199) we conclude that for all gauges which satisfy (195) we have that

$$h_k^G(k \gg \mathcal{H}) = h_k^N(k \gg \mathcal{H}) + O(\mathcal{H}^2/k^2)\,, \tag{200}$$

which proves the approximate gauge independence of $\rho_{\rm GW}$ (189). It should be noted that while the work of [161] is not a solution per se, it shows when and why the predictions for induced GWs are well-defined and gives strong evidence that the calculations in this class of gauges are meaningful. In any case, the final word would be to find the GW detector response at second order.

Before ending this section, let us place some more attention to point (ii), i.e., that the source term to induced GWs is not active. This means that the amplitude of scalar fluctuations has decayed enough so that they are no longer a significant source of induced GWs. This is often implied by taking the subhorizon limit $k \gg \mathcal{H}$ since induced GWs are mainly generated at horizon crossing, i.e., $k \sim \mathcal{H}$, and not afterwards. However, there is one exception: the dust dominated universe studied in Section 6. In a dust dominated universe, the Newtonian potential Φ is constant on all scales and, therefore, is constantly sourcing induced GWs. This is the reason why Ref. [149] finds such large gauge dependence of induced GWs in a dust dominated universe. Yet, we learned in Section 6 that the final spectrum of induced GWs is very sensitive to the transition from dust to radiation, where Φ changes from a constant to an oscillating function. Thus, one must follow the induced tensor modes generated in the dust dominated universe until they are deep inside the horizon in the radiation dominated era. It is from that moment on that the conditions (i)–(iii) of [161] apply. Then we know that the spectrum of induced GWs is approximately gauge invariant.

8. Other GW Sources Related to PBH Formation

So far, we have focused on the GW induced by large primordial fluctuations. These large fluctuations collapse to form PBH if their rms amplitude is large enough. Once PBH form they may source other GWs independent of the induced GWs. Let us anticipate that the other GW counterparts are from PBHs mergers and graviton emission by Hawking evaporation. We briefly comment on these possibilities below.

On one hand, PBHs will form binaries in the early universe and eventually merge. The formation of such binaries is mainly due to the three body interaction with the nearest PBH [309]. These PBH binaries can be detected by the observation of resolved or unresolved merger events. In the former we would see the full GW waveform, e.g., as in the LIGO detections [310], and the latter will appear as a stochastic GW background. These make two additional possible GW counterparts to PBHs. Note that if PBHs are lighter than 10^{15} g they evaporated by today and, therefore, we most likely cannot observe any resolved merger. PBHs heavier than 10^{15} g have not yet evaporated and their binaries could be merging in the nearby universe. The GW spectrum from the unresolved mergers is studied in Refs. [142,311,312] while the estimates for the merger rates can be found in Refs. [174,309,313–316]. A very rough order of magnitude estimate of the frequency at which the GWs from the mergers of PBH binaries will show up is given by the frequency of the GWs emitted at the Innermost Stable Circular Orbit [185]

$$f_{\rm GW,max} \approx 2 f_{\rm ISCO} \approx 4.4 \text{ kHz}\left(\frac{M_\odot}{M}\right), \tag{201}$$

where M is the total mass of the binary and $M_\odot$ is a solar mass. For a monochromatic PBH mass function then $M = 2M_{\rm PBH,i}$. This may change by a factor of a few depending on the redshift at the time of merger, so that $f_{\rm GW,max,0} = f_{\rm GW,max}/(1+z)$. For PBHs with $M_{\rm PBH,i} \gg 10^{15}$ g we can assume that most of the mergers occur in the nearby universe and so the frequency (201) already gives a good intuition of the position of the peak in the SGWB from mergers and the chirp frequency of the GW waveform, see for example Ref. [142]. However, PBHs with $M_{\rm PBH,i} < 5 \times 10^{8}$ g have merged and evaporated long before BBN. In this case, we can assume that most of the mergers occur close to the evaporation time and, if PBH dominated the universe as in Section 6.2, we can take into account the redshift of the frequency (201) until today. A quick estimate yields

$$f^{\rm merger}_{\rm peak,0} \approx 10^{15}\,\text{Hz}\left(\frac{M_{\rm PBH,i}}{10^4 \text{ g}}\right)^{1/2}\left(\frac{g_*(T_{\rm eva})}{106.75}\right)^{1/4}\left(\frac{g_{s*}(T_{\rm eva})}{106.75}\right)^{-1/3}\left(\frac{g_H(T_{\rm PBH})}{108}\right)^{-1/2}, \tag{202}$$

which gives the right order of magnitude estimate for the position of the peak. For the detailed spectral shape in this case see Ref. [68]. The peak frequency of this SGWB is unfortunately too high to be observed by current laser interferometers. Note that the PBH mass range $10^9 \text{ g} < M_{\rm PBH,i} < 10^{14}$ g is subject to tight constraints from entropy production, changes in the abundance of light elements, the extragalactic photon background and damping of CMB temperature anisotropies. For a detailed review see Ref. [176].

On the other hand, when PBHs evaporate they also emit "gravitons" by Hawking radiation. The typical frequency of such gravitons can be estimated by assuming that Hawking radiation follows a black body spectrum. Then, the peak frequency is roughly at $f_{\rm peak} \approx 2.82\, T_{\rm PBH,i}$ where $T_{\rm PBH,i}$ is given in Equation (164) in terms of the initial PBH mass $M_{\rm PBH,i}$. Note that the frequency $f_{\rm peak}$ is the peak frequency of the spectrum when PBH evaporate. Thus, we have to take into account that frequencies redshift proportional to a^{-1} to calculate the frequency of these gravitons today. To do that, we will use the calculations of Section 6.2 which considers the case that PBH dominate the universe. Then, we can compute the evaporation temperature $T_{\rm eva}$ in terms of $M_{\rm PBH,i}$ by Equation (165). Inserting all numerical coefficients, we find that the graviton peak frequency today is given by

$$f^{\rm graviton}_{\rm peak,0} \approx 10^{16}\,\text{Hz}\left(\frac{M_{\rm PBH,i}}{10^4 \text{ g}}\right)^{1/2}\left(\frac{g_*(T_{\rm eva})}{106.75}\right)^{1/4}\left(\frac{g_{s*}(T_{\rm eva})}{106.75}\right)^{-1/3}\left(\frac{g_H(T_{PBH})}{108}\right)^{-1/2}. \tag{203}$$

A more detailed derivation of the graviton spectrum and the frequency range can be found in Ref. [68]. Unfortunately, these are very high frequency GWs which are not observable by direct detection with the current capabilities. Nevertheless, they might be within range of future CMB experiments by looking at the contribution of these gravitons to the effective number of relativistic species at BBN [288–292]. Very interestingly if PBHs have a large spin, their imprint on the effective number is within range of the CMB-S4 experiment [291,292].

It is also important to notice that in some inflationary models involving axions coupled to gauge fields the enhancement in the primordial curvature perturbation is accompanied by an enhancement in the primordial tensor spectrum [88,100,317]. In this case the primordial GW signal is in general larger than the induced GW one. Thus, in this scenario the induced GW signal may be buried inside the primordial GWs. It is possible that a detailed observation of the spectrum and the PBH counterpart helps to distinguish such scenario.

9. Current and Future Observational Prospects

In this section, we briefly discuss the future observational prospects regarding induced GWs. We will not focus on the prospects for PBHs, although they are equally interesting. Nevertheless, let us mention the most interesting PBH windows as they provide good motivation to look for induced GWs in particular ranges. The first one is the LIGO binary black hole mergers [309,313–315]. Notably, although some models are already ruled out by current data [318] there seems to be evidence for a small fraction of PBHs in addition to astrophysical ones [319,320]. Interestingly, the corresponding induced GWs to the PBHs in the LIGO window fall in the PTA band.[24] Coincidentally, the NANOGrav collaboration reported a possible SGWB in such range [18], but is yet to be confirmed. We discuss more on NANOGrav a bit later. Another interesting window is the planet mass PBHs [321] where the OGLE collaboration reported few microlensing events of planet mass objects [322]. Future microlensing observations such as with Subaru/HSC [323] will be able to confirm the results from OGLE. If we assume that radiation dominated early universe, planet mass PBHs have an induced GW counterpart that falls in between PTA and LISA. The last promising window is for PBHs in the asteroid mass range where there are practically no observational constraints yet [174,176] and therefore PBHs can make up for all dark matter. In this case, the induced GW counterpart falls right inside the LISA band. Thus, induced GW will be an essential probe of PBH as dark matter. For a review see Ref. [179]. Additionally, see Ref. [142] for a detailed discussion.

Let us now turn to the observational prospects of induced GWs. Although there are weak constraints from LIGO [324] on the induced SGWB [312,325,326], the most interesting part is the future prospects. On one hand, the absence of induced GWs directly translates into bounds on the primordial spectrum [30]. In this way, assuming radiation domination, we see from Ref. [30] that the bounds are roughly $\mathcal{P}_\mathcal{R} \sim 10^{-5}$ for $k \sim 10^6 - 10^8\,\mathrm{Mpc}^{-1}$ in the PTA band, $\mathcal{P}_\mathcal{R} \sim 10^{-4}$ for $k \sim 10^{11} - 10^{14}\,\mathrm{Mpc}^{-1}$ in the LISA band and $\mathcal{P}_\mathcal{R} \sim 10^{-3}$ for $k \sim 10^{15} - 10^{18}\,\mathrm{Mpc}^{-1}$ in the ET band. CMB spectral distortions will probe down to $\mathcal{P}_\mathcal{R} \sim 10^{-8}$ for $k \sim 10 - 10^5\,\mathrm{Mpc}^{-1}$ [32,33]. However, one of the problems is that such future bounds depend on the shape of the induced GW spectrum. For example, if we only see part of the induced GW spectrum, e.g., the UV or IR tail but not the peak, then in order to extract information on the amplitude of primordial fluctuations we must extrapolate the results beyond the range of the GW detector by assuming some template [172]. On top of that, we have the issue that some induced GW templates might degenerate with other sources. As a good example for these potential problems, we proceed to discuss the various induced GW explanations of the NANOGrav results [18].

The NANOGrav collaboration reported a common process signal in the time residuals of the frequencies of pulsars [18]. While no quadrupolar nature of the signal has been found, it is interesting to consider the possibility that the signal measured is in fact the SGWB of induced GWs. The main features of the reported spectrum [18], in cosmologists terms, are that it has an amplitude of $\Omega_{\mathrm{GW},0}h^2 \sim 10^{-9}$ at a frequency of 2×10^{-9} Hz and a spectral index, that is $\Omega_{GW,0} \propto k^n$, that ranges from $n \sim 1/2$ to $n \sim -3/2$ within the 1-σ bounds. This means that the NANOGrav data seems to prefer a rather flat GW spectrum. This is illustrated in Figure 9. Using the very rough estimate of Section 2, that is $\Omega_{GW,0}h^2 \sim 10^{-6}\mathcal{P}_{\mathcal{R}}^2$, we find that if the peak of the induced GW spectrum lies around $f \sim 2 \times 10^{-9}$ Hz then the amplitude of primordial fluctuations that sourced such induced GWs is about $\mathcal{P}_{\mathcal{R}} \sim O(10^{-2})$. Such amplitude of primordial fluctuations is right at the value when PBH formation is interesting, and their masses are in the LIGO window.

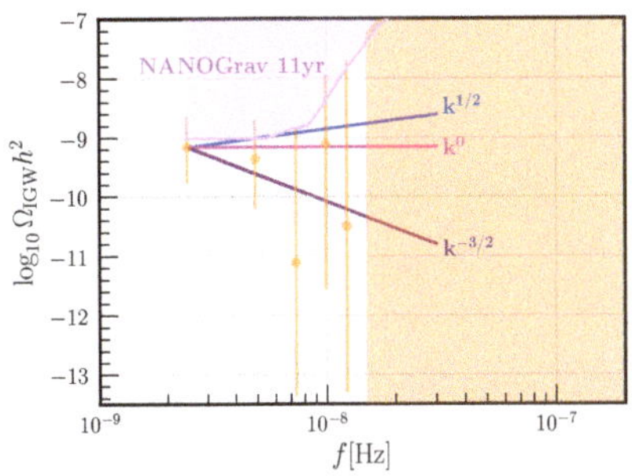

Figure 9. NANOGrav results on the possible SGWB [18] in terms of spectral density. The orange dots are the data points from the timing residuals with their error bars. We only show the first five data points which are the most relevant. The points in the orange shaded region were compatible with white noise and not considered for the fit. In light purple we also show the NANOGrav 11-yr sensitivity curve from [327]. In solid lines we show the allowed slopes within the 1-σ contours. In blue and purple we respectively show the upper and lower limits on the slope, $k^{1/2}$ and $k^{-3/2}$. In magenta we give the approximate best fit for the data, which is a scale invariant spectrum.

The main complication is that the NANOGrav GW spectrum is rather flat and the induced GWs from a sharp peak are very peaked and decay as k^3 in the IR. This means that there are two possibilities: either the GWs were induced by a broad peak in the primordial spectrum (including a flat spectrum) [129–131,133,135] or the GWs were induced in non-radiation dominated universe with $w < 1/3$ [132]. Below we list the main results of these works:

- Ref. [129] found a negligible amount of PBH in LIGO region but the PBH could be the seeds of supermassive black holes.
- Ref. [131] found a larger amount of PBH in the LIGO band than [129] and also pointed out that the SGWB from unresolved mergers could detectable in the future by LISA and ET.;
- Ref. [130] showed that the induced GWs from a flat primordial spectrum could explain all dark matter;
- Ref. [134] proposed a particular inflationary model with an axion-like curvaton which could explain the NANOGrav result as well as the LIGO events due to primordial non-gaussianities. Similar claims appear in Ref. [135] with a Higgs-like inflation with non-canonical kinetic terms;
- In a different direction, Refs. [132,133] studied the induced GWs from a peaked primordial spectrum in a non-radiation dominated universe. They both find that a soft equation of state seems to be preferred. In particular, Ref. [132] use the IR tail of the induced GW spectrum from a very peaked primordial spectrum (see Section 5) to fit the NANOGrav results. They found that the 1σ contours on the spectral tilt translate to bounds for the equation state parameter as $-0.1 < w < 0.05$;
- Ref. [81] uses the UV tail of the induced GW spectrum from a broken power-law primordial spectrum. They concluded that the NANOGrav results imply a small non-gaussianity parameter $f_{NL} < 0.7$;

It is important to note that the induced GWs are not the only explanations of the NANOGrav results. For example, among many others, possible candidates are cosmic strings [328–331] and first order phase transitions [332–339]. Within the next decades, we expect to have more information from PTA experiments. We also expect the GW detector LISA to launch. In order to distinguish these candidates to the NANOGrav data, we will need better resolution for a better Bayesian analysis [336] but crucially we will need other discriminators of cosmic GWs. In the case of induced GWs, if some of the LIGO binary black hole mergers are found to be PBHs, the NANOGrav signal would have a strong case to be induced GWs.

10. Summary of Main Formulas

We dedicate this section to recollect the main formulas scattered around the review and present them so that they are ready for use. We first give the GW spectrum measured today $\Omega_{\rm GW,0}$ in terms of the spectral density of GWs at a pivot scale $\Omega_{\rm GW,rh}$. The subscript "rh" stands for reheating and refers to the time of instantaneous transition to the standard radiation dominated era. In the case where GWs are induced during radiation domination, the pivot time "rh" refers to the time when induced GWs of given wavenumber k are sufficiently inside the cosmological horizon to be treated as a radiation fluid in an expanding universe. With these clarifications, we can use the results of Sections 4 and 5 to write

$$\Omega_{\rm GW,0}h^2 = 1.62\times10^{-5}\left(\frac{\Omega_{r,0}h^2}{4.18\times10^{-5}}\right)\left(\frac{g_*(T_{\rm rh})}{106.75}\right)\left(\frac{g_{*s}(T_{\rm rh})}{106.75}\right)^{-4/3}\Omega_{\rm GW,rh}\,. \tag{204}$$

We give the detailed expression of $\Omega_{\rm GW,rh}$ for a general equation of state in Section 10.1 and for the particular case of a dust universe with an instantaneous transition to radiation domination in Section 10.2.

10.1. General Equation of State

We studied in detail the derivation of $\Omega_{\rm GW,rh}$ for general expansion histories in Section 4. The main approximations used are the following: (i) there is an epoch of constant w and c_s after inflation, (ii) the peak (or plateau) in the primordial spectrum enters the horizon during such era and (iii) the transition to radiation domination is instantaneous. Note that in the case of standard radiation domination these assumptions are unnecessary. The presence of a transition to radiation domination introduces another relevant scale $k_{\rm rh}$, which is the last co-moving scale that entered the horizon at the transition. The GW spectrum for scales $k \gg k_{\rm rh}$ takes a general form given by

$$\Omega_{\rm GW,rh} = \left(\frac{k}{k_{\rm rh}}\right)^{-2b}\int_0^\infty dv\int_{|1-v|}^{1+v} du\,\mathcal{T}(u,v,b,c_s)\mathcal{P}_{\mathcal{R}}(ku)\mathcal{P}_{\mathcal{R}}(kv)\,, \tag{205}$$

where we defined the transfer function as

$$\begin{aligned}\mathcal{T}(u,v,b,c_s) =&\mathcal{N}(b,c_s)\left(\frac{4v^2-(1-u^2+v^2)^2}{4u^2v^2}\right)^2|1-y^2|^b\\ &\times\left\{\left(\mathsf{P}_b^{-b}(y)+\frac{b+2}{b+1}\mathsf{P}_{b+2}^{-b}(y)\right)^2\Theta[c_s(u+v)-1]\right.\\ &+\frac{4}{\pi^2}\left(\mathsf{Q}_b^{-b}(y)+\frac{b+2}{b+1}\mathsf{Q}_{b+2}^{-b}(y)\right)^2\Theta[c_s(u+v)-1]\\ &\left.+\frac{4}{\pi^2}\left(\mathcal{Q}_b^{-b}(-y)+2\frac{b+2}{b+1}\mathcal{Q}_{b+2}^{-b}(-y)\right)^2\Theta[1-c_s(u+v)]\right\},\end{aligned} \tag{206}$$

and the numerical coefficient is given by

$$\mathcal{N}(b,c_s) \equiv \frac{4^{2b}}{3c_s^4}\Gamma^4\left[b+\tfrac{3}{2}\right]\left(\frac{b+2}{2b+3}\right)^2(1+b)^{-2(1+b)}. \tag{207}$$

In Equation (206) we have introduced the following notation for convenience:

$$b = \frac{1-3w}{1+3w} \quad \text{and} \quad y = 1 - \frac{1-c_s^2(u-v)^2}{2c_s^2 uv}. \tag{208}$$

These equations are general enough for $k \gg k_{\rm rh}$ that can be used for any shape of the primordial spectrum as long as the main feature, whether it is a peak or a plateau, enters during the $w =$ constant epoch. If the transition to radiation domination is gradual, then we only expect a change in the IR tail of the spectrum for modes which entered during the relevant stages of the transition. Following with the instantaneous transition, it is important to note that for $b > 1$ the IR tail has a red tilt and the spectrum for $k \ll k_{\rm rh}$ becomes important. Nevertheless, a good approximation is to stop the spectrum at $k \sim k_{\rm rh}$ and match it to the typical IR scaling in radiation domination, i.e., $\Omega_{\rm GW,rh} \propto k^3$. Details on the spectrum for $k \ll k_{\rm rh}$ can be found in Section 4. It should be noted that (206) is not valid for $c_s = 0$. The case of $c_s = 0$ is studied in detail in Section 6 and a quick estimate is presented below in Section 10.2.

The transfer function (206) is expressed in terms of Associated Legendre functions which are provided in Appendix H in terms of Hypergeometric functions. This general form, while very useful, might be a hindrance to a quick implementation. For this reason, we present below particular examples of the transfer function (206). We choose the cases where $b \in \mathbb{Z}$ and $b \in \mathbb{Z} + \frac{1}{2}$ since then the associated Legendre functions (and the Hypergeometric functions) reduce to polynomials. Note that only when $b \in \mathbb{Z}$ logarithmic terms appear. We will first check the standard case of radiation domination ($b = 0$, $w = 1/3$) and we will later turn to other situations. We give the transfer function for a stiff fluid $b = -1/2$ ($w = 1$), a soft fluid $b = 1/2$ ($w = 1/9$), a pressureless fluid $b = 1$ ($w = 0$) and negative equation of state fluid $b = 2$ ($w = -1/9$). We keep the sound speed c_s arbitrary. For an adiabatic perfect fluid, we have $c_s^2 = w$ and for a canonical scalar field $c_s^2 = 1$.

10.1.1. Radiation Domination

The transfer function (206) for $b = 0$ ($w = 1/3$) reads

$$\begin{aligned}\mathcal{T}_{RD}(u,v,c_s,w=1/3) = &\frac{1y^2}{3c_s^4}\left(\frac{4v^2-(1-u^2+v^2)^2}{4u^2v^2}\right)^2 \\ &\times\left\{\frac{\pi^2}{4}y^2\Theta[c_s(u+v)-1] + \left(1-\frac{1}{2}y\ln\left|\frac{1+y}{1-y}\right|\right)^2\right\}, \end{aligned} \tag{209}$$

where we kept the variable y as it simplifies considerably the expressions. This recovers the result of Kohri and Terada [63] after taking into account all numerical factors in (7).

10.1.2. Stiff Fluid (Kinetic) Domination

Another notable case of interest is a stiff fluid with $b = -1/2$ ($w = 1$). This is for example typical of quintessential inflation scenarios [340–343]. For a kinetic dominated period, the transfer function (206) is given by

$$\mathcal{T}_{\rm KD}(u,v,c_s,w=1) = \frac{4}{3\pi c_s^4|1-y^2|}\left(\frac{4v^2-(1-u^2+v^2)^2}{4u^2v^2}\right)^2$$
$$\times\left\{\left(1+3y^2\right)\Theta[c_s(u+v)-1]+\left(1-3y^2+3y\sqrt{|1-y^2|}\right)^2\Theta[1-c_s(u+v)]\right\}. \quad (210)$$

This coincides with the formulas derived in Ref. [64,74]. It has also been studied numerically in Ref. [75].

10.1.3. Soft Fluid Domination

The next case is that of a soft fluid with $b = 1/2$ ($w = 1/9$). This could be due to a scalar field rolling down an exponential potential. After some algebra, we find that

$$\begin{aligned}\mathcal{T}(u,v,c_s,w=1/9) = {} & \frac{2^8}{3^8\pi c_s^4}\left(\frac{4v^2-(1-u^2+v^2)^2}{4u^2v^2}\right)^2 \\ & \times\left\{\left(4+45y^2\right)\Theta[c_s(u+v)-1]\right. \\ & \left.+\left(y(3-10y^2)+(2+10y^2)\sqrt{|1-y^2|}\right)^2\Theta[1-c_s(u+v)]\right\}. \quad (211)\end{aligned}$$

10.1.4. Pressure-Less Fluid Domination

We turn now to the case of a pressure-less fluid with $b = 1$ ($w = 0$). Note that this is not dust in the sense that $c_s \neq 0$. So that even though the fluid has no pressure, perturbations still propagate at a finite speed. This is the case of a scalar field rolling down an exponential potential and might also be close to the coherent oscillations of a scalar field around the bottom of the potential [344]. The kernel for $b = 1$ is given by

$$\begin{aligned}\mathcal{T}(u,v,c_s,w=0) = {} & \frac{3^3\,5^2}{2^{14}c_s^4}\left(\frac{4v^2-(1-u^2+v^2)^2}{4u^2v^2}\right)^2 \\ & \times\left\{\frac{\pi^2}{4}(1-y^2)^2(1+3y^2)^2\Theta[c_s(u+v)-1]\right. \\ & \left.+\left(y(1-3y^2)-\frac{1}{2}(1+2y^2-3y^4)\ln\left|\frac{1+y}{1-y}\right|\right)^2\right\}. \quad (212)\end{aligned}$$

10.1.5. Negative EoS Fluid Domination

The last case we are interested in is a negative EoS parameter. When $-1/3 < w < 0$ the universe is decelerating very slowly, which yields to interesting slopes in the IR tail. For convenience we present the kernel for the case $b = 2$ ($w = -1/9$), which reads

$$\begin{aligned}\mathcal{T}(u,v,c_s,w=-1/9) = {} & \frac{5^2\,7^2}{2^8\,3^9c_s^4}\left(\frac{4v^2-(1-u^2+v^2)^2}{4u^2v^2}\right)^2 \\ & \times\left\{\frac{225\pi^2}{4}(1-y^2)^2(1+y^2-2y^4)^2\Theta[c_s(u+v)-1]\right. \\ & \left.+\left(y(9+35y^2-30y^4)+\frac{15}{2}(1-3y^4+2y^6)\ln\left|\frac{1+y}{1-y}\right|\right)^2\right\}. \quad (213)\end{aligned}$$

10.2. Dust Domination

Here we provide a useful order of magnitude estimate for the following situation: (i) the universe is dominated by dust $w = c_s^2 = 0$ when induced GWs are generated, (ii) the transition to radiation domination is instantaneous and (iii) the spectrum of fluctuations for the Newtonian potential Φ is given by

$$\mathcal{P}_\Phi = \mathcal{A}_\Phi \left(\frac{k}{k_{\rm UV}}\right)^{-n} \Theta(k_{UV} - k)\,, \tag{214}$$

where $n < 2$. This power spectrum includes a power-law generated during inflation and the case of PBH density fluctuations. The details of the derivation can be found in Section 6. With these assumptions, the spectrum of induced GWs is approximately given by

$$\Omega_{\rm GWs,rh}(k \sim k_{\rm UV}) \approx \Omega_{\rm GWs}^{\rm peak} \left(\frac{k_{\rm UV}}{k_{\rm rh}}\right)^{7-2n} \Theta(k_{\rm UV} - k)\,, \tag{215}$$

where the amplitude of the peak reads

$$\Omega_{\rm GWs}^{\rm peak} \approx \frac{\pi}{9216\sqrt{3}} \left(\frac{4}{3}\right)^n \left(\frac{k_{\rm UV}}{k_{\rm rh}}\right)^7 \mathcal{A}_\Phi^2\,. \tag{216}$$

We can then use Equation (157) together with (204) to estimate the induced GW spectrum today. Note that if the UV cut-off $k_{\rm UV}$ is at very small scales, i.e., $k_{\rm UV} \gg k_{\rm rh}$, the amplitude of induced GW is very much enhanced and strong constraints apply. It is important to clarify that if $k_{\rm UV}$ is very large (larger than k_{NL} in Equation (139)), some density fluctuations would enter the non-linear regime. Thus, the above result must be considered as a rough estimate. For detailed discussion we refer the reader to Section 6.

11. Conclusions

Gravitational waves induced by primordial fluctuations, so-called induced GWs, have been shown to be a very promising tool to complete our picture of inflation and the early universe. Although this effect was first pointed out 50 years ago, induced GWs together with primordial black holes are receiving serious attention since the first GW detection by LIGO [61]. They are even more interesting after the NANOGrav collaboration recently reported a possible stochastic GW signal [18]. Within the next couple of decades, LISA will launch, and we will have a new data on stochastic backgrounds of gravitational waves in a wide frequency range. Even in the worst-case scenario, where no conclusive induced GW signal is found, we will obtain new constraints on the physics of inflation and the early universe in completely unexplored regimes.

The field of cosmology with induced GWs is relatively new and still developing. In this review we have focused on the current analytical techniques and estimates to compute the induced GW spectrum. We believe they will be most useful in analysing future SGWB data. We have not dwelled on the details of PBH formation nor on the observational forecasts for future GW detectors, both of which deserve a separate review. In the next years there will be more studies on induced GWs; new and better analytical and numerical results might be developed. Until then, particularly interesting and important directions are: (i) the impact of primordial non-gaussianities, not necessarily in the local shape, (ii) possible discriminators of induced GWs to distinguish them from other sources, (iii) the GW detector response at second order and (iv) GWs from the non-linear regime in an early matter dominated stage. We hope this review will be useful to the next advancements in the cosmology of induced GWs.

Funding: G.D. as a Fellini fellow was supported by the European Union's Horizon 2020 research and innovation programme under the Marie Skłodowska-Curie grant agreement No. 754496.

Institutional Review Board Statement: Not applicable.

Informed Consent Statement: Not applicable.

Data Availability Statement: No data is reported.

Acknowledgments: I have benefited from many helpful and insightful discussions and collaborations over the years while carrying out the works in which this review is based. In particular, I would like to thank Misao Sasaki for his constant support and encouragement. I thank Jinn-Ouk Gong and Shi Pi for awaking my interest on induced GWs some years ago. I also thank Sabino Matarrese for insightful discussions and for pointing out the first papers on induced GWs. I have also learned a lot from discussions with Vicente Atal, Nicola Bartolo, Albert Escrivá, Joseph Fedrow, Jacopo Fumagalli, Jaume Garriga, Cristiano Germani, Marc Kamionkowski, David Langlois, Chunshan Lin, Atsushi Naruko, Sébastien Renaux-Petel, Javier Rubio, Volodymyr Takhistov, Vicent Vennin and Lukas T. Witkowski. I acknowledge very helpful correspondence with Jai-chan Hwang, Keisuke Inomata, Kazunori Kohri and Zach Weiner. Last but not least, I would like to thank D. Rojas and A. D. Rojas for their constant support.

Conflicts of Interest: The author declares no conflict of interest.

Abbreviations

The following abbreviations are used in this manuscript:

GW	Gravitational Wave
PBH	Primordial Black Hole
SGWB	Stochastic Gravitational Wave Background
CMB	Cosmic Microwave Background
FLRW	Friedmann-Lemaître-Robertson-Walker
ADM	Arnowitt-Deser-Misner
WKB	Wentzel–Kramers–Brillouin
LIGO	Laser Interferometer Gravitational-Wave Observatory
KAGRA	Kamioka Gravitational Wave Detector
LISA	Laser Interferometer Space Antenna
PTA	Pulsar Timing Array
DECIGO	Deci-hertz Interferometer Gravitational wave Observatory
AION	Atom Interferometer Observatory and Network
MAGIS	Matter-wave Atomic Gradiometer Interferometric Sensor
ET	Einstein Telescope
OGLE	Optical Gravitational Lensing Experiment
rms	root mean squared
NG	Non-Gaussianity

Appendix A. Useful Formulas and Numerical Values

In this appendix we present the numerical values and formulas used to derive the parameters in the main text. First, we used the following values from the Planck results [2]: $k_{\rm eq} \approx 0.0104\,{\rm Mpc}^{-1}$, $z_{\rm eq} \approx 3400$, $T_0 = 2.7255\,{\rm K}$. We also used the value for the reduced Planck mass $M_{\rm pl} \approx 4.235 \times 10^{18}\,{\rm GeV} \approx 4.34 \times 10^{-6}\,{\rm g}$ and the solar mass $M_\odot \approx 2 \times 10^{33}\,{\rm g}$. A useful relation between units (in the Planck units) is the following:

$$1\,{\rm K} \approx 8.62 \times 10^{-5}\,{\rm eV} \approx 4.34\,{\rm cm}^{-1} \approx 1.31 \times 10^{11}\,{\rm Hz}. \tag{A1}$$

On the effective degrees of freedom that we used for radiation at temperature T, the energy density is given by

$$\rho = \frac{\pi^2}{30} g_*(T) T^4 \quad \text{with} \quad g_*(T) = \sum g_b + \frac{7}{8} \sum g_f\,, \tag{A2}$$

where g_b and g_f are respectively the degrees of freedom of bosons and fermions. The entropy can then be calculated by

$$s = \frac{2\pi^2}{45} g_{*s}(T) T^3 \,. \tag{A3}$$

The different values of g_* and g_{*s} in the standard model are reviewed in Ref. [293]. At very high temperatures, i.e., $T \gg 100\,\mathrm{GeV}$ one has $g_* \approx g_{*s} \approx 106.75$. At the time of matter-radiation equality and at present we took $g_*(T_{\rm eq}) \approx g_*(T_0) \approx 3.38$ and $g_{*s}(T_{\rm eq}) \approx g_{*s}(T_0) \approx 3.94$.

The Hubble parameter in the radiation dominated universe can be related by the Friedmann equations to the temperature as

$$H = \frac{\pi}{3\sqrt{10} M_{\rm pl}} g_*^{1/2}(T) T^2 \,, \tag{A4}$$

Another important relation is that since the entropy is conserved, which in an expanding universe means $s \propto a^{-3}$, there is a direct relation between the scale factor and the temperature. In particular we have that

$$\frac{a}{a_\star} = \frac{T_\star}{T} \left(\frac{g_{*s}(T_\star)}{g_{*s}(T)} \right)^{1/3} \,. \tag{A5}$$

More detailed explanations can be found, e.g., in Baumann's lecture notes as of August 2021 http://cosmology.amsterdam/education/cosmology/.

Appendix B. Green's Function Method

In this appendix we present the Green's method to find particular solutions to a differential equation where the homogeneous solutions are known. In the main text we found that the induced GWs have the following equations of motion:

$$h''_{\mathbf{k},\lambda} + 2\mathcal{H} h'_{\mathbf{k},\lambda} + k^2 h_{\mathbf{k},\lambda} = \mathcal{S}_{\mathbf{k},\lambda} \,. \tag{A6}$$

Assuming that we know the two homogeneous solution, say h_1 and h_2, we can check that the following Green's function,

$$G(\tau, \tilde{\tau}) = \frac{1}{W(h_1, h_2, \tilde{\tau})} \left(h_1(\tau) h_2(\tilde{\tau}) - h_1(\tilde{\tau}) h_2(\tau) \right) , \tag{A7}$$

is a solution to

$$G''_{\mathbf{k},\lambda}(\tau, \tilde{\tau}) + 2\mathcal{H} G'_{\mathbf{k},\lambda}(\tau, \tilde{\tau}) + k^2 G_{\mathbf{k},\lambda}(\tau, \tilde{\tau}) = \delta(\tau - \tilde{\tau}) \,. \tag{A8}$$

In Equation (A7) we have defined the Wronskian as

$$W(h_1, h_2, \tilde{\tau}) = h'_1(\tilde{\tau}) h_2(\tilde{\tau}) - h_1(\tilde{\tau}) h'_2(\tilde{\tau}) \,. \tag{A9}$$

Then with the Green's function (A7) we have that the particular solution to $h_{\mathbf{k},\lambda}$ with initial conditions $h_{\mathbf{k},\lambda}(\tau_i) = h'_{\mathbf{k},\lambda}(\tau_i) = 0$ is given by

$$h_{\mathbf{k},\lambda} = \int_{\tau_i}^{\tau} d\tilde{\tau} G(\tau, \tilde{\tau}) \mathcal{S}_{\mathbf{k},\lambda}(\tilde{\tau}) \,. \tag{A10}$$

Appendix C. ADM Formalism

In this appendix we briefly give the ADM or (3 + 1)-decomposition of the Einstein-Hilbert action, that is

$$S = \int d^4x \sqrt{-g} \frac{M_{pl}^2}{2} R . \tag{A11}$$

More details can be found in Ref. [223]. First, the line element is parametrised by

$$ds^2 = -N^2 dt^2 + h_{ij} \left(dx^i + N^i dt \right) \left(dx^j + N^j dt \right) , \tag{A12}$$

where N is the lapse and N^i is the shift vector and describe the foliation of our spacetime. In this decomposition the Einstein-Hilbert action, after integrating out boundary terms, is given by

$$S = \int d^3x dt N \sqrt{h} \, \frac{M_{pl}^2}{2} \left(R^{(3)}[h] + K_{ij} K^{ij} - K^2 \right) , \tag{A13}$$

where

$$K_{ij} \equiv \frac{1}{2N} \left(\dot{h}_{ij} - D_i N_j - D_j N_i \right) , \tag{A14}$$

is the extrinsic curvature, $K = h^{ij} K_{ij}$ and D_i is the covariant derivative of h_{ij}. Explicit expressions of the metric, its inverse and all the Christoffel symbols can be found in Appendix A of Ref. [345].

Appendix D. Fourier Conventions and Polarization Tensors

In this appendix we specify the conventions used for Fourier transforms and the polarization tensors. First, we define the Fourier transform of a scalar mode as

$$\Phi(\tau, \mathbf{k}) = \int \frac{d^3k}{(2\pi)^3} \, \Phi_{\mathbf{k}}(\tau) e^{i\mathbf{k}\mathbf{x}} , \tag{A15}$$

and of a tensor mode as

$$h_{ij}(\tau, \mathbf{k}) = \sum_{\lambda} \int \frac{d^3k}{(2\pi)^3} \, e_{ij}^{\lambda}(\mathbf{k}) \, h_{\mathbf{k},\lambda}(\tau) e^{i\mathbf{k}\mathbf{x}} , \tag{A16}$$

where λ are the polarization, e.g., $+$ and $\times$, and $e_{ij}^{\lambda}(\mathbf{k})$ are the polarization tensors. Such expansion is valid for real polarization tensors. Then, reality conditions[25] of h_{ij} imply [29] $e_{ij}^{\lambda}(\mathbf{k}) = e_{ij}^{\lambda}(-\mathbf{k})$ and $h_{\mathbf{k},\lambda}^* = h_{-\mathbf{k},\lambda}$. We define the polarization tensors in terms of polarization vectors as follows:

$$e_{ij}^{+}(\mathbf{k}) = \frac{1}{\sqrt{2}} \left[e_i(\mathbf{k}) e_j(\mathbf{k}) - \bar{e}_i(\mathbf{k}) \bar{e}_j(\mathbf{k}) \right] , \tag{A17}$$

$$e_{ij}^{\times}(\mathbf{k}) = \frac{1}{\sqrt{2}} \left[e_i(\mathbf{k}) \bar{e}_j(\mathbf{k}) + \bar{e}_i(\mathbf{k}) e_j(\mathbf{k}) \right] . \tag{A18}$$

Then, using that $e_i(\mathbf{k}) k^i = \bar{e}_i(\mathbf{k}) k^i = 0$ and $e_i(\mathbf{k}) e^i(\mathbf{k}) = \bar{e}_i(\mathbf{k}) \bar{e}^i(\mathbf{k}) = 1$, we arrive at the usual properties of the polarization tensors, namely

$$e_{ij}^{+}(\mathbf{k}) e^{+ij}(\mathbf{k}) = 1 \quad , \quad e_{ij}^{\times}(\mathbf{k}) e^{\times ij}(\mathbf{k}) = 1 \quad , \quad e_{ij}^{+}(\mathbf{k}) e^{\times ij}(\mathbf{k}) = 0 , \tag{A19}$$
$$\delta_{ij} e^{+ij}(\mathbf{k}) = \delta_{ij} e^{\times ij}(\mathbf{k}) = k_i e^{+ij}(\mathbf{k}) = k_i e^{\times ij}(\mathbf{k}) = 0 .$$

Appendix D.1. Spherical Parametrisation

When calculating the projection of polarization tensors with the scalar momenta we choose spherical coordinates for convenience. Then, one can write in general that the wavenumber of a tensor mode is given by

$$\mathbf{k} = k(\sin\theta_k \cos\varphi_k, \sin\theta_k \sin\varphi_k, \cos\theta_k)\,, \tag{A20}$$

where θ_k and φ_k respectively are the polar and azimuthal angles. With this choice, a pair of orthonormal polarization vectors are

$$\mathbf{e}(\mathbf{k}) = (\cos\theta_k \cos\varphi_k, \cos\theta_k \sin\varphi_k, -\sin\theta_k)\,, \tag{A21}$$

$$\bar{\mathbf{e}}(\mathbf{k}) = (-\sin\varphi_k, \cos\varphi_k, 0)\,. \tag{A22}$$

We parametrize two scalar momenta $\mathbf{q}$ and $\mathbf{l}$ as

$$\mathbf{q} = q(\sin\theta_q \cos\varphi_q, \sin\theta_q \sin\varphi_q, \cos\theta_q)\,, \tag{A23}$$

and

$$\mathbf{l} = l(\sin\theta_l \cos\varphi_l, \sin\theta_l \sin\varphi_l, \cos\theta_l)\,. \tag{A24}$$

In the calculations of the tensor spectrum of Section 3.1 we considered general local primordial non-gaussianity. In this case, due to the symmetry between $\mathbf{l}$ and $\mathbf{q}$, we find it the most convenient to choose $\mathbf{k}$ in the z-axis, namely we set

$$\theta_k = 0 \quad , \quad \varphi_k = 0\,. \tag{A25}$$

Then one has that

$$|\mathbf{k}-\mathbf{q}|^2 = k^2 + q^2 - 2kq\cos\theta_k\,, \tag{A26}$$

$$|\mathbf{k}-\mathbf{l}|^2 = k^2 + l^2 - 2kl\cos\theta_l\,, \tag{A27}$$

$$|\mathbf{q}-\mathbf{l}|^2 = q^2 + l^2 - 2lq\left(\cos\theta_q \cos\theta_l + \cos(\varphi_q - \varphi_l)\sin\theta_q \sin\theta_l\right). \tag{A28}$$

Note that the azimuthal dependence is only in $|\mathbf{q}-\mathbf{l}|$. The projections of $\mathbf{q}$ and $\mathbf{l}$ with the polarization tensors then read

$$e^{+}_{ij}(\mathbf{k})q^i q^j = \frac{1}{\sqrt{2}}q^2 \sin^2\theta_k \cos(2\varphi_q) \quad , \quad e^{\times}_{ij}(\mathbf{k})q^i q^j = \frac{1}{\sqrt{2}}q^2 \sin^2\theta_k \sin(2\varphi_q) \tag{A29}$$

$$e^{+}_{ij}(\mathbf{k})l^i l^j = \frac{1}{\sqrt{2}}l^2 \sin^2\theta_l \cos(2\varphi_l)\,, \quad , \quad e^{\times}_{ij}(\mathbf{k})l^i l^j = \frac{1}{\sqrt{2}}l^2 \sin^2\theta_l \sin(2\varphi_l)\,. \tag{A30}$$

Lastly, what we need to derive the induced GW spectrum are the following results:

$$\left(e^{+}_{ij}(\mathbf{k})q^i q^j\right)^2 + \left(e^{\times}_{ij}(\mathbf{k})q^i q^j\right)^2 = \frac{q^4}{2}\sin^4\theta_k\,, \tag{A31}$$

and

$$e^{+}_{ij}(\mathbf{k})q^i q^j e^{+}_{ij}(\mathbf{k})l^i l^j + e^{\times}_{ij}(\mathbf{k})q^i q^j e^{\times}_{ij}(\mathbf{k})l^i l^j = \frac{q^2 l^2}{2}\sin^2\theta_k \sin^2\theta_l \cos(2(\varphi_q - \varphi_l))\,. \tag{A32}$$

Appendix E. Formulas in a General Gauge

In this appendix we present the explicit expressions for Einstein equations in a perturbed flat FLRW universe filled with a perfect fluid. We present the results in a general gauge. The convention for the perturbed metric is given in Equation (183) and the energy momentum tensor in Equation (44).

Appendix E.1. Background

At zeroth order in perturbation theory we have the two Friedmann equations and energy conservation equations. These are respectively given by

$$3M_{\rm pl}^2\mathcal{H}^2 = a^2\rho\,, \tag{A33}$$

$$M_{\rm pl}^2\left(\mathcal{H}^2 + 2\mathcal{H}'\right) = -a^2P\,, \tag{A34}$$

and

$$\rho' + 3\mathcal{H}(\rho + P) = 0\,. \tag{A35}$$

Appendix E.2. First Order

At first order we find that the Hamiltonian and momentum constraints yield [164]

$$6\mathcal{H}(\phi' - \mathcal{H}\alpha) + 2\Delta(\phi + \mathcal{H}\sigma) = M_{\rm pl}^2a^2\delta\rho\,, \tag{A36}$$

$$2\phi' - 2\mathcal{H}\alpha = M_{\rm pl}^2a^2(\rho + P)(v + \beta)\,. \tag{A37}$$

The remaining Einstein equations are

$$2\phi'' + 4\mathcal{H}\phi' - 2\mathcal{H}\alpha' - (4\mathcal{H}' + 2\mathcal{H}^2)\alpha = -M_{\rm pl}^2a^2\delta P\,, \tag{A38}$$

$$\sigma' + 2\mathcal{H}\sigma + \phi + \alpha = 0\,. \tag{A39}$$

We assumed that there is no anisotropic stress for matter and we defined

$$\sigma \equiv \beta - E'\,, \tag{A40}$$

which corresponds to the scalar shear of the metric [164]. The energy and momentum conservation equations can be derived from the Einstein equations.

Appendix E.3. Second Order

Here we only present the second order equations for the tensor modes with a scalar squared source. In the source terms, there are in general vector and tensor components as well, although they are often subdominant [151]. After a long algebra, the equations of motion for tensor modes are given by [150]

$$(\hat{D}_\tau - \Delta)h_{ij} = \widehat{TT}_{ij}^{ab}S_{ab}\,, \tag{A41}$$

where $\hat{D}_\tau \equiv a^{-2}\partial_\tau(a^2\partial_\tau)$, $\Delta \equiv \delta^{ij}\partial_i\partial_j$ is the flat three-dimensional Laplacian. $\widehat{TT}_{ij}^{ab}$ is the transverse-traceless projector, which is given by

$$\widehat{TT}_{ij}{}^{ab} \equiv \left(\delta_i^{(a} - \partial_i\partial^{(a}\Delta^{-1}\right)\left(\delta_j^{b)} - \partial_j\partial^{b)}\Delta^{-1}\right) - \frac{1}{2}\left(\delta_{ij} - \partial_i\partial_j\Delta^{-1}\right)\left(\delta^{ab} - \partial^a\partial^b\Delta^{-1}\right), \tag{A42}$$

where the parenthesis in the indexes denote normalised symmetrisation. The source term in a general gauge can be written as

$$S_{ab} = 4\partial_a\Phi\partial_b\Phi + 2a^2(\rho + p)\partial_aV\partial_bV - (\hat{D}_\tau - \Delta)\left[\partial_a\sigma\partial_b\sigma + \partial_a\partial^kE\partial_k\partial_bE\right], \tag{A43}$$

where we defined for convenience

$$\Phi \equiv \phi - \frac{1}{3}\Delta E + \sigma \quad , \quad V \equiv v + E'\,. \tag{A44}$$

The source term (A43) reduces to the result of Section 3.1 in the shear-free or Newtonian gauge, that is $\sigma = E = 0$.

Appendix F. Bessel Functions

In this appendix we write useful formulas and relations of the Bessel functions. First, the asymptotic expansion for small argument is given by

$$J_\nu(x \ll 1) \approx x^\nu \frac{2^{-\nu}}{\Gamma[1+\nu]} + O(x^{\nu+1}) \quad , \quad Y_\nu(x \ll 1) \approx -\frac{2^\nu}{\pi}\Gamma[\nu]x^{-\nu} + O(x^{-\nu+1}) \,. \tag{A45}$$

For large arguments we have that the Bessel functions oscillate periodically as

$$J_\nu(x \gg 1) \approx \sqrt{\frac{2}{\pi x}} \cos\left(x - \frac{\nu\pi}{2} - \frac{\pi}{4}\right) + O(x^{-1}) \,, \tag{A46}$$

and

$$Y_\nu(x \gg 1) \approx \sqrt{\frac{2}{\pi x}} \sin\left(x - \frac{\nu\pi}{2} - \frac{\pi}{4}\right) + O(x^{-1}) \,. \tag{A47}$$

Lastly, useful relations between derivative and Bessel functions of similar order are respectively given by

$$\partial_x J_\nu(x) = J_{\nu-1}(x) - (\nu/x) J_\nu(x) \,, \tag{A48}$$

and

$$J_{\nu-1}(x) + J_{\nu+1}(x) = (2\nu/x) J_\nu(x) \,. \tag{A49}$$

Appendix G. Integrals with Two and Three Bessel Functions

In this appendix we give the expression for the integrals used in Section 4. Most of the formulae can be found in [233,346].

Appendix G.1. Superhorizon Approximation

The integrals (80) in the kernel (79) for the superhorizon approximation ($x \ll 1$) of Section 4.3 involve an integral with two Bessel functions of the same order. Then, the relevant integrals necessary for Section 4.3 are

$$\int dx x J_\mu^2(ax) = \tfrac{1}{2}x^2\left(J_\mu^2(ax) - J_{\mu-1}(ax)J_{\mu+1}(ax)\right) \,, \tag{A50}$$

and

$$\int dx x^{-2\mu+1} J_\mu^2(ax) = \frac{x^{-2\mu+2}}{2(1-2\mu)}\left(J_\mu^2(ax) + J_{\mu-1}^2(ax)\right) . \tag{A51}$$

By using the above formulas it is straightforward to show that the integrals (80) reduce after integration to

$$\mathcal{I}_J \approx \frac{3+2b}{1+b}\frac{2^{-b-1/2}}{\Gamma[b+3/2]}\frac{x}{\pi v} \,, \tag{A52}$$

$$\mathcal{I}_Y \approx \frac{2b+3}{b(1+b)c_s\pi}\left(2^{b-1/2}\Gamma[b+1/2]\frac{x^{-2b}}{\pi v} - 2^{-b-3/2}\frac{c_s^{2b}}{\Gamma[b+3/2]}\frac{1+b+b^2}{1+b}\right) . \tag{A53}$$

Appendix G.2. Subhorizon Approximation

The integrals (80) in the kernel (79) for the subhorizon approximation ($x \gg 1$) of Section 4.2 involve definite integral of three Bessel functions. The relevant integrals necessary for Section 4.2 can be found in Ref. [233]. Here we review their main results. On one hand, they find that for $|a-b| < c < a+b$ the integrals (80) read

$$\begin{aligned}\mathcal{I}_{J/Y} &= \int_0^\infty d\tilde{x}\tilde{x}^{1-\rho}\begin{Bmatrix} J_\rho(c\tilde{x}) \\ Y_\rho(c\tilde{x}) \end{Bmatrix} J_\nu(a\tilde{x})J_\nu(b\tilde{x}) \\ &= \frac{1}{\pi}\sqrt{\frac{2}{\pi}}\frac{(ab)^{\rho-1}}{c^\rho}(\sin\varphi)^{\rho-1/2}\begin{Bmatrix} \frac{\pi}{2}\mathsf{P}_{\nu-1/2}^{-\rho+1/2}(\cos\varphi) \\ -\mathsf{Q}_{\nu-1/2}^{-\rho+1/2}(\cos\varphi) \end{Bmatrix},\end{aligned} \tag{A54}$$

where

$$16\Delta^2 \equiv \left(c^2-(a-b)^2\right)\left((a+b)^2-c^2\right), \tag{A55}$$

and

$$\cos\varphi = \frac{a^2+b^2-c^2}{2ab} \quad , \quad \sin\varphi = \frac{2\Delta}{ab}. \tag{A56}$$

On the other hand, if $c > a+b$ the integrals are instead given by

$$\begin{aligned}\mathcal{I}_{J/Y} &= \int_0^\infty d\tilde{x}\tilde{x}^{1-\rho}\begin{Bmatrix} J_\rho(c\tilde{x}) \\ Y_\rho(c\tilde{x}) \end{Bmatrix} J_\nu(a\tilde{x})J_\nu(b\tilde{x}) \\ &= \frac{1}{\pi}\sqrt{\frac{2}{\pi}}\frac{(ab)^{\rho-1}}{c^\rho}(\sinh\phi)^{\rho-1/2}\Gamma[\nu-\rho+1]\mathcal{Q}_{\nu-1/2}^{-\rho+1/2}(\cosh\phi)\begin{Bmatrix} -\sin[(\nu-\rho)\pi] \\ \cos[(\nu-\rho)\pi] \end{Bmatrix},\end{aligned} \tag{A57}$$

where

$$16\tilde{\Delta}^2 \equiv \left(c^2-(a-b)^2\right)\left(c^2-(a+b)^2\right), \tag{A58}$$

and

$$\cosh\phi = \frac{c^2-(a^2+b^2)}{2ab} \quad , \quad \sinh\phi = \frac{2\tilde{\Delta}}{ab}. \tag{A59}$$

To use these formulas in the main text, we identify

$$c = 1 \quad , \quad a = c_s v \quad , \quad b = c_s u. \tag{A60}$$

Then, the range $|1-v| < u < 1+v$ can be split into $1 > c_s(u+v)$ ($c > a+b$) and $1 < c_s(u+v)$ ($c < a+b$). In the latter case we always have that $c_s|u-v| < 1$ by momentum conservation and so the formulas presented above cover all ranges of interest.

H. Associated Legendre Functions

In this appendix, we present useful formulae for the associated Legendre functions that can be found in Ref. [346]. We begin with the definitions of the associated Legendre functions of the first and second kind in terms of hypergeometric functions, which read

$$\mathsf{P}_\nu^\mu(x) = \left(\frac{1+x}{1-x}\right)^{\mu/2}\mathbf{F}\left(\nu+1,-\nu;1-\mu;\tfrac{1}{2}-\tfrac{1}{2}x\right), \tag{A61}$$

$$\mathsf{Q}^{\mu}_{\nu}(x) = \frac{\pi}{2\sin(\mu\pi)}\left(\cos(\mu\pi)\left(\frac{1+x}{1-x}\right)^{\mu/2}\mathbf{F}\left(\nu+1,-\nu;1-\mu;\tfrac{1}{2}-\tfrac{1}{2}x\right)\right.$$
$$\left. - \frac{\Gamma(\nu+\mu+1)}{\Gamma(\nu-\mu+1)}\left(\frac{1-x}{1+x}\right)^{\mu/2}\mathbf{F}\left(\nu+1,-\nu;1+\mu;\tfrac{1}{2}-\tfrac{1}{2}x\right)\right), \tag{A62}$$

and

$$\mathcal{Q}^{\mu}_{\nu}(x) = \frac{\pi}{2\sin(\mu\pi)\Gamma(\nu+\mu+1)}\left(\left(\frac{x+1}{x-1}\right)^{\mu/2}\mathbf{F}\left(\nu+1,-\nu;1-\mu;\tfrac{1}{2}-\tfrac{1}{2}x\right)\right.$$
$$\left. - \frac{\Gamma(\nu+\mu+1)}{\Gamma(\nu-\mu+1)}\left(\frac{x-1}{x+1}\right)^{\mu/2}\mathbf{F}\left(\nu+1,-\nu;1+\mu;\tfrac{1}{2}-\tfrac{1}{2}x\right)\right). \tag{A63}$$

Note that $\mathsf{P}^{\mu}_{\nu}(x)$ and $\mathsf{Q}^{\mu}_{\nu}(x)$ are also called the Ferrer's functions and are valid for $|x| < 1$. Then, $\mathcal{Q}^{\mu}_{\nu}(x)$ is also known as Olver's function, which is real for $|x| > 1$. In the above definitions we have used the compact notation

$$\mathbf{F}(a,b;c;x) = \frac{1}{\Gamma[c]}F(a,b;c;x), \tag{A64}$$

with $F(a,b;c;x)$ being the Gauss's Hypergeometric function.

H.1. Asymptotic Limits

We write down the asymptotic behaviour of the Associated Legendre functions near the singular points $x \sim 1$ and $x \to \infty$. It should be noted that the formulas below are valid for $\mu \notin \mathbb{Z}$. Relevant cases with $\mu \notin \mathbb{Z}$ are given in Section 10.

The limit $x \to 1^-$: First, the Ferrer's function of the first kind asymptotically goes as

$$\mathsf{P}^{\mu}_{\nu}(x) \sim \frac{1}{\Gamma(1-\mu)}\left(\frac{2}{1-x}\right)^{\mu/2}. \tag{A65}$$

Second, the Ferrer's function of the second kind goes for $\mu > 0$ as

$$\mathsf{Q}^{-\mu}_{\nu}(x) \sim \frac{\Gamma(\mu)\Gamma(\nu-\mu+1)}{2\Gamma(\nu+\mu+1)}\left(\frac{2}{1-x}\right)^{\mu/2}, \tag{A66}$$

while for $\mu < 0$ behaves as

$$\mathsf{Q}^{\mu}_{\nu}(x) \sim \frac{1}{2}\cos(\mu\pi)\Gamma(\mu)\left(\frac{2}{1-x}\right)^{\mu/2} \qquad \mu \neq 1/2. \tag{A67}$$

The limit $x \to 1^+$: Since $|x| > 1$, this case only concerns the Olver's function of the second kind. In the limit $x \to 1^+$ we have that

$$\mathcal{Q}^{\mu}_{\nu}(x) \sim \frac{\Gamma(\mu)}{2\Gamma(\nu+\mu+1)}\left(\frac{2}{x-1}\right)^{\mu/2}. \tag{A68}$$

The limit $x \to \infty$: This case also applies only to the Olver's function of the second kind. The asymptotic behaviour is then given by

$$\mathcal{Q}^{\mu}_{\nu}(x) \sim \frac{\pi^{1/2}}{\Gamma(\nu+\frac{3}{2})(2x)^{\nu+1}}. \tag{A69}$$

H.1.1. Useful Relations

In order to apply the asymptotic limits of Appendix H.1 we have to evaluate the functions for negative argument and negative order. This can be done by the useful relations below:

$$\mathrm{P}_b^{-b}(-x) = \mathrm{P}_b^{-b}(x) \quad , \quad \mathrm{P}_{b+2}^{-b}(-x) = \mathrm{P}_{b+2}^{-b}(x) \, , \tag{A70}$$

$$\mathrm{Q}_b^{-b}(-x) = -\mathrm{Q}_b^{-b}(x) \quad , \quad \mathrm{Q}_{b+2}^{-b}(-x) = -\mathrm{Q}_{b+2}^{-b}(x) \, , \tag{A71}$$

and

$$\mathcal{Q}_b^{-b}(x) = \mathcal{Q}_b^{b}(x) \quad , \quad \mathcal{Q}_{b+2}^{-b}(x) = \mathcal{Q}_{b+2}^{b}(x) \, . \tag{A72}$$

H.1.2. The Resonance Limit

Using the formulas in the Appendix H.1 we can compute the terms that appear in Section 4.2 in Equation (83) near the resonant point $y \sim 1$. In this way, we have that on one hand for $y \to -1^+$

$$\mathrm{P}_b^{-b}(y) + \frac{b+2}{b+1}\mathrm{P}_{b+2}^{-b}(y) \sim \frac{3+2b}{1+b}\frac{1}{\Gamma[1+b]}\left(\frac{2}{1+y}\right)^{-b/2} \, , \tag{A73}$$

and

$$\mathrm{Q}_b^{-b}(y) + \frac{b+2}{b+1}\mathrm{Q}_{b+2}^{-b}(y) \sim -\frac{3+2b}{1+b}\begin{cases} (1+b+b^2)\dfrac{\Gamma[b]}{\Gamma[2b+3]}\left(\dfrac{2}{1+y}\right)^{b/2} & b>0 \\ \dfrac{1}{2}\cos(b\pi)\Gamma[-b]\left(\dfrac{2}{1+y}\right)^{-b/2} & b<0 \end{cases} . \tag{A74}$$

On the other hand, for $y \to 1^+$ we find

$$\mathcal{Q}_b^{-b}(y) + 2\frac{b+2}{b+1}\mathcal{Q}_{b+2}^{-b}(y) \sim \frac{3+2b}{1+b}\begin{cases} (1+b+b^2)\dfrac{\Gamma[b]}{\Gamma[2b+3]}\left(\dfrac{2}{1-y}\right)^{b/2} & b>0 \\ \dfrac{1}{2}\Gamma[-b]\left(\dfrac{2}{1-y}\right)^{-b/2} & b<0 \end{cases} . \tag{A75}$$

Notes

1 Observations of the first photons that decoupled from the thermal plasma after neutral hydrogen formed.

2 To have a quantitative idea, the comoving wavenumber corresponding to the size of Hubble horizon at the time of neutrino decoupling is roughly $k \sim 10^4$ Mpc^{-1}. CMB observations constraint the primordial fluctuations roughly on scales $k \sim 10^{-4} - 5\times10^{-1}$Mpc^{-1}. Any information on much smaller scales, that is for $k \gg 10^4$ Mpc^{-1}, is erased by complicated astrophysical processes. On the largest scales we are limited by cosmic variance. CMB spectral distortions might probe down to $k \sim 10^5$ Mpc^{-1}.

3 Take one source per Hubble patch and count how many Hubble patches at a given redshift fit the current universe. The number actually grows as $(1+z)^3$ and we are talking about GWs generated at $z > 10^{10}$. On top of that, the angular resolution of GW detectors is not enough to pinpoint cosmic GWs generated in tiny Hubble patches.

4 For example, GWs generated when the universe had a temperature of around 1 GeV and 10^{10} GeV, respectively have a typical (peak) frequency roughly around 10^{-8} Hz and 100 Hz.

5 We have that the linear terms in the perturbative expansion in $h_{\mu\nu}$ vanish after averaging.

6 Note that sometimes in the literature there is an additional factor 1/2 in front of h_{ij}, e.g., in Refs. [38,63]. This slightly changes the numerical factors in different steps of the calculations but, of course, yields the same results in the end.

7 We are assuming that the tensor modes are drawn from a Gaussian distribution. This is essentially the case when they are generated by quantum fluctuations during inflation. Then, the expectation of a two-point correlation function of a random variable that follows a Gaussian distribution is proportional to the Dirac delta.

8 Here we define the critical threshold in terms of $\delta\rho/\rho$. However, it may also be defined through the peak of the so-called compaction function [199]. See Refs. [200–202] for recent studies.

9 From the critical collapse we know there will be some dispersion in the masses. For simplicity, we neglect it here.

10 Although we will abuse the notation "Newtonian gauge", the Newtonian gauge is actually defined as the shear-free slicing at the linear level. Our gauge choice reduces to the definition of the Newtonian gauge at first order but there might be some subtleties if one wants to relate them at second order.

11 As in the tensor modes we have that the dimensionless power spectrum is given by

$$\langle \Phi_k \Phi_{k'} \rangle = \frac{2\pi^2}{k^3} \mathcal{P}_\Phi(k) \times (2\pi)^3 \delta^3(\boldsymbol{k} + \boldsymbol{k}') \,.$$

12 Although the name is not informative at all, as strictly speaking any distribution which is not a Gaussian is non-Gaussian, in cosmology the term "non-Gaussianity" is often used to indicate small departures from a Gaussian distribution, e.g., by a small but non-vanishing 3-point correlation function. For a detailed explanation, I recommend to read E. A. Lim's notes on primordial NG as of August 2021: https://nms.kcl.ac.uk/eugene.lim/AdvCos/lecture2.pdf.

13 This factor is actually to compensate the factor in (61) for $w = 0$ since the original definition [230] was in terms of Φ not $\mathcal{R}$.

14 Since there are more possible contractions in the non-gaussian case with respect to the gaussian one, the numerical factors of the NG contribution will be larger.

15 For reference note that the radiation-matter equality corresponds to $k \sim 10^{-2}\,\text{Mpc}^{-1}$. This means that at the transition to the cold dark matter dominated epoch such induced GWs are already freely propagating GWs and can be treated as radiation.

16 For instance, one has arbitrary constant w and $c_s = 1$ for a canonical scalar field in an exponential potential [231].

17 Isocurvature fluctuations are fluctuations that leave the total energy density of matter homogeneous. Thus, they are only possible in multi-fluid systems where the density fluctuations of one fluid can be compensated by the density fluctuations of another.

18 Actually to derive Equation (70) we used the equations of motion for the curvature perturbation $\mathcal{R}$ for general w and c_s as can be found, e.g., in [183,232] and then we changed variables according to Equation (61).

19 We used that $\int_{-\infty}^{\infty} dz\, \text{Ci}^2[z] = \pi$.

20 For example, this is also the case of a matter inhomogeneity in the universe, such as a galaxy, where the matter density is clearly larger than the mean density of the universe but the gravitational potential can be considered as a perturbation. In this perturbative expansion one recovers Newtonian gravity which is very accurate in galactic scales.

21 It is parametrised as the effective number of neutrinos below the electron-positron annihilation temperature. In this way, the factor 7/8 is the relative factor of the energy density of fermions with respect to bosons. The factor 4/11 is the relative factor of the entropy of neutrinos and photons. Since $T \propto s^{1/3}$ and $\rho \propto T^4$ we get a power of 4/3. This is well explained in Baumann's lecture notes as of August 2021 http://cosmology.amsterdam/education/cosmology/.

22 Note that one obtains similar bounds from studies of the CMB [301,302]. However, these CMB constraints consider gravitational waves as a dark radiation component and depend on the initial conditions of such dark radiation fluctuations. As an order of magnitude estimate we limit ourselves to the BBN constraints [299,300] on the fraction of extra relativistic particles. We thank an anonymous referee for clarifying this point.

23 Without the symmetries of an FLRW background the definition of tensor modes is more subtle. For example, in anisotropic inflation they are constructed from the 2D scalar and vector perturbations [304,305].

24 That is assuming an early universe dominated by radiation and using the estimates of Section 2.3.

25 In case of doubt about conventions and normalisation conditions, it is advisable to treat h_{ij} as a field operator and express the Fourier expansion in the Fock representation. If the polarization tensors are complex, such as in circular polarization, the general normalization conditions are $e^{\lambda}_{ij}(\mathbf{k})\left(e^{\lambda ij}(\mathbf{k})\right)^* = \delta^{\lambda,\lambda'}$, where an asterisk refers to complex conjugate. We thank M. Sasaki for explaining this point.

References

1. Allahverdi, R.; Amin, M.A.; Berlin, A.; Bernal, N.; Byrnes, C.T.; Delos, M.S.; Erickcek, A.L.; Escudero, M.; Figueroa, D.G.; Freese, K.; et al. The First Three Seconds: A Review of Possible Expansion Histories of the Early Universe. *Open J. Astrophys.* **2021**, *4*. [CrossRef]
2. Aghanim, N. et al. [Planck Collaboration]. Planck 2018 results. VI. Cosmological parameters. *Astron. Astrophys.* **2020**, *641*, A6. [CrossRef]
3. Akrami, Y. et al. [Planck Collaboration]. Planck 2018 results. X. Constraints on inflation. *Astron. Astrophys.* **2020**, *641*, A10. [CrossRef]
4. Brout, R.; Englert, F.; Gunzig, E. The Creation of the Universe as a Quantum Phenomenon. *Ann. Phys.* **1978**, *115*, 78. [CrossRef]
5. Starobinsky, A.A. Spectrum of relict gravitational radiation and the early state of the universe. *JETP Lett.* **1979**, *30*, 682–685.
6. Guth, A.H. The Inflationary Universe: A Possible Solution to the Horizon and Flatness Problems. *Phys. Rev. D* **1981**, *23*, 347–356. [CrossRef]

7. Sato, K. First Order Phase Transition of a Vacuum and Expansion of the Universe. *Mon. Not. R. Astron. Soc.* **1981**, *195*, 467–479. [CrossRef]
8. Mukhanov, V.F.; Chibisov, G.V. Quantum Fluctuations and a Nonsingular Universe. *JETP Lett.* **1981**, *33*, 532–535.
9. Linde, A.D. A New Inflationary Universe Scenario: A Possible Solution of the Horizon, Flatness, Homogeneity, Isotropy and Primordial Monopole Problems. *Phys. Lett. B* **1982**, *108*, 389–393. [CrossRef]
10. Albrecht, A.; Steinhardt, P.J. Cosmology for Grand Unified Theories with Radiatively Induced Symmetry Breaking. *Phys. Rev. Lett.* **1982**, *48*, 1220–1223. [CrossRef]
11. Sasaki, M. Gauge Invariant Scalar Perturbations in the New Inflationary Universe. *Prog. Theor. Phys.* **1983**, *70*, 394. [CrossRef]
12. Kodama, H.; Sasaki, M. Cosmological Perturbation Theory. *Prog. Theor. Phys. Suppl.* **1984**, *78*, 1–166. [CrossRef]
13. Lentati, L.; Taylor, S.R.; Mingarelli, C.M.F.; Sesana, A.; Sanidas, S.A.; Vecchio, A.; Caballero, R.N.; Lee, K.J.; van Haasteren, R.; Babak, S.; et al. European Pulsar Timing Array Limits On An Isotropic Stochastic Gravitational-Wave Background. *Mon. Not. R. Astron. Soc.* **2015**, *453*, 2576–2598. [CrossRef]
14. Shannon, R.M.; Ravi, V.; Lentati, L.T.; Lasky, P.D.; Hobbs, G.; Kerr, M.; Manchester, R.N.; Coles, W.A.; Levin, Y.; Bailes, M.; et al. Gravitational waves from binary supermassive black holes missing in pulsar observations. *Science* **2015**, *349*, 1522–1525. [CrossRef]
15. Arzoumanian, Z. et al. [NANOGrav Collaboration]. The NANOGrav Nine-year Data Set: Limits on the Isotropic Stochastic Gravitational Wave Background. *Astrophys. J.* **2016**, *821*, 13. [CrossRef]
16. Qin, W.; Boddy, K.K.; Kamionkowski, M.; Dai, L. Pulsar-timing arrays, astrometry, and gravitational waves. *Phys. Rev.* **2019**, *D99*, 063002. [CrossRef]
17. Aggarwal, K.; Arzoumanian, Z.; Baker, P.T.; Brazier, A.; Brinson, M.R.; Brook, P.R.; Burke-Spolaor, S.; Chatterjee, S.; Cordes, J.M.; Cornish, N.J.; et al. The NANOGrav 11-Year Data Set: Limits on Gravitational Waves from Individual Supermassive Black Hole Binaries. *Astrophys. J.* **2019**, *880*, 2. [CrossRef]
18. Arzoumanian, Z. et al. [NANOGrav Collaboration]. The NANOGrav 12.5 yr Data Set: Search for an Isotropic Stochastic Gravitational-wave Background. *Astrophys. J. Lett.* **2020**, *905*, L34. [CrossRef]
19. Maggiore, M.; Broeck, C.V.D.; Bartolo, N.; Belgacem, E.; Bertacca, D.; Bizouard, M.A.; Branchesi, M.; Clesse, S.; Foffa, S.; García-Bellido, J.; et al. Science Case for the Einstein Telescope. *JCAP* **2020**, *2020*, 050. [CrossRef]
20. Amaro-Seoane, P. et al. [LISA Collaboration]. Laser Interferometer Space Antenna. *arXiv* **2017**, arXiv:1702.00786.
21. Barausse, E.; Berti, E.; Hertog, T.; Hughes, S.A.; Jetzer, P.; Pani, P.; Sotiriou, T.P.; Tamanini, N.; Witek, H.; Yagi, K.; et al. Prospects for Fundamental Physics with LISA. *Gen. Rel. Grav.* **2020**, *52*, 81. [CrossRef]
22. Seto, N.; Kawamura, S.; Nakamura, T. Possibility of direct measurement of the acceleration of the universe using 0.1-Hz band laser interferometer gravitational wave antenna in space. *Phys. Rev. Lett.* **2001**, *87*, 221103. [CrossRef] [PubMed]
23. Yagi, K.; Seto, N. Detector configuration of DECIGO/BBO and identification of cosmological neutron-star binaries. *Phys. Rev.* **2011**, *D83*, 044011. Erratum in *Phys. Rev.* **2017**, *D95*, 109901. [CrossRef]
24. Kawamura, S.; Ando, M.; Seto, N.; Sato, S.; Musha, M.; Kawano, I.; Yokoyama, J.; Tanaka, T.; Ioka, K.; Akutsu, T. et al. Current status of space gravitational wave antenna DECIGO and B-DECIGO. *arXiv* **2020**, arXiv:2006.13545.
25. Badurina, L.; Bentine, E.; Blas, D.; Bongs, K.; Bortoletto, D.; Bowcock, T.; Bridges, K.; Bowden, W.; Buchmueller, O.; Burrage, C.; et al. AION: An Atom Interferometer Observatory and Network. *arXiv* **2019**, arXiv:1911.11755.
26. Ruan, W.H.; Guo, Z.K.; Cai, R.G.; Zhang, Y.Z. Taiji Program: Gravitational-Wave Sources. *arXiv* **2018**, arXiv:1807.09495.
27. Luo, J. et al. [TianQin Collaboration]. TianQin: A space-borne gravitational wave detector. *Class. Quant. Grav.* **2016**, *33*, 035010. [CrossRef]
28. Thrane, E.; Romano, J.D. Sensitivity curves for searches for gravitational-wave backgrounds. *Phys. Rev.* **2013**, *D88*, 124032. [CrossRef]
29. Caprini, C.; Figueroa, D.G. Cosmological Backgrounds of Gravitational Waves. *Class. Quant. Grav.* **2018**, *35*, 163001. [CrossRef]
30. Gow, A.D.; Byrnes, C.T.; Cole, P.S.; Young, S. The power spectrum on small scales: Robust constraints and comparing PBH methodologies. *JCAP* **2021**, *2021*, 002. [CrossRef]
31. Chluba, J.; Kogut, A.; Patil, S.P.; Abitbol, M.H.; Aghanim, N.; Ali-Haïmoud, Y.; Amin, M.A.; Aumont, J.; Bartolo, N.; Basu, K.; et al. Spectral Distortions of the CMB as a Probe of Inflation, Recombination, Structure Formation and Particle Physics: Astro2020 Science White Paper. *Bull. Am. Astron. Soc.* **2019**, *51*, 184.
32. Kite, T.; Ravenni, A.; Patil, S.P.; Chluba, J. Bridging the gap: Spectral distortions meet gravitational waves. *arXiv* **2020**, arXiv:2010.00040.
33. Unal, C.; Kovetz, E.D.; Patil, S.P. Multi-messenger Probes of Inflationary Fluctuations and Primordial Black Holes. *arXiv* **2020**, arXiv:2008.11184.
34. Tomita, K. Non-Linear Theory of Gravitational Instability in the Expanding Universe. *Progress of Theoretical Physics* **1967**, *37*, 831–846. [CrossRef]
35. Matarrese, S.; Pantano, O.; Saez, D. A General relativistic approach to the nonlinear evolution of collisionless matter. *Phys. Rev. D* **1993**, *47*, 1311–1323. [CrossRef] [PubMed]
36. Matarrese, S.; Pantano, O.; Saez, D. General relativistic dynamics of irrotational dust: Cosmological implications. *Phys. Rev. Lett.* **1994**, *72*, 320–323. [CrossRef]

37. Matarrese, S.; Mollerach, S.; Bruni, M. Second order perturbations of the Einstein-de Sitter universe. *Phys. Rev. D* **1998**, *58*, 043504. [CrossRef]
38. Ananda, K.N.; Clarkson, C.; Wands, D. The Cosmological gravitational wave background from primordial density perturbations. *Phys. Rev. D* **2007**, *75*, 123518. [CrossRef]
39. Baumann, D.; Steinhardt, P.J.; Takahashi, K.; Ichiki, K. Gravitational Wave Spectrum Induced by Primordial Scalar Perturbations. *Phys. Rev. D* **2007**, *76*, 084019. [CrossRef]
40. Mangilli, A.; Bartolo, N.; Matarrese, S.; Riotto, A. The impact of cosmic neutrinos on the gravitational-wave background. *Phys. Rev. D* **2008**, *78*, 083517. [CrossRef]
41. Sarkar, D.; Serra, P.; Cooray, A.; Ichiki, K.; Baumann, D. Cosmic shear from scalar-induced gravitational waves. *Phys. Rev. D* **2008**, *77*, 103515. [CrossRef]
42. Martineau, P.; Brandenberger, R. A Back-reaction Induced Lower Bound on the Tensor-to-Scalar Ratio. *Mod. Phys. Lett. A* **2008**, *23*, 727–735. [CrossRef]
43. Bartolo, N.; Matarrese, S.; Riotto, A.; Vaihkonen, A. The Maximal Amount of Gravitational Waves in the Curvaton Scenario. *Phys. Rev. D* **2007**, *76*, 061302. [CrossRef]
44. Boubekeur, L.; Creminelli, P.; Norena, J.; Vernizzi, F. Action approach to cosmological perturbations: The 2nd order metric in matter dominance. *JCAP* **2008**, *2008*, 028. [CrossRef]
45. Saito, R.; Yokoyama, J. Gravitational wave background as a probe of the primordial black hole abundance. *Phys. Rev. Lett.* **2009**, *102*, 161101. Erratum in *Phys. Rev. Lett.* **2011**, *107*, 069901. [CrossRef] [PubMed]
46. Saito, R.; Yokoyama, J. Gravitational-Wave Constraints on the Abundance of Primordial Black Holes. *Prog. Theor. Phys.* **2010**, *123*, 867–886. Erratum in *Prog. Theor. Phys.* **2011**, *126*, 351–352. [CrossRef]
47. Bugaev, E.; Klimai, P. Induced gravitational wave background and primordial black holes. *Phys. Rev. D* **2010**, *81*, 023517. [CrossRef]
48. Bugaev, E.V.; Klimai, P.A. Bound on induced gravitational wave background from primordial black holes. *JETP Lett.* **2010**, *91*, 1–5. [CrossRef]
49. Bugaev, E.; Klimai, P. Constraints on the induced gravitational wave background from primordial black holes. *Phys. Rev. D* **2011**, *83*, 083521. [CrossRef]
50. Assadullahi, H.; Wands, D. Constraints on primordial density perturbations from induced gravitational waves. *Phys. Rev. D* **2010**, *81*, 023527. [CrossRef]
51. Assadullahi, H.; Wands, D. Gravitational waves from an early matter era. *Phys. Rev. D* **2009**, *79*, 083511. [CrossRef]
52. Arroja, F.; Assadullahi, H.; Koyama, K.; Wands, D. Cosmological matching conditions for gravitational waves at second order. *Phys. Rev. D* **2009**, *80*, 123526. [CrossRef]
53. Alabidi, L.; Kohri, K.; Sasaki, M.; Sendouda, Y. Observable Spectra of Induced Gravitational Waves from Inflation. *JCAP* **2012**, *2012*, 017. [CrossRef]
54. Alabidi, L.; Kohri, K.; Sasaki, M.; Sendouda, Y. Observable induced gravitational waves from an early matter phase. *JCAP* **2013**, *2013*, 033. [CrossRef]
55. Kawasaki, M.; Kitajima, N.; Yokoyama, S. Gravitational waves from a curvaton model with blue spectrum. *JCAP* **2013**, *2013*, 042. [CrossRef]
56. Nakama, T.; Suyama, T. Primordial black holes as a novel probe of primordial gravitational waves. *Phys. Rev. D* **2015**, *92*, 121304. [CrossRef]
57. Nakama, T.; Suyama, T. Primordial black holes as a novel probe of primordial gravitational waves. II: Detailed analysis. *Phys. Rev. D* **2016**, *94*, 043507. [CrossRef]
58. Suyama, T.; Yokoyama, J. Temporal enhancement of super-horizon curvature perturbations from decays of two curvatons and its cosmological consequences. *Phys. Rev. D* **2011**, *84*, 083511. [CrossRef]
59. Saga, S.; Ichiki, K.; Sugiyama, N. Impact of anisotropic stress of free-streaming particles on gravitational waves induced by cosmological density perturbations. *Phys. Rev. D* **2015**, *91*, 024030. [CrossRef]
60. Fidler, C.; Pettinari, G.W.; Beneke, M.; Crittenden, R.; Koyama, K.; Wands, D. The intrinsic B-mode polarisation of the Cosmic Microwave Background. *JCAP* **2014**, *2014*, 011. [CrossRef]
61. Abbott, B.P. et al. [LIGO Scientific and Virgo Collaborations]. GW151226: Observation of Gravitational Waves from a 22-Solar-Mass Binary Black Hole Coalescence. *Phys. Rev. Lett.* **2016**, *116*, 241103. [CrossRef] [PubMed]
62. Espinosa, J.R.; Racco, D.; Riotto, A. A Cosmological Signature of the SM Higgs Instability: Gravitational Waves. *JCAP* **2018**, *2018*, 012. [CrossRef]
63. Kohri, K.; Terada, T. Semianalytic calculation of gravitational wave spectrum nonlinearly induced from primordial curvature perturbations. *Phys. Rev. D* **2018**, *97*, 123532. [CrossRef]
64. Domènech, G. Induced gravitational waves in a general cosmological background. *Int. J. Mod. Phys. D* **2020**, *29*, 2050028. [CrossRef]
65. Inomata, K.; Kohri, K.; Nakama, T.; Terada, T. Gravitational Waves Induced by Scalar Perturbations during a Gradual Transition from an Early Matter Era to the Radiation Era. *JCAP* **2019**, *2019*, 071. [CrossRef]
66. Inomata, K.; Kohri, K.; Nakama, T.; Terada, T. Enhancement of Gravitational Waves Induced by Scalar Perturbations due to a Sudden Transition from an Early Matter Era to the Radiation Era. *Phys. Rev. D* **2019**, *100*, 043532. [CrossRef]

67. Dalianis, I.; Kouvaris, C. Gravitational Waves from Density Perturbations in an Early Matter Domination Era. *J. Cosmol. Astropart. Phys.* **2021**, *7*, 46. [CrossRef]
68. Inomata, K.; Kawasaki, M.; Mukaida, K.; Terada, T.; Yanagida, T.T. Gravitational Wave Production right after a Primordial Black Hole Evaporation. *Phys. Rev. D* **2020**, *101*, 123533. [CrossRef]
69. Papanikolaou, T.; Vennin, V.; Langlois, D. Gravitational waves from a universe filled with primordial black holes. *arXiv* **2020**, arXiv:2010.11573.
70. Domènech, G.; Lin, C.; Sasaki, M. Gravitational wave constraints on the primordial black hole dominated early universe. *arXiv* **2020**, arXiv:2012.08151.
71. Domènech, G.; Takhistov, V.; Sasaki, M. Exploring Evaporating Primordial Black Holes with Gravitational Waves. *arXiv* **2021**, arXiv:2105.06816.
72. Hajkarim, F.; Schaffner-Bielich, J. Thermal History of the Early Universe and Primordial Gravitational Waves from Induced Scalar Perturbations. *Phys. Rev. D* **2020**, *101*, 043522. [CrossRef]
73. Bhattacharya, S.; Mohanty, S.; Parashari, P. Primordial black holes and gravitational waves in nonstandard cosmologies. *Phys. Rev. D* **2020**, *102*, 043522. [CrossRef]
74. Domènech, G.; Pi, S.; Sasaki, M. Induced gravitational waves as a probe of thermal history of the universe. *JCAP* **2020**, *2020*, 017. [CrossRef]
75. Dalianis, I.; Kritos, K. Exploring the Spectral Shape of Gravitational Waves Induced by Primordial Scalar Perturbations and Connection with the Primordial Black Hole Scenarios. *Phys. Rev. D* **2021**, *103*, 023505. [CrossRef]
76. Abe, K.T.; Tada, Y.; Ueda, I. Induced gravitational waves as a cosmological probe of the sound speed during the QCD phase transition. *arXiv* **2020**, arXiv:2010.06193.
77. Hook, A.; Marques-Tavares, G.; Racco, D. Causal gravitational waves as a probe of free streaming particles and the expansion of the Universe. *JHEP* **2021**, *02*, 117. [CrossRef]
78. Cai, R.G.; Pi, S.; Sasaki, M. Universal infrared scaling of gravitational wave background spectra. *Phys. Rev. D* **2020**, *102*, 083528. [CrossRef]
79. Yuan, C.; Chen, Z.C.; Huang, Q.G. Log-dependent slope of scalar induced gravitational waves in the infrared regions. *Phys. Rev. D* **2020**, *101*, 043019. [CrossRef]
80. Liu, J.; Guo, Z.K.; Cai, R.G. Analytical approximation of the scalar spectrum in the ultraslow-roll inflationary models. *Phys. Rev. D* **2020**, *101*, 083535. [CrossRef]
81. Atal, V.; Domènech, G. Probing non-Gaussianities with the high frequency tail of induced gravitational waves. *arXiv* **2021**, arXiv:2103.01056.
82. Pi, S.; Sasaki, M. Gravitational Waves Induced by Scalar Perturbations with a Lognormal Peak. *JCAP* **2020**, *2020*, 037. [CrossRef]
83. Cai, R.G.; Pi, S.; Wang, S.J.; Yang, X.Y. Resonant multiple peaks in the induced gravitational waves. *JCAP* **2019**, *2019*, 013. [CrossRef]
84. Fumagalli, J.; Renaux-Petel, S.; Witkowski, L.T. Oscillations in the stochastic gravitational wave background from sharp features and particle production during inflation. *arXiv* **2020**, arXiv:2012.02761.
85. Braglia, M.; Hazra, D.K.; Finelli, F.; Smoot, G.F.; Sriramkumar, L.; Starobinsky, A.A. Generating PBHs and small-scale GWs in two-field models of inflation. *JCAP* **2020**, *2020*, 001. [CrossRef]
86. Braglia, M.; Chen, X.; Hazra, D.K. Probing Primordial Features with the Stochastic Gravitational Wave Background. *arXiv* **2020**, arXiv:2012.05821.
87. Fumagalli, J.; Renaux-Petel, S.; Witkowski, L.T. Resonant features in the stochastic gravitational wave background. *arXiv* **2021**, arXiv:2105.06481.
88. Garcia-Bellido, J.; Peloso, M.; Unal, C. Gravitational Wave signatures of inflationary models from Primordial Black Hole Dark Matter. *JCAP* **2017**, *2017*, 013. [CrossRef]
89. Cai, R.; Pi, S.; Sasaki, M. Gravitational Waves Induced by non-Gaussian Scalar Perturbations. *Phys. Rev. Lett.* **2019**, *122*, 201101. [CrossRef] [PubMed]
90. Unal, C. Imprints of Primordial Non-Gaussianity on Gravitational Wave Spectrum. *Phys. Rev. D* **2019**, *99*, 041301. [CrossRef]
91. Yuan, C.; Huang, Q.G. Gravitational waves induced by the local-type non-Gaussian curvature perturbations. *arXiv* **2020**, arXiv:2007.10686.
92. Adshead, P.; Lozanov, K.D.; Weiner, Z.J. Non-Gaussianity and the induced gravitational wave background. *arXiv* **2021**, arXiv:2105.01659.
93. Ragavendra, H.V. Accounting for scalar non-Gaussianity in secondary gravitational waves. *arXiv* **2021**, arXiv:2108.04193.
94. Ota, A. Induced superhorizon tensor perturbations from anisotropic non-Gaussianity. *Phys. Rev. D* **2020**, *101*, 103511. [CrossRef]
95. Cai, Y.F.; Chen, C.; Tong, X.; Wang, D.G.; Yan, S.F. When Primordial Black Holes from Sound Speed Resonance Meet a Stochastic Background of Gravitational Waves. *Phys. Rev. D* **2019**, *100*, 043518. [CrossRef]
96. Cai, R.G.; Guo, Z.K.; Liu, J.; Liu, L.; Yang, X.Y. Primordial black holes and gravitational waves from parametric amplification of curvature perturbations. *JCAP* **2020**, *2020*, 013. [CrossRef]
97. Zhou, Z.; Jiang, J.; Cai, Y.F.; Sasaki, M.; Pi, S. Primordial black holes and gravitational waves from resonant amplification during inflation. *Phys. Rev. D* **2020**, *102*, 103527. [CrossRef]

98. Cai, Y.F.; Jiang, J.; Sasaki, M.; Vardanyan, V.; Zhou, Z. Beating the Lyth bound by parametric resonance during inflation. *arXiv* **2021**, arXiv:2105.12554.
99. Ragavendra, H.V.; Sriramkumar, L.; Silk, J. Could PBHs and secondary GWs have originated from squeezed initial states? *arXiv* **2020**, arXiv:2011.09938.
100. Garcia-Bellido, J.; Peloso, M.; Unal, C. Gravitational waves at interferometer scales and primordial black holes in axion inflation. *JCAP* **2016**, *2016*, 031. [CrossRef]
101. Di, H.; Gong, Y. Primordial black holes and second order gravitational waves from ultra-slow-roll inflation. *JCAP* **2018**, *2018*, 007. [CrossRef]
102. Ando, K.; Kawasaki, M.; Nakatsuka, H. Formation of primordial black holes in an axionlike curvaton model. *Phys. Rev. D* **2018**, *98*, 083508. [CrossRef]
103. Byrnes, C.T.; Cole, P.S.; Patil, S.P. Steepest growth of the power spectrum and primordial black holes. *JCAP* **2019**, *2019*, 028. [CrossRef]
104. Gao, T.J.; Yang, X.Y. Gravitational waves induced from string axion model of inflation. *Int. J. Mod. Phys. A* **2019**, *34*, 1950213. [CrossRef]
105. Xu, W.T.; Liu, J.; Gao, T.J.; Guo, Z.K. Gravitational waves from double-inflection-point inflation. *Phys. Rev. D* **2020**, *101*, 023505. [CrossRef]
106. Mahbub, R. Primordial black hole formation in inflationary α-attractor models. *Phys. Rev. D* **2020**, *101*, 023533. [CrossRef]
107. Mishra, S.S.; Sahni, V. Primordial Black Holes from a tiny bump/dip in the Inflaton potential. *JCAP* **2020**, *2020*, 007. [CrossRef]
108. Fu, C.; Wu, P.; Yu, H. Scalar induced gravitational waves in inflation with gravitationally enhanced friction. *Phys. Rev. D* **2020**, *101*, 023529. [CrossRef]
109. Özsoy, O.; Tasinato, G. On the slope of the curvature power spectrum in non-attractor inflation. *JCAP* **2020**, *2020*, 048. [CrossRef]
110. Özsoy, O.; Lalak, Z. Primordial black holes as dark matter and gravitational waves from bumpy axion inflation. *JCAP* **2021**, *2021*, 040. [CrossRef]
111. Gao, Q.; Gong, Y.; Yi, Z. Primordial black holes and secondary gravitational waves from natural inflation. *arXiv* **2020**, arXiv:2012.03856.
112. Bhaumik, N.; Jain, R.K. Stochastic induced gravitational waves and lowest mass limit of primordial black holes with the effects of reheating. *arXiv* **2020**, arXiv:2009.10424.
113. Lin, J.; Gao, Q.; Gong, Y.; Lu, Y.; Zhang, C.; Zhang, F. Primordial black holes and secondary gravitational waves from k and G inflation. *Phys. Rev. D* **2020**, *101*, 103515. [CrossRef]
114. Ragavendra, H.V.; Saha, P.; Sriramkumar, L.; Silk, J. PBHs and secondary GWs from ultra slow roll and punctuated inflation. *arXiv* **2020**, arXiv:2008.12202.
115. Yi, Z.; Gao, Q.; Gong, Y.; Zhu, Z.h. Primordial black holes and secondary gravitational waves from inflationary model with a non-canonical kinetic term. *arXiv* **2020**, arXiv:2011.10606.
116. Gao, Q. Primordial black holes and secondary gravitational waves from chaotic inflation. *arXiv* **2021**, arXiv:2102.07369.
117. Gao, T.J.; Yang, X.Y. Double peaks of gravitational wave spectrum induced from inflection point inflation. *arXiv* **2021**, arXiv:2101.07616.
118. Solbi, M.; Karami, K. Primordial black holes and induced gravitational waves in Galileon inflation. *arXiv* **2021**, arXiv:2102.05651.
119. Drees, M.; Xu, Y. Overshooting, Critical Higgs Inflation and Second Order Gravitational Wave Signatures. *Eur. Phys. J. C* **2021**, *81*, 182. [CrossRef]
120. Yi, Z.; Gong, Y.; Wang, B.; Zhu, Z.h. Primordial Black Holes and Secondary Gravitational Waves from Higgs field. *arXiv* **2020**, arXiv:2007.09957.
121. Kohri, K.; Terada, T. Primordial Black Hole Dark Matter and LIGO/Virgo Merger Rate from Inflation with Running Spectral Indices: Formation in the Matter- and/or Radiation-Dominated Universe. *Class. Quant. Grav.* **2018**, *35*, 235017. [CrossRef]
122. Bartolo, N.; De Luca, V.; Franciolini, G.; Lewis, A.; Peloso, M.; Riotto, A. Primordial Black Hole Dark Matter: LISA Serendipity. *Phys. Rev. Lett.* **2019**, *122*, 211301. [CrossRef] [PubMed]
123. Bartolo, N.; De Luca, V.; Franciolini, G.; Peloso, M.; Racco, D.; Riotto, A. Testing primordial black holes as dark matter with LISA. *Phys. Rev. D* **2019**, *99*, 103521. [CrossRef]
124. Tada, Y.; Yokoyama, S. Primordial black hole tower: Dark matter, earth-mass, and LIGO black holes. *Phys. Rev. D* **2019**, *100*, 023537. [CrossRef]
125. Ballesteros, G.; Rey, J.; Taoso, M.; Urbano, A. Primordial black holes as dark matter and gravitational waves from single-field polynomial inflation. *JCAP* **2020**, *2020*, 025. [CrossRef]
126. Inomata, K.; Kawasaki, M.; Mukaida, K.; Tada, Y.; Yanagida, T.T. Inflationary primordial black holes for the LIGO gravitational wave events and pulsar timing array experiments. *Phys. Rev. D* **2017**, *95*, 123510. [CrossRef]
127. Nakama, T.; Silk, J.; Kamionkowski, M. Stochastic gravitational waves associated with the formation of primordial black holes. *Phys. Rev. D* **2017**, *95*, 043511. [CrossRef]
128. Ando, K.; Inomata, K.; Kawasaki, M.; Mukaida, K.; Yanagida, T.T. Primordial black holes for the LIGO events in the axionlike curvaton model. *Phys. Rev. D* **2018**, *97*, 123512. [CrossRef]
129. Vaskonen, V.; Veermäe, H. Did NANOGrav see a signal from primordial black hole formation? *Phys. Rev. Lett.* **2021**, *126*, 051303. [CrossRef]

130. De Luca, V.; Franciolini, G.; Riotto, A. NANOGrav Data Hints at Primordial Black Holes as Dark Matter. *Phys. Rev. Lett.* **2021**, *126*, 041303. [CrossRef]
131. Kohri, K.; Terada, T. Solar-Mass Primordial Black Holes Explain NANOGrav Hint of Gravitational Waves. *Phys. Lett. B* **2021**, *813*, 136040. [CrossRef]
132. Domènech, G.; Pi, S. NANOGrav Hints on Planet-Mass Primordial Black Holes. *arXiv* **2020**, arXiv:2010.03976.
133. Bhattacharya, S.; Mohanty, S.; Parashari, P. Implications of the NANOGrav result on primordial gravitational waves in nonstandard cosmologies. *arXiv* **2020**, arXiv:2010.05071.
134. Inomata, K.; Kawasaki, M.; Mukaida, K.; Yanagida, T.T. NANOGrav results and LIGO-Virgo primordial black holes in axion-like curvaton model. *arXiv* **2020**, arXiv:2011.01270.
135. Yi, Z.; Zhu, Z.H. NANOGrav signal and LIGO-Virgo Primordial Black Holes from Higgs inflation. *arXiv* **2021**, arXiv:2105.01943.
136. Orlofsky, N.; Pierce, A.; Wells, J.D. Inflationary theory and pulsar timing investigations of primordial black holes and gravitational waves. *Phys. Rev. D* **2017**, *95*, 063518. [CrossRef]
137. Cai, R.G.; Pi, S.; Wang, S.J.; Yang, X.Y. Pulsar Timing Array Constraints on the Induced Gravitational Waves. *JCAP* **2019**, *2019*, 059. [CrossRef]
138. Chen, Z.C.; Yuan, C.; Huang, Q.G. Pulsar Timing Array Constraints on Primordial Black Holes with NANOGrav 11-Year Dataset. *Phys. Rev. Lett.* **2020**, *124*, 251101. [CrossRef]
139. Chen, Z.C.; Huang, Q.G. Distinguishing Primordial Black Holes from Astrophysical Black Holes by Einstein Telescope and Cosmic Explorer. *JCAP* **2020**, *2020*, 039. [CrossRef]
140. Inomata, K.; Nakama, T. Gravitational waves induced by scalar perturbations as probes of the small-scale primordial spectrum. *Phys. Rev. D* **2019**, *99*, 043511. [CrossRef]
141. Clesse, S.; García-Bellido, J.; Orani, S. Detecting the Stochastic Gravitational Wave Background from Primordial Black Hole Formation. *arXiv* **2018**, arXiv:1812.11011.
142. Wang, S.; Terada, T.; Kohri, K. Prospective constraints on the primordial black hole abundance from the stochastic gravitational-wave backgrounds produced by coalescing events and curvature perturbations. *Phys. Rev. D* **2019**, *99*, 103531. Erratum in *Phys. Rev. D* **2020**, *101*, 069901. [CrossRef]
143. Yuan, C.; Chen, Z.C.; Huang, Q.G. Probing primordial–black-hole dark matter with scalar induced gravitational waves. *Phys. Rev. D* **2019**, *100*, 081301. [CrossRef]
144. Lu, Y.; Gong, Y.; Yi, Z.; Zhang, F. Constraints on primordial curvature perturbations from primordial black hole dark matter and secondary gravitational waves. *JCAP* **2019**, *2019*, 031. [CrossRef]
145. Zhang, F.; Ali, A.; Gong, Y.; Lin, J.; Lu, Y. On the waveform of the scalar induced gravitational waves. *arXiv* **2020**, arXiv:2008.12961.
146. Wang, Y.T.; Cai, Y.; Liu, Z.G.; Piao, Y.S. Probing the primordial universe with gravitational waves detectors. *JCAP* **2017**, *2017*, 010. [CrossRef]
147. Bartolo, N.; Domcke, V.; Figueroa, D.G.; García-Bellido, J.; Peloso, M.; Pieroni, M.; Ricciardone, A.; Sakellariadou, M.; Sorbo, L.; Tasinato, G. Probing non-Gaussian Stochastic Gravitational Wave Backgrounds with LISA. *JCAP* **2018**, *2018*, 034. [CrossRef]
148. Bartolo, N.; Bertacca, D.; De Luca, V.; Franciolini, G.; Matarrese, S.; Peloso, M.; Ricciardone, A.; Riotto, A.; Tasinato, G. Gravitational wave anisotropies from primordial black holes. *JCAP* **2020**, *2020*, 028. [CrossRef]
149. Hwang, J.C.; Jeong, D.; Noh, H. Gauge dependence of gravitational waves generated from scalar perturbations. *Astrophys. J.* **2017**, *842*, 46. [CrossRef]
150. Domènech, G.; Sasaki, M. Hamiltonian approach to second order gauge invariant cosmological perturbations. *Phys. Rev. D* **2018**, *97*, 023521. [CrossRef]
151. Gong, J.O. Analytic integral solutions for induced gravitational waves. *arXiv* **2019**, arXiv:1909.12708.
152. Tomikawa, K.; Kobayashi, T. Gauge dependence of gravitational waves generated at second order from scalar perturbations. *Phys. Rev. D* **2020**, *101*, 083529. [CrossRef]
153. De Luca, V.; Franciolini, G.; Kehagias, A.; Riotto, A. On the Gauge Invariance of Cosmological Gravitational Waves. *JCAP* **2020**, *2020*, 014. [CrossRef]
154. Inomata, K.; Terada, T. Gauge Independence of Induced Gravitational Waves. *Phys. Rev. D* **2020**, *101*, 023523. [CrossRef]
155. Yuan, C.; Chen, Z.C.; Huang, Q.G. Scalar induced gravitational waves in different gauges. *Phys. Rev. D* **2020**, *101*, 063018. [CrossRef]
156. Chang, Z.; Wang, S.; Zhu, Q.H. Gauge Invariant Second Order Gravitational Waves. *arXiv* **2020**, arXiv:2009.11994.
157. Chang, Z.; Wang, S.; Zhu, Q.H. Gauge invariance of the second order cosmological perturbations. *arXiv* **2020**, arXiv:2009.11025.
158. Lu, Y.; Ali, A.; Gong, Y.; Lin, J.; Zhang, F. Gauge transformation of scalar induced gravitational waves. *Phys. Rev. D* **2020**, *102*, 083503. [CrossRef]
159. Ali, A.; Gong, Y.; Lu, Y. Gauge transformation of scalar induced tensor perturbation during matter domination. *Phys. Rev. D* **2021**, *103*, 043516. [CrossRef]
160. Chang, Z.; Wang, S.; Zhu, Q.H. On the Gauge Invariance of Scalar Induced Gravitational Waves: Gauge Fixings Considered. *arXiv* **2020**, arXiv:2010.01487.
161. Domènech, G.; Sasaki, M. Approximate gauge independence of the induced gravitational wave spectrum. *arXiv* **2020**, arXiv:2012.14016.

162. Gurian, J.; Jeong, D.; Hwang, J.C.; Noh, H. Gauge-Invariant Tensor Perturbations Induced from Baryon-CDM Relative Velocity and the B-mode Polarization of the CMB. *arXiv* **2021**, arXiv:2104.03330.
163. Mukhanov, V.F.; Feldman, H.A.; Brandenberger, R.H. Theory of cosmological perturbations. Part 1. Classical perturbations. Part 2. Quantum theory of perturbations. Part 3. Extensions. *Phys. Rept.* **1992**, *215*, 203–333. [CrossRef]
164. Malik, K.A.; Wands, D. Cosmological perturbations. *Phys. Rept.* **2009**, *475*, 1–51. [CrossRef]
165. Durrer, R. Cosmological perturbation theory. *Lect. Notes Phys.* **2004**, *653*, 31–70. [CrossRef]
166. Langlois, D. Lectures on inflation and cosmological perturbations. *Lect. Notes Phys.* **2010**, *800*, 1–57. [CrossRef]
167. Baumann, D. Physics of the Large and the Small: TASI 09. In Proceedings of the Theoretical Advanced Study Institute in Elementary Particle Physics, Boulder, CO, USA, 1–26 June 2009; pp 523–686. [CrossRef]
168. Piattella, O.F. *Lecture Notes in Cosmology*; UNITEXT for Physics; Springer: Cham, Switzerland, 2018. [CrossRef]
169. Gong, J.O. Multi-field inflation and cosmological perturbations. *Int. J. Mod. Phys. D* **2016**, *26*, 1740003. [CrossRef]
170. Chen, X. Primordial Features as Evidence for Inflation. *JCAP* **2012**, *2012*, 038. [CrossRef]
171. Christensen, N. Stochastic Gravitational Wave Backgrounds. *Rept. Prog. Phys.* **2019**, *82*, 016903. [CrossRef]
172. Kuroyanagi, S.; Chiba, T.; Takahashi, T. Probing the Universe through the Stochastic Gravitational Wave Background. *JCAP* **2018**, *2018*, 038. [CrossRef]
173. Guzzetti, M.C.; Bartolo, N.; Liguori, M.; Matarrese, S. Gravitational waves from inflation. *Riv. Nuovo Cim.* **2016**, *39*, 399–495. [CrossRef]
174. Sasaki, M.; Suyama, T.; Tanaka, T.; Yokoyama, S. Primordial black holes—Perspectives in gravitational wave astronomy. *Class. Quant. Grav.* **2018**, *35*, 063001. [CrossRef]
175. Khlopov, M.Y. Primordial Black Holes. *Res. Astron. Astrophys.* **2010**, *10*, 495–528. [CrossRef]
176. Carr, B.; Kohri, K.; Sendouda, Y.; Yokoyama, J. Constraints on Primordial Black Holes. *arXiv* **2020**, arXiv:2002.12778.
177. Carr, B.; Kuhnel, F. Primordial Black Holes as Dark Matter: Recent Developments. *Ann. Rev. Nucl. Part. Sci.* **2020**, *70*, 355–394. [CrossRef]
178. Green, A.M.; Kavanagh, B.J. Primordial Black Holes as a dark matter candidate. *J. Phys. G* **2021**, *48*, 4. [CrossRef]
179. Yuan, C.; Huang, Q.G. A topic review on probing primordial black hole dark matter with scalar induced gravitational waves. *arXiv* **2021**, arXiv:2103.04739.
180. Komatsu, E. Hunting for Primordial Non-Gaussianity in the Cosmic Microwave Background. *Class. Quant. Grav.* **2010**, *27*, 124010. [CrossRef]
181. Bartolo, N.; Matarrese, S.; Riotto, A. Non-Gaussianity and the Cosmic Microwave Background Anisotropies. *Adv. Astron.* **2010**, *2010*, 157079. [CrossRef]
182. Byrnes, C.T.; Choi, K.Y. Review of local non-Gaussianity from multi-field inflation. *Adv. Astron.* **2010**, *2010*, 724525. [CrossRef]
183. Koyama, K. Non-Gaussianity of quantum fields during inflation. *Class. Quant. Grav.* **2010**, *27*, 124001. [CrossRef]
184. Misner, C.W.; Thorne, K.S.; Wheeler, J.A. *Gravitation*; W. H. Freeman: San Francisco, CA, USA, 1973.
185. Maggiore, M. *Gravitational Waves. Vol. 1: Theory and Experiments*; Oxford Master Series in Physics; Oxford University Press: Oxford, UK, 2007.
186. Isaacson, R.A. Gravitational Radiation in the Limit of High Frequency. I. The Linear Approximation and Geometrical Optics. *Phys. Rev.* **1968**, *166*, 1263–1271. [CrossRef]
187. Isaacson, R.A. Gravitational Radiation in the Limit of High Frequency. II. Nonlinear Terms and the Ef fective Stress Tensor. *Phys. Rev.* **1968**, *166*, 1272–1279. [CrossRef]
188. Kawamura, S.; Ando, M.; Seto, N.; Sato, S.; Nakamura, T.; Tsubono, K.; Kanda, N.; Tanaka, T.; Yokoyama, J.; Funaki, I. et al. The Japanese space gravitational wave antenna: DECIGO. *Class. Quant. Grav.* **2011**, *28*, 094011. [CrossRef]
189. Hawking, S.W. Particle Creation by Black Holes. *Commun. Math. Phys.* **1975**, *43*, 199–220. Erratum in *Commun. Math. Phys.* **1976**, *46*, 206. [CrossRef]
190. Carr, B.J.; Hawking, S. Black holes in the early Universe. *Mon. Not. R. Astron. Soc.* **1974**, *168*, 399–415. [CrossRef]
191. Crawford, M.; Schramm, D.N. Spontaneous Generation of Density Perturbations in the Early Universe. *Nature* **1982**, *298*, 538–540. [CrossRef]
192. Kodama, H.; Sasaki, M.; Sato, K. Abundance of Primordial Holes Produced by Cosmological First Order Phase Transition. *Prog. Theor. Phys.* **1982**, *68*, 1979. [CrossRef]
193. Garriga, J.; Vilenkin, A.; Zhang, J. Black holes and the multiverse. *JCAP* **2016**, *2016*, 064. [CrossRef]
194. Cotner, E.; Kusenko, A. Primordial black holes from supersymmetry in the early universe. *Phys. Rev. Lett.* **2017**, *119*, 031103. [CrossRef]
195. Cotner, E.; Kusenko, A.; Sasaki, M.; Takhistov, V. Analytic Description of Primordial Black Hole Formation from Scalar Field Fragmentation. *JCAP* **2019**, *2019*, 077. [CrossRef]
196. Amendola, L.; Rubio, J.; Wetterich, C. Primordial black holes from fifth forces. *Phys. Rev. D* **2018**, *97*, 081302. [CrossRef]
197. Savastano, S.; Amendola, L.; Rubio, J.; Wetterich, C. Primordial dark matter halos from fifth forces. *Phys. Rev. D* **2019**, *100*, 083518. [CrossRef]
198. Flores, M.M.; Kusenko, A. Primordial Black Holes from Long-Range Scalar Forces and Scalar Radiative Cooling. *Phys. Rev. Lett.* **2021**, *126*, 041101. [CrossRef]

199. Shibata, M.; Sasaki, M. Black hole formation in the Friedmann universe: Formulation and computation in numerical relativity. *Phys. Rev. D* **1999**, *60*, 084002. [CrossRef]
200. Nakama, T.; Harada, T.; Polnarev, A.G.; Yokoyama, J. Identifying the most crucial parameters of the initial curvature profile for primordial black hole formation. *JCAP* **2014**, *2014*, 037. [CrossRef]
201. Harada, T.; Yoo, C.M.; Nakama, T.; Koga, Y. Cosmological long-wavelength solutions and primordial black hole formation. *Phys. Rev. D* **2015**, *91*, 084057. [CrossRef]
202. Escrivà, A.; Romano, A.E. Effects of the shape of curvature peaks on the size of primordial black holes. *JCAP* **2021**, *2021*, 066. [CrossRef]
203. Zel'dovich, Y.B.; Novikov, I.D. The Hypothesis of Cores Retarded during Expansion and the Hot Cosmological Model. *Sov. Astron.* **1967**, *10*, 602.
204. Hawking, S. Gravitationally collapsed objects of very low mass. *Mon. Not. R. Astron. Soc.* **1971**, *152*, 75. [CrossRef]
205. Meszaros, P. The behaviour of point masses in an expanding cosmological substratum. *Astron. Astrophys.* **1974**, *37*, 225–228.
206. Carr, B.J. The Primordial black hole mass spectrum. *Astrophys. J.* **1975**, *201*, 1–19. [CrossRef]
207. Khlopov, M.; Malomed, B.; Zeldovich, I. Gravitational instability of scalar fields and formation of primordial black holes. *Mon. Not. R. Astron. Soc.* **1985**, *215*, 575–589. [CrossRef]
208. Niemeyer, J.C.; Jedamzik, K. Dynamics of primordial black hole formation. *Phys. Rev. D* **1999**, *59*, 124013. [CrossRef]
209. Musco, I.; Miller, J.C.; Rezzolla, L. Computations of primordial black hole formation. *Class. Quant. Grav.* **2005**, *22*, 1405–1424. [CrossRef]
210. Musco, I.; Miller, J.C.; Polnarev, A.G. Primordial black hole formation in the radiative era: Investigation of the critical nature of the collapse. *Class. Quant. Grav.* **2009**, *26*, 235001. [CrossRef]
211. Musco, I.; Miller, J.C. Primordial black hole formation in the early universe: Critical behaviour and self-similarity. *Class. Quant. Grav.* **2013**, *30*, 145009. [CrossRef]
212. Harada, T.; Yoo, C.M.; Kohri, K. Threshold of primordial black hole formation. *Phys. Rev. D* **2013**, *88*, 084051. Erratum in *Phys. Rev. D* **2014**, *89*, 029903. [CrossRef]
213. Escrivà, A.; Germani, C.; Sheth, R.K. Universal threshold for primordial black hole formation. *Phys. Rev. D* **2020**, *101*, 044022. [CrossRef]
214. Musco, I.; De Luca, V.; Franciolini, G.; Riotto, A. Threshold for primordial black holes. II. A simple analytic prescription. *Phys. Rev. D* **2021**, *103*, 063538. [CrossRef]
215. Young, S.; Musso, M. Application of peaks theory to the abundance of primordial black holes. *JCAP* **2020**, *2020*, 022. [CrossRef]
216. Escrivà, A.; Germani, C.; Sheth, R.K. Analytical thresholds for black hole formation in general cosmological backgrounds. *JCAP* **2021**, *2021*, 030. [CrossRef]
217. Kehagias, A.; Musco, I.; Riotto, A. Non-Gaussian Formation of Primordial Black Holes: Effects on the Threshold. *JCAP* **2019**, *2019*, 029. [CrossRef]
218. Atal, V.; Cid, J.; Escrivà, A.; Garriga, J. PBH in single field inflation: The effect of shape dispersion and non-Gaussianities. *JCAP* **2020**, *2020*, 022. [CrossRef]
219. Yoo, C.M.; Gong, J.O.; Yokoyama, S. Abundance of primordial black holes with local non-Gaussianity in peak theory. *JCAP* **2019**, *2019*, 033. [CrossRef]
220. Riccardi, F.; Taoso, M.; Urbano, A. Solving peak theory in the presence of local non-gaussianities. *arXiv* **2021**, arXiv:2102.04084.
221. Young, S.; Musco, I.; Byrnes, C.T. Primordial black hole formation and abundance: Contribution from the non-linear relation between the density and curvature perturbation. *JCAP* **2019**, *2019*, 012. [CrossRef]
222. Gourgoulhon, E. 3 + 1 formalism and bases of numerical relativity. *arXiv* **2007**, arXiv:0703035.
223. Poisson, E. *A Relativist's Toolkit: The Mathematics of Black-Hole Mechanics*; Cambridge University Press: Cambridge, UK, 2009. [CrossRef]
224. Arnowitt, R.L.; Deser, S.; Misner, C.W. The Dynamics of general relativity. *Gen. Rel. Grav.* **2008**, *40*, 1997–2027. [CrossRef]
225. Wald, R.M. *General Relativity*; University of Chicago Press: Chicago, IL, USA, 1984. [CrossRef]
226. Maldacena, J.M. Non-Gaussian features of primordial fluctuations in single field inflationary models. *JHEP* **2003**, *05*, 013. [CrossRef]
227. Arkani-Hamed, N.; Maldacena, J. Cosmological Collider Physics. *arXiv* **2015**, arXiv:1503.08043.
228. Lyth, D.H.; Malik, K.A.; Sasaki, M. A General proof of the conservation of the curvature perturbation. *JCAP* **2005**, *2005*, 004. [CrossRef]
229. Mukhanov, V. *Physical Foundations of Cosmology*; Cambridge University Press: Oxford, UK, 2005.
230. Komatsu, E.; Spergel, D.N.; Wandelt, B.D. Measuring primordial non-Gaussianity in the cosmic microwave background. *Astrophys. J.* **2005**, *634*, 14–19. [CrossRef]
231. Lucchin, F.; Matarrese, S. Power Law Inflation. *Phys. Rev. D* **1985**, *32*, 1316. [CrossRef] [PubMed]
232. Garriga, J.; Mukhanov, V.F. Perturbations in k-inflation. *Phys. Lett. B* **1999**, *458*, 219–225. [CrossRef]
233. Gervois, A.; Navelet, H. Integrals of three Bessel functions and Legendre functions. I. *J. Math. Phys.* **1985**, *26*, 633–644. [CrossRef]
234. Carrilho, P.; Malik, K.A.; Mulryne, D.J. Dissecting the growth of the power spectrum for primordial black holes. *Phys. Rev. D* **2019**, *100*, 103529. [CrossRef]

235. Kawasaki, M.; Sugiyama, N.; Yanagida, T. Primordial black hole formation in a double inflation model in supergravity. *Phys. Rev. D* **1998**, *57*, 6050–6056. [CrossRef]
236. Frampton, P.H.; Kawasaki, M.; Takahashi, F.; Yanagida, T.T. Primordial Black Holes as All Dark Matter. *JCAP* **2010**, *2010*, 023. [CrossRef]
237. Kawasaki, M.; Kitajima, N.; Yanagida, T.T. Primordial black hole formation from an axionlike curvaton model. *Phys. Rev. D* **2013**, *87*, 063519. [CrossRef]
238. Inomata, K.; Kawasaki, M.; Mukaida, K.; Tada, Y.; Yanagida, T.T. Inflationary Primordial Black Holes as All Dark Matter. *Phys. Rev. D* **2017**, *96*, 043504. [CrossRef]
239. Pi, S.; Zhang, Y.l.; Huang, Q.G.; Sasaki, M. Scalaron from R^2-gravity as a heavy field. *JCAP* **2018**, *2018*, 042. [CrossRef]
240. Cai, Y.F.; Tong, X.; Wang, D.G.; Yan, S.F. Primordial Black Holes from Sound Speed Resonance during Inflation. *Phys. Rev. Lett.* **2018**, *121*, 081306. [CrossRef]
241. Chen, C.; Cai, Y.F. Primordial black holes from sound speed resonance in the inflaton-curvaton mixed scenario. *JCAP* **2019**, *2019*, 068. [CrossRef]
242. A. Ashoorioon, A. Rostami and J. T. Firouzjaee, EFT compatible PBHs: effective spawning of the seeds for primordial black holes during inflation. *JHEP* **2021**, *07*, 087. [CrossRef]
243. Chen, C.; Ma, X.H.; Cai, Y.F. Dirac-Born-Infeld realization of sound speed resonance mechanism for primordial black holes. *Phys. Rev. D* **2020**, *102*, 063526. [CrossRef]
244. Garcia-Bellido, J.; Linde, A.D.; Wands, D. Density perturbations and black hole formation in hybrid inflation. *Phys. Rev. D* **1996**, *54*, 6040–6058. [CrossRef]
245. Yokoyama, J. Chaotic new inflation and formation of primordial black holes. *Phys. Rev. D* **1998**, *58*, 083510. [CrossRef]
246. Kohri, K.; Lin, C.M.; Matsuda, T. Primordial black holes from the inflating curvaton. *Phys. Rev. D* **2013**, *87*, 103527. [CrossRef]
247. Clesse, S.; García-Bellido, J. Massive Primordial Black Holes from Hybrid Inflation as Dark Matter and the seeds of Galaxies. *Phys. Rev. D* **2015**, *92*, 023524. [CrossRef]
248. Cheng, S.L.; Lee, W.; Ng, K.W. Production of high stellar-mass primordial black holes in trapped inflation. *JHEP* **2017**, *02*, 008. [CrossRef]
249. Espinosa, J.R.; Racco, D.; Riotto, A. Cosmological Signature of the Standard Model Higgs Vacuum Instability: Primordial Black Holes as Dark Matter. *Phys. Rev. Lett.* **2018**, *120*, 121301. [CrossRef]
250. Kannike, K.; Marzola, L.; Raidal, M.; Veermäe, H. Single Field Double Inflation and Primordial Black Holes. *JCAP* **2017**, *2017*, 020. [CrossRef]
251. Garcia-Bellido, J.; Ruiz Morales, E. Primordial black holes from single field models of inflation. *Phys. Dark Univ.* **2017**, *18*, 47–54. [CrossRef]
252. Cheng, S.L.; Lee, W.; Ng, K.W. Primordial black holes and associated gravitational waves in axion monodromy inflation. *JCAP* **2018**, *2018*, 001. [CrossRef]
253. Inomata, K.; Kawasaki, M.; Mukaida, K.; Yanagida, T.T. Double inflation as a single origin of primordial black holes for all dark matter and LIGO observations. *Phys. Rev. D* **2018**, *97*, 043514. [CrossRef]
254. Palma, G.A.; Sypsas, S.; Zenteno, C. Seeding primordial black holes in multifield inflation. *Phys. Rev. Lett.* **2020**, *125*, 121301. [CrossRef]
255. Fumagalli, J.; Renaux-Petel, S.; Ronayne, J.W.; Witkowski, L.T. Turning in the landscape: A new mechanism for generating Primordial Black Holes. *arXiv* **2020**, arXiv:2004.08369.
256. Atal, V.; Germani, C. The role of non-gaussianities in Primordial Black Hole formation. *Phys. Dark Univ.* **2019**, *24*, 100275. [CrossRef]
257. Chen, X. Primordial Non-Gaussianities from Inflation Models. *Adv. Astron.* **2010**, *2010*, 638979. [CrossRef]
258. Chluba, J.; Hamann, J.; Patil, S.P. Features and New Physical Scales in Primordial Observables: Theory and Observation. *Int. J. Mod. Phys. D* **2015**, *24*, 1530023. [CrossRef]
259. Slosar, A.; Abazajian, K.N.; Abidi, M.; Adshead, P.; Ahmed, Z.; Alonso, D.; Amin, M.A.; Ansarinejad, B.; Armstrong, R.; Baccigalupi, C. et al. Scratches from the Past: Inflationary Archaeology through Features in the Power Spectrum of Primordial Fluctuations. *arXiv* **2019**, arXiv:1903.09883.
260. Starobinsky, A.A. Spectrum of adiabatic perturbations in the universe when there are singularities in the inflation potential. *JETP Lett.* **1992**, *55*, 489–494.
261. Adams, J.A.; Cresswell, B.; Easther, R. Inflationary perturbations from a potential with a step. *Phys. Rev. D* **2001**, *64*, 123514. [CrossRef]
262. Bean, R.; Chen, X.; Hailu, G.; Tye, S.H.H.; Xu, J. Duality Cascade in Brane Inflation. *JCAP* **2008**, *2008*, 026. [CrossRef]
263. Adshead, P.; Dvorkin, C.; Hu, W.; Lim, E.A. Non-Gaussianity from Step Features in the Inflationary Potential. *Phys. Rev. D* **2012**, *85*, 023531. [CrossRef]
264. Bartolo, N.; Cannone, D.; Matarrese, S. The Effective Field Theory of Inflation Models with Sharp Features. *JCAP* **2013**, *2013*, 038. [CrossRef]
265. Palma, G.A. Untangling features in the primordial spectra. *JCAP* **2015**, *2015*, 035. [CrossRef]
266. Ballesteros, G.; Beltran Jimenez, J.; Pieroni, M. Black hole formation from a general quadratic action for inflationary primordial fluctuations. *JCAP* **2019**, *2019*, 016. [CrossRef]

267. Kefala, K.; Kodaxis, G.P.; Stamou, I.D.; Tetradis, N. Features of the inflaton potential and the power spectrum of cosmological perturbations. *arXiv* **2020**, arXiv:2010.12483.
268. Achucarro, A.; Gong, J.O.; Hardeman, S.; Palma, G.A.; Patil, S.P. Features of heavy physics in the CMB power spectrum. *JCAP* **2011**, *2011*, 030. [CrossRef]
269. Shiu, G.; Xu, J. Effective Field Theory and Decoupling in Multi-field Inflation: An Illustrative Case Study. *Phys. Rev. D* **2011**, *84*, 103509. [CrossRef]
270. Gao, X.; Langlois, D.; Mizuno, S. Influence of heavy modes on perturbations in multiple field inflation. *JCAP* **2012**, *2012*, 040. [CrossRef]
271. Pahud, C.; Kamionkowski, M.; Liddle, A.R. Oscillations in the inflaton potential? *Phys. Rev. D* **2009**, *79*, 083503. [CrossRef]
272. Chen, X.; Easther, R.; Lim, E.A. Generation and Characterization of Large Non-Gaussianities in Single Field Inflation. *JCAP* **2008**, *2008*, 010. [CrossRef]
273. Silverstein, E.; Westphal, A. Monodromy in the CMB: Gravity Waves and String Inflation. *Phys. Rev. D* **2008**, *78*, 106003. [CrossRef]
274. Flauger, R.; McAllister, L.; Pajer, E.; Westphal, A.; Xu, G. Oscillations in the CMB from Axion Monodromy Inflation. *JCAP* **2010**, *2010*, 009. [CrossRef]
275. Chen, X.; Namjoo, M.H. Standard Clock in Primordial Density Perturbations and Cosmic Microwave Background. *Phys. Lett. B* **2014**, *739*, 285–292. [CrossRef]
276. Chen, X.; Namjoo, M.H.; Wang, Y. Models of the Primordial Standard Clock. *JCAP* **2015**, *2015*, 027. [CrossRef]
277. Gao, X.; Gong, J.O. Towards general patterns of features in multi-field inflation. *JHEP* **2015**, *08*, 115. [CrossRef]
278. Huang, Q.G.; Pi, S. Power-law modulation of the scalar power spectrum from a heavy field with a monomial potential. *JCAP* **2018**, *2018*, 001. [CrossRef]
279. Domènech, G.; Rubio, J.; Wons, J. Mimicking features in alternatives to inflation with interacting spectator fields. *Phys. Lett. B* **2019**, *790*, 263–269. [CrossRef]
280. Chen, X.; Loeb, A.; Xianyu, Z.Z. Unique Fingerprints of Alternatives to Inflation in the Primordial Power Spectrum. *Phys. Rev. Lett.* **2019**, *122*, 121301. [CrossRef]
281. Young, S.; Byrnes, C.T. Primordial black holes in non-Gaussian regimes. *JCAP* **2013**, *2013*, 052. [CrossRef]
282. Atal, V.; Garriga, J.; Marcos-Caballero, A. Primordial black hole formation with non-Gaussian curvature perturbations. *JCAP* **2019**, *2019*, 073. [CrossRef]
283. De Luca, V.; Franciolini, G.; Kehagias, A.; Peloso, M.; Riotto, A.; Ünal, C. The Ineludible non-Gaussianity of the Primordial Black Hole Abundance. *JCAP* **2019**, *2019*, 048. [CrossRef]
284. Kawasaki, M.; Nakatsuka, H. Effect of nonlinearity between density and curvature perturbations on the primordial black hole formation. *Phys. Rev. D* **2019**, *99*, 123501. [CrossRef]
285. Kodama, H.; Sasaki, M. Evolution of Isocurvature Perturbations. 1. Photon - Baryon Universe. *Int. J. Mod. Phys. A* **1986**, *1*, 265. [CrossRef]
286. Kodama, H.; Sasaki, M. Evolution of Isocurvature Perturbations. 2. Radiation Dust Universe. *Int. J. Mod. Phys. A* **1987**, *2*, 491. [CrossRef]
287. Hooper, D.; Krnjaic, G.; McDermott, S.D. Dark Radiation and Superheavy Dark Matter from Black Hole Domination. *JHEP* **2019**, *2019*, 001. [CrossRef]
288. Dong, R.; Kinney, W.H.; Stojkovic, D. Gravitational wave production by Hawking radiation from rotating primordial black holes. *JCAP* **2016**, *2016*, 034. [CrossRef]
289. Arbey, A.; Auffinger, J.; Silk, J. Evolution of primordial black hole spin due to Hawking radiation. *Mon. Not. R. Astron. Soc.* **2020**, *494*, 1257–1262. [CrossRef]
290. Kuhnel, F. Enhanced Detectability of Spinning Primordial Black Holes. *Eur. Phys. J. C* **2020**, *80*, 243. [CrossRef]
291. Arbey, A.; Auffinger, J.; Sandick, P.; Shams Es Haghi, B.; Sinha, K. Precision Calculation of Dark Radiation from Spinning Primordial Black Holes and Early Matter Dominated Eras. *arXiv* **2021**, arXiv:2104.04051.
292. Masina, I. Dark matter and dark radiation from evaporating Kerr primordial black holes. *arXiv* **2021**, arXiv:2103.13825.
293. Husdal, L. On Effective Degrees of Freedom in the Early Universe. *Galaxies* **2016**, *4*, 78. [CrossRef]
294. Saikawa, K.; Shirai, S. Primordial gravitational waves, precisely: The role of thermodynamics in the Standard Model. *JCAP* **2018**, *2018*, 035. [CrossRef]
295. Kawasaki, M.; Kohri, K.; Sugiyama, N. Cosmological constraints on late time entropy production. *Phys. Rev. Lett.* **1999**, *82*, 4168. [CrossRef]
296. Kawasaki, M.; Kohri, K.; Sugiyama, N. MeV scale reheating temperature and thermalization of neutrino background. *Phys. Rev. D* **2000**, *62*, 023506. [CrossRef]
297. Hannestad, S. What is the lowest possible reheating temperature? *Phys. Rev. D* **2004**, *70*, 043506. [CrossRef]
298. Hasegawa, T.; Hiroshima, N.; Kohri, K.; Hansen, R.S.L.; Tram, T.; Hannestad, S. MeV-scale reheating temperature and thermalization of oscillating neutrinos by radiative and hadronic decays of massive particles. *JCAP* **2019**, *2019*, 012. [CrossRef]
299. Cyburt, R.H.; Fields, B.D.; Olive, K.A.; Skillman, E. New BBN limits on physics beyond the standard model from ^{4}He. *Astropart. Phys.* **2005**, *23*, 313–323. [CrossRef]

300. Fields, B.D.; Olive, K.A.; Yeh, T.H.; Young, C. Big-Bang Nucleosynthesis after Planck. *JCAP* **2020**, *2020*, 010. Erratum in JCAP **2020**, *11*, E02. [CrossRef]
301. Sendra, I.; Smith, T.L. Improved limits on short-wavelength gravitational waves from the cosmic microwave background. *Phys. Rev. D* **2012**, *85*, 123002. [CrossRef]
302. Pagano, L.; Salvati, L.; Melchiorri, A. New constraints on primordial gravitational waves from Planck 2015. *Phys. Lett. B* **2016**, *760*, 823–825. [CrossRef]
303. Abazajian, K.N. et al. [CMB-S4 Collaboration]. CMB-S4 Science Book, First Edition. *arXiv* **2016**, arXiv:1610.02743.
304. Watanabe, M.; Kanno, S.; Soda, J. The Nature of Primordial Fluctuations from Anisotropic Inflation. *Prog. Theor. Phys.* **2010**, *123*, 1041–1068. [CrossRef]
305. Soda, J. Statistical Anisotropy from Anisotropic Inflation. *Class. Quant. Grav.* **2012**, *29*, 083001. [CrossRef]
306. Naruko, A.; Pitrou, C.; Koyama, K.; Sasaki, M. Second-order Boltzmann equation: Gauge dependence and gauge invariance. *Class. Quant. Grav.* **2013**, *30*, 165008. [CrossRef]
307. Saito, R.; Naruko, A.; Hiramatsu, T.; Sasaki, M. Geodesic *curve*-of-sight formulae for the cosmic microwave background: A unified treatment of redshift, time delay, and lensing. *JCAP* **2014**, *2014*, 051. [CrossRef]
308. Namikawa, T.; Naruko, A.; Saito, R.; Taruya, A.; Yamauchi, D. Unified approach to secondary effects on the CMB B-mode polarization. *arXiv* **2021**, arXiv:2103.10639.
309. Nakamura, T.; Sasaki, M.; Tanaka, T.; Thorne, K.S. Gravitational waves from coalescing black hole MACHO binaries. *Astrophys. J. Lett.* **1997**, *487*, L139–L142. [CrossRef]
310. Abbott, B. et al. [LIGO Scientific and Virgo Collaborations]. GW170817: Observation of Gravitational Waves from a Binary Neutron Star Inspiral. *Phys. Rev. Lett.* **2017**, *119*, 161101. [CrossRef]
311. Mandic, V.; Bird, S.; Cholis, I. Stochastic Gravitational-Wave Background due to Primordial Binary Black Hole Mergers. *Phys. Rev. Lett.* **2016**, *117*, 201102. [CrossRef]
312. Wang, S.; Wang, Y.F.; Huang, Q.G.; Li, T.G.F. Constraints on the Primordial Black Hole Abundance from the First Advanced LIGO Observation Run Using the Stochastic Gravitational-Wave Background. *Phys. Rev. Lett.* **2018**, *120*, 191102. [CrossRef]
313. Sasaki, M.; Suyama, T.; Tanaka, T.; Yokoyama, S. Primordial Black Hole Scenario for the Gravitational-Wave Event GW150914. *Phys. Rev. Lett.* **2016**, *117*, 061101. Erratum in *Phys. Rev. Lett.* **2018**, *121*, 059901. [CrossRef]
314. Bird, S.; Cholis, I.; Muñoz, J.B.; Ali-Haïmoud, Y.; Kamionkowski, M.; Kovetz, E.D.; Raccanelli, A.; Riess, A.G. Did LIGO detect dark matter? *Phys. Rev. Lett.* **2016**, *116*, 201301. [CrossRef]
315. Ali-Haïmoud, Y.; Kovetz, E.D.; Kamionkowski, M. Merger rate of primordial black-hole binaries. *Phys. Rev. D* **2017**, *96*, 123523. [CrossRef]
316. Garriga, J.; Triantafyllou, N. Enhanced cosmological perturbations and the merger rate of PBH binaries. *JCAP* **2019**, *2019*, 043. [CrossRef]
317. Peloso, M.; Sorbo, L.; Unal, C. Rolling axions during inflation: Perturbativity and signatures. *JCAP* **2016**, *2016*, 001. [CrossRef]
318. Hall, A.; Gow, A.D.; Byrnes, C.T. Bayesian analysis of LIGO-Virgo mergers: Primordial vs. astrophysical black hole populations. *Phys. Rev. D* **2020**, *102*, 123524. [CrossRef]
319. De Luca, V.; Franciolini, G.; Pani, P.; Riotto, A. Bayesian Evidence for Both Astrophysical and Primordial Black Holes: Mapping the GWTC-2 Catalog to Third-Generation Detectors. *JCAP* **2021**, *2021*, 003. [CrossRef]
320. Franciolini, G.; Baibhav, V.; De Luca, V.; Ng, K.K.Y.; Wong, K.W.K.; Berti, E.; Pani, P.; Riotto, A.; Vitale, S. Quantifying the evidence for primordial black holes in LIGO/Virgo gravitational-wave data. *arXiv* **2021**, arXiv:2105.03349.
321. Niikura, H.; Takada, M.; Yokoyama, S.; Sumi, T.; Masaki, S. Constraints on Earth-mass primordial black holes from OGLE 5-year microlensing events. *Phys. Rev. D* **2019**, *99*, 083503. [CrossRef]
322. Mróz, P.; Udalski, A.; Skowron, J.; Poleski, R.; Kozłowski, S.; Szymański, M.K.; Soszyński, I.; Wyrzykowski, L.; Pietrukowicz, P.; Ulaczyk, K.; et al. No large population of unbound or wide-orbit Jupiter-mass planets. *Nature* **2017**, *548*, 183–186. [CrossRef]
323. Sugiyama, S.; Takhistov, V.; Vitagliano, E.; Kusenko, A.; Sasaki, M.; Takada, M. Testing Stochastic Gravitational Wave Signals from Primordial Black Holes with Optical Telescopes. *Phys. Lett. B* **2021**, *814*, 136097. [CrossRef]
324. Abbott, B. et al. [LIGO Scientific and Virgo Collaborations]. Search for the isotropic stochastic background using data from Advanced LIGO's second observing run. *Phys. Rev. D* **2019**, *100*, 061101. [CrossRef]
325. Kapadia, S.J.; Pandey, K.L.; Suyama, T.; Kandhasamy, S.; Ajith, P. Search for the stochastic gravitational-wave background induced by primordial curvature perturbations in LIGO's second observing run. *arXiv* **2020**, arXiv:2009.05514.
326. Romero-Rodriguez, A.; Martinez, M.; Pujolàs, O.; Sakellariadou, M.; Vaskonen, V. Search for a scalar induced stochastic gravitational wave background in the third LIGO-Virgo observing run. *arXiv* **2021**, arXiv:2107.11660.
327. Arzoumanian, Z. et al. [NANOGrav Collaboration]. The NANOGrav 11-year Data Set: Pulsar-timing Constraints On The Stochastic Gravitational-wave Background. *Astrophys. J.* **2018**, *859*, 47. [CrossRef]
328. Ellis, J.; Lewicki, M. Cosmic String Interpretation of NANOGrav Pulsar Timing Data. *Phys. Rev. Lett.* **2021**, *126*, 041304. [CrossRef]
329. Blasi, S.; Brdar, V.; Schmitz, K. Has NANOGrav found first evidence for cosmic strings? *Phys. Rev. Lett.* **2021**, *126*, 041305. [CrossRef]
330. Buchmuller, W.; Domcke, V.; Schmitz, K. From NANOGrav to LIGO with metastable cosmic strings. *Phys. Lett. B* **2020**, *811*, 135914. [CrossRef]
331. Samanta, R.; Datta, S. Gravitational wave complementarity and impact of NANOGrav data on gravitational leptogenesis. *JHEP* **2021**, *05*, 211. [CrossRef]

332. Nakai, Y.; Suzuki, M.; Takahashi, F.; Yamada, M. Gravitational Waves and Dark Radiation from Dark Phase Transition: Connecting NANOGrav Pulsar Timing Data and Hubble Tension. *Phys. Lett. B* **2021**, *816*, 136238. [CrossRef]
333. Addazi, A.; Cai, Y.F.; Gan, Q.; Marciano, A.; Zeng, K. NANOGrav results and dark first order phase transitions. *Sci. China Phys. Mech. Astron.* **2021**, *64*, 290411. [CrossRef]
334. Neronov, A.; Roper Pol, A.; Caprini, C.; Semikoz, D. NANOGrav signal from magnetohydrodynamic turbulence at the QCD phase transition in the early Universe. *Phys. Rev. D* **2021**, *103*, 041302. [CrossRef]
335. Ratzinger, W.; Schwaller, P. Whispers from the dark side: Confronting light new physics with NANOGrav data. *SciPost Phys.* **2021**, *10*, 047. [CrossRef]
336. Bian, L.; Cai, R.G.; Liu, J.; Yang, X.Y.; Zhou, R. Evidence for different gravitational-wave sources in the NANOGrav dataset. *Phys. Rev. D* **2021**, *103*, L081301. [CrossRef]
337. Li, H.H.; Ye, G.; Piao, Y.S. Is the NANOGrav signal a hint of dS decay during inflation? *Phys. Lett. B* **2021**, *816*, 136211. [CrossRef]
338. Liu, J.; Cai, R.G.; Guo, Z.K. Large Anisotropies of the Stochastic Gravitational Wave Background from Cosmic Domain Walls. *Phys. Rev. Lett.* **2021**, *126*, 141303. [CrossRef]
339. Paul, A.; Mukhopadhyay, U.; Majumdar, D. Gravitational Wave Signatures from Domain Wall and Strong First-Order Phase Transitions in a Two Complex Scalar extension of the Standard Model. *JHEP* **2021**, *05*, 223. [CrossRef]
340. Spokoiny, B. Deflationary universe scenario. *Phys. Lett. B* **1993**, *315*, 40–45. [CrossRef]
341. Peebles, P.J.E.; Vilenkin, A. Quintessential inflation. *Phys. Rev. D* **1999**, *59*, 063505. [CrossRef]
342. Brax, P.; Martin, J. Coupling quintessence to inflation in supergravity. *Phys. Rev. D* **2005**, *71*, 063530. [CrossRef]
343. Hossain, M.W.; Myrzakulov, R.; Sami, M.; Saridakis, E.N. Variable gravity: A suitable framework for quintessential inflation. *Phys. Rev. D* **2014**, *90*, 023512. [CrossRef]
344. Martin, J.; Papanikolaou, T.; Pinol, L.; Vennin, V. Metric preheating and radiative decay in single-field inflation. *JCAP* **2020**, *2020*, 003. [CrossRef]
345. Deffayet, C.; Esposito-Farese, G.; Steer, D.A. Counting the degrees of freedom of generalized Galileons. *Phys. Rev. D* **2015**, *92*, 084013. [CrossRef]
346. NIST Digital Library of Mathematical Functions. Release 1.0.24 of 2019-09-15. Available online: http://dlmf.nist.gov/ (accessed on 2 August 2021).

MDPI AG
Grosspeteranlage 5
4052 Basel
Switzerland
Tel.: +41 61 683 77 34

Universe Editorial Office
E-mail: universe@mdpi.com
www.mdpi.com/journal/universe

www.ingramcontent.com/pod-product-compliance
Lightning Source LLC
Chambersburg PA
CBHW061604220926
51850CB00005B/416
* 9 7 8 3 7 2 5 8 7 1 5 6 8 *